[3.1]	a^0	1 $(a \neq 0)$
	a^{-n}	the reciprocal of a^n, $\dfrac{1}{a^n}$ $(a \neq 0)$
[3.2]	$a^{1/n}$	the nth real root of a for $a \in R$, or the positive one if there are two
	$a^{m/n}$	$(a^{1/n})^m$
[3.3]	$\sqrt[n]{a}$	the nth real root of a for $a \in R$, or the positive one if there are two
[3.5]	C	the set of complex numbers
	i	the imaginary unit, $\sqrt{-1}$
	$\sqrt{-b}, b > 0$	$i\sqrt{b}$
	$\bar{z}$	the conjugate of z
[5.1]	(a, b)	the ordered pair of numbers whose first component is a and whose second component is b
	$R \times R$, or R^2	the Cartesian product of R and R
	f, g, h, F, etc.	symbols denoting functions
	$f(x)$	f of x, or the value of f at x
	$f(x)\vert_a^b$	$f(b) - f(a)$
[5.2]	d	the distance between two points
	m	the slope of a line
[5.4]	f^{-1}	the inverse function of f
[5.5]	$[x]$	the greatest integer not greater than x
[7.1]	e	an irrational number, approximately equal to 2.7182818
[7.2]	$\log_b x$	the logarithm to the base b of x, or the logarithm of x to the base b
[7.3]	$\text{antilog}_b x$	the antilogarithm to the base b of x, or the antilogarithm of x to the base b for $b = 10$ and $b = e$
[8.2]	(x, y, z)	the ordered triple of numbers whose first component is x, second component is y, and third component is z

The Fifth Edition of College Algebra

Edwin F. Beckenbach
University of California, Los Angeles

Irving Drooyan
Los Angeles Pierce College

William Wooton

Wadsworth Publishing Company
Belmont, California

A division of Wadsworth, Inc.

Mathematics Editor: Richard Jones
Production: Greg Hubit Bookworks

© 1982 by Wadsworth, Inc.

© 1978, 1973, 1968, 1964 by Wadsworth Publishing Company, Inc. All rights reserved. No part of this book may be reproduced, stored in a retrieval system, or transcribed, in any form or by any means, electronic, mechanical, photocopying, recording, or otherwise, without the prior written permission of the publisher, Wadsworth Publishing Company, Belmont, California 94002, a division of Wadsworth, Inc.

Printed in the United States of America
1 2 3 4 5 6 7 8 9 10—86 85 84 83 82

Library of Congress Cataloging in Publication Data

Beckenbach, Edwin F.
  College Algebra.

  Includes index.
  1. Algebra. I. Drooyan, Irving. II. Wooton, William. III. Title.
QA154.2.B42   1982   512.9   81-13109
ISBN 0-543-01007-5   AACR2

Contents

1 Real Numbers and Their Properties — 1

    1.1 Preliminary Concepts — 1
    1.2 Real Numbers — 6
    1.3 Field Properties — 11
    1.4 Order in R — 15
    Chapter Review — 22

2 Polynomials; Rational Expressions — 24

    2.1 Definitions; Sums of Polynomials — 24
    2.2 Products of Polynomials — 30
    2.3 Factoring Polynomials — 34
    2.4 Quotients of Polynomials — 38
    2.5 Synthetic Division — 42
    2.6 Equivalent Fractions — 46
    2.7 Sums and Differences of Rational Expressions — 50
    2.8 Products and Quotients of Rational Expressions — 53
    Chapter Review — 57

3 Rational Exponents—Radicals — 60

    3.1 Powers with Integral Exponents — 60
    3.2 Powers with Rational Exponents — 66
    3.3 Radical Expressions — 71
    3.4 Operations on Radical Expressions — 76
    3.5 Complex Numbers — 79
    Chapter Review — 86

4 Equations and Inequalities in One Variable — 89

- 4.1 Equivalent Equations; First-Degree Equations — 89
- 4.2 Second-Degree Equations — 95
- 4.3 The Quadratic Formula — 99
- 4.4 Equations Involving Radicals — 102
- 4.5 Substitution in Solving Equations — 105
- 4.6 Solution of Linear Inequalities — 107
- 4.7 Solution of Nonlinear Inequalities — 112
- 4.8 Equations and Inequalities Involving Absolute Value — 119
- 4.9 Word Problems — 122
- Chapter Review — 129

Supplemental Exercises for Chapters 1–4 — 131

5 Functions and Graphs — 134

- 5.1 Pairings of Real Numbers — 134
- 5.2 Linear Equations in Two Variables — 142
- 5.3 Forms for Linear Equations — 148
- 5.4 Inverse Functions — 153
- 5.5 Special Functions — 156
- 5.6 Graphs of First-Degree Inequalities — 160
- Chapter Review — 163

6 Nonlinear Relations and Functions — 165

- 6.1 Quadratic Functions — 165
- 6.2 Conic Sections — 173
- 6.3 Variation as a Functional Relationship — 180
- 6.4 Zeros of a Polynomial Function — 186
- 6.5 Rational Zeros of Polynomial Functions — 193
- 6.6 Graphing Polynomial Functions — 197
- 6.7 Rational Functions — 201
- Chapter Review — 208

7 Exponential and Logarithmic Functions — 211

- 7.1 Exponential Functions — 211
- 7.2 Logarithmic Functions — 214
- 7.3 Special Logarithms and Powers — 219
- 7.4 Solution of Exponential Equations; Applications — 228
- Chapter Review — 236

Supplemental Exercises for Chapters 5–7 — 239

Contents

8 Systems of Equations and Inequalities — **241**

- 8.1 Systems of Linear Equations in Two Variables — 241
- 8.2 Systems of Linear Equations in Three Variables — 249
- 8.3 Partial Fractions — 255
- 8.4 Systems of Nonlinear Equations — 262
- 8.5 Systems of Inequalities — 272
- 8.6 Convex Sets—Polygonal Regions — 274
- 8.7 Linear Programming — 277
- Chapter Review — 280

9 Matrices and Determinants — **282**

- 9.1 Definitions; Matrix Addition — 282
- 9.2 Matrix Multiplication — 288
- 9.3 Solution of Linear Systems by Using Row-Equivalent Matrices — 295
- 9.4 The Determinant Function — 304
- 9.5 Properties of Determinants — 310
- 9.6 The Inverse of a Square Matrix — 315
- 9.7 Solution of Linear Systems Using Matrix Inverses — 322
- 9.8 Cramer's Rule — 326
- Chapter Review — 330

Supplemental Exercises for Chapters 8 and 9 — ***332***

10 Sequences and Series — **334**

- 10.1 Sequences — 334
- 10.2 Series — 341
- 10.3 Limits of Sequences and Series — 347
- 10.4 The Binomial Theorem — 353
- 10.5 Mathematical Induction — 361
- Chapter Review — 365

11 Counting and Probability — **367**

- 11.1 Basic Counting Principles; Permutations — 367
- 11.2 Combinations — 374
- 11.3 Probability Functions — 377
- 11.4 Probability of the Union of Events — 380
- 11.5 Probability of the Intersection of Events — 383
- Chapter Review — 388

Supplemental Exercises for Chapters 10 and 11 — ***390***

Appendices — 393

A *Translation of Axes; Parametric Equations* — 394

    A.1 Translation of Axes — 394
    A.2 Parametric Representations of Relations — 397

B *Mathematical Structure* — 401

    Axioms for Real Numbers — 401
    Properties of Numbers — 402
    Mathematical Systems — 411

C *Tables* — 413

    Table I   Exponential Functions, Base e — 413
    Table II   Common Logarithms — 414
    Table III Natural Logarithms — 416
    Table IV Squares, Square Roots, and Prime Factors — 417

Answers to Odd-Numbered Exercises — 418

Index — 465

Preface

The Fifth Edition of College Algebra is designed, as was its predecessors, to provide a contemporary mathematics course for first-year college students.

In preparing this edition, we have sought to clarify and simplify the material. The number of examples in each chapter has been increased. Each exercise set has been revised and graded for difficulty, and most sets have been expanded. In addition to these revisions we have made the following changes:

> Chapter 1 has been revised to include interval notation with the order axioms for real numbers.
>
> Chapter 6 now includes material on graphing polynomials by finding turning points analytically.
>
> Chapter 7 has been revised to allow the use of a calculator in teaching exponents and logarithms.
>
> Chapter 8 now includes material on systems of equations with complex-number solutions.
>
> Chapter 9 now includes material on finding the inverse of a matrix by row reduction.
>
> Appendix A has been added to include a treatment of translation of axes and a discussion of parametric equations.
>
> Supplementary exercise sets, containing exercises which typify the algebraic problems encountered in more advanced courses, have been added after Chapters 4, 7, 9, and 11.

Portions of the preface to the fourth edition that are applicable to the fifth edition follow.

The first chapter, which introduces the student to the complete ordered field of real numbers, is followed by three chapters—polynomials, rational exponents, and equations and inequalities in one variable—that may be optional for the student who shows a mastery of second-year high-school algebra. Beginning with

Chapter 5, discussions generally center around the function concept. Polynomial and rational functions (and their graphs) are covered in detail; variation is treated from the function standpoint; logarithms are developed from a consideration of exponential functions; determinants are presented as functions of matrices; sequences are treated as functions having sets of positive integers as domain; and probability is discussed from a set-function standpoint. Appendix B includes the basic substance of the text in compact form and is a handy guide for following the development of the course. It also includes a brief treatment of mathematical systems as related to the number systems considered in the text.

We have organized the contents with an eye to providing material for a variety of courses of varying scope and depth. The book can be used for a semester course of three, four, or five units, or for two three-unit quarter courses. Chapters 7–11 are sufficiently independent of each other that any may be omitted for a short course.

A section at the end of the book provides answers for odd-numbered exercises, along with graphs.

As in the fourth edition, a second color is used functionally to highlight key procedures in routine manipulations and to focus attention on key elements of figures. Marginal annotations direct the reader's attention to important ideas.

A solutions manual and a study guide covering topics in the text are available for student use. Ancillary materials that relate to the text are also available to instructors.

We sincerely thank Professor Michael D. Grady of Loyola Marymount University for extensive help in preparing this edition.

Edwin F. Beckenbach
Irving Drooyan
William Wooton

1 Real Numbers and Their Properties

In earlier mathematics courses, you explored the basic properties of real numbers and saw how some of these properties apply to real-life situations. In elementary algebra, you manipulated symbols which represented numbers. Such manipulations will continue in this course, but the symbols may represent objects which belong to systems other than the system of real numbers. However, the purpose of this first chapter is to review, briefly, some facts about the real-number system.

1.1 Preliminary Concepts

In algebra, we are interested in various *sets* of numbers and in their relations to *sets* of points on a line or in a plane or in space. In this section we shall review some basic concepts and terminology in order to provide a convenient mathematical language for what follows.

Sets

A **set** is simply a collection of objects. Any one of the objects of a collection is called a **member** or an **element** of the set. For example, the collection of numbers 2, 4, 6, 8 is a set, and the number 2 is a member or an element of that set. Sets are usually designated by means of capital letters, A, B, C, etc. Sets are also identified by means of **braces**, { }, with the members either listed or described. For example, the set $\{1, 2, 3, \ldots\}$ (where the three dots indicate that the pattern established continues indefinitely) is denoted by N. This set is called the set of **natural numbers**.

The symbol $\in$ (read "is a member of" or "is an element of") is used to denote membership in a set. Thus, for example,

$$2 \in \{1, 2, 3\} \quad \text{and} \quad 127 \in N.$$

Definition 1.1 Two sets A and B are **equal**, $A = B$, if and only if they have the same elements.

The order in which the elements of a set are named is of no importance in determining its membership, nor is the fact that an element might be named more than once. Thus, if A denotes $\{1, 2, 3\}$, B denotes $\{3, 2, 1\}$, and C denotes $\{1, 2, 1, 3, 1\}$, then $A = B = C$.

Definition 1.2 If every element of a set A is an element of a set B, then A is a **subset** of B.

The symbol $\subset$ (read "is a subset of" or "is contained in") will be used to denote the subset relationship. Thus,

$$\{1, 2, 3\} \subset \{1, 2, 3, 4\} \quad \text{and} \quad \{1, 2, 3\} \subset \{1, 2, 3\}.$$

Observe that, by definition, every set is a subset of itself. Note that we write

$$\{2\} \subset \{1, 2, 3\} \quad \text{and} \quad 2 \in \{1, 2, 3\},$$

since $\{2\}$ is a *subset*, whereas 2 is an *element*, of $\{1, 2, 3\}$.

The set that contains no elements is called the **empty set**, or **null set**, and is denoted by the symbol $\emptyset$ (read "the empty set" or "the null set"); $\emptyset$ is a subset of every set.

If a set S is the null set or contains exactly n elements for some fixed natural number n, then S is said to be **finite**. A set that is not finite is said to be **infinite**. For example, the set of *all* natural numbers, $\{1, 2, 3, \ldots\}$, is an infinite set.

Union of two sets

Definition 1.3 The **union** of two sets A and B is the set of all elements that belong either to A or to B or to both.

The symbol $\cup$ is used to denote the union of sets. Thus $A \cup B$ (read "the union of A and B") is the set of all elements that are in A or B or both. For example, if $A = \{1, 2, 3, 4\}$ and $B = \{3, 4, 5\}$, then

$$A \cup B = \{1, 2, 3, 4, 5\}.$$

Notice that although 3 and 4 are elements of both A and B, it is sufficient to list them only once when identifying $A \cup B$.

Intersection of two sets

Definition 1.4 The **intersection** of two sets A and B is the set of all elements common to both A and B.

The symbol $\cap$ is used to denote the intersection of sets. Thus $A \cap B$ (read "the intersection of A and B") is the set of all elements that are in both A and B. For example, if $A = \{1, 2, 3, 4\}$ and $B = \{3, 4, 5\}$, then

$$A \cap B = \{3, 4\}.$$

1.1 Preliminary Concepts

If two sets A and B contain no element in common, A and B are said to be **disjoint**. That is, A and B are disjoint if and only if $A \cap B = \emptyset$. For example, if $A = \{1, 2, 3\}$ and $B = \{5, 6, 7\}$, then A and B are disjoint.

Complements

When discussing sets, it is often helpful to have in mind some general set from which the elements of all sets under consideration are drawn. For example, if we wish to talk about sets of integers, we may use as a general set the set of all integers; or taking a larger view we might use the set of all real numbers, or any one of many possible general sets. Such a general set is called the **universal set** and is usually denoted simply by the capital letter U.

Definition 1.5   The **complement** of a set A in a universal set U is the set of all elements of U not in A.

The symbol A' denotes the complement of A. Thus if $U = \{1, 2, 3, 4, 5, 6\}$ and $A = \{1, 2, 3, 4\}$, then

$$A' = \{5, 6\}.$$

Variables

When discussing an individual but unspecified element of a set we usually denote the element by a lowercase italic letter (for example, a, d, s, x), or sometimes by a letter from the Greek alphabet: α (alpha), β (beta), γ (gamma), and so on. Symbols used in this way are called *variables*.

Definition 1.6   A **variable** is a symbol representing an unspecified element of a given set.

The given set is called the **replacement set**, or **domain**, of the variable. For example,

$$x \in A$$

means that the variable x represents an (unspecified) element of the set A. The elements of the replacement set are called the **values** of the variable. A symbol which represents a fixed value in a given discussion is called a **constant**.

Negation

The slant bar $/$, drawn through certain symbols of relation, is used to indicate negation. Thus $\neq$ is read "is not equal to," $\not\subset$ is read "is not a subset of," and $\notin$ is read "is not an element of." For example,

$$\{1, 2\} \neq \{1, 2, 3\}, \quad \{1, 2, 3\} \not\subset \{1, 2\}, \quad \text{and} \quad 3 \notin \{1, 2\}.$$

Set-builder notation

Another symbolism useful in discussing sets is $\{x \mid x \text{ has a certain property}\}$; for example,

$$\{x \mid x \in A \quad \text{and} \quad x \notin B\}$$

(read "the set of all x such that x is a member of A and is not a member of B"). This symbolism, called **set-builder notation**, is used in several places in this book. What it does is to specify a variable (in this case, x) and, at the same time, state a condition on the variable (in this case, that x is contained in the set A and is not contained in the set B).

Exercise 1.1

A *Designate each of the following sets using braces and listing the members.*

Examples a. {natural numbers less than 7}   b. {natural numbers between 2 and 4}

Solutions a. {1, 2, 3, 4, 5, 6}   b. {3}

1. {natural numbers between 3 and 11 inclusive}
2. {natural numbers between 31 and 35 inclusive}
3. {letters in the word "mathematics"}
4. {letters in the words "college algebra"}
5. {days in the week}
6. {months in the year}

In Exercises 7–36, let $U = \{1, 2, 3, 4, 5, 6, 7, 8\}$, $A = \{1, 2, 3, 4\}$, $B = \{5, 6, 7, 8\}$, $C = \{2, 4, 6, 8\}$, *and* $D = \{1, 3, 5, 7\}$.

Replace the asterisk * *with either* $\in$ *or* $\notin$ *to make a true statement.*

7. $3 * A$
8. $3 * B$
9. $\{3\} * C$
10. $\{3\} * D$

Replace the asterisk * *with either* $=$ *or* $\neq$ *to make a true statement.*

11. $A * \{12, 34\}$
12. $A * \{4, 3, 2, 1\}$
13. $A * \{1, 1, 2, 2, 3, 4\}$
14. $A * \{1234\}$

Replace the asterisk * *with either* $\subset$ *or* $\not\subset$ *to make a true statement.*

15. $B * \{4, 5, 6, 7\}$
16. $B * \{5, 6, 7\}$
17. $B * \{5, 6, 7, 8\}$
18. $B * \{4, 5, 6, 7, 8\}$

1.1 Preliminary Concepts

List the elements of the following sets.

19. $A \cup C$
20. $B \cup D$
21. $B \cap C$
22. $A \cap D$
23. A'
24. C'
25. $(A \cap C)'$
26. $(B \cap D)'$
27. $(A \cup C)'$
28. $(B \cup D)'$
29. $A' \cup D$
30. $B' \cup C$
31. $(A' \cup C) \cap D$
32. $(B' \cap C) \cup D$
33. $(A \cap C') \cap D$
34. $(B \cap C) \cap D$
35. $A \cup (D \cap \emptyset)$
36. $B \cap (C \cup \emptyset)$

Designate the given set in set-builder notation. Use N to denote the set of natural numbers.

Example {even natural numbers}

Solution $\{x \mid x = 2n, n \in N\}$

37. {odd natural numbers}
38. {negative integers}
39. {multiples of 3 in N}
40. {multiples of 5 in N}
41. {natural numbers less than 100}
42. {natural numbers greater than 5}

Describe the following sets in terms of A, B, and C, and unions, intersections, and complements.

Examples   a. $\{x \mid x \in A \text{ and } x \notin B\}$ b. $\{x \mid x \in A \text{ or } x \in B\}$

Solutions   a. $A \cap B'$ b. $A \cup B$

43. $\{x \mid x \in B \text{ and } x \in C\}$
44. $\{x \mid x \in A \text{ or } x \in C'\}$
45. $\{x \mid x \in A \text{ and } x \text{ is an element of } B \text{ or } C\}$
46. $\{x \mid x \notin A \text{ and } x \notin B\}$
47. $\{x \mid x \in A \text{ but } x \text{ is not an element of either } B \text{ or } C\}$
48. $\{x \mid x \in A \text{ and } x \in B \text{ and } x \notin C\}$

B 49. Let U be a set with three elements. Count all the subsets of U. Repeat for sets with four and five elements and use these results to find a formula for the number of subsets of a set with n elements.

50. Explain why, for any two sets A and B, $A \subset A \cup B$.
51. Explain why, for any two sets A and B, $A \cap B \subset A$.
52. Explain why $(A')' = A$.

1.2 Real Numbers

We shall frequently refer to the following five sets of numbers.

1. The set N of **natural numbers**, whose elements are the counting numbers:
$$N = \{1, 2, 3, \ldots\}.$$

2. The set J of **integers**, whose elements are the counting numbers, their negatives, and zero:
$$J = \{\ldots, -2, -1, 0, 1, 2, \ldots\}.$$

3. The set Q of **rational numbers**, whose elements are all those numbers that can be represented as the quotient of two integers $\dfrac{a}{b}$ (or a/b, or $a \div b$), where b is not 0. Among the elements of Q are such numbers as $-3/4$, $18/27$, $3/1$, and $-6/1$. In symbols,
$$Q = \left\{x \,\middle|\, x = \frac{a}{b}, \quad a, b \in J, \quad b \neq 0\right\}.$$

Every rational number can be represented by a terminating or repeating decimal numeral, such as 3.2, 1.975, 6.3333..., and 2.171717... .

4. The set H of **irrational numbers**, whose elements are the numbers with decimal representations that are nonterminating and nonrepeating. Among the elements of this set are such numbers as $\sqrt{2}$, $\sqrt{3}$, $\sqrt{5}$, $\sqrt{7}$, π, and $-\sqrt{7}$. An irrational number cannot be represented in the form a/b, where a and b are integers.

5. The set R of **real numbers**, which contains all of the elements of the set of rational numbers and all of the elements of the set of irrational numbers:
$$R = Q \cup H.$$

The foregoing sets of numbers are related as indicated in Figure 1.1. Thus we have
$$H \subset R, \quad Q \subset R, \quad \text{and} \quad N \subset J \subset Q.$$

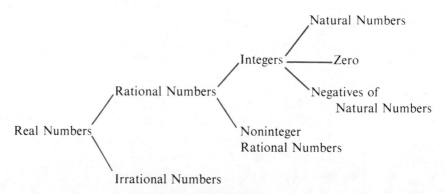

Figure 1.1

1.2 Real Numbers

Axioms

For a given mathematical system, there are some properties of that system we accept without proof. These properties are called **axioms** or **postulates**. The words **law** and **principle** are sometimes used in referring to axioms or postulates, although these words may also be applied to certain consequences of the axioms or postulates of the system.

Equality axioms

The first properties to be considered here have to do with equality. We accept the following statements about the equality relation as axioms. For any elements a, b, and c of a set S, the equality ($=$) relationship satisfies the following laws:

E-1 $a = a$ *Reflexive law for equality.*

E-2   If $a = b$, then $b = a$.   *Symmetric law for equality.*

E-3   If $a = b$ and $b = c$, then $a = c$.   *Transitive law for equality.*

E-4   If $a = b$, then a may be replaced by b and b by a in any mathematical statement without altering the truth or falsity of the statement.†   *Substitution law for equality.*

The equality axioms indicate how we use symbols and warn that we must neither change the meaning of a symbol in the middle of a discussion nor use the same symbol for two different things in the same context.

Other axioms, somewhat different conceptually, are used to characterize mathematical systems. They are axioms concerning the behavior of elements of sets under **binary operations**, that is, operations which involve using two elements of a set to produce a third element in the set. Thus, for example, addition and multiplication are binary operations on the set R of real numbers. To state that the element produced by a binary operation on two elements of a set A is also in the set A, we say that A is **closed** under the operation.

Addition and multiplication axioms

Since the set R of real numbers is the set of greatest interest to us at the moment, we shall state the properties of addition and multiplication in R which we shall accept as axioms. You are probably familiar with these laws from your earlier work. They are listed on page 8 for reference. Parentheses are used in stating some of the relations to indicate that symbols within parentheses are to be viewed as representing a single entity.

If, for a given set and a given pair of binary operations (not necessarily the set of real numbers or ordinary addition and multiplication, Axioms F-1 through F-11 are satisfied by the elements of the set, then the set is called a **field** under these operations. Thus, we speak of the **field R of real numbers**.

---

† We have adopted a very powerful axiom in E-4, one that includes E-2 and E-3 as special cases. By wording E-4 as we have, we can eliminate a great deal of detail in later arguments; and, because E-2 and E-3 are fundamental properties of the equality relation, we have elected to retain them as axioms.

Field axioms of the set R of real numbers

Let a, b, c be arbitrary elements of R.

F-1 $a + b$ is a unique element of R. — *Closure law for addition.*

F-2 $(a + b) + c = a + (b + c)$. — *Associative law for addition.*

F-3 There exists an element $0 \in R$ (called the **identity element for addition**) with the property
$$a + 0 = a \quad \text{and} \quad 0 + a = a$$
for each $a \in R$. — *Additive-identity law.*

F-4 For each $a \in R$, there exists an element $-a \in R$ (called the **additive inverse**, or **negative**, of a) with the property
$$a + (-a) = 0 \quad \text{and} \quad (-a) + a = 0.$$
— *Additive-inverse law.*

F-5 $a + b = b + a$. — *Commutative law for addition.*

F-6 $a \times b$ is a unique element of R. — *Closure law for multiplication.*

F-7 $(a \times b) \times c = a \times (b \times c)$. — *Associative law for multiplication.*

F-8 There exists an element $1 \in R$ (called the **identity element for multiplication**), $1 \neq 0$, with the property
$$a \times 1 = a \quad \text{and} \quad 1 \times a = a$$
for each $a \in R$. — *Multiplicative-identity law.*

F-9 For each element $a \in R$, $a \neq 0$, there exists an element $\frac{1}{a} \in R$ (called the **multiplicative inverse**, or **reciprocal**, of a) with the property
$$a \times \frac{1}{a} = 1 \quad \text{and} \quad \frac{1}{a} \times a = 1.$$
— *Multiplicative-inverse law.*

F-10 $a \times b = b \times a$. — *Commutative law for multiplication.*

F-11 $a \times (b + c) = (a \times b) + (a \times c)$
and
$(b + c) \times a = (b \times a) + (c \times a)$. — *Distributive law.*

1.2 Real Numbers

To give meaning to expressions $a + b + c$, $a \times b \times c$, $a + b + c + d$, and so on, let us make the following agreement.

Definition 1.7 If $a, b, c, d, \ldots \in R$, then
$$a + b + c = a + (b + c), \qquad a + b + c + d = a + (b + c + d), \ldots,$$
and
$$a \times b \times c = a \times (b \times c), \qquad a \times b \times c \times d = a \times (b \times c \times d), \ldots.$$

The associative properties F-2 and F-7 ensure that $a + b + c$, $a \times b \times c$, etc., can actually be evaluated in any order.

Inverse operations

In terms of the operations of addition and multiplication, we can define two additional operations on real numbers: **subtraction** (finding a *difference*) and **division** (finding a *quotient*).

Definition 1.8 The **difference** of elements $a \in R$ and $b \in R$, denoted by $a - b$, is given by
$$a - b = a + (-b).$$

Definition 1.9 The **quotient** of elements $a \in R$ and $b \in R$, $b \neq 0$, denoted by $\frac{a}{b}$, a/b, or $a \div b$, is given by
$$\frac{a}{b} = a \times \frac{1}{b}.$$

In all that follows, we shall adopt the customary practice of writing ab or $a \cdot b$ for $a \times b$. Observe that the quotient $\frac{a}{b}$ requires $b \neq 0$; division by 0 is not defined.

Exercise 1.2

A *State whether each statement is true or false. The sets N, J, Q, H, and R are as given in the text.*

1. $-1 \in N$
2. $0 \in Q$
3. $\sqrt{3} \in R$
4. $\frac{1}{2} \notin H$
5. $0.125 \in Q$
6. $0.1717\ldots \in Q$
7. $Q \cap H = \{0\}$
8. $Q \cap H = \emptyset$
9. $J \subset H$
10. $J \cap Q = N$
11. $N \not\subset Q$
12. $\{1, 2\} \not\subset J$

Each variable in Exercises 13–50 denotes a real number. Each of the statements 13–20 is an application of one of the axioms E-1 through E-4. Justify the statement by citing an appropriate axiom. (*There may be more than one correct justification.*)

Examples a. If $x + y = 8$ and $y = 3$, then $x + 3 = 8$.

b. If $x + y = x + z$, then $x + z = x + y$.

Solutions a. Substitution law for equality, E-4.

b. Symmetric law for equality, E-2.

13. If $x = 3z$, then $3z = x$.
14. $3 + x = 3 + x$
15. If $a + b = b + c$ and $b + c = c + d$, then $a + b = c + d$.
16. If $x = 3y$ and $3y = 6z$, then $x = 6z$.
17. If $x = 6$ and $3x = y$, then $3 \cdot 6 = y$.
18. If $a + b = 1$ and $c - (a + b) = d$, then $c - 1 = d$.
19. $2a - (b + c) = 2a - (b + c)$
20. If $a + b = b + c$, then $b + c = a + b$.

Each statement in Exercises 21–36 is an application of one of the axioms F-1 through F-11. Justify the statement by citing the appropriate axiom.

Examples a. $3 \cdot (x + 5) = (x + 5) \cdot 3$

b. $3(x + 5) = 3 \cdot x + 3 \cdot 5$

Solutions a. Commutative law for multiplication, F-10.

b. Distributive law, F-11.

21. $x + (y + 0) = (x + y) + 0$
22. $(3x)y = y(3x)$
23. $(p + q)(r + s) = (r + s)(p + q)$
24. $(a + 1) \cdot 1 = a + 1$
25. $3a + 0 = 3a$
26. $(a + 1) \cdot 1 = a \cdot 1 + 1 \cdot 1$
27. $(kl)(m + n) = k[l(m + n)]$
28. $1 + (1 + 1) = (1 + 1) + 1$
29. $(p + q)(r + s) = p(r + s) + q(r + s)$
30. $0 + (3 + x) = 3 + x$
31. $1 \cdot (1 + 1) = 1 + 1$
32. $[5(x + y)] \cdot 2 = 5[(x + y) \cdot 2]$
33. $(a + 3) + [-(a + 3)] = 0$
34. $(0 + k) \cdot 1 = 0 \cdot 1 + k \cdot 1$
35. $(a + 1)(b + 1) = (a + 1) \cdot b + (a + 1) \cdot 1$
36. $5x + (-5x) = 0$

1.3 Field Properties

Use the distributive laws to rewrite any product among the following as a sum and any sum as a product.

Examples

a. $3(7 + p)$ **b.** $8 + 4m$

Solutions

a. $3(7 + p) = 3 \cdot 7 + 3p$ **b.** $8 + 4m = 4 \cdot 2 + 4m$
$\qquad\qquad\; = 21 + 3p$ $\qquad\qquad\qquad\; = 4(2 + m)$

37. $a(b + 1)$
38. $(a + 1)b$
39. $p(q + r)$
40. $ab + cb$
41. $(d + 3)d$
42. $(18 + 2)10$
43. $(de)f + gf$
44. $xy + xyz$
45. $(13 + q)r$
46. $a(2 + b)$
47. $st + su + sv$
48. $ab + (-1)b$
49. $3(x + y + z)$
50. $xyz + xy + xz$

1.3 Field Properties

The field axioms together with the axioms for equality imply other properties of the real numbers. Such implications are generally stated as theorems. A **theorem** is simply a statement of a fact that follows logically from the axioms (or postulates) and other theorems. We shall list (without proof) some of the theorems ordinarily encountered in lower-level algebra courses. We have numbered these theorems to provide an efficient way to refer to them later and have also named those that have commonly accepted names. In most cases, illustrative examples are provided.

Addition law for equality

First, consider the following result, which reaffirms the uniqueness of the sum of two real numbers.

Theorem 1.1 If $a, b, c \in R$ and $a = b$, then
$$a + c = b + c \quad \text{and} \quad c + a = c + b.$$

Example If $x - 3 = 5$, then $(x - 3) + 3 = 5 + 3$.

Multiplication law for equality

A theorem closely analogous to Theorem 1.1 can be stated as follows.

Theorem 1.2 If $a, b, c \in R$ and $a = b$, then
$$ac = bc \quad \text{and} \quad ca = cb.$$

Example

If $5z = 15$, then $\frac{1}{5}(5z) = \frac{1}{5}(15)$.

Additional properties

The next theorem states that the additive inverse of a real number and the multiplicative inverse of a nonzero real number are unique—that is, that a given real number has only one additive inverse and only one multiplicative inverse.

Theorem 1.3

I If $a, b \in R$ and $a + b = 0$, then
$$b = -a \quad \text{and} \quad a = -b.$$

II If $a, b \in R$ and $a \cdot b = 1$, then
$$a = \frac{1}{b} \quad \text{and} \quad b = \frac{1}{a} \quad (a, b \neq 0).$$

Examples

a. If $x + 3 = 0$, then $x = -3$ and $3 = -x$.

b. If $4x = 1$, then $x = \frac{1}{4}$ and $4 = \frac{1}{x}$.

The following theorems assert other useful properties of real numbers.

Theorem 1.4 If $a, b, c \in R$ and $a + c = b + c$, then $a = b$.

Theorem 1.5 If $a, b, c \in R$, $c \neq 0$, and $ac = bc$, then $a = b$.

Theorem 1.6 For every $a \in R$, $a \cdot 0 = 0$.

Theorem 1.7 If $a, b \in R$ and $a \cdot b = 0$, then either $a = 0$ or $b = 0$ or both.

Examples

a. If $y + 2 = 3 + 2$, then $y = 3$.   b. If $5z = 5 \cdot 3$, then $z = 3$.

c. $3 \cdot 0 = 0$ d. If $5(x + y) = 0$, then $x + y = 0$.

Combining Theorems 1.6 and 1.7, we see that for $a, b \in R$ we have $ab = 0$ *if and only if* either $a = 0$ or $b = 0$ or both.

Law of signs

The following theorem concerns the familiar "laws of signs" for operating with real numbers.

1.3 Field Properties

Theorem 1.8   If $a, b \in R$, then

I $-(-a) = a,$ II $(-a) + (-b) = -(a + b),$

III $(-a)(b) = -(ab),$ IV $(-a)(-b) = ab,$

V $\dfrac{-a}{b} = \dfrac{a}{-b} = -\dfrac{a}{b}$ $(b \neq 0),$ VI $\dfrac{-a}{-b} = \dfrac{a}{b}$ $(b \neq 0).$

Examples

a. $-(-10) = 10$

b. $(-1) + (-3) = -(1 + 3) = -4$

c. $\left(-\dfrac{1}{2}\right) \cdot \dfrac{1}{3} = -\dfrac{1}{6}$

d. $(-2)\left(-\dfrac{1}{4}\right) = 2\left(\dfrac{1}{4}\right) = \dfrac{1}{2}$

e. $\dfrac{-6}{3} = \dfrac{6}{-3} = -\dfrac{6}{3} = -2$

f. $\dfrac{-8}{-2} = \dfrac{8}{2} = 4$

Properties of Quotients

The following theorem gives a condition for the equality of quotients.

Theorem 1.9   If $a, b, c \in R$, then

$$\dfrac{a}{b} = \dfrac{c}{d} \quad \text{if and only if} \quad ad = bc \quad (b, d \neq 0).$$

Example $\dfrac{x}{5} = \dfrac{2}{3}$ if and only if $3x = 2 \cdot 5 = 10.$

As a direct consequence of the characterization of equal quotients given in Theorem 1.9, we have a theorem that is sometimes referred to as the **fundamental principle of fractions**.

Theorem 1.10   If $a, b, c \in R$, then

$$\dfrac{ac}{bc} = \dfrac{a}{b} \quad (b, c \neq 0).$$

Examples

a. $\dfrac{4}{12} = \dfrac{1 \cdot 4}{3 \cdot 4} = \dfrac{1}{3}$

b. $\dfrac{3}{5} = \dfrac{3 \cdot 2}{5 \cdot 2} = \dfrac{6}{10}$

Note that this theorem can be used either to "reduce" fractions as shown in Example (a) above or to "build up" fractions as shown in Example (b) above.

Theorem 1.11   If $a, b, c, d \in R$, then

I $\dfrac{1}{a} \cdot \dfrac{1}{b} = \dfrac{1}{ab}$ $(a, b \neq 0)$,

II $\dfrac{a}{b} \cdot \dfrac{c}{d} = \dfrac{ac}{bd}$ $(b, d \neq 0)$,

III $\dfrac{a}{c} + \dfrac{b}{c} = \dfrac{a+b}{c}$ $(c \neq 0)$,

IV $\dfrac{a}{b} + \dfrac{c}{d} = \dfrac{ad + bc}{bd}$ $(b, d \neq 0)$,

V $\dfrac{a}{b} - \dfrac{c}{d} = \dfrac{ad - bc}{bd}$ $(b, d \neq 0)$,

VI $\dfrac{1}{\frac{a}{b}} = \dfrac{b}{a}$ $(a, b \neq 0)$,

VII $\dfrac{\frac{a}{b}}{\frac{c}{d}} = \dfrac{a}{b} \div \dfrac{c}{d} = \dfrac{ad}{bc}$ $(b, c, d \neq 0)$.

Examples

a. $\dfrac{1}{2} \cdot \dfrac{1}{5} = \dfrac{1}{2 \cdot 5} = \dfrac{1}{10}$

b. $\dfrac{2}{3} \cdot \dfrac{5}{7} = \dfrac{2 \cdot 5}{3 \cdot 7} = \dfrac{10}{21}$

c. $\dfrac{2}{7} + \dfrac{3}{7} = \dfrac{2 + 3}{7} = \dfrac{5}{7}$

d. $\dfrac{1}{5} + \dfrac{3}{8} = \dfrac{1 \cdot 8 + 5 \cdot 3}{5 \cdot 8} = \dfrac{23}{40}$

e. $\dfrac{1}{\frac{5}{8}} = \dfrac{8}{5}$

f. $\dfrac{\frac{2}{3}}{\frac{7}{11}} = \dfrac{2}{3} \div \dfrac{7}{11} = \dfrac{2 \cdot 11}{3 \cdot 7} = \dfrac{22}{21}$

Exercise 1.3

A   In Exercises 1–20, each statement is justified by one part of Theorems 1.1–1.11. Cite an appropriate justification. All variables denote elements of the set R of real numbers. Assume no denominators are zero.

Examples

a. $-[x + (-y)] = -x + [-(-y)]$

b. If $x - 3 = 1$, then $(x - 3) + 3 = 1 + 3$.

Solutions

a. Theorem 1.8-II.

b. Theorem 1.1.

1. $\dfrac{x}{3} \cdot \dfrac{y}{5} = \dfrac{xy}{15}$

2. If $5(xy) = 15$, then $xy = 3$.

3. If $x + 5 = 15$, then $x = 10$.

4. $\left(-\dfrac{1}{3}\right)\left(\dfrac{1}{5}\right) = -\dfrac{1}{15}$

5. If $3(x + y) = 0$, then $x + y = 0$.

6. If $p + \dfrac{1}{2} = 0$, then $p = -\dfrac{1}{2}$.

1.4 Order in R

7. $\dfrac{2}{5} \div \dfrac{3}{7} = \dfrac{2 \cdot 7}{5 \cdot 3}$

8. $\dfrac{a}{2} + \dfrac{b}{3} = \dfrac{3a + 2b}{6}$

9. $\dfrac{xy}{xz} = \dfrac{y}{z}$

10. $x - (-y) = x + y$

11. If $3x = 5y$, then $\dfrac{x}{5} = \dfrac{y}{3}$.

12. $(-8)(-2) = 16$

13. $\dfrac{a+b}{2} + \dfrac{c}{3} = \dfrac{3(a+b) + 2c}{6}$

14. If $x - 2 = 5$, then $(x - 2) + 2 = 7$.

15. If $\dfrac{x}{2} = 5$, then $x = 10$.

16. $\dfrac{1}{x} + \dfrac{1}{y} = \dfrac{y + x}{xy}$

17. If $\dfrac{x}{5} = \dfrac{3}{10}$, then $10x = 15$.

18. If $10x = 15$, then $x = \dfrac{15}{10}$.

19. $\dfrac{2x+1}{2} + \dfrac{x}{3} = \dfrac{3(2x+1) + 2x}{6}$

20. $\dfrac{\dfrac{2x+1}{2}}{\dfrac{x}{3}} = \dfrac{(2x+1) \cdot 3}{2x}$

1.4 Order in R

In the preceding section we discussed some of the properties of multiplication and addition of real numbers. In this section we shall state axioms of order for the real numbers and discuss some of the properties of the real numbers under these axioms.

The real number line

We shall accept as an axiom the statement that a geometric line can be scaled in such a way that to each real number there corresponds one and only one point on the line. To illustrate this, we imagine the line scaled in convenient units, with the positive direction (from 0 toward 1) denoted by an arrowhead. The line is then called a **number line**; the real number corresponding to a point on the line is called the **coordinate** of the point, and the point is called the **graph** of the number. The point corresponding to 0 is called the **origin**. For example, number-line representations of 1, 3, and 5 are shown in Figure 1.2. Figure 1.2 may also be interpreted as the graph of the set {1, 3, 5}.

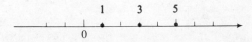

Figure 1.2

Positive and negative numbers

A horizontal number line with positive direction to the right can be separated into three disjoint subsets: {the points to the left of the origin}, {the origin}, {the points to the right of the origin}. The set of numbers whose elements are associated with the points on the right-hand side of the origin belong to the set R_+ of **positive real numbers**, and the set whose elements are associated with the points on the left-hand side belong to the set R_- of **negative real numbers**.

Order axioms

We now formally state the two order properties that we shall accept as axioms for the real number system.

O-1 If a is a real number, then exactly one of the following statements is true: a is a negative number, a is 0, or a is a positive number.   *Trichotomy law.*

O-2 If a and b are positive real numbers, then $a + b$ and $a \cdot b$ are positive real numbers.   *Closure law for positive real numbers.*

The first of these axioms asserts that every real number belongs to one of the sets R_+, {0}, or R_-, but to only one of them. The second asserts that the set R_+ of positive real numbers is closed with respect to the binary operations of addition and multiplication. These axioms along with the field axioms can be used to show that a real number a is positive if and only if $-a$ is negative.

Since the set R of real numbers satisfies axioms O-1 and O-2 as well as axioms F-1 through F-11, we say that R is an **ordered field**.

By Axiom O-2, if a and b are positive real numbers, then ab is a positive real number. This fact, together with Parts III and IV of Theorem 1.8, is sufficient to establish that the product of a positive real number and a negative real number is a negative real number, while the product of two negative real numbers is a positive real number.

Less than and greater than

The addition of a positive real number d to a real number a can be visualized on a number-line graph as the process of locating the point corresponding to a on the line and then moving along the line d units to the right to arrive at the point corresponding to $a + d$ (Figure 1.3). With this idea in mind, we define what is meant by "less than."

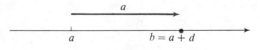

Figure 1.3

Definition 1.10 If $a, b \in R$, then a is **less than** b if and only if there exists a positive real number d such that $a + d = b$.

1.4 Order in R

The fact that d is positive implies, for any two real numbers a and b, that if the graph of a lies to the left of the graph of b, then a is less than b. The inequality symbol $<$ is used to denote the phrase "is less than," and $a < b$ is read "a is less than b." The inequality symbol $>$ means "is greater than." The statements $a < b$ and $b > a$ are taken as equivalent.

Definition 1.10 and the field axioms have the following implications, which we shall not prove.

Theorem 1.12 For any $a, b, c \in R$:

I   If $a < b$ and $b < c$, then $a < c$.

II If $a < b$, then $a + c < b + c$.

III If $a < b$ and $c > 0$, then $ac < bc$.

IV If $a < b$ and $c < 0$, then $ac > bc$.

Examples
a. $2 < 5$ and $5 < 7$, so $2 < 7$
b. $3 < 8$, so $3 + 5 < 8 + 5$, or $8 < 13$
c. $3 < 6$ and $2 > 0$, so $2 \cdot 3 < 2 \cdot 6$, or $6 < 12$
d. $2 < 7$ and $-2 < 0$, so $-2(2) > -2(7)$, or $-4 > -14$

As you can see, Theorem 1.12-I is comparable to the transitive law for equality. That is, we can say that "less than" is a transitive relationship.

Each of the symbols $\leq$ and $\geq$ (read "is less than or equal to" and "is greater than or equal to," respectively) is a contraction for two symbols, one of equality and one of inequality, connected by the word "or." For example, $x \leq 7$ is the statement that x is less than 7 *or* x equals 7.

Absolute value The graphs of the numbers a and $-a$ on a number line lie the same distance from the origin, but on opposite sides of it. If we wish to refer to the *distance* of the graph of a number from the origin, and not to the side of the origin on which it is located, then we use the term **absolute value**. Thus, the absolute value of a and the absolute value of $-a$ are the same nonnegative number. The symbol $|a|$ is used to denote the absolute value of a. We formalize the definition as follows:

Definition 1.11 If $a \in R$, then the **absolute value** of a is given by

$$|a| = \begin{cases} a, & \text{if } a \geq 0, \\ -a, & \text{if } a < 0. \end{cases}$$

For example,

$$|-3| = -(-3) = 3, \quad |7| = 7, \quad \text{and} \quad |0| = 0.$$

Geometric Interpretation of Order

We have noted in the foregoing discussion that certain algebraic statements concerning the order of real numbers can be interpreted geometrically. We summarize some of the more common correspondences in the following table, where in each case $a, b, c \in R$.

Algebraic statement	Geometric statement	Graph
1. a is positive	1. The graph of a lies to the right of the origin.	1. ———•——→ $\quad\quad\quad\;$ 0 $\;\;$ a
2. a is negative	2. The graph of a lies to the left of the origin.	2. ←——•———→ $\;\;$ a $\quad\;$ 0
3. $a > b$	3. The graph of a lies to the right of the graph of b.	3. ———•——•——→ $\quad\quad$ b $\;\;$ a
4. $a < b$	4. The graph of a lies to the left of the graph of b.	4. ———•——•——→ $\quad\quad$ a $\;\;$ b
5. $a < c < b$	5. The graph of c is to the right of the graph of a and to the left of the graph of b.	5. ——•——•——•——→ $\;\;$ a $\;\;$ c $\;\;$ b
6. $\lvert a \rvert < c$	6. The graph of a is less than c units from the origin.	6. ——┼——•——┼——→ $\;\;$ −c $\;$ 0 a $\;\;$ c
7. $\lvert a - b \rvert < c$	7. The graphs of a and b are less than c units from each other.	7. $d < c$ $\quad$ ⊢─d─⊣ $\quad\quad\quad\quad\quad$ a $\;\;$ b
8. $\lvert a \rvert < \lvert b \rvert$	8. The graph of a is closer to the origin than the graph of b.	8. ——┼——•——┼——→ $\;\;$ −|b| a 0 $\;\;$ |b|

Notice that statement 5, $a < c < b$, is a contraction of the two inequalities $a < b$ and $b < c$ and is read "a is less than b and b is less than c." Similarly, $0 < x$ and $x \le 1$ can be written together as $0 < x \le 1$.

Interval notation

A notation called **interval notation** is sometimes used to denote sets such as $\{x \mid -2 < x \le 5\}$, which can be called *intervals* of real numbers. Such notation involves the use of a parenthesis to denote an open endpoint and a bracket to denote a closed endpoint. Thus,

$$(-2, 5] = \{x \mid -2 < x \le 5\}.$$

For an infinite interval such as $\{x \mid x < 6\}$ we would write $(-\infty, 6)$, where the symbol $-\infty$ denotes the inclusion of all real numbers less than 6 in the interval. Similarly, $\{x \mid x \ge 4\}$ can be written as $[4, +\infty)$.

Intervals of real numbers can be graphed on a number line as shown in the following table.

1.4 Order in R

Interval notation	Graph
1. (a, b)	1. open circle at a, open circle at b
2. $[a, b)$	2. closed at a, open at b
3. $(a, b]$	3. open at a, closed at b
4. $[a, b]$	4. closed at a, closed at b
5. $(-\infty, a)$	5. ray to open at a
6. $(-\infty, a]$	6. ray to closed at a
7. (a, ∞)	7. open at a, ray right
8. $[a, \infty)$	8. closed at a, ray right

Note that an open circle indicates the number corresponding to the indicated point is *not* in the set.

Exercise 1.4

A *For $x, y \in R$, justify the given statement by citing one part of Theorem 1.12.*

Example If $-2x > 1$, then $x < -\dfrac{1}{2}$.

Solution Part IV. Each member of $-2x > 1$ is multiplied by $-\dfrac{1}{2}$.

1. If $x - y > 0$, then $x > y$.
2. If $x < 5$, then $2x < 10$.
3. If $x < y$ and $y < 5$, then $x < 5$.
4. If $-x > 1$, then $x < -1$.
5. If $3x < 18$, then $x < 6$.
6. If $x + 3 < 7$, then $x < 4$.
7. If $-2 < -x$, then $x < 2$.
8. If $2 < x$ and $x < z$, then $2 < z$.

Express the given statement by means of the symbols $<$, $\leq$, $>$, and $\geq$.

Examples   a. x is at least 7   b. x is between 1 and 3

Solutions   a. $x \geq 7$ b. $1 < x < 3$

9. x is greater than 3
10. x is less than 3
11. x is no greater than 3
12. x is no smaller than 3
13. x is at least as great as 3
14. x is at most 3
15. x is between 1 and 3 inclusive
16. x is not greater than 3
17. x is at least 1 and less than 3
18. x is greater than 1 but no greater than 3

Order the numbers in each list from least to greatest.

Examples   a. $\frac{3}{4}, 0, -1, 2$ b. $-1, -4, 2, \frac{1}{2}$

Solutions   Plot the values on a number line and list the values from the left to right.

a.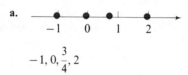

$-1, 0, \frac{3}{4}, 2$

b.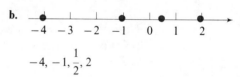

$-4, -1, \frac{1}{2}, 2$

19. $2, -3, 1, -1$
20. $-4, 1, 2, -2$
21. $\frac{1}{2}, 0, \frac{2}{3}, 1$
22. $\frac{2}{5}, \frac{1}{3}, \frac{3}{10}, 1$
23. $-\frac{1}{4}, \frac{1}{8}, -\frac{2}{3}, -1$
24. $-\frac{5}{2}, -\frac{2}{5}, 0, -1$

For $x, y \in R$ rewrite the given expression without using absolute-value notation.

Examples   a. $|-13|$ b. $|x - 1|$

Solutions   a. Since -13 is negative,

$|-13| = -(-13) = 13.$

b. $|x - 1| = \begin{cases} x - 1 & \text{if } x - 1 \geq 0 \\ -(x - 1) & \text{if } x - 1 < 0 \end{cases}$

so $|x - 1| = \begin{cases} x - 1 & \text{if } x \geq 1 \\ 1 - x & \text{if } x < 1 \end{cases}$

1.4 Order in R

25. $|-(-11)|$ 26. $|-11|$ 27. $|11|$ 28. $-|-11|$
29. $-|11|$ 30. $|x|$ 31. $|-x|$ 32. $|1-x|$
33. $-|1-x|$ 34. $-|1+x|$ 35. $|1+x|$ 36. $|x^2|$

From each of the following lists of numbers, select the greatest.

37. $-1, |-2|, 0$ 38. $|-3|, |-4|, 2$ 39. $|-3|, 4, |2|$
40. $3, |-4|, 2$ 41. $|-1|, -2, 0$ 42. $1, -|2|, 0$

Write each set in interval notation and graph.

Examples   a. $\{x | 2 < x \leq 4\}$ b. $\{x | 3 < x\}$

Solutions   a. $(2, 4]$ b. $(3, \infty)$

43. $\{x | -2 \leq x \leq 0\}$ 44. $\{x | 2 \leq x < 5\}$
45. $\{x | x \leq 4\}$ 46. $\{x | x \geq 0\}$

Write each set in set-builder notation and graph.

Examples   a. $[-1, 1)$ b. $(-\infty, 0)$

Solutions   a. $\{x | -1 \leq x < 1\}$ b. $\{x | x < 0\}$

47. $(-3, 1]$ 48. $\left(-1, \dfrac{1}{2}\right)$ 49. $(-\infty, 0]$ 50. $[2, \infty)$

Graph each of the following sets on a number line. Describe each set as a single interval.

Example $(1, 4] \cap [-3, 2)$

Solution   Graph each set separately as shown. The values for which the graphs overlap is the intersection.

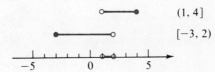

The set is $(1, 2)$.

51. $(-7, 4] \cap (0, 5)$ 52. $[-5, 2) \cap [-1, 4]$ 53. $(-\infty, 1] \cap (0, 2)$
54. $[1, 3) \cap [0, +\infty)$ 55. $[-1, 2) \cap [2, 3]$ 56. $(-\infty, -1) \cap (0, 2)$

Graph each of the following sets on a number line.

Example $[-2, 3) \cup (4, +\infty)$

Solution Graph each set on the same line. The entire region graphed is the union.

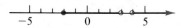

57. $(-1, 1) \cup [3, 5]$
58. $(-1, +\infty) \cup [-4, -1)$
59. $[-3, 2) \cup (2, 3]$
60. $[1, +\infty) \cup (-\infty, 4)$
61. $[-4, -2) \cup [0, 1) \cup [3, 4]$
62. $(-\infty, 1] \cup [2, 3) \cup [4, +\infty)$

Chapter Review

[1.1] Let $A = \{2, 3, 5, 7, 11\}$, $B = \{3, 6, 9, 12\}$, and $U = \{1, 2, 3, \ldots, 11, 12\}$. List the elements of the following sets.

1. A'
2. $A' \cup B'$
3. $A \cup B$
4. $A \cap B'$
5. $A \cap (B \cup \emptyset)$
6. $A' \cup (B \cap \emptyset)$

State whether each of the following is true or false.

7. $A' \neq \{1, 4, 6, 8, 10\}$
8. $B \subset \{x \mid x \in U, \ x \geq 3\}$
9. $9 \in A \cap B'$
10. $\{x \mid x \in U, \ x > 1\} \subset A$
11. $3 \notin A \cup B$
12. $B = \{x \mid x \in U, \ x \text{ is a multiple of } 3\}$

[1.2] Let $A = \left\{-2, 0, 2, -\dfrac{1}{2}, \dfrac{1}{2}, \sqrt{2}, -\sqrt{2}, \dfrac{1}{\sqrt{2}}\right\}$. List the elements of the following sets.

13. {natural numbers in A}
14. {integers in A}
15. {rational numbers in A}
16. {irrational numbers in A}
17. {negative real numbers in A}
18. {positive real numbers in A}

For $a, b, c, d \in R$, justify each statement by citing an appropriate field axiom.

19. $ab + ac = a(b + c)$
20. $ab + ac = ac + ab$
21. $ab + ac = ab + ca$
22. $a(b + c) = (b + c)a$

Review Exercises

[1.3] For $a, b, c \in R$, justify each statement by citing a statement from Theorems 1.1–1.11.

23. If $a + 2 = 7$, then $a = 5$.

24. $(-a)(-b) = ab$

25. $2 \div \dfrac{3}{4} = \dfrac{8}{3}$

26. $-(a - 2) = -a + 2$

27. $\dfrac{a}{2} + \dfrac{b}{3} = \dfrac{3a + 2b}{6}$

28. If $(a + b)c = 1$, then $c = \dfrac{1}{a + b}$.

[1.4] For $a, b \in R$, justify each statement by citing part of Theorem 1.12.

29. If $a < b$, then $3a < 3b$.

30. If $-a < b$, then $a > -b$.

31. If $a < 5$, then $a - 1 < 4$.

32. If $a + 1 < 3b$, then $\dfrac{a + 1}{3} < b$.

Express each of the following using inequality symbols.

33. a is less than b

34. a is between 3 and 7 inclusive

35. a is not less than b

36. a is at least 2

Order the numbers in each list from least to greatest.

37. $-\dfrac{2}{3}, 1, \dfrac{1}{2}, -1$

38. $-3, \dfrac{2}{3}, \dfrac{1}{8}, 0$

Rewrite these expressions without using absolute values.

39. $|-5|$

40. $|x - 5|$

Write each set in interval notation and graph the set.

41. $\left\{x \mid -\dfrac{1}{2} \leq x < 2\right\}$

42. $\left\{x \mid x \geq \dfrac{1}{2}\right\}$

2 Polynomials; Rational Expressions

In this chapter we investigate the simplest algebraic expressions—polynomials and rational expressions. We consider their behavior under the basic arithmetic operations of addition, subtraction, multiplication, and division, and discuss simplification of the results of these operations.

2.1 Definitions; Sums of Polynomials

Any grouping of constants and variables obtained by applying a finite number of the elementary operations—addition, subtraction, multiplication, division, or the extraction of roots—is called an **algebraic expression**. For example,

$$\frac{3x^2 + \sqrt{2x-1}}{3x^3 + 7} \quad \text{and} \quad xy + 3x^2z - \sqrt[5]{z}$$

are algebraic expressions. If two expressions have equal values for all values of the variables for which both expressions are defined, then we say that the expressions are **equivalent**.

Natural-number powers

You should recall that an expression of the form a^n is called a **power** of a, where a is the **base** of the power and n is the **exponent** of the power.

Definition 2.1 If $n \in N$ and $a \in R$, then

$$a^n = \underbrace{a \cdot a \cdot a \cdots \cdot a}_{n \text{ factors}}.$$

2.1 Definitions; Sums of Polynomials

Monomial expressions

An algebraic expression of the form cx^n, where n is a nonnegative integer, is called a **monomial** in the variable x. The number c is called the **coefficient** of the monomial, and the number n is called the **degree** of the monomial in x. A constant other than 0 is said to be a monomial of degree 0; no degree is assigned to the special monomial 0.

An algebraic expression of the form $cx^m y^n$ where m and n are nonnegative integers, is called a monomial in the variables x and y. As in the case of a monomial in one variable, the number c is called the coefficient of the monomial, m is the degree of the monomial in x, and n is the degree of the monomial in y. The degree of the monomial in both x and y is the sum $m + n$.

Example

The degree of the monomial $-4x^2 y^3$ is 2 in x, 3 in y, and 5 in x and y.

We define a monomial in any number of variables in a similar way.

Polynomial expressions

In any algebraic expression of the form $A + B + C + \cdots$, where $A, B, C, \ldots$ are algebraic expressions, $A, B, C, \ldots$ are called **terms** of the expression. For example, in $x + (y + 3)$ the terms are x and $(y + 3)$, but in $x + y + 3$ the terms are x, y, and 3.

A **polynomial** is any algebraic expression that can be written as one which contains only terms which are monomials. The **degree of a polynomial** is the same as the degree of its term of highest degree. Since no degree is assigned to the monomial 0, no degree is assigned to the polynomial 0, either.

Example

The degree of the polynomial $3xy^2 + 4xyz^2$ is 1 in x, 2 in y, 3 in x and y, and 4 in x, y, and z.

Because $a - b$ is defined to be $a + (-b)$, we shall view the signs in any polynomial as signs denoting numbers or their negatives, and the operation involved as addition. Thus,

$$3x - 5y + 4z = (3x) + (-5y) + (4z),$$

and $3x - 5y + 4z$ is a polynomial with terms $3x$, $-5y$, and $4z$. Again, an expression such as

$$a - (bx + cx^2),$$

in which a set of parentheses is preceded by a negative sign, can be written

$$a + [-(bx + cx^2)],$$

or

$$a + [-bx - cx^2],$$

or, finally,

$$a + (-bx) + (-cx^2).$$

Thus, $a - (bx + cx^2)$ is equivalent to a polynomial with terms a, $-bx$, and $-cx^2$.

Also, since
$$\frac{a}{b} = a\left(\frac{1}{b}\right),$$
we can view division by a constant as multiplication by its reciprocal (multiplicative inverse), and, for example, write
$$\frac{3x^2}{4} + \frac{x}{2} \quad \text{as} \quad \frac{3}{4}x^2 + \frac{1}{2}x.$$

Accordingly, since a polynomial can be considered to involve only the operations of addition and multiplication, and since the set R of real numbers is closed with respect to these operations, it follows that, for any specific real value of x, a polynomial with real coefficients represents a real number. Therefore, the properties for the real numbers are applicable to the terms in such polynomials and to the polynomials themselves.

By applying the commutative, associative, and distributive properties in various ways, we can frequently rewrite polynomials and sums of polynomials in what might be termed "simpler" forms.

Examples

a. $(2x^2 + 3x + 5) + 2x + (6x^2 + 7)$
$= 2x^2 + 6x^2 + 3x + 2x + 5 + 7$
$= (2 + 6)x^2 + (3 + 2)x + 5 + 7$
$= 8x^2 + 5x + 12$

b. $(3x^2 - 2x) - x^2 - x$
$= 3x^2 - x^2 - 2x - x$
$= (3 + (-1))x^2 + (-2 - 1)x$
$= 2x^2 - 3x$

We shall often be concerned with polynomials in one variable. A polynomial of degree n, $n \geq 0$, in x can be represented—when its terms are rearranged, if need be— by an expression in the **standard form**
$$a_n x^n + a_{n-1} x^{n-1} + a_{n-2} x^{n-2} + \cdots + a_1 x + a_0 \quad (a_n \neq 0),$$
where it is understood that the a's are the (constant) coefficients of the powers of x in the polynomial.

The term $a_n x^n$ is called the **leading term** and the coefficient a_n is called the **leading coefficient** in the polynomial. It is often convenient to have the leading term of the form x^n. In this case (that is, when $a_n = 1$), the polynomial is said to be **monic**.

If the coefficients in a polynomial are real numbers, then the polynomial is called a **polynomial over the real-number field**, or simply a **polynomial over R**. If the variable is restricted to represent only real numbers, then the polynomial is said to be a **polynomial in the real variable** x. If *both* the coefficients and the variable are restricted to real values, we say that the polynomial is a **real polynomial**.

Symbols for polynomials

Polynomials are frequently represented by symbols such as
$$P(x), \quad D(y), \quad \text{and} \quad Q(z),$$
where the symbol in parentheses designates the variable. Thus, we might write
$$P(x) = 2x^3 - 3x + 2,$$
$$D(y) = y^6 - 2y^2 + 3y - 2,$$
$$Q(z) = 8z^4 + 3z^3 - 2z^2 + z - 1.$$

2.1 Definitions; Sums of Polynomials

Value of a polynomial

The notation $P(x)$ can be used to denote values of the polynomial for specific values of x. Thus, $P(2)$ means the value of the polynomial $P(x)$ when x is replaced by 2. For example, if

$$P(x) = x^2 - 2x + 1,$$

then

$$P(2) = 2^2 - 2(2) + 1 = 1, \quad P(3) = 3^2 - 2(3) + 1 = 4,$$
$$P(-4) = (-4)^2 - 2(-4) + 1 = 25.$$

In some applications, the notation $P(x)|_a^b$ denotes $P(b) - P(a)$. Thus, if

$$P(x) = \frac{x^2}{2} - 4x,$$

then

$$P(x)|_2^3 = P(3) - P(2) = \left[\frac{3^2}{2} - 4(3)\right] - \left[\frac{2^2}{2} - 4(2)\right] = -\frac{3}{2}.$$

Exercise 2.1

A *Write each polynomial in standard form.*

Examples

a. $(x^2 - 3x + 1) - (x^2 - 3x + 4)$ b. $(x^2 - 2x + 4) - (2x^2 - 2)$

Solutions

a. $(x^2 - 3x + 1) - (x^2 - 3x + 4)$
$= x^2 - 3x + 1 - x^2 + 3x - 4$
$= (x^2 - x^2) + (-3x + 3x) + (1 - 4)$
$= -3$

b. $(x^2 - 2x + 4) - (2x^2 - 2)$
$= x^2 - 2x + 4 - 2x^2 + 2$
$= (x^2 - 2x^2) + (-2x) + (4 + 2)$
$= -x^2 - 2x + 6$

1. $(x^2 + 5x - 4) + (x^2 - 4x + 4)$
2. $(x^3 + 2x^2 + x) + (x^3 - x^2 + 2x)$
3. $(x^2 - 7x - 10) - (3x^2 - x + 5)$
4. $(x^3 - x + 4) - (2x^3 - 3x^2 + 4x - 5)$
5. $(x^3 - x^2 + x - 1) - 2x^3$
6. $(x^5 - x^3 + 1) + (x^5 - 2x^3 + 3)$
7. $(2x^4 + 2x^2 + 1) + (3x^3 - x) - (4x^4 + 6x^3)$
8. $(3x^2 - x + 5) - (2x^2 + x - 5) + 10$
9. $(x^2 + 3x + 2) - (x^2 + x + 1)$
10. $(x^3 - 1) - (x^3 + 3x^2 + 3x + 1)$

Simplify each polynomial by combining like terms.

Example $(3x^2 + 2xy - y^2) - (x^2y - xy + x^2)$

Solution $(3x^2 + 2xy - y^2) - (x^2y - xy + x^2) = 3x^2 + 2xy - y^2 - x^2y + xy - x^2$
$= (3x^2 - x^2) + (2xy + xy) - y^2 - x^2y$
$= 2x^2 + 3xy - y^2 - x^2y$

11. $(2x^2y + xy^2 + y^3) - (x^3 + xy^2 - 3x^2y) - (y^3 - x^3)$
12. $(2xy + x^2 - y^2) + (xy - x^2) - (y^2 - 3xy)$
13. $(x^2yz + xy^2z) - (2x^3y + 3x^2yz) + (x^3y + xy^2z)$
14. $(x^2y + x^2z) - (xyz + 2x^2z + 3xz^2) - (x^2y - xyz + xz^2)$

Let $P(x) = x^2 - 2x - 3$, $Q(x) = 2x^2 - x + 1$, and $R(x) = -x^2 + 3x + 4$.
Write each polynomial in standard form.

Examples   **a.** $P(x) + Q(x)$ **b.** $R(x) - Q(x)$

Solutions   **a.** $P(x) + Q(x)$ **b.** $R(x) - Q(x)$
$= (x^2 - 2x - 3) + (2x^2 - x + 1)$ $= (-x^2 + 3x + 4) - (2x^2 - x + 1)$
$= x^2 - 2x - 3 + 2x^2 - x + 1$ $= -x^2 + 3x + 4 - 2x^2 + x - 1$
$= (x^2 + 2x^2) + (-2x - x)$ $= (-x^2 - 2x^2) + (3x + x) + (4 - 1)$
$\qquad\qquad + (-3 + 1)$ $= -3x^2 + 4x + 3$
$= 3x^2 - 3x - 2$

15. $P(x) - Q(x)$ 16. $P(x) + [Q(x) - R(x)]$
17. $P(x) - [Q(x) + R(x)]$ 18. $P(x) - [Q(x) - R(x)]$
19. $[P(x) - Q(x)] - R(x)$ 20. $Q(x) + Q(x)$
21. $R(x) + R(x)$ 22. $[P(x) - P(x)] + [Q(x) - Q(x)]$

Write each polynomial in standard form.

Examples   **a.** $x - [2x - (3x - 1)]$ **b.** $3x - [2 - x - (x + 5)]$

Solutions   **a.** $x - [2x - (3x - 1)]$ **b.** $3x - [2 - x - (x + 5)]$
$= x - [2x - 3x + 1]$ $= 3x - [2 - x - x - 5]$
$= x - [-x + 1]$ $= 3x - [-2x - 3]$
$= x + x - 1$ $= 3x + 2x + 3$
$= 2x - 1$ $= 5x + 3$

(Note in both examples how inner grouping symbols are removed first.)

2.1 Definitions; Sums of Polynomials

23. $2x - [3 + x - (4x - 5)]$
24. $4x - [(x + 2) - (3x + 1)]$
25. $x - [2x - \{3x - (4x - 1)\}]$
26. $x^2 - [1 - (1 - x^2)]$
27. $x^2 - 1 - [x - (2x^2 - 3x + 1)]$
28. $2x^2 - x + 5 - [x^2 - x - (2x^2 - 3x + 1)]$
29. $-\{-[(x^2 - 2x) + (x^3 - x^2)] + 3x\} - \{-(x - 2) + (x^2 - 4) - 3\}$
30. $x^2 - \{x - (x^3 - [x^2 - (x + 1) - x^3] + x^2) - x\}$
31. $-\{[(x^2 + x + 1) - (x^3 + x - 1)] - x + [-(x - \{x - (2x + 1)\}) + x]\}$
32. $2x^3 - \{x - [x^2 - (x^3 + 1)] - (x + 1) - [x^4 - (x^4 + x^2 - 1)]\}$

Example Given $P(x) = x^2 - 5x + 4$, find $P(0), P(1), P(2), P(x)|_0^2$.

Solution
$P(0) = 0^2 - 5 \cdot 0 + 4 = 4$
$P(1) = 1^2 - 5 \cdot 1 + 4 = 0$
$P(2) = 2^2 - 5 \cdot 2 + 4 = -2$
$P(x)|_0^2 = P(2) - P(0) = -2 - 4 = -6$

33. Given $P(x) = x^3 - x + 2$, find $P(-1), P(0), P(2), P(x)|_{-1}^0$.
34. Given $P(x) = x^5 - 2x^3 + x$, find $P(-1), P(1), P(3), P(x)|_1^3$.
35. Given $P(x) = x^7$, find $P(-1), P(0), P(1), P(x)|_{-1}^1$.
36. Given $P(x) = x^{14}$, find $P(-1), P(0), P(1), P(x)|_{-1}^1$.

Example Given $P(x) = x^2 + x$ and $Q(x) = x - 1$, find $P(Q(2))$.

Solution $Q(2) = 2 - 1 = 1; \quad P(Q(2)) = P(1) = 1^2 + 1 = 2$

37. Given $P(x) = x + 3$ and $Q(x) = x^2$, find $P(Q(2))$ and $Q(P(-1))$.
38. Given $P(x) = x + 2$ and $Q(x) = x^2 - 1$, find $P(Q(3))$ and $Q(P(0))$.
39. Given $P(x) = x^2 + 2x + 1$ and $Q(x) = x + 3$, find $P(Q(1))$ and $Q(P(3))$.
40. Given $P(x) = x^2 - 3x$ and $Q(x) = x^2 + 1$, find $P(Q(1))$ and $Q(P(3))$.

B Given $P(x) = x^3 - 4x^2 + 1$, $Q(x) = x^2 - 1$ and $R(x) = x^2 + 1$, compute each of the following.

41. $P(Q(x))|_0^1$
42. $P(R(x))|_1^3$
43. $P(Q(R(1)))$
44. $Q(P(R(-1)))$

45. If $P(x)$ is of degree n and $Q(x)$ is of degree $n - 2$, what is the degree of
$$P(x) + Q(x)? \quad \text{Of } P(x) - Q(x)? \quad \text{Of } Q(x) + Q(x)? \quad \text{Of } Q(x) - Q(x)?$$

46. If $P(x)$ and $Q(x)$ are each of degree n, what can be said about the degree of
$$P(x) + Q(x)?$$

2.2 Products of Polynomials

Laws of exponents for natural-number exponents

By Definition 2.1 (page 24), we have
$$a^n = \underbrace{a \cdot a \cdot a \cdot \cdots \cdot a}_{n \text{ factors}}.$$

The product $a^m \cdot a^n$, where m and n are natural numbers, is then given by
$$a^m \cdot a^n = \underbrace{(a \cdot a \cdot a \cdot \cdots \cdot a)}_{m \text{ factors}} \underbrace{(a \cdot a \cdot a \cdot \cdots \cdot a)}_{n \text{ factors}}$$
$$= \underbrace{a \cdot a \cdot a \cdot \cdots \cdot a}_{(m+n) \text{ factors}} = a^{m+n}.$$

We state the result formally, together with two related properties whose proofs are similar to the one above and are omitted.

Theorem 2.1   If $a \in R$ and $m, n \in N$, then

    I $a^m \cdot a^n = a^{m+n}$,
    II $(a^m)^n = a^{mn}$,
    III $(ab)^n = a^n b^n$.

This theorem justifies writing the product of two natural-number powers of the same base as a power (of the same base) with an exponent equal to the sum of the two exponents.

Examples

Rewrite the given product as a power.

    a. $x^2 x^3$ **b.** $y^3 y^4 y^2$ **c.** $(xy)^3 (xy)$

Solutions

    a. $x^2 x^3 = x^{2+3}$ **b.** $y^3 y^4 y^2 = y^{3+4+2}$ **c.** $(xy)^3 (xy) = (xy)^{3+1}$
            $= x^5$ $= y^9$ $= (xy)^4$

Products of monomials

In rewriting the product of two monomials, we use the commutative and associative laws and Theorem 2.1 to simplify the product.

2.2 Products of Polynomials

Examples Simplify each product.

a. $(3x^2y)(2xy^2)$

b. $(-4xy)\left(\dfrac{1}{2}y^2\right)$

Solutions

a. $(3x^2y)(2xy^2)$
$= 3 \cdot 2 \cdot x^2 \cdot x \cdot y \cdot y^2$
$= 6 \cdot x^{2+1} \cdot y^{2+1}$
$= 6x^3y^3$

b. $(-4xy)\left(\dfrac{1}{2}y^2\right)$
$= -4 \cdot \dfrac{1}{2} \cdot x \cdot y \cdot y^2$
$= -2 \cdot x \cdot y^{1+2}$
$= -2xy^3$

Products of polynomials

The distributive law (page 8) can be generalized to

$$a(b_1 + b_2 + \cdots + b_n) = ab_1 + ab_2 + \cdots + ab_n$$

and can be applied to write as a polynomial the product of a monomial and a polynomial containing more than one term.

Examples

a. $3xy(x^2 + y + z^2)$
$= 3xy(x^2) + 3xy(y) + 3xy(z^2)$
$= 3x^3y + 3xy^2 + 3xyz^2$

b. $4x(x^2 + 2x + 1)$
$= 4x(x^2) + 4x(2x) + 4x(1)$
$= 4x^3 + 8x^2 + 4x$

The distributive law can be applied successively to simplify the products of polynomials containing more than one term.

Examples

a. $(3x + 2y)(x - y)$
$= 3x(x - y) + 2y(x - y)$
$= 3x^2 - 3xy + 2xy - 2y^2$
$= 3x^2 - xy - 2y^2$

b. $(2x + y)(x + 2y)$
$= 2x(x + 2y) + y(x + 2y)$
$= 2x^2 + 4xy + xy + 2y^2$
$= 2x^2 + 5xy + 2y^2$

Of course, products of polynomials should ordinarily be simplified mentally if it is convenient to do so. The following binomial products are types so frequently encountered that you should learn to recognize them on sight:

$$(x + a)(x + b) = x^2 + (a + b)x + ab$$
$$(x + a)^2 = x^2 + 2ax + a^2$$
$$(x + a)(x - a) = x^2 - a^2$$
$$(ax + b)(cx + d) = acx^2 + (bc + ad)x + bd$$

Examples

a. $(x + 2)(x + 3)$
$= x^2 + (2 + 3)x + (2)(3)$
$= x^2 + 5x + 6$

b. $(x + 4)(x - 4)$
$= x^2 - (4)^2$
$= x^2 - 16$

Exercise 2.2

A *Simplify each product by combining all constants and all powers of each variable.*

Examples   **a.** $(2x^2y)(5xyz^2)$ **b.** $(3a^n)(2a^{n+3})$

Solutions   **a.** $(2x^2y)(5xyz^2)$ **b.** $(3a^n)(2a^{n+3})$
$= 2 \cdot 5 \cdot x^2 \cdot x \cdot y \cdot y \cdot z^2$ $= 3 \cdot 2 \cdot a^n \cdot a^{n+3}$
$= 10x^3y^2z^2$ $= 6a^{n+n+3}$
$= 6a^{2n+3}$

1. $(-x^2y)(2xy^3)$
2. $(x^4y^2)(3xy^2)$
3. $(2x^2y)(-3x^2y^2)(xy^3)$
4. $(-2x^2)(xy^3)(-3xy)$
5. $x^{n+2}(3x^n)$
6. $(2x^2)(4x)(x^n)$
7. $(-x^ny)(2x^2y^n)$
8. $(3x^2y^n)(-3x^{2n}y^n)$

Examples   **a.** $3(x^3 - x)$ **b.** $x(2x^2 + x - 1)$

Solutions   **a.** $3(x^3 - x) = 3x^3 - 3x$ **b.** $x(2x^2 + x - 1) = 2x^3 + x^2 - x$

9. $5(x - 2)$
10. $2(x^2 + 1)$
11. $x(2x - 4)$
12. $x(x^2 + 1)$
13. $x(x^3 + x + 1)$
14. $x^2(x^2 + x + 1)$

Examples   **a.** $(x - 3)(x + 1)$ **b.** $-(x - 1)(x + 2)$

Solutions   **a.** $(x - 3)(x + 1)$ **b.** $-(x - 1)(x + 2)$
$= x^2 + x - 3x - 3$ $= -[x^2 + 2x - x - 2]$
$= x^2 - 2x - 3$ $= -[x^2 + x - 2]$
$= -x^2 - x + 2$

15. $(x + 2)(x + 3)$
16. $(x - 1)(x + 5)$
17. $(3x - 1)(x + 3)$
18. $(x + 4)(2x + 1)$
19. $3(2x - 1)(x + 1)$
20. $5(x + 2)(3x - 1)$
21. $-(x - 1)(x + 1)$
22. $-(x + 1)(x + 2)$
23. $-(2x - 1)(x + 3)$
24. $-(x + 3)(3x - 1)$

2.2 Products of Polynomials

Example

$(x + 2)(x^2 + 3x + 4)$

Solution

$$(x + 2)(x^2 + 3x + 4) = x(x^2 + 3x + 4) + 2(x^2 + 3x + 4)$$
$$= x^3 + 3x^2 + 4x + 2x^2 + 6x + 8$$
$$= x^3 + 5x^2 + 10x + 8$$

An alternative format:

$$\begin{array}{r} x^2 + 3x + 4 \\ x + 2 \\ \hline x^3 + 3x^2 + 4x \\ 2x^2 + 6x + 8 \\ \hline x^3 + 5x^2 + 10x + 8 \end{array}$$

← This is $x^2 + 3x + 4$ times x.
← This is $x^2 + 3x + 4$ times 2.

25. $(x - 1)(x^2 + x + 1)$
26. $(x + 1)(x^2 - x - 1)$
27. $(x + 2)(3x^2 - x + 4)$
28. $(x + 5)(x^2 - 2x + 3)$
29. $(2x + 1)(x^2 + 2x + 1)$
30. $(2x - 1)(2x^2 - x + 1)$
31. $x(x + 1)(x^2 + x + 1)$
32. $x(x - 2)(x^2 - 2)$
33. $(x^2 + x + 1)(x^2 - x + 2)$
34. $(x^2 + x)(x^2 - 2x + 3)$

Write each product as a single polynomial.

35. $(x + 1)[(x + 1)^2 - (2x - 1)]$
36. $(x - 3)[(2x + 1)^2 - (x^2 - x - 1)]$
37. $(2x - 3)[(2x + 1) - (x^2 + x)](x^2 - 4)$
38. $(x^2 - x + 4)[(x^2 + x + 4) - (x^2 + x - 4)](x - 1)$

Example

Given $P(x) = x^2 + 2x + 3$, find $P(a + 1)$, $P(a + h)$, and $P(x)|_a^{a+h}$.

Solution

$P(a + 1) = (a + 1)^2 + 2(a + 1) + 3$
$\qquad = a^2 + 2a + 1 + 2a + 2 + 3$
$\qquad = a^2 + 4a + 6$

$P(a + h) = (a + h)^2 + 2(a + h) + 3$
$\qquad = a^2 + 2ah + h^2 + 2a + 2h + 3$

$P(x)|_a^{a+h} = P(a + h) - P(a)$
$\qquad = (a^2 + 2ah + h^2 + 2a + 2h + 3) - (a^2 + 2a + 3)$
$\qquad = 2ah + h^2 + 2h$

39. Given $P(x) = x^2 + x - 2$, find $P(c)$, $P(c + h)$, $P(x)|_c^{c+h}$.
40. Given $P(x) = x^2 - 2x - 3$, find $P(a)$, $P(a - h)$, $P(x)|_{a-h}^a$.
41. Given $P(x) = x^2 + 1$, find $P(x + 2)$, $P(x^2)$, $(P(x))^2$.
42. Given $P(x) = x^2 + x + 3$, find $P(x + 2)$, $P(x^2)$, $(P(x))^2$.

B 43. If $P(x)$ and $Q(x)$ have degrees m and n respectively, what is the degree of $P(x) \cdot Q(x)$?

44. If $P(x)$ has degree n, what is the degree of $(P(x))^2$? Of $[P(x)]^k$, $k \in N$?

45. If $P(x)$ has degree n, what is the degree of $P(x^2)$? Of $P(x^k)$, $k \in N$?

46. If $P(x)$ has degree n and $Q(x)$ has degree k, what is the degree of $P(Q(x))$? Of $Q(P(x))$?

2.3 Factoring Polynomials

What do we mean when we say that we have *factored* an integer or a polynomial? It is true, for example, that

$$2 = 4\left(\frac{1}{2}\right),$$

but we would not ordinarily say that 4 and 1/2 are factors of 2. On the other hand, since

$$10 = (2)(5),$$

we do say that 2 and 5 are factors of 10 in the domain J of integers.

Now consider the polynomial

$$2x^2 - 10,$$

which is completely factored as

$$2x^2 - 10 = 2(x^2 - 5)$$

if we are limited to **integral** coefficients—that is, coefficients that are integers; but if we consider polynomials in x having real numbers as coefficients, its complete factorization is

$$2x^2 - 10 = 2(x - \sqrt{5})(x + \sqrt{5}).$$

Thus the result depends in part on the coefficients we consider permissible. In this book, we are primarily concerned with polynomials whose coefficients are integers.

In factoring polynomials having integer coefficients, we consider as factors only polynomials having integer coefficients with no common integer factor other than 1 or -1, and we say that such a polynomial is **prime** if it is not the product of two polynomials of this sort and its leading coefficient is positive. For example, $x + 1$, $2x + 1$, and $x^2 + 1$ are prime, while $2x + 2 = 2(x + 2)$ and $x^2 - 1 = (x + 1)(x - 1)$ are not.

We say that a polynomial other than 0 or 1 is **completely factored** if it is written equivalently as a product of prime polynomials or as such a product times -1.

Examples a. $6x^4 - 6 = 6(x^4 - 1)$ b. $3x^2 + 18x + 27 = 3(x^2 + 6x + 9)$
$ = 6(x^2 - 1)(x^2 + 1)$ $ = 3(x + 3)^2$
$ = 6(x - 1)(x + 1)(x^2 + 1)$

2.3 Factoring Polynomials

Factors of quadratics

One very common type of factoring is that involving quadratic (second-degree) binomials or trinomials with integer coefficients. From Section 2.2, we recall that

$$(x + a)(x + b) = x^2 + (a + b)x + ab, \qquad (1)$$

$$(x + a)^2 = x^2 + 2ax + a^2, \qquad (2)$$

$$(x + a)(x - a) = x^2 - a^2, \qquad (3)$$

$$(ax + b)(cx + d) = acx^2 + (bc + ad)x + bd. \qquad (4)$$

These four forms, each of which involves prime polynomial factors, are those most commonly encountered in the chapters that follow. In this section, we are interested in viewing these relationships from right to left—that is, from polynomial to factored form.

A few other polynomials occur frequently enough to justify a study of their factorization. In particular, the forms

$$(a + b)(x + y) = ax + ay + bx + by, \qquad (5)$$

$$(x + a)(x^2 - ax + a^2) = x^3 + a^3, \qquad (6)$$

$$(x - a)(x^2 + ax + a^2) = x^3 - a^3 \qquad (7)$$

are often encountered in one or another part of mathematics. We are again interested in viewing these relationships from right to left. Expressions such as the right-hand member of form (5) are factorable by grouping. For example, to factor

$$3x^2y + 2y + 3xy^2 + 2x,$$

we write it in the form

$$3x^2y + 2x + 3xy^2 + 2y$$

and factor the common monomial x from the first group of two terms and y from the second group of two terms, obtaining

$$x(3xy + 2) + y(3xy + 2).$$

If we now factor the common binomial $(3xy + 2)$ from each term, we have

$$(3xy + 2)(x + y),$$

in which both factors are prime.

The application of forms (6) and (7) is direct.

Examples

a. $a^3 - 1 = (a - 1)(a^2 + a + 1)$

b. $8a^3 + b^3 = (2a)^3 + b^3$
$= (2a + b)[(2a)^2 - 2ab + b^2]$
$= (2a + b)[4a^2 - 2ab + b^2]$

Exercise 2.3

A Factor completely into products of polynomials with integer coefficients.

Examples a. $3xy^2 + 15x^2y + 6xy$ b. $x^2 + 10x + 25$

Solutions a. $3xy^2 + 15x^2y + 6xy$
           $= 3xy(y + 5x + 2)$
           b. $x^2 + 10x + 25$
           $= (x + 5)^2$ by formula (2)

1. $2x^3y^2 + 8x^2y + 16x^2y^2$
2. $7y^4z^2 + 14y^3z + 7y^2z^2$
3. $x^2 - 16$
4. $2x^2 - 18$
5. $3x^2y^4 - 12x^2$
6. $x^2y^2z^2 - 4x^2z^2$
7. $2x^2 + 12x + 18$
8. $x^2 + 14x + 49$
9. $3y^3x^2 + 6y^3x + 3y^3$
10. $2x^2y + 16xy + 32y$
11. $5x^2yz + 10xyz + 15x$
12. $2x^3y^2 + 16x^2y^3 + 32x^2y^2$

Examples a. $x^2 + 5x + 6$ b. $xy + 3x + 4y + 12$

Solutions a. $x^2 + 5x + 6 = (x + 2)(x + 3)$
           by formula (1)
           b. $xy + 3x + 4y + 12$
           $= x(y + 3) + 4(y + 3)$
           $= (x + 4)(y + 3)$
           by formula (5)

13. $x^2 + 3x - 10$
14. $x^2 - 5x - 6$
15. $x^2 - 8x + 15$
16. $2x^2 + 6x + 4$
17. $2x^2 + 3x + 1$
18. $3x^2 + 5x + 1$
19. $3x^2 - 7x + 2$
20. $2x^2 - 5x + 1$
21. $2x^2 - 3x - 2$
22. $3x^2 - x - 2$
23. $6x^2 - 11x + 3$
24. $8x^2 + 14x + 3$
25. $16x^2 + 10x + 1$
26. $12x^2 - 17x + 6$
27. $12x^2 + 2x - 2$
28. $12x^2 - 8x - 15$
29. $3x^2 + 12x - 15$
30. $x^3 - 2x^2 - 8x$
31. $xy^2 + 5xy - 14x$
32. $x^2y^2 + 5xy^2 - 14y^2$
33. $xy - y + 3x - 3$
34. $xy + 3y - x - 3$
35. $2xy + 6x - 4y - 12$
36. $xy^2 - 2y^2 + 2xy - 4y$

2.3 Factoring Polynomials

Examples a. $x^3 + 8$ b. $x^3 - 8y^3$

Solutions
a. $x^3 + 8 = x^3 + 2^3$
$= (x + 2)(x^2 - 2x + 4)$
by formula (6)

b. $x^3 - 8y^3 = x^3 - (2y)^3$
$= (x - 2y)(x^2 + 2xy + 4y^2)$
by formula (7)

37. $27 + x^3$
38. $x^3y^3 - 8$
39. $1 + (x + 1)^3$
40. $(x + 1)^3 - x^3$
41. $64x^3 - 8y^3$
42. $64x^3 - y^3$

Factor completely into products of polynomials with integer coefficients.

43. $x^3 + 3x^2 + yx + 3y$
44. $x^2y^3 - 8x^2$
45. $x^2 + 11x + 30$
46. $x^4 + 6x^2 + 8$
47. $x^2y^2 + 3x^2 + y^2 + 3$
48. $x - xy^2$
49. $x^5 - x$
50. $1 - xy + x - y$

B 51. Consider the polynomial $x^4 + x^2y^2 + 25y^4$. If $9x^2y^2$ is both added to and subtracted from this expression, we have

$$x^4 + x^2y^2 + 25y^4 + 9x^2y^2 - 9x^2y^2,$$
$$(x^4 + 10x^2y^2 + 25y^4) - 9x^2y^2,$$
$$(x^2 + 5y^2)^2 - (3xy)^2,$$
$$[(x^2 + 5y^2) - 3xy][(x^2 + 5y^2) + 3xy],$$
$$(x^2 - 3xy + 5y^2)(x^2 + 3xy + 5y^2).$$

By adding and subtracting an appropriate monomial, factor $x^4 + x^2y^2 + y^4$.

52. Use the method of Exercise 51 to factor $x^4 - 3x^2y^2 + y^4$.
53. Use the method of Exercise 51 to factor $x^4 + x^2 + 1$.
54. Use the method of Exercise 51 to factor $x^4 + 3x^2 + 4$.

Factor completely into products of polynomials with integer coefficients. Assume all exponents are natural numbers.

Examples a. $x^{2n} + 3x^n + 2$ b. $x^{3n} - 1$

Solutions
a. $x^{2n} + 3x^n + 2$
$= (x^n)^2 + 3x^n + 2$
$= (x^n + 1)(x^n + 2)$

b. $x^{3n} - 1$
$= (x^n)^3 - 1^3$
$= (x^n - 1)(x^{2n} + x^n + 1)$
$= (x - 1)(x^{n-1} + x^{n-2} + \cdots$
$+ x + 1)(x^{2n} + x^n + 1)$

55. $x^n - y^n$

56. $x^{2n} - y^{2n}$

57. $x^{2n} - 1$

58. $3x^{2n} - 9x^n - 12$

59. $x^{4n} - y^{2n}$

60. $x^{3n} + 8$

61. $x^{6n} + 2x^{4n} + x^{2n}$

62. $x^{4n} - 8x^{3n} + 16x^{2n}$

63. $x^{6n} + 1$

64. $x^{6n} + 2x^{3n} + 1$

2.4 Quotients of Polynomials

The fractions 2/3, 1/5, 8/9, and −3/4 are all examples of noninteger quotients of integers; therefore the set of integers is *not* closed with respect to division. Similarly, we can see that the set of polynomials is not closed with respect to division, because $1/x$ is the quotient of the polynomials 1 and x but is not, itself, a polynomial.

Laws of exponents for quotients

Let us examine some ways in which we can rewrite quotients of polynomials, even if the resulting expressions are not always, themselves, polynomials. We begin with the simplest case, that in which the polynomials are monomials and their quotient is a monomial. Consider

$$\frac{a^m}{a^n} \quad (a \neq 0,\, m, n \in N, \text{ and } m > n).$$

We have

$$\frac{a^m}{a^n} = a^m \cdot \frac{1}{a^n} = (a^{m-n} \cdot a^n) \cdot \frac{1}{a^n}$$

$$= a^{m-n} \cdot \left(a^n \cdot \frac{1}{a^n}\right) = a^{m-n} \cdot 1.$$

Hence,

$$\frac{a^m}{a^n} = a^{m-n}.$$

This establishes the first part of the following result. Proof of the second part is similar to the proof of Theorem 2.1 and is omitted.

Theorem 2.2 *If $a, b \in R\,(a, b \neq 0)$, $m, n \in N$, $m > n$, then*

$$\text{I} \quad \frac{a^m}{a^n} = a^{m-n},$$

$$\text{II} \quad \left(\frac{a}{b}\right)^m = \frac{a^m}{b^m}.$$

2.4 Quotients of Polynomials

The following examples illustrate the use of this theorem.

Examples

a. $\dfrac{12a^5 b^3}{4a^2 b^2} = \dfrac{12}{4} a^{5-2} b^{3-2}$

$= 3a^3 b, \quad (a, b \neq 0).$

b. $\dfrac{4(x-1)^2 x^3}{(x-1)x^2} = 4(x-1)^{2-1} x^{3-2}$

$= 4(x-1)x, \quad (x \neq 0, 1).$

Note that in Example (a) the variables a and b are not permitted to take the value 0, because if they were, $3a^3 b$ would represent a real number, 0, although $(12a^5 b^3)/(4a^2 b^2)$ would not be defined. In Example (b), the variable x can take neither the value 0 nor the value 1, because if x takes either of these values $4(x-1)x$ would represent the value 0 although $[4(x-1)^2 x^3]/[(x-1)x^2]$ would not be defined.

Quotients of polynomials

Theorem 1.11-III (together with the closure laws) permits us to rewrite quotients of polynomials whose numerators are not monomials. For example,

$$\frac{2x^3 + 4x^2 + 8x}{2x} = \frac{2x^3}{2x} + \frac{4x^2}{2x} + \frac{8x}{2x},$$

and, by an application of Theorem 2.2-I, the right-hand member can then be denoted by the expression

$$x^2 + 2x + \frac{8x}{2x} \quad (x \neq 0).$$

Though Theorem 2.2-I is not applicable to the variable factors in expressions such as $(8x)/(2x)$, by Theorem 1.11-II and the fact that $x/x = 1$ for all $x \neq 0$, we have

$$\frac{8x}{2x} = \frac{8}{2} \cdot \frac{x}{x} = \frac{8}{2} \cdot 1 = 4,$$

for every $x \neq 0$. Thus,

$$\frac{2x^3 + 4x^2 + 8x}{2x} = x^2 + 2x + 4 \quad (x \neq 0).$$

As we have noted, quotients of polynomials cannot always be represented by polynomials. For example, we have

$$\frac{2x^3 + 4x + 1}{x} = \frac{2x^3}{x} + \frac{4x}{x} + \frac{1}{x}$$

$$= 2x^2 + 4 + \frac{1}{x} \quad (x \neq 0),$$

where the resulting expression is not a polynomial, but rather an expression of the form

$$Q(x) + \frac{R(x)}{D(x)},$$

where the polynomial $Q(x)$ (equal to $2x^2 + 4$) may be called a "partial quotient" and $R(x)$ (equal to 1) is called a "remainder."

If the divisor of a quotient contains more than one term, the familiar **long-division algorithm** involving successive subtractions can be used to rewrite the quotient. Notice that the algorithm can be continued until the degree of the remainder is less than that of the divisor.

Example Write $\dfrac{x^4 + x^2 + 2x - 1}{x + 3}$ in the form $Q(x) + \dfrac{R(x)}{x + 3}$, where $Q(x)$ is a polynomial and $R(x)$ is a constant.

Solution Using the long-division algorithm, we have

$$\begin{array}{r}
x^3 - 3x^2 + 10x - 28 \\
x+3 \overline{\smash{\big)}\, x^4 + 0x^3 + x^2 + 2x - 1} \\
\underline{x^4 + 3x^3 } \\
-3x^3 + x^2 \\
\underline{-3x^3 - 9x^2 } \\
10x^2 + 2x \\
\underline{10x^2 + 30x } \\
-28x - 1 \\
\underline{-28x - 84} \\
83 \ (\text{remainder})
\end{array}$$

Thus, for $x + 3 \neq 0$,

$$\frac{x^4 + x^2 + 2x - 1}{x + 3} = x^3 - 3x^2 + 10x - 28 + \frac{83}{x + 3}.$$

Observe in the preceding example that $0x^3$ is used in the dividend for a term involving x^3, even though the dividend contains no such term explicitly.

Exercise 2.4

A *Write each quotient as a polynomial.*

Examples **a.** $\dfrac{10x^4 y^2}{2x^2 y}$ **b.** $\dfrac{x^3 + 3x^2 + 4x}{x}$

Solutions **a.** $\dfrac{10x^4 y^2}{2x^2 y} = \dfrac{10}{2} x^{4-2} y^{2-1}$ **b.** $\dfrac{x^3 + 3x^2 + 4x}{x} = \dfrac{x^3}{x} + \dfrac{3x^2}{x} + \dfrac{4x}{x}$

$ = 5x^2 y \quad (x, y \neq 0)$ $ = x^2 + 3x + 4 \quad (x \neq 0)$

2.4 Quotients of Polynomials

1. $\dfrac{15y^4z^3}{3yz^2}$
2. $\dfrac{8x^4y^2}{2xy}$
3. $\dfrac{6x^4y^2}{xy^2}$
4. $\dfrac{3x^5y^6z^7}{x^2y^4z^6}$
5. $\dfrac{4x^4 + 2x^2}{2x}$
6. $\dfrac{x^3 + 5x^2 - 6x}{x}$
7. $\dfrac{3x^2y - 2xy^2 + xy}{xy}$
8. $\dfrac{x^2yz - xy^2z + xyz^2}{xyz}$

Write each quotient $\dfrac{P(x)}{D(x)}$ either in the form $Q(x)$ or $Q(x) + \dfrac{R(x)}{D(x)}$, where $Q(x)$ and $R(x)$ are polynomials and the degree of $R(x)$ is less than that of $D(x)$.

Examples

a. $\dfrac{x^3 + 3x + 5}{x}$

b. $\dfrac{x^3 + 3x + 5}{x + 1}$

Solutions

a. $\dfrac{x^3 + 3x + 5}{x}$

$= \dfrac{x^3}{x} + \dfrac{3x}{x} + \dfrac{5}{x}$

$= x^2 + 3 + \dfrac{5}{x} \quad (x \neq 0)$

b.
$$
\begin{array}{r}
x^2 - x + 4 \\
x + 1 \,\overline{\smash{)}\, x^3 + 0\cdot x^2 + 3x + 5} \\
\underline{x^3 + x^2 } \\
-x^2 + 3x \\
\underline{-x^2 - x } \\
4x + 5 \\
\underline{4x + 4} \\
1
\end{array}
$$

$\dfrac{x^3 + 3x + 5}{x + 1}$

$= x^2 - x + 4 + \dfrac{1}{x + 1} \quad (x \neq -1).$

9. $\dfrac{x^2 + 2x + 1}{x}$
10. $\dfrac{2x^2 - 3x + 4}{x}$
11. $\dfrac{4x^3 - 2x + 6}{2x}$
12. $\dfrac{z^3 + z^2 + z + 1}{z^3}$
13. $\dfrac{x^5 + 2x^3 + x - 4}{x^2}$
14. $\dfrac{3x^2 + 8x + 4}{x^3}$
15. $\dfrac{x^2 + 2x + 1}{x + 1}$
16. $\dfrac{2x^2 - 3x + 4}{x + 1}$
17. $\dfrac{4x^3 - 2x + 6}{2x - 3}$
18. $\dfrac{z^3 + z^2 + z + 1}{z - 1}$
19. $\dfrac{x^5 + 2x^3 + x - 4}{x - 4}$
20. $\dfrac{3x^2 + 8x + 4}{x - 2}$
21. $\dfrac{x^5 + 2x^3 + x - 4}{x^3 - x}$
22. $\dfrac{x^5 - 4x^3 + x}{x^2 + x + 1}$
23. $\dfrac{4x^4 + 3x^3 + 2x^2 + x}{2x^2 + 1}$
24. $\dfrac{x^5 - 1}{x - 1}$
25. $\dfrac{x^5 + 1}{x + 1}$
26. $\dfrac{x^5}{x^2 + 1}$

B 27. $\dfrac{x^4 + x^2 - 1}{x^3 + x - 1}$

28. $\dfrac{x^5 + 2x^4 - x + 1}{x^3 + 1}$

29. $\dfrac{(x+1)^2 - (x+1) - 1}{(x+1)^2 - 1}$

30. $\dfrac{(2x-1)^3 - (2x-1) + 4}{(2x-1)^2 + 1}$

Let $P(x)$ and $D(x)$ be polynomials with degrees k and n, respectively. Assume $n \le k$.

31. If $\dfrac{P(x)}{D(x)} = Q(x) + \dfrac{R(x)}{D(x)}$ with the degree of $R(x)$ less than n, what is the degree of $Q(x)$?

2.5 Synthetic Division

If the divisor is of the form $x + c$, then the process of dividing one polynomial by another can be simplified by a process called **synthetic division**. Consider the example on page 40, where $x^4 + x^2 + 2x - 1$ is divided by $x + 3$. If we omit the variables, writing only the coefficients of the terms, and use zero for the coefficient of any missing power, we have

$$\begin{array}{r}
1 - 3 + 10 - 28 \\
1 + 3 \,\overline{\big)\, 1 + 0 + 1 + 2 - 1}\\
\underline{1 + 3 }\\
-3 + (1) \\
\underline{-3 - 9 }\\
10 + (2) \\
\underline{10 + 30 }\\
-28 - (1)\\
\underline{-28 - 84}\\
83 \text{ (remainder).}
\end{array}$$

Now, observe that the numerals shown in color are repetitions of the numerals written immediately above and are also repetitions of the coefficients of the associated variables in the quotient; the numerals in parentheses, ( ), are repetitions of the coefficients of the dividend. Therefore, the whole process can be written in compact form as

$$\begin{array}{rl}
(1) & \underline{3\,\big|}\,1 0 1 2 -1 \\
(2) & 3 -9 30 -84 \\
(3) & 1 -3 10 -28 83 \quad \text{(remainder: 83)}
\end{array}$$

where the repetitions are omitted and where 1, the coefficient of x in the divisor, has also been omitted.

The entries in line (3), which are the coefficients of the variables in the quotient and the remainder, have been obtained by *subtracting* the **detached coefficients** in line (2) from the detached coefficients of terms of the same degree in line (1).

2.5 Synthetic Division

We could obtain the same result by replacing 3 with -3 in the divisor and *adding* instead of subtracting at each step, and this is what is done in the *synthetic-division* process. The final form then appears:

$$
\begin{array}{r|rrrrr}
(1) \quad -3 & 1 & 0 & 1 & 2 & -1 \\
(2) & & -3 & 9 & -30 & 84 \\
\hline
(3) & 1 & -3 & 10 & -28 & 83
\end{array}
\quad \text{(remainder: 83)}.
$$

Comparing the results of using synthetic division with those obtained on the same example using long division, on page 40, we observe that the entries in line (3) are the coefficients of the polynomial $x^3 - 3x^2 + 10x - 28$, and that there is a remainder of 83.

Example

Write $\dfrac{3x^3 - 4x - 1}{x - 2}$ in the form $Q(x) + \dfrac{r}{D(x)}$, where $Q(x)$ and $D(x)$ are polynomials and r is a constant.

Solution

Using synthetic division, we write

$$\underline{2\,|\,3} \quad 0 \quad -4 \quad -1,$$

where 0 has been inserted in the position that would be occupied by the coefficient of a second-degree term if such a term were present in the dividend. Our divisor is the negative of -2, or 2.

$$
\begin{array}{r|rrrr}
(1) \quad 2 & 3 & 0 & -4 & -1 \\
(2) & & 6 & 12 & 16 \\
\hline
(3) & 3 & 6 & 8 & 15
\end{array}
\quad \text{(remainder: 15)}.
$$

This process employs these steps:

1. 3 is "brought down" from line (1) to line (3).
2. 6, the product of 2 and 3, is written in the next position on line (2).
3. 6, the sum of 0 and 6, is written on line (3).
4. 12, the product of 2 and 6, is written in the next position on line (2).
5. 8, the sum of -4 and 12, is written on line (3).
6. 16, the product of 2 and 8, is written in the next position on line (2).
7. 15, the sum of -1 and 16, is written on line (3).

We use the first three entries on line (3) as coefficients to write a polynomial of degree one less than the degree of the dividend. This polynomial is the quotient lacking the remainder. The last number is the remainder. Thus, for $x - 2 \neq 0$, $3x^3 - 4x - 1$ divided by $x - 2$ is $3x^2 + 6x + 8$ with a remainder of 15; that is,

$$\frac{3x^3 - 4x - 1}{x - 2} = 3x^2 + 6x + 8 + \frac{15}{x - 2} \quad (x \neq 2).$$

Example

Find a polynomial $Q(x)$ and a real number r such that

$$x^3 + 2x^2 + 1 = (x - 1)Q(x) + r.$$

Solution on overleaf

Solution Using synthetic division to compute $x^3 + 2x^2 + 1/(x - 1)$, we write

$$\underline{1|}\ 1\ \ 2\ \ 0\ \ 1$$
$$\phantom{\underline{1|}\ 1\ \ }1\ \ 3\ \ 3$$
$$\phantom{\underline{1|}\ }1\ \ 3\ \ 3\ \ 4$$

Thus,

$$\frac{x^3 + 2x^2 + 1}{x - 1} = x^2 + 3x + 3 + \frac{4}{x - 1} \quad (x \neq 1).$$

Now multiplying both members of the above equation by $x - 1$ yields

$$x^3 + 2x^2 + 1 = (x^2 + 3x + 3)(x - 1) + 4.$$

Thus, $Q(x) = x^2 + 3x + 3$ and $r = 4$.

The foregoing example illustrates a theorem that we shall state without proof.

Theorem 2.3 *If $P(x)$ is a real polynomial and c is any real number, then there exists a unique real polynomial $Q(x)$ and a real number r, such that*

$$P(x) = (x - c)Q(x) + r.$$

Although this theorem does not directly involve the quotient $\dfrac{P(x)}{x - c}$, it does assure us that, for $x \neq c$,

$$\frac{P(x)}{x - c} = Q(x) + \frac{r}{x - c}.$$

Exercise 2.5

A Use synthetic division to write each quotient $\dfrac{P(x)}{x - c}$ in the form $Q(x)$ or in the form $Q(x) + \dfrac{r}{x - c}$, where $Q(x)$ is a polynomial and r is a constant.

Examples

a. $\dfrac{x^4 + x^3 - 3x^2 - 8x + 4}{x - 2}$

b. $\dfrac{x^4 - x^3 + 3x - 7}{x + 3}$

Solutions

a. $\underline{2|}\ 1\ \ \ \ -3\ \ -8\ \ \ \ 4$
$\phantom{\underline{2|}\ 1\ \ \ }2\ \ \ \ \ \ 6\ \ \ \ \ \ 6\ \ -4$
$\phantom{\underline{2|}\ }1\ \ \ \ \ 3\ \ \ \ \ \ 3\ \ -2\ \ \ \ 0$

Thus,

$$\frac{x^4 + x^3 - 3x^2 - 8x + 4}{x - 2}$$

$$= x^3 + 3x^2 + 3x - 2$$

$(x \neq 2)$.

b. $\underline{-3|}\ 1\ \ -1\ \ \ \ \ 0\ \ \ \ \ \ 3\ \ \ \ -7$
$\phantom{\underline{-3|}\ 1\ \ }-3\ \ \ \ 12\ \ -36\ \ \ \ 99$
$\phantom{\underline{-3|}\ }1\ \ -4\ \ \ \ 12\ \ -33\ \ \ \ 92$

Thus,

$$\frac{x^4 - x^3 + 3x - 7}{x + 3}$$

$$= x^3 - 4x^2 + 12x - 33 + \frac{92}{x + 3}$$

$(x \neq -3)$.

2.5 Synthetic Division

1. $\dfrac{x^2 + 4x - 21}{x - 3}$
2. $\dfrac{3x^2 - 4x - 7}{x + 1}$
3. $\dfrac{x^2 - 30}{x + 5}$
4. $\dfrac{x^2 + 30}{x - 5}$
5. $\dfrac{2x^3 + x + 18}{x + 2}$
6. $\dfrac{2x^3 + x + 18}{x - 2}$
7. $\dfrac{3x^4 + x^3 - 2x^2 + 6}{x - 2}$
8. $\dfrac{x^5 + x^3 + 1}{x + 1}$
9. $\dfrac{x^6 - 1}{x - 1}$
10. $\dfrac{x^5 + 1}{x + 1}$
11. $\dfrac{4x^5 + x^4 - 2x^3 + x^2}{x - 3}$
12. $\dfrac{3x^3 + 4x^2 - 2x + 4}{x + 2}$
13. $\dfrac{x^3 - 3x^2 + 3x - 1}{x + 1}$
14. $\dfrac{x^3 - 3x^2 + 3x - 1}{x - 1}$
15. $\dfrac{x^4 + x - 14}{x + 3}$
16. $\dfrac{x^4 + x - 14}{x - 2}$
17. $\dfrac{x^3 + 11}{x + 5}$
18. $\dfrac{x^6 - 2x^3 + 1}{x + 1}$
19. $\dfrac{x^5 + 2x^4 + 3x^3 + 9x^2 - 7}{x + 2}$
20. $\dfrac{x^4 - 3x^3 + 5x^2 - x + 2}{x - 3}$

B Use synthetic division to write each quotient $\dfrac{P(x)}{ax + b}$ in the form $Q(x) + \dfrac{r}{ax + b}$, where $Q(x)$ is a polynomial and r is a constant.

Example $\dfrac{x^3 + 2x^2 + x - 1}{2x - 4}$

Solution We first write

$$\dfrac{x^3 + 2x^2 + x - 1}{2x - 4} = \dfrac{x^3 + 2x^2 + x - 1}{2(x - 2)}$$

$$= \dfrac{1}{2}\left(\dfrac{x^3 + 2x^2 + x - 1}{x - 2}\right).$$

Using synthetic division on the quotient $(x^3 + 2x^2 + x - 1)/(x - 2)$, we write

$$\underline{2|}\;1\quad 2\quad 1\quad -1$$
$$\phantom{\underline{2|}\;1\quad}\;2\quad 8\quad 18$$
$$\phantom{\underline{2|}\;}1\quad 4\quad 9\quad 17.$$

Thus,

$$\dfrac{x^3 + 2x^2 + x - 1}{2x - 4} = \dfrac{1}{2}\left(x^2 + 4x + 9 + \dfrac{17}{x - 2}\right)$$

$$= \dfrac{1}{2}x^2 + 2x + \dfrac{9}{2} + \dfrac{17}{2x - 4}.$$

21. $\dfrac{x^2 + x + 1}{2x - 2}$

22. $\dfrac{x^2 - x - 4}{2x - 4}$

23. $\dfrac{3x^3 + 2x^2 + x - 1}{3x + 6}$

24. $\dfrac{4x^3 - 2x^2 + 3x - 1}{2x - 6}$

25. $\dfrac{2x^4 - 3x^3 + x - 1}{2x - 2}$

26. $\dfrac{3x^4 + x^2 + 1}{3x + 3}$

27. $\dfrac{x^4 - 2x^3 + x + 2}{2x - 1}$

28. $\dfrac{x^4 - x^2 - 1}{2x + 1}$

29. $\dfrac{x^3 - x^2 - 1}{-2x + 2}$

30. $\dfrac{x^3 - 2x^2 - x - 1}{-3x + 6}$

2.6 Equivalent Fractions

Rational expressions

A fraction is an expression denoting a quotient. If the numerator (dividend) and the denominator (divisor) are polynomials, then the fraction is said to be a **rational expression**. Trivially, any polynomial can be considered as being a rational expression, since it is the quotient of itself and 1. For each replacement of the variable(s) for which the numerator and denominator of a fraction represent real numbers and for which the denominator is not zero, the fraction represents a real number. Of course, for any value of the variable(s) for which the denominator vanishes (is equal to zero), the fraction does not represent a real number and its value is undefined.

Since for each replacement of the variable(s) for which its denominator is not zero, a rational expression represents a real number, the following relationships for rational expressions follow directly from Theorems 1.8, 1.9, and 1.10 and the laws of closure.

For values of the variables for which the denominators do not vanish,

I $\quad \dfrac{P(x)}{Q(x)} = \dfrac{R(x)}{S(x)} \quad$ if and only if $\quad P(x) \cdot S(x) = Q(x) \cdot R(x)$,

II $\quad -\dfrac{P(x)}{Q(x)} = \dfrac{-P(x)}{Q(x)} = \dfrac{P(x)}{-Q(x)} = -\dfrac{-P(x)}{-Q(x)}$,

III $\quad \dfrac{P(x)}{Q(x)} = \dfrac{-P(x)}{-Q(x)} = -\dfrac{-P(x)}{Q(x)} = -\dfrac{P(x)}{-Q(x)}$,

IV $\quad \dfrac{P(x) \cdot R(x)}{Q(x) \cdot R(x)} = \dfrac{P(x)}{Q(x)} \quad$ (*fundamental principle of fractions*).

Reducing fractions

A fraction is said to be in **lowest terms** when the numerator and denominator do not contain prescribed types of factors in common. The arithmetic fraction a/b, where a and b are integers and $b \neq 0$, is in lowest terms provided a and b contain no

2.6 Equivalent Fractions

common positive integral factors other than 1. If the numerator and denominator of a fraction are polynomials with integral coefficients, then the fraction is said to be in lowest terms if the numerator and denominator have no common factors other than ± 1.

To express a given fraction in lowest terms (called **reducing** the fraction), we can factor the numerator and denominator and apply the fundamental principle of fractions.

Examples

a. $\dfrac{y}{y^2} = \dfrac{1 \cdot y}{y \cdot y}$

$= \dfrac{1}{y} \quad (y \neq 0)$

b. $\dfrac{(x-1)^2}{x-1} = \dfrac{(x-1) \cdot (x-1)}{1 \cdot (x-1)}$

$= x - 1 \quad (x \neq 1)$

Diagonal lines are sometimes used to abbreviate the procedure of reducing a fraction. Thus, in Example (**a**) above, we may write

$$\dfrac{y}{y^2} = \dfrac{\cancel{y}^{\,1}}{\cancel{y^2}_{\,y}} = \dfrac{1}{y} \quad (y \neq 0).$$

Reducing a fraction to lowest terms should be accomplished mentally whenever convenient.

To reduce fractions with polynomial numerators and denominators, you should, when possible, write them in factored form. Common factors are then evident by inspection.

Examples

a. $\dfrac{2x^2 + x - 15}{2x + 6} = \dfrac{(2x - 5)(x + 3)}{2(x + 3)}$

$= \dfrac{2x - 5}{2} \quad (x \neq -3)$

b. $\dfrac{2x + 4}{2x^2 + 8x + 8} = \dfrac{2(x + 2)}{2(x + 2)(x + 2)}$

$= \dfrac{1}{x + 2} \quad (x \neq -2)$

Building fractions

We can also change fractions to equivalent fractions in higher terms by applying the fundamental principle in the form

$$\dfrac{P(x)}{Q(x)} = \dfrac{P(x) \cdot R(x)}{Q(x) \cdot R(x)}.$$

In general, to change a fraction $P(x)/Q(x)$ to an equivalent fraction with $Q(x) \cdot R(x)$ as a denominator, we can determine the factor $R(x)$ by inspection, and then can multiply the numerator and the denominator of the original fraction by this factor.

Example

Express $\dfrac{x}{x + 1}$ as a fraction with denominator $x^2 + 3x + 2$.

Solution on overleaf

Solution

We observe that $x^2 + 3x + 2 = (x + 1)(x + 2)$ and write

$$\frac{x}{x + 1} = \frac{x(x + 2)}{(x + 1)(x + 2)} = \frac{x^2 + 2x}{x^2 + 3x + 2}.$$

Exercise 2.6

A Reduce the fractions to lowest terms. Specify restrictions on the variables for which the reduction is not valid.

Examples

a. $\dfrac{5x^3 y}{15xy^2}$

b. $\dfrac{a + 3}{a^2 + 6a + 9}$

Solutions

a. $\dfrac{5x^3 y}{15xy^2} = \dfrac{(5xy)(x^2)}{(5xy)(3y)}$

$= \dfrac{x^2}{3y} \quad (x, y \neq 0)$

b. $\dfrac{a + 3}{a^2 + 6a + 9} = \dfrac{(a + 3) \cdot 1}{(a + 3)(a + 3)}$

$= \dfrac{1}{a + 3} \quad (a \neq -3)$

1. $\dfrac{8x^3 y^4}{2x^2 y}$

2. $\dfrac{3x^{11} y^3}{33y^2}$

3. $\dfrac{4x^2 + 2x}{2x + 1}$

4. $\dfrac{x^2 - x}{(x - 1)^2}$

5. $\dfrac{x^2 - 1}{x - 1}$

6. $\dfrac{1 - 4x^2}{2x - 1}$

7. $\dfrac{a^2 - b^2}{a + b}$

8. $\dfrac{c^2 - cd}{(1 - c)(1 - d)}$

9. $\dfrac{ab^3 - a^3 b^5}{ab + a^2 b^2}$

10. $\dfrac{xy^3 - xy}{xy - x}$

11. $\dfrac{x^2 - 2x + 1}{x - 1}$

12. $\dfrac{x^2 + 3x - 4}{x^2 - 1}$

13. $\dfrac{x^2 - 3x - 4}{x^2 - 1}$

14. $\dfrac{x^2 + 5x + 6}{x^2 + 4x + 4}$

15. $\dfrac{x^3 + 1}{(x + 1)^3}$

Examples

a. $\dfrac{x^3 - 1}{x^2 - 1}$

b. $\dfrac{x^2 + 2xy + y^2}{(x + y)^2}$

a. $\dfrac{x^3 - 1}{x^2 - 1} = \dfrac{(x - 1)(x^2 + x + 1)}{(x - 1)(x + 1)}$

$= \dfrac{x^2 + x + 1}{x + 1} \quad (x \neq 1, -1)$

b. $\dfrac{x^2 + 2xy + y^2}{(x + y)^2} = \dfrac{(x + y)^2}{(x + y)^2}$

$= 1 \quad (x \neq -y)$

16. $\dfrac{x^2 - 4x - 5}{x^2 + 4x + 3}$

17. $\dfrac{x^3 + 8}{x^2 + 5x + 6}$

18. $\dfrac{x + 5}{x^2 + 14x + 45}$

2.6 Equivalent Fractions

19. $\dfrac{x^4 - 1}{2x^2 + 2}$ 20. $\dfrac{3x^2 - 6x - 45}{9x^2 + 27x}$ 21. $\dfrac{x^3 - 1}{(x - 1)(x^2 + x + 1)}$

22. $\dfrac{a^4 - x^4}{x^2 - a^2}$ 23. $\dfrac{1 - 9x^2}{3x - 1}$ 24. $\dfrac{2x - 6}{x^3 - 27}$

25. $\dfrac{x^4 - 4x^2 - 21}{x^3 + 3x}$ 26. $\dfrac{xy - 2x + 3y - 6}{x^2 + x - 6}$ 27. $\dfrac{b^2 - 4c^2}{ab + 2ac}$

28. $\dfrac{x^3 - y^3}{(x - y)^2}$ 29. $\dfrac{x^6 - 1}{x^2 - 1}$ 30. $\dfrac{x^3 - x^2 + x - 1}{x^4 - 1}$

31. $\dfrac{-y^3 + 2xy^2 + 6x - 3y}{2x^2 - 3xy + y^2}$ 32. $\dfrac{x^4 + 3x^2y^2 + 4y^4}{x^3y - x^2y^2 + 2xy^3}$ 33. $\dfrac{x^4 + x^2y^2 + y^4}{x^3 - y^3}$

Express each fraction as an equivalent fraction with the indicated denominator. Specify the values of the variables for which the fractions are equivalent.

Examples a. $\dfrac{2}{5}, \dfrac{}{35}$ b. $\dfrac{a}{bc}, \dfrac{}{ab^2c}$

Solutions a. $35 = 5 \cdot 7$, so
$$\dfrac{2}{5} = \dfrac{2 \cdot 7}{5 \cdot 7} = \dfrac{14}{35}$$

b. $ab^2c = (ab)(bc)$, so
$$\dfrac{a}{bc} = \dfrac{a(ab)}{(bc)(ab)} = \dfrac{a^2b}{ab^2c} \quad (a, b, c \neq 0)$$

34. $\dfrac{2}{9}, \dfrac{}{54}$ 35. $\dfrac{3}{7}, \dfrac{}{49}$ 36. $\dfrac{a^2}{ab^2c}, \dfrac{}{a^2b^2c^2}$

37. $\dfrac{xyz}{xz}, \dfrac{}{x^2yz^2}$ 38. $\dfrac{x}{x + 1}, \dfrac{}{x^2 + x}$ 39. $\dfrac{x + 1}{x - 1}, \dfrac{}{x^2 - 1}$

40. $\dfrac{3}{x - y}, \dfrac{}{x^2 - y^2}$ 41. $\dfrac{1}{x + 1}, \dfrac{}{x^3 + 1}$ 42. $\dfrac{x + 1}{4}, \dfrac{}{8x^2 + 8x}$

B Reduce each fraction to lowest terms.

43. $\dfrac{6x^2(3x - 4) - 2x(3x - 4)^2}{x^4}$

44. $\dfrac{2(x + 2)(x^2 - 1)^3 - 6x(x^2 - 1)^2(x + 2)^2}{(x^2 - 1)^6}$

45. $\dfrac{(x + 1)^3(x - 4) - 4(x + 1)(x - 4)}{4x^2 - 4x - 48}$

46. $\dfrac{(2x - 3)^4(x^4 - 1) - x^3(2x - 3)(x^4 - 1)}{x^4 - x^3 + x^2 - x}$

2.7 Sums and Differences of Rational Expressions

Since rational expressions represent real numbers for each replacement of the variable(s) for which the denominators are not zero, the following relationships are a direct consequence of Theorem 1.11-III and the laws of closure.

For values of the variables for which $Q(x) \neq 0$,

$$\frac{P(x)}{Q(x)} + \frac{R(x)}{Q(x)} = \frac{P(x) + R(x)}{Q(x)},$$

and

$$\frac{P(x)}{Q(x)} - \frac{R(x)}{Q(x)} = \frac{P(x) - R(x)}{Q(x)}.$$

Examples **a.** $\dfrac{3x^2}{5} + \dfrac{x}{5} = \dfrac{3x^2 + x}{5}$ **b.** $\dfrac{3}{y} - \dfrac{x}{y} = \dfrac{3 - x}{y}$ $(y \neq 0)$

This principle, of course, extends to any number of fractions. If the fractions in a sum or difference have unlike denominators, we can replace the fractions with equivalent fractions having common denominators and then write the sum or difference as a single fraction.

For values for which denominators do not equal zero,

$$\frac{P(x)}{Q(x)} + \frac{R(x)}{S(x)} = \frac{P(x) \cdot S(x)}{Q(x) \cdot S(x)} + \frac{R(x) \cdot Q(x)}{S(x) \cdot Q(x)}$$

$$= \frac{P(x) \cdot S(x) + R(x) \cdot Q(x)}{Q(x) \cdot S(x)}$$

and

$$\frac{P(x)}{Q(x)} - \frac{R(x)}{S(x)} = \frac{P(x) \cdot S(x) - R(x) \cdot Q(x)}{Q(x) \cdot S(x)}.$$

Least common multiple

In rewriting fractions in a sum or difference so that they share a common denominator, any such denominator may be used. If the **least common multiple** of the denominators (called the **least common denominator**) is used, however, the resulting fraction will be in simpler form than if any other common denominator is employed. The least common multiple of two or more natural numbers is the least natural number that is exactly divisible by each of the given numbers (that is, each quotient is a natural number).

Example Find the least common multiple of 24, 45, and 60.

2.7 Sums and Differences of Rational Expressions

Solution Factor 24, 45, 60
as $2^3 \cdot 3$, $3^2 \cdot 5$, $2^2 \cdot 3 \cdot 5$.
The least common multiple is $2^3 \cdot 3^2 \cdot 5 = 360$.

The notion of a least common multiple among several polynomial expressions is, in general, meaningless without further specification of what is desired. We can, however, define the least common multiple of a set of polynomials with integer coefficients to be the polynomial of lowest degree with integer coefficients which has each of the given polynomials as a factor and which, among all such polynomials, has the least possible positive leading coefficient.

Very often, the least common multiple of a set of natural numbers or polynomials can be determined by inspection. When inspection fails us, however, we can find the least common multiple of a set of polynomials with integer coefficients as follows:

1. Express each polynomial in completely factored form.
2. Write as factors of a product each *different* factor occurring in any of the polynomials, including each factor the greatest number of times it occurs in any one of the given polynomials.

Example Find the least common multiple of $x^3 - x$, $x^2 + x$, and $x^2 - 3x + 2$.

Solution Factor $x^3 - x$, $x^3 + x^2$, $x^2 - 3x + 2$
as $x(x-1)(x+1)$, $x^2(x+1)$, $(x-1)(x-2)$.
The least common multiple is $x^2(x-1)(x+1)(x-2)$.

To simplify sums of fractions having different denominators, we can ascertain the least common denominator of the fractions, determine the factor necessary to express each of the fractions as a fraction having this common denominator, write the fractions accordingly, and then express the sum as a single fraction.

Example Write $\dfrac{1}{x} + \dfrac{-2}{x+1} + \dfrac{2x}{(x+1)^2}$ as a single fraction.

Solution The least common denominator of the fractions is $x(x+1)^2$. Thus

$$\frac{1}{x} + \frac{-2}{x+1} + \frac{2x}{(x+1)^2} = \frac{1(x+1)^2}{x(x+1)^2} + \frac{-2x(x+1)}{x(x+1)^2} + \frac{2x \cdot x}{x(x+1)^2}$$

$$= \frac{x^2 + 2x + 1}{x(x+1)^2} + \frac{-2x^2 - 2x}{x(x+1)^2} + \frac{2x^2}{x(x+1)^2}$$

$$= \frac{x^2 + 1}{x(x+1)^2} \quad (x \neq 0, -1).$$

Exercise 2.7

A Find the least common multiple.

Examples a. 54, 20, 63 b. $x^3, x^2 + x, x^4 + 2x^3 + x^2$

Solutions
a. Factor 54, 20, 63 as $2 \cdot 3^3$, $2^2 \cdot 5$, $3^2 \cdot 7$; least common multiple is $2^2 \cdot 3^3 \cdot 5 \cdot 7 = 3780$.

b. Factor x^3, $x^2 + x$, $x^4 + 2x^3 + x^2$ as x^3, $x(x + 1)$, $x^2(x + 1)^2$; least common multiple is $x^3(x + 1)^2$.

1. 18, 27, 35
2. 15, 20, 45
3. 21, 56, 48
4. $x^2 - 1, x^3 + x^2$
5. $x^4 - x^3, x^2 + 3x - 4$
6. $2x^3, 4x + 2, 4x^2 + 8x + 1$

Write each sum or difference as a single fraction in lowest terms. Assume no variable in a denominator takes a value for which the denominator vanishes.

Examples
a. $\dfrac{x + 1}{x} + \dfrac{x - 1}{x + 1}$
b. $\dfrac{x}{x^2 - 1} + \dfrac{1}{x - 1} - \dfrac{1}{x + 1}$

Solutions
a. $\dfrac{x + 1}{x} + \dfrac{x - 1}{x + 1}$
$= \dfrac{(x + 1)^2}{x(x + 1)} + \dfrac{x(x - 1)}{x(x + 1)}$
$= \dfrac{x^2 + 2x + 1 + x^2 - x}{x(x + 1)}$
$= \dfrac{2x^2 + x + 1}{x(x + 1)}$

b. $\dfrac{x}{x^2 - 1} + \dfrac{1}{x - 1} - \dfrac{1}{x + 1}$
$= \dfrac{x}{x^2 - 1} + \dfrac{1 \cdot (x + 1)}{(x - 1)(x + 1)} - \dfrac{1 \cdot (x - 1)}{(x + 1)(x - 1)}$
$= \dfrac{x + (x + 1) - (x - 1)}{x^2 - 1}$
$= \dfrac{x + 2}{x^2 - 1}$

7. $\dfrac{x - 1}{x^2} + \dfrac{1}{x^2 + x}$
8. $\dfrac{1}{x^2 - 1} + \dfrac{x}{x - 1}$
9. $\dfrac{y + 3}{4y} - \dfrac{1}{2y^2}$
10. $\dfrac{x - 2}{x^2} + \dfrac{1}{x}$
11. $\dfrac{1}{x - 1} - \dfrac{1}{x + 1}$
12. $\dfrac{y - 1}{y + 2} + \dfrac{y - 3}{y + 4}$

13. $\dfrac{4x-1}{x+2} + \dfrac{3x+1}{x-1}$

14. $\dfrac{x-1}{x+3} + \dfrac{x-2}{x+6}$

15. $\dfrac{-4x+2}{2x-1} - \dfrac{2x+1}{x-3}$

16. $\dfrac{3x-1}{x+1} - \dfrac{x}{x+2}$

17. $\dfrac{1}{2-x} - \dfrac{x}{x-2}$

18. $\dfrac{2}{x} - \dfrac{1}{x^2+x}$

19. $\dfrac{1}{w^2-3w+2} + \dfrac{1}{w^2-1}$

20. $\dfrac{1}{x^2+5x+4} - \dfrac{1}{x^2-1}$

21. $\dfrac{x+1}{x^2-5x+6} - \dfrac{1}{x^2-6x+9}$

22. $\dfrac{2}{u^2+u-2} + \dfrac{u}{(u-1)^2}$

23. $\dfrac{1}{x+1} - \dfrac{1}{x^2-6x+9}$

24. $\dfrac{1}{x} - \dfrac{1}{x+1}$

25. $\dfrac{1}{x+1} - \dfrac{1}{x^2-1} + \dfrac{x}{x-1}$

26. $\dfrac{2}{x} + \dfrac{x+1}{x^2+x} + \dfrac{1}{x+1}$

27. $z - \dfrac{1}{z} - \dfrac{1}{z^2}$

28. $1 + \dfrac{x}{1+x} + \dfrac{x^2}{1+2x+x^2}$

29. $1 - \dfrac{1}{x-1} + \dfrac{x}{x^2-1}$

30. $a + 1 + \dfrac{1}{a} + \dfrac{1-a^3}{a^2+a}$

31. $\dfrac{x+3}{3x^2+7x+4} - \dfrac{x-7}{3x^2+13x+12}$

32. $\dfrac{z+4}{2z^2-5z-3} + \dfrac{2z-1}{2z^2+3z+1}$

33. $\dfrac{x}{x-a} - \dfrac{a}{x+a}$

34. $\dfrac{4ax}{x^2-a^2} + \dfrac{x-a}{x+a}$

B 35. $\dfrac{x+1}{x^3-x^2+x-1} + \dfrac{x-1}{x^4-x^3-x+1}$

36. $\dfrac{2x+1}{2x^3-3x^2-2x+3} - \dfrac{x+2}{2x^2-5x+1}$

37. $\dfrac{x}{x^3-1} - \dfrac{x}{x^4+x^2+1}$

38. $\dfrac{x}{x^4+2x^2+9} - \dfrac{x^2+1}{x^3-2x^2+3x}$

2.8 Products and Quotients of Rational Expressions

By Parts II and VII of Theorem 1.11 and the laws of closure, we have the following result.

For values of the variables for which the denominators do not vanish,

$$\frac{P(x)}{Q(x)} \cdot \frac{R(x)}{S(x)} = \frac{P(x) \cdot R(x)}{Q(x) \cdot S(x)},$$

and

$$\frac{P(x)}{Q(x)} \div \frac{R(x)}{S(x)} = \frac{P(x) \cdot S(x)}{Q(x) \cdot R(x)}.$$

We can use the above relationships to rewrite a product or quotient of fractions as a single fraction in lowest terms.

Example

$$\frac{x^2 - 4}{x^2 + x - 12} \cdot \frac{x - 3}{x^2 + 4x + 4} = \frac{(x - 2)(x + 2)}{(x + 4)(x - 3)} \cdot \frac{x - 3}{(x + 2)^2}$$

$$= \frac{(x - 2)(x + 2)(x - 3)}{(x + 4)(x - 3)(x + 2)^2}$$

$$= \frac{x - 2}{(x + 4)(x + 2)} = \frac{x - 2}{x^2 + 6x + 8} \quad (x \neq -2, 3, -4)$$

Since the factors of the numerator and denominator of the product of two fractions are just the factors of the numerators and denominators (respectively) of the fractions, we can divide common factors out of the numerators and denominators before writing the product as a single fraction. Thus, in the example above, we could write

$$\frac{x^2 - 4}{x^2 + x - 12} \cdot \frac{x - 3}{x^2 + 4x + 4} = \frac{(x - 2)\cancel{(x + 2)}}{(x + 4)\cancel{(x - 3)}} \cdot \frac{\cancel{x - 3}}{(x + 2)\cancel{(x + 2)}}$$

$$= \frac{x - 2}{(x + 4)(x + 2)} \quad (x \neq -2, 3, -4).$$

Of course this procedure can be used to simplify the expression obtained by rewriting a quotient as a product.

Example

$$\frac{x^2 - 5x + 4}{x^2 + 5x + 6} \div \frac{x - 1}{x + 2} = \frac{\cancel{(x - 1)}(x - 4)}{\cancel{(x + 2)}(x + 3)} \cdot \frac{\cancel{x + 2}}{\cancel{x - 1}} = \frac{x - 4}{x + 3} \quad (x \neq -2, -3, 1)$$

The quotient $\dfrac{P(x)}{Q(x)} \div \dfrac{R(x)}{S(x)}$ can also be denoted by

$$\frac{\dfrac{P(x)}{Q(x)}}{\dfrac{R(x)}{S(x)}}.$$

This latter form is called a **complex fraction**; that is, it is a fraction containing a fraction in either the numerator or denominator or both.

Reduction of complex fractions

When the quotient of two fractions is given in the form of a complex fraction, we have a choice of procedures available to us for writing the quotient in the form of a simple (not complex) fraction.

2.8 Products and Quotients of Rational Expressions

Example Write $\dfrac{\dfrac{x+1}{4}}{x - \dfrac{2}{3}}$ as a simple fraction in lowest terms.

Solution 1 We can apply the fundamental principle of fractions to multiply numerator and denominator by the least common denominator of the simple fractions involved. Then, we have

$$\frac{\left(\dfrac{x+1}{4}\right)12}{\left(x - \dfrac{2}{3}\right)12} = \frac{3x + 3}{12x - 8} \quad \left(x \neq \frac{2}{3}\right).$$

Solution 2 Alternatively, we can rewrite the complex fraction as

$$\frac{\dfrac{x+1}{4}}{x - \dfrac{2}{3}} = \left(\frac{x+1}{4}\right) \div \left(x - \frac{2}{3}\right) = \left(\frac{x+1}{4}\right) \div \left(\frac{3x-2}{3}\right)$$

$$= \frac{x+1}{4} \cdot \frac{3}{3x-2} = \frac{3x+3}{12x-8} \quad \left(x \neq \frac{2}{3}\right).$$

In the event we have a more complicated expression involving a complex fraction, we can rewrite the expression by simplifying small parts of it at a time.

Example Write $\dfrac{1}{x + \dfrac{1}{x + \dfrac{1}{x}}}$ as a simple fraction in lowest terms.

Solution We can begin by concentrating on the lower right-hand expression, $\dfrac{1}{x + \dfrac{1}{x}}$. We have

$$\frac{1}{x + \dfrac{1}{x}} = \frac{(1)\,x}{\left(x + \dfrac{1}{x}\right)x} = \frac{x}{x^2 + 1} \quad (x \neq 0).$$

Thus,

$$\frac{1}{x + \dfrac{1}{x + \dfrac{1}{x}}} = \frac{1}{x + \dfrac{x}{x^2 + 1}} \quad (x \neq 0).$$

Solution continued on overleaf

From this point, we can apply either of the methods shown in the previous example to the right-hand member above. Using the first method, we have

$$\frac{1}{x + \dfrac{1}{x + \dfrac{1}{x}}} = \frac{1(x^2+1)}{\left(x + \dfrac{x}{x^2+1}\right)(x^2+1)}$$

$$= \frac{x^2+1}{x^3+x+x} = \frac{x^2+1}{x^3+2x} \quad (x \neq 0).$$

Exercise 2.8

A Write each product or quotient as a simple fraction in lowest terms. Assume that no denominator vanishes.

Examples

a. $\dfrac{3a^2b}{2c^2} \div \dfrac{6a}{b^2c}$

b. $\dfrac{x^2-1}{x} \cdot \dfrac{x^2}{x-1}$

Solutions

a. $\dfrac{3a^2b}{2c^2} \div \dfrac{6a}{b^2c} = \dfrac{3a^2b}{2c^2} \cdot \dfrac{b^2c}{6a}$

$= \dfrac{3a^2b^3c}{12ac^2} = \dfrac{ab^3}{4c}$

b. $\dfrac{x^2-1}{x} \cdot \dfrac{x^2}{x-1} = \dfrac{(x-1)(x+1)x^2}{x(x-1)}$

$= (x+1) \cdot x$

1. $\dfrac{x^2yz}{ab} \cdot \dfrac{a^2bc}{xy^2z}$

2. $\dfrac{xy}{a} \div \dfrac{ay}{x}$

3. $\dfrac{4xyz}{3a} \div \dfrac{6xy^2}{10ab}$

4. $\dfrac{x^2}{yz} \cdot \dfrac{2x^2y}{z}$

5. $\dfrac{x+1}{x} \cdot \dfrac{x^2}{x^2-1}$

6. $\dfrac{x^2-2x+1}{x+1} \cdot \dfrac{x^2+2x+1}{x-1}$

7. $\dfrac{x^2-1}{x^2-4} \div \dfrac{x^2+2x+1}{x^2+4x+4}$

8. $\dfrac{2x^2-2}{x+1} \div \dfrac{x}{6x+6}$

9. $\dfrac{a^2b^2}{a^2} \cdot \dfrac{a^2-ab+b^2}{a^2} \div \dfrac{a^3+b^3}{a^4}$

10. $\dfrac{x^3+8}{x^2-9} \cdot \dfrac{x^3-27}{x^2+5x+6} \div \dfrac{x^2-2x+4}{x^2+6x+9}$

11. $\left(1 - \dfrac{1}{z}\right)\left(1 + \dfrac{1}{z}\right)$

12. $\left(\dfrac{1}{z} + 1\right) \div \left(\dfrac{1}{z} - 1\right)$

2.8 Products and Quotients of Rational Expressions

Examples

a. $\left(\dfrac{1}{x} - \dfrac{1}{x+1}\right)x$

b. $a - \dfrac{a}{a + \dfrac{1}{a}}$

Solutions

a. $\left(\dfrac{1}{x} - \dfrac{1}{x+1}\right)x$

$= \left[\dfrac{(x+1) - x}{x(x+1)}\right]x$

$= \dfrac{1}{x(x+1)} \cdot x$

$= \dfrac{1}{x+1}$

b. $a - \dfrac{a}{a + \dfrac{1}{a}}$

$= a - \dfrac{a}{\dfrac{a^2 + 1}{a}} = a - a \cdot \dfrac{a}{a^2 + 1}$

$= a - \dfrac{a^2}{a^2 + 1} = \dfrac{a^3 + a - a^2}{a^2 + 1}$

$= \dfrac{a^3 - a^2 + a}{a^2 + 1}$

---

13. $\left(\dfrac{5}{x+5} - \dfrac{4}{x+4}\right) \cdot \dfrac{x+4}{x}$

14. $\left(\dfrac{x}{x^2+1} - \dfrac{1}{x+1}\right) \cdot \dfrac{x+1}{x-1}$

15. $\left(\dfrac{5}{x^2-9} + 1\right) \div \dfrac{x+2}{x-3}$

16. $\left(\dfrac{y}{y+1} + \dfrac{1}{y-1}\right) \div \dfrac{y^2+1}{y+1}$

17. $\dfrac{1}{1 + \dfrac{1}{x}}$

18. $\dfrac{3}{2 - \dfrac{3}{x^2}}$

19. $2 - \dfrac{2}{2 - \dfrac{1}{x}}$

20. $3 + \dfrac{1}{1 + \dfrac{2}{x}}$

21. $\dfrac{1 + \dfrac{1}{2x+1}}{1 + \dfrac{1}{x}}$

22. $\dfrac{1 - \dfrac{2}{x+1}}{\dfrac{x-1}{2}}$

23. $1 + \dfrac{2 + \dfrac{1}{x}}{x - \dfrac{1}{x}}$

24. $\dfrac{a + \dfrac{2}{a+3}}{a + 3 - \dfrac{1}{a+3}}$

25. $\dfrac{a + 3 + \dfrac{12}{a-5}}{a + 5 + \dfrac{16}{a-5}}$

26. $\dfrac{x - \dfrac{5}{x+1}}{x + 2 - \dfrac{3}{x-1}}$

27. $\dfrac{x - 7 + \dfrac{36}{x+6}}{x - 1 + \dfrac{12}{x+6}}$

28. $\dfrac{x - 2 + \dfrac{4}{x+3}}{x^2 - 3x + 2}$

29. $\dfrac{1 + \dfrac{1}{1 - \dfrac{x}{y}}}{1 - \dfrac{3}{1 + \dfrac{x}{y}}}$

30. $\dfrac{1 - \dfrac{1}{\dfrac{x}{y} + 2}}{1 + \dfrac{3}{\dfrac{x}{y} + 2}}$

Chapter Review

[2.1] Simplify.

1. $(x^2 - 5x + 1) - (2 - 3x - x^2)$
2. $3\{x^2 + 1 - [x^2 - 2(x^2 - 1)] + 1\}$

Given $P(x) = 2x^2 - 3x + 4$, find the following.

3. $P(3)$
4. $P(x)|_{-1}^{2}$

[2.2] Simplify.

5. $2(x + 1)(3x + 2)$
6. $(x + 2)(3x^2 + x + 1)$

[2.3] Factor completely.

7. $y^2 - 8y + 15$
8. $x^3 + 4x^2 + 4x$
9. $x^4 - 16$
10. $z^3 - 64$

[2.4] Write each quotient as a polynomial.

11. $\dfrac{30x^2y^3z^4}{5xyz^3}$
12. $\dfrac{18y^3 + 21y^2 + 3y}{3y}$
13. $\dfrac{2x^2 + x - 15}{x + 3}$
14. $\dfrac{6x^2 + x - 2}{2x - 1}$

[2.5] Use synthetic division to write each quotient $\dfrac{P(x)}{D(x)}$ in the form $Q(x)$ or in the form $Q(x) + \dfrac{r}{D(x)}$, where $Q(x)$ is a polynomial and r is a real number.

15. $\dfrac{x^4 - 5x^2 + 7x - 10}{x - 2}$
16. $\dfrac{3x^3 + 6x^2 - 4x - 5}{x + 2}$

[2.6] Reduce to lowest terms and state all restrictions on the variables.

17. $\dfrac{6y^2z + 33yz^2}{3yz}$
18. $\dfrac{x^2 + 10x + 21}{x + 3}$
19. $\dfrac{x^2 + 4x - 21}{x^2 - 9}$
20. $\dfrac{3x^2 + 12x + 12}{3x^2 - 12}$

Review Exercises

[2.7] Simplify. Assume that no denominator vanishes.

21. $\dfrac{x}{3} + \dfrac{x+1}{4}$

22. $\dfrac{x+1}{x-1} - \dfrac{x-2}{x+2}$

23. $\dfrac{x^2+3}{x+1} - 2$

24. $\dfrac{y+2}{y^2-y-2} + \dfrac{y-3}{y+1}$

[2.8] Simplify. Assume that no denominator vanishes.

25. $\dfrac{x^3 y}{z} \cdot \dfrac{xz^2}{y^3}$

26. $\dfrac{x^2+2x+1}{x-3} \cdot \dfrac{x^2+4x-21}{x^2+6x+5}$

27. $\dfrac{x^2-1}{x+2} \div \dfrac{x^2+4x+3}{x^2+4x+4}$

28. $\dfrac{x - \dfrac{1}{x+1}}{x+4 + \dfrac{11}{x-3}}$

3 Rational Exponents—Radicals

In Chapter 2, powers of real numbers were defined for natural-number exponents, and some simple properties of products and quotients of powers were examined. In this chapter, powers with integral and rational exponents are defined in a manner consistent with these properties.

3.1 Powers with Integral Exponents

Reason for defining a^0 to be 1

In Section 2.4 we saw that for $a \in R$, $a \neq 0$, and $m, n \in N$, $m > n$, we have

$$\frac{a^m}{a^n} = a^{m-n}. \tag{1}$$

If (1) is to hold also for $m = n$, then we must have

$$\frac{a^n}{a^n} = a^{n-n} = a^0.$$

Since $a^n/a^n = 1$ for $a \neq 0$, for consistency we can state the following definition.

Definition 3.1   If $a \in R$, $a \neq 0$, then

$$a^0 = 1.$$

Examples   **a.** $2^0 = 1$ **b.** $(x - 3)^0 = 1$ for $x \neq 3$

3.1 Powers with Integral Exponents

Reason for defining a^{-n} to be $1/a^n$

In a similar way, if (1) is to hold for $m = 0$, then we must have

$$\frac{a^0}{a^n} = a^{0-n} = a^{-n}.$$

Since $a^0 = 1$, for consistency we can make the following definition.

Definition 3.2   If $a \in R$, $a \neq 0$, and $n \in N$, then

$$a^{-n} = \frac{1}{a^n}.$$

Examples

a. $3^{-2} = \dfrac{1}{3^2}$ b. $(x + 1)^{-1} = \dfrac{1}{x + 1}$ for $x \neq -1$

Laws of exponents for integral exponents

The laws of exponents for positive-integer exponents given by Theorems 2.1 and 2.2 are also valid for negative integers. For example, if m and n are both negative integers then $-m$ and $-n$ are both positive integers and we can write

$$a^m \cdot a^n = \frac{1}{a^{-m}} \cdot \frac{1}{a^{-n}} \qquad \text{By Definition 3.2.}$$

$$= \frac{1}{a^{-m} \cdot a^{-n}} \qquad \text{By Theorem 1.11-II.}$$

$$= \frac{1}{a^{(-m)+(-n)}} \qquad \text{By Theorem 2.1-I.}$$

$$= \frac{1}{a^{-(m+n)}} = a^{m+n} \qquad \text{By Definition 3.2.}$$

Thus, $a^m \cdot a^n = a^{m+n}$ for m and n negative integers. A complete discussion of this law and of all of the others given by Theorems 2.1 and 2.2 requires consideration of several cases. The arguments are similar to the one given above, and we shall not give them here. We now restate these theorems for integral exponents $m, n \in J$.

Theorem 3.1   If $a, b \in R$ ($a, b \neq 0$) and $m, n \in J$, then

    I $a^m \cdot a^n = a^{m+n}$,   IV $(ab)^n = a^n b^n$,

   II $\dfrac{a^m}{a^n} = a^{m-n}$,   V $\left(\dfrac{a}{b}\right)^n = \dfrac{a^n}{b^n}.$

  III $(a^m)^n = a^{mn}$,

Some applications of this theorem should prove enlightening.

Examples a. $(x^2)^3 = x^{2 \cdot 3} = x^6$ By Theorem 3.1-III.
b. $(xy)^3 = x^3 y^3$ By Theorem 3.1-IV.
c. $\left(\dfrac{x^2}{y}\right)^{-4} = \dfrac{(x^2)^{-4}}{(y)^{-4}}$ By Theorem 3.1-V.

$= \dfrac{x^{-8}}{y^{-4}}$ By Theorem 3.1-III.

$= \dfrac{1/x^8}{1/y^4}$ By Definition 3.2.

$= \dfrac{y^4}{x^8} \quad (x, y \neq 0)$ By Theorem 1.11-VII.

Although we shall not do anything about proving it here, Parts IV, and V of Theorem 3.1 can be shown to apply to expressions involving additional factors.

Examples a. $\left(\dfrac{a^3 b}{3c^2}\right)^{-2} = \dfrac{a^{-6} b^{-2}}{3^{-2} c^{-4}}$ b. $\dfrac{x^{-1} - 1}{x^{-2} y} = \dfrac{\dfrac{1}{x} - 1}{\dfrac{y}{x^2}}$

$= \dfrac{3^2 c^4}{a^6 b^2}$

$= \dfrac{9c^4}{a^6 b^2} \quad (a, b, c \neq 0)$

$= \left(\dfrac{1}{x} - 1\right) \cdot \dfrac{x^2}{y}$

$= \dfrac{1 - x}{x} \cdot \dfrac{x^2}{y}$

$= \dfrac{(1 - x) \cdot x}{y} \quad (x, y \neq 0)$

Scientific notation

In many applications of mathematics, we deal with numbers which are very close to zero or which are very large. These numbers are usually expressed as a product of a number between 1 and 10 (excluding 10) and a power of 10. When a number is expressed in this way, we say it is expressed in **scientific notation**.

To express a number in scientific notation, we can factor the given number by simply moving the decimal point to a position where the resulting number is between 1 and 10 (excluding 10) and multiplying by the power of 10, where the exponent on 10 corresponds to the number of decimal places the point has been moved. The exponent is negative if the point was moved to the right and positive if the point was moved to the left.

Examples a. 7,450,000,000 b. 0.00000612
$= 7{,}450{,}000{,}000.$ $= 0.00000612$
$= 7.45 \times 10^9$ $= 6.12 \times 10^{-6}$

3.1 Powers with Integral Exponents

Many calculators display results of operations in the form of a number between 1 and 10 (excluding 10) and an integer which represents a power of 10, that is, essentially in scientific notation.

To translate a number from scientific notation to standard decimal notation, we simply move the decimal point the number of places indicated by the exponent on 10, moving it to the right if the power is positive and to the left if the power is negative.

Examples

a. 6.7345×10^2
$= 6.7345$
$= 673.45$

b. 4.3571×10^{-5}
$= 00004.3571$
$= 0.000043571$

In order to avoid the necessity of constantly noting exceptions, we shall assume that in the exercises and examples in this chapter the variables are restricted so that no denominator vanishes.

Exercise 3.1

A For each of the following expressions, write an equivalent basic numeral—that is, a numeral with exponent 1.

Examples

a. $4 \cdot 3^{-2}$

b. $\dfrac{3^{-2}}{6^{-1}}$

c. $2^{-1} + 2^0 + 2^1$

Solutions

a. $4 \cdot 3^{-2} = 4 \cdot \dfrac{1}{3^2}$
$= \dfrac{4}{9}$

b. $\dfrac{3^{-2}}{6^{-1}} = \dfrac{\frac{1}{3^2}}{\frac{1}{6}} = \dfrac{1}{9} \cdot \dfrac{6}{1}$
$= \dfrac{6}{9} = \dfrac{2}{3}$

c. $2^{-1} + 2^0 + 2^1 = \dfrac{1}{2} + 1 + 2$
$= \dfrac{7}{2}$

1. 3^{-1}
2. 4^{-2}
3. $(-2)^{-3}$
4. $(-2)^{-4}$
5. $\dfrac{1}{2^{-2}}$
6. $\dfrac{1}{3^{-1}}$
7. $\dfrac{2^0 \cdot 3^{-1}}{4}$
8. $\dfrac{3^0}{4^{-1}5^{-2}}$
9. $\left(\dfrac{1}{3}\right)^{-1}$
10. $\left(\dfrac{2}{5}\right)^{-2}$
11. $4^{-1} + 2^{-2}$
12. $3^{-2} - 7^0$

Write each expression as a product or quotient in which each variable occurs at most once in the expression and all exponents are positive.

Examples

a. $x^4 x^{-5}$

b. $(x^{-2}y^3)^{-2}$

c. $\left(\dfrac{x^{-1}yz^0}{xy^{-1}z}\right)^{-1}$

Solutions on overleaf

Solutions

a. $x^4 x^{-5} = x^{-1}$
$= \dfrac{1}{x}$

b. $(x^{-2}y^3)^{-2} = x^4 y^{-6}$
$= \dfrac{x^4}{y^6}$

c. $\left(\dfrac{x^{-1}yz^0}{xy^{-1}z}\right)^{-1} = \dfrac{x^1 y^{-1} z^0}{x^{-1} yz^{-1}}$
$= \dfrac{x \cdot x \cdot z}{y \cdot y}$
$= \dfrac{x^2 z}{y^2}$

13. $x^8 x^{-5}$
14. $y^{-3} y^2$
15. $\dfrac{y^{-1}}{x^2}$
16. $\dfrac{x^3}{y^{-2}}$

17. $(x^2 y)^5$
18. $(xy^{-3})^2$
19. $\left(\dfrac{x^2}{2y}\right)^3$
20. $\left(\dfrac{2x}{y^2}\right)^2 \cdot \dfrac{y}{x^2}$

21. $\left(\dfrac{x^{-1}y^2}{2z}\right)^{-2}$
22. $\left(\dfrac{x^0 y^2}{z^2}\right)^{-1} \cdot \dfrac{x}{y^{-1}}$

23. $\left(\dfrac{xy^2}{z^3}\right)^{-2} \cdot \left(\dfrac{xy^0}{z^2}\right)^{-1}$
24. $\left(\dfrac{2x^{-1}}{y^2}\right)^{-1} \cdot \dfrac{x^2}{y}$

Represent each expression as a single fraction involving positive exponents only.

Examples

a. $x^{-1} + y^{-1}$

b. $\dfrac{x^{-1}}{(x^{-1} + y^{-2})^{-1}}$

Solutions

a. $x^{-1} + y^{-1} = \dfrac{1}{x} + \dfrac{1}{y}$
$= \dfrac{y+x}{xy}$

b. $\dfrac{x^{-1}}{(x^{-1} + y^{-2})^{-1}} = \dfrac{\dfrac{1}{x}}{\left(\dfrac{1}{x} + \dfrac{1}{y^2}\right)^{-1}}$
$= \dfrac{1}{x}\left(\dfrac{1}{x} + \dfrac{1}{y^2}\right)$
$= \dfrac{1}{x}\left(\dfrac{y^2 + x}{xy^2}\right) = \dfrac{y^2 + x}{x^2 y^2}$

25. $x^{-1} + y^{-2}$
26. $x^{-2} - y^{-1}$
27. $xy^{-1} - x^{-1}y$

28. $\dfrac{x}{y^{-1}} + \dfrac{y^{-1}}{x}$
29. $\dfrac{x^2}{y} - \dfrac{x^{-1}}{y^2}$
30. $x(x+1)^{-1}$

31. $xy(x^{-1} + y^{-1})$
32. $x^{-1}(x^{-1} + 1)$
33. $\dfrac{x^{-1} + y^{-1}}{x^{-1} y^{-1}}$

34. $\dfrac{x^{-1} - y^{-2}}{(x^2 y)^0}$
35. $(x^{-2} + y^{-1})^{-2}$
36. $\dfrac{(x^{-1} + y^{-1})^{-1}}{x^{-1} + y^{-1}}$

3.1 Powers with Integral Exponents

Write each expression as a rational expression, that is, as the quotient of two polynomials.

Examples
a. $(x + 3x^{-1})(x^{-1} + 1)$
b. $x^{-2}y + x^2y^{-3}$

Solutions
a. $(x + 3x^{-1})(x^{-1} + 1)$
$$= xx^{-1} + 3x^{-1}x^{-1} + x + 3x^{-1}$$
$$= 1 + 3x^{-2} + x + 3x^{-1}$$
$$= 1 + \frac{3}{x^2} + x + \frac{3}{x}$$
$$= \frac{x^2 + 3 + x^3 + 3x}{x^2}$$
$$= \frac{x^3 + x^2 + 3x + 3}{x^2}$$

b. $x^{-2}y + x^2y^{-3}$
$$= \frac{y}{x^2} + \frac{x^2}{y^3}$$
$$= \frac{y^3y + x^2x^2}{x^2y^3}$$
$$= \frac{y^4 + x^4}{x^2y^3}$$

37. $(x^{-1} + 1)(x^{-1} - 1)$
38. $x^{-1}(x + 2x^2 + 3x^3)$
39. $(x^{-1} + y^{-1})(x + y)$
40. $(3y + y^{-1})(y^{-1} + 3)$
41. $(2x - x^{-1})(x^{-2} + x)$
42. $(3y + y^{-1})(y^{-1} - 2)$

Factor as indicated.

Examples
a. $x^{-2}y + x^2y^{-3} = x^{-2}y^{-3}(?)$
b. $x^2y^{-2} + y^2x^{-2} = x^{-3}y^{-2}(?)$

Solutions
a. $x^{-2}y^{-3}(y^4 + x^4)$
b. $x^{-3}y^{-2}(x^5 + y^4x)$

43. $x^{-3}y^2 + 4xy^{-1} = x^{-3}y^{-1}(?)$
44. $xy^{-1} + x^{-1}y^{-2} = x^{-1}y^{-2}(?)$
45. $3x^{-3}y^{-4} + 4x^{-4}y^{-3} = x^{-4}y^{-4}(?)$
46. $4y^{-1} + x^2y^{-3} = y^{-3}(?)$
47. $xy^2z^3 - x^2y^{-1}z^{-3} = xy^{-1}z^{-3}(?)$
48. $x^{-1}z^{-3} + 2y^{-2}z = x^{-1}y^{-2}z^{-3}(?)$

Write each number in scientific notation.

Examples
a. 13400
b. 0.00532

Solutions
a. $13400 = 1.34 \times 10^4$
b. $0.00532 = 5.32 \times 10^{-3}$

49. 19711
50. 36
51. 1976
52. 6140
53. 0.0586
54. 0.0002
55. 0.01001
56. 0.9

Write each number in decimal notation.

Examples a. 2.31×10^6 b. 4.34×10^{-2}

Solutions a. 2,310,000 b. 0.0434

57. 2.48×10^8
58. 5.76×10^3
59. -6.542×10^6
60. -7.361×10^2
61. 3.65×10^{-4}
62. 7.61×10^{-6}
63. -4.625×10^{-8}
64. -6.571×10^{-3}

Write each product or quotient as a single number in scientific notation.

65. $(0.00043)(2000)$
66. $(2,100,000)(8,000,000)$
67. $\dfrac{201,000,000}{0.0003}$
68. $\dfrac{0.000000046}{0.00002}$

B For each expression, write an equivalent product in which each variable occurs only once.

Examples a. $\dfrac{x^n x^{n+3}}{x^{n-1}}$ b. $\dfrac{(x^{n-1})^2}{x^n}$

Solutions a. $\dfrac{x^n x^{n+3}}{x^{n-1}} = x^{n+(n+3)-(n-1)}$ b. $\dfrac{(x^{n-1})^2}{x^n} = x^{2(n-1)-n}$

$\qquad\qquad\qquad = x^{n+4}$ $\qquad\qquad\qquad = x^{n-2}$

69. $x^n x^{n-3}$
70. $x^n x^{n+1}$
71. $y^{n+1} y^{-n+1}$
72. $\dfrac{x^{n+1} x^{n+2}}{x^{n+3}}$
73. $\dfrac{y^{3n} y^{2n-1}}{y^{-2}}$
74. $\dfrac{x^2}{x^{n-2}}$
75. $\left(\dfrac{x^n}{x^{-1}}\right)^2$
76. $\dfrac{(x^2)^n}{(x^n)^{-1}}$
77. $\dfrac{x^n y^{n+2}}{(xy)^{-1}}$
78. $\dfrac{x^{2n+1} y^{2n-1}}{x^n y^n}$
79. $\left(\dfrac{xy}{x^n}\right)^{-1}$
80. $\left(\dfrac{x^{n+1} y}{xy^{n+1}}\right)^{-1}$

3.2 Powers with Rational Exponents

In Section 3.1, we defined powers of real numbers with 0 and negative-integer exponents so that the law of exponents given in Theorem 2.1-I would remain valid. It followed that the remaining laws of positive-integer exponents hold for all

3.2 Powers with Rational Exponents

integer exponents also. Now we want to give meaning to powers of real numbers with rational numbers as exponents. As before, we shall want any such definition to be consistent with all of the laws of exponents for integers.

We observed in Chapter 2 that for $a \in R$ and $m, n \in N$,

$$(a^m)^n = a^{mn}. \tag{1}$$

If (1) is to hold for $m = 1/n$, and $a^{1/n}$ is a real number, then we must have

$$(a^{1/n})^n = a^{(1/n)(n)} = a^{n/n} = a^1 = a,$$

so that the nth power of $a^{1/n}$ must be a. A number having a as its nth power is called an **nth root** of a. In particular, for n equal to 2 or 3, respectively, an nth root is called a **square root** or a **cube root**.

Number of roots

For n odd, each $a \in R$ has just one real nth root. Thus

$$(-2)^3 = -8 \quad \text{and} \quad 2^3 = 8,$$

so that -2 is the cube root of -8, and 2 is the cube root of 8.

For n even and $a > 0$, a has two real nth roots. Thus,

$$(-2)^4 = 16 \quad \text{and} \quad 2^4 = 16,$$

so that -2 and 2 are both fourth roots of 16.

For n even and $a < 0$, a has no real nth root. Thus -1 has no real square root, since the square of each real number is nonnegative.

If $a = 0$, then a has exactly one real nth root, for n either odd or even, namely 0. We therefore make the following definition.

Definition 3.3 If $a \in R, n \in N$, then $\mathbf{a^{1/n}}$ is the real number if only one exists, and is the positive real number if two exist, such that

$$(\mathbf{a^{1/n}})^n = \mathbf{a}.$$

Examples

a. $25^{1/2} = 5$
b. $-25^{1/2} = -5$
c. $(-25)^{1/2}$ is not a real number.
d. $27^{1/3} = 3$
e. $-27^{1/3} = -3$
f. $(-27)^{1/3} = -3$

Notice in parts **b** and **e** that $-25^{1/2}$ and $-27^{1/3}$ denote $-(25^{1/2})$ and $-(27^{1/3})$, respectively.

To generalize from rational exponents of the form $1/n$, for $n \in N$, to rational exponents of the form m/n, for $m \in J, n \in N$, we need the results expressed in the following two theorems, which will be stated without proof.

Theorem 3.2 If $a^{1/n} \in R, m \in J,$ and $n \in N$, with $a \neq 0$ for $m \leq 0$, then

$$(a^{1/n})^m = (a^m)^{1/n}.$$

Examples **a.** $(9^{1/2})^3 = 3^3 = 27$
$(9^3)^{1/2} = (729)^{1/2} = 27$
Hence, $(9^{1/2})^3 = (9^3)^{1/2}$

b. $(8^{1/3})^{-2} = 2^{-2} = \dfrac{1}{4}$

$(8^{-2})^{1/3} = \left(\dfrac{1}{64}\right)^{1/3} = \dfrac{1}{4}$

Hence, $(8^{1/3})^{-2} = (8^{-2})^{1/3}$

Observe that Theorem 3.2 requires that $a^{1/n}$ be a real number. This requirement is not satisfied if n is even and a is negative.

Examples **a.** $[(-3)^{1/2}]^4$ is undefined because $(-3)^{1/2} \notin R$

b. $[(-3)^4]^{1/2} = 81^{1/2} = 9$

Theorem 3.3 If $a^{1/np} \in R$, $m \in J$, and $n, p \in N$, with $a \neq 0$ for $m \leq 0$, then
$$(a^{1/np})^{mp} = (a^{1/n})^m.$$

Theorems 3.2 and 3.3 show that if $a^{1/np}$ is a real number, then $(a^{1/n})^m$, $(a^m)^{1/n}$, $(a^{1/np})^{mp}$, and $(a^{mp})^{1/np}$ all represent the same number.

Examples **a.** $(16^{1/2})^3 = 4^3 = 64$
c. $(16^{1/4})^6 = 2^6 = 64$

b. $(16^3)^{1/2} = 4096^{1/2} = 64$
d. $(16^6)^{1/4} = 16{,}777{,}216^{1/4} = 64$

We make the following definition.

Definition 3.4 If $a^{1/n} \in R$, $m \in J$, and $n \in N$, then
$$a^{m/n} = (a^{1/n})^m.$$

Observe that requiring $n \in N$ does not alter the fact that m/n can represent every rational number, since all that is done is to restrict the denominator of the fraction representing the rational number to be positive, which is always possible in light of Theorems 1.8-V and 1.8-VI.

Examples **a.** $8^{2/3} = (8^{1/3})^2 = 2^2 = 4$

b. $16^{-3/4} = (16^{1/4})^{-3} = 2^{-3} = \dfrac{1}{8}$

Because we define $a^{1/n}$ to be the positive nth root of a for a positive and n an even natural number, and since a^m is positive for a negative and m an even natural number, it follows that, for m and n *even* natural numbers and a *any* real number,
$$(a^m)^{1/n} = |a|^{m/n}.$$

3.2 Powers with Rational Exponents

For the special case $m = n$ (m and n even),

$$(a^n)^{1/n} = |a|.$$

Examples a. $(a^2)^{1/2} = |a|^{2/2} = |a|$ b. $(a^6)^{1/2} = |a|^{6/2} = |a|^3$

Observe that in the last two examples it was necessary to use absolute-value notation because the expressions had to be positive. For example, if $a = -2$ then

$$((-2)^2)^{1/2} = 4^{1/2} = 2 \neq (-2)^1$$

and

$$((-2)^6)^{1/2} = 64^{1/2} = 8 \neq (-2)^3.$$

Powers with rational exponents have the same fundamental properties as powers with integral exponents, as long as the base is positive. Theorem 3.1 can be invoked to rewrite exponential expressions involving rational exponents whenever $a, b > 0$. If the base is negative, however, considerable care must be exercised in dealing with any exponents m/n where m and n are even.

Examples a. $\dfrac{x^{3/5}}{x^{2/5}} = x^{3/5 - 2/5} = x^{1/5}$ b. $(x^6 y^3)^{1/3} = (x^6)^{1/3}(y^3)^{1/3} = x^2 y$

c. $x^{2/3} y^{1/2} \cdot x^{1/3} y^{1/4}$
$= x^{2/3 + 1/3} y^{1/2 + 1/4}$
$= xy^{3/4}$

d. $(x^{3/2} y^{1/4})^{2/3}$
$= x^{3/2 \cdot 2/3} y^{1/4 \cdot 2/3}$
$= xy^{1/6}$

Exercise 3.2

A For each expression, write an equivalent basic numeral—that is, a numeral not involving exponents.

Examples a. $36^{1/2}$ b. $(-8)^{5/3}$ c. $\left(\dfrac{1}{27}\right)^{-2/3}$

Solutions a. $36^{1/2}$
$= 6$

b. $(-8)^{5/3}$
$= ((-8)^{1/3})^5$
$= (-2)^5$
$= -32$

c. $\left(\dfrac{1}{27}\right)^{-2/3}$
$= \left(\left(\dfrac{1}{27}\right)^{1/3}\right)^{-2} = \left(\dfrac{1}{3}\right)^{-2}$
$= \dfrac{1}{\left(\dfrac{1}{3}\right)^2} = \dfrac{1}{\dfrac{1}{9}} = 9$

1. $9^{1/2}$ 2. $9^{-1/2}$ 3. $4^{-3/2}$ 4. $4^{4/2}$
5. $8^{2/3}$ 6. $8^{-2/3}$ 7. $(-8)^{2/3}$ 8. $(-8)^{-2/3}$
9. $(-32)^{-2/5}$ 10. $(-32)^{2/5}$ 11. $\left(\dfrac{9}{4}\right)^{3/2}$ 12. $\left(\dfrac{9}{4}\right)^{-3/2}$

Write each expression as a product or quotient in which each variable occurs only once and all exponents are positive. Assume all variable bases are positive.

Examples a. $\dfrac{x^{1/2}}{x^{1/3}}$ b. $\dfrac{(x^{1/2}y)^2}{(xy^{1/3})^3}$ c. $\dfrac{(x^2y^3)^{1/2}}{xy^2}$

Solutions a. $\dfrac{x^{1/2}}{x^{1/3}} = x^{1/2-1/3}$ b. $\dfrac{(x^{1/2}y)^2}{(xy^{1/3})^3} = \dfrac{xy^2}{x^3y}$ c. $\dfrac{(x^2y^3)^{1/2}}{xy^2} = \dfrac{xy^{3/2}}{xy^2}$
$\qquad\qquad\; = x^{1/6}$ $\qquad\qquad\qquad\qquad = \dfrac{y}{x^2}$ $\qquad\qquad\qquad\qquad = \dfrac{1}{y^{1/2}}$

13. $x^{1/2} \cdot x^{1/3}$ 14. $x^{-1/2} \cdot x^{2/3}$ 15. $y \cdot y^{1/3}$
16. $y^{-1/3} \cdot y^2$ 17. $\dfrac{y^{1/2}}{y^{1/6}}$ 18. $\dfrac{y^{3/5}}{y^{2/5}}$
19. $\dfrac{(4x)^{1/2}}{(16x)^{-1/4}}$ 20. $\dfrac{(8x)^{1/3}}{(8x)^{-2/3}}$ 21. $\left(\dfrac{x^3y^{1/2}}{x^{1/3}}\right)^6$
22. $\left(\dfrac{x^{1/2}y^{1/3}}{x^{1/4}}\right)^2$ 23. $\left(\dfrac{125x^3y^4}{27x^{-6}y}\right)^{1/3}$ 24. $\left(\dfrac{8x^3y^2}{x^{1/2}}\right)^{1/3}$

Apply the distributive law to write each product as a sum.

Examples a. $x^{1/4}(x^{3/4} - x)$ b. $(x^{1/2} - x)(x^{1/2} + x)$

Solutions a. $x^{1/4}x^{3/4} - x^{1/4}x$ b. $x^{1/2}x^{1/2} - x^2$
$\qquad\quad x - x^{5/4}$ $\qquad\qquad\qquad x - x^2$

25. $x(x^{1/2} + x)$ 26. $x^{1/2}(1 + x)$
27. $x^{1/2}(x^{1/2} - 1)$ 28. $x^{1/3}(x^{1/3} + 2)$
29. $y^{2/3}(y - y^{1/3})$ 30. $y^{1/2}(y^2 - y^{1/2})$
31. $(x^{1/2} - y^{1/2})(x^{1/2} + y^{1/2})$ 32. $(x^{-1/2} + y^{1/2})(x^{-1/2} - y^{1/2})$
33. $(x + y)^{1/2}[(x + y)^{1/2} - (x + y)]$ 34. $(x - y)^{2/3}[(x - y)^{-1/3} + (x - y)]$

3.3 Radical Expressions

Factor as indicated. (Observe that in each expression the least power of the variable is factored out.)

Examples

a. $y^{-1/2} + y^{1/2} = y^{-1/2}(?)$
b. $x^{3/2} - x^{-1/2} = x^{-1/2}(?)$

Solutions

a. $y^{-1/2}(1 + y)$
b. $x^{-1/2}(x^2 - 1)$

35. $x^{3/2} + x = x(?)$
36. $y - y^{2/3} = y^{2/3}(?)$
37. $x^{-3/2} + x^{-1/2} = x^{-3/2}(?)$
38. $y^{3/4} - y^{-1/4} = y^{-1/4}(?)$
39. $(x + 1)^{1/2} - (x + 1)^{-1/2} = (x + 1)^{-1/2}(?)$
40. $(y + 2)^{1/5} - (y + 2)^{-4/5} = (y + 2)^{-4/5}(?)$

B Write each expression as a product or quotient in which each variable occurs only once and all exponents are positive. Assume all variable bases are positive and all variables in exponents are natural numbers.

41. $(x^{1/n}y^n)^{1/n}(z^2)^{n/2}$
42. $(x^{3n})^2(y^4)^{n/2}$
43. $(x^{n/2}y^{n+1})^3$
44. $\left(\dfrac{a^n}{b^{2n}}\right)^{1/2}$
45. $\dfrac{(a^n)^3}{a^{3-2n}}$
46. $(x^{2n}y^{-n})^{-1/n}$

In the preceding exercises the variables were restricted to positive numbers. In the remaining exercises, consider each variable base to be any real number. Simplify.

Examples

a. $[(-5)^2]^{1/2}$
b. $(u^2)^{1/2}$
c. $[x^2(x + 1)]^{1/2}$

Solutions

a. $[(-5)^2]^{1/2} = 25^{1/2}$
   $= 5$
b. $(u^2)^{1/2} = |u|$
c. $(x^2)^{1/2}(x + 1)^{1/2}$
   $= |x|(x + 1)^{1/2}$

47. $((-3)^6)^{1/3}$
48. $((-2)^8)^{1/4}$
49. $((-2)^6)^{1/2}$
50. $((-3)^{12})^{1/4}$
51. $(9x^2)^{1/2}$
52. $(x^2(x^2 + 1))^{1/2}$
53. $(x^4(1 - x))^{-1/2}$
54. $(x^{-2}(x + 1))^{-1/2}$

3.3 Radical Expressions

Powers of real numbers with rational numbers for exponents are frequently denoted by symbols involving the use of the radical sign, $\sqrt{}$.

Definition 3.5 If $a^{1/n} \in R$ and $n \in N$, then

$$\sqrt[n]{a} = a^{1/n}.$$

Naturally, the radical expression on the left is not defined if the power on the right is not. In the symbolism $\sqrt[n]{a}$, a is called the **radicand** and n the **index** of the radical, and the expression is called a **radical expression of order** n. If no index is shown with a radical expression, as, for example, in the case $\sqrt{a}$, then the index 2 is understood to apply. Thus the symbol $\sqrt{a}$ equals $a^{1/2}$, which is the *nonnegative square root of a*, where, of course, a cannot be negative if $\sqrt{a}$ is to be real. The symbol $\sqrt{x^2}$, where $x \in R$, therefore provides us with an alternative means of writing $|x|$. That is, $\sqrt{x^2} = |x|$.

Examples **a.** $\sqrt{5^2} = |5| = 5$ **b.** $\sqrt{(-3)^2} = |-3| = 3$

Properties of radicals

An immediate consequence of the foregoing definition and the theorems pertaining to exponents is the following.

Theorem 3.4 *For real values of a and b for which all the radical expressions denote real numbers,*

I_A $\sqrt[n]{a^n} = a$ (*n* an odd natural number),

I_B $\sqrt[n]{a^n} = |a|$ (*n* an even natural number),

II $\sqrt[n]{a^m} = (\sqrt[n]{a})^m$ ($n \in N, m \in J$),

III $\sqrt[n]{a} \cdot \sqrt[n]{b} = \sqrt[n]{ab}$ ($n \in N$),

IV $\dfrac{\sqrt[n]{a}}{\sqrt[n]{b}} = \sqrt[n]{\dfrac{a}{b}}$ ($b \neq 0, n \in N$),

V $\sqrt[cn]{a^{cm}} = \sqrt[n]{a^m}$ ($m \in J, n, c \in N$).

Note in III and IV that if $a, b < 0$ and n is even, then the radicals in the left-hand members are not defined, even though the radical in the right-hand member is. Thus in the case $a, b < 0$ and n even, parts III and IV are *not* valid.

The several parts of Theorem 3.4 can be used to rewrite radical expressions in various ways, and, in particular, to write them in what is called **standard form**. A radical expression is said to be in standard form if:

a. the radicand contains no polynomial factor raised to a power equal to or greater than the index of the radical,
b. the radicand contains no fractions,
c. no radical expressions are contained in denominators of fractions, and
d. the index of the radical is as small as possible.

3.3 Radical Expressions

Examples

a. $\sqrt[3]{x^4} = \sqrt[3]{x^3}\sqrt[3]{x} = x\sqrt[3]{x}$

b. $\sqrt[3]{\dfrac{1}{x^3}} = \dfrac{\sqrt[3]{1}}{\sqrt[3]{x^3}} = \dfrac{1}{x}$

c. $\dfrac{1}{\sqrt{x}} = \dfrac{1\sqrt{x}}{\sqrt{x}\sqrt{x}} = \dfrac{\sqrt{x}}{x}$

d. $\sqrt[6]{a^3} = \sqrt{a}$

The process employed in simplifying the expressions in parts **b** and **c** in the foregoing examples is called "rationalizing the denominator," because the result is a fraction with the denominator free of radicals. This does not exclude the possibility that the denominator is an irrational number.

It should be noted, though, that it is not *always* preferable, in working with fractions, to have their denominators rationalized. Sometimes, in fact, it is desirable to rationalize the numerator. Thus, for example,

$$\dfrac{\sqrt[3]{2}}{\sqrt{3}} = \dfrac{\sqrt[3]{2}\sqrt[3]{4}}{\sqrt{3}\sqrt[3]{4}} = \dfrac{2}{\sqrt{3}\sqrt[3]{4}}.$$

Exercise 3.3

A *(In Exercises 1–64, assume that all variables represent positive real numbers and that all radicands are positive.)*

Write in radical form.

Examples

a. $4^{1/3}$

b. $x^{1/2}y$

c. $(x^2 + y^2)^{-1/2}$

Solutions

a. $4^{1/3} = \sqrt[3]{4}$

b. $x^{1/2}y = \sqrt{x}\,y$
   $= y\sqrt{x}$

c. $(x^2 + y^2)^{-1/2} = \dfrac{1}{\sqrt{x^2 + y^2}}$

1. $5^{1/2}$
2. $2^{1/4}$
3. $2x^{1/3}$
4. $(2x)^{1/3}$
5. $(9y)^{1/2}$
6. $9y^{1/2}$
7. $x^{3/5}$
8. $x^{5/4}$
9. $xy^{1/3}$
10. $x^{1/3}y$
11. $(xy)^{1/3}$
12. $(x + y)^{1/3}$
13. $(x^2 + y^2)^{1/2}$
14. $(1 + 4x^2)^{-1/2}$
15. $(x^2 - y^2)^{-3/4}$
16. $(x^3 + y^3)^{2/3}$

Write in exponential form.

Examples

a. $\sqrt{y^5}$

b. $\sqrt[3]{3x^2}$

c. $\sqrt[5]{(x^2 + y^2)^t}$

Solutions on overleaf

Solutions a. $\sqrt{y^5} = (y^5)^{1/2}$ b. $\sqrt[3]{3x^2} = (3x^2)^{1/3}$ c. $\sqrt[5]{x^2 + y^2} = (x^2 + y^2)^{1/5}$
            $= y^{5/2}$ $= 3^{1/3}x^{2/3}$

17. $\sqrt[3]{8x^4}$ 18. $8\sqrt[3]{x^4}$ 19. $\sqrt[4]{x^2y^3}$ 20. $\sqrt{4y^4}$
21. $\sqrt{(x+y)^3}$ 22. $\sqrt[5]{x^6y^4}$ 23. $\sqrt[3]{x^2+1}$ 24. $\sqrt{1-x^2}$

Find the indicated root.

Examples a. $\sqrt[3]{8}$ b. $\sqrt{x^4}$ c. $\sqrt[3]{8x^6}$

Solutions a. $\sqrt[3]{8} = 2$ b. $\sqrt{x^4} = x^{4/2}$ c. $\sqrt[3]{8x^6} = \sqrt[3]{8}\sqrt[3]{x^6}$
                                     $= x^2$ $= 2x^{6/3}$
                                                                    $= 2x^2$

25. $\sqrt[3]{-27}$ 26. $\sqrt[4]{81}$ 27. $\sqrt[5]{-32}$ 28. $\sqrt{100}$
29. $\sqrt{x^6y^2}$ 30. $\sqrt[3]{x^6y^3}$ 31. $\sqrt{9x^2y^4}$ 32. $\sqrt[3]{\dfrac{125x^6}{8y^3}}$

Write in standard form.

Examples a. $\sqrt{8x^5}$ b. $\sqrt{3xy}\sqrt{6xy^3}$

Solutions a. $\sqrt{8x^5} = \sqrt{4x^4}\sqrt{2x}$ b. $\sqrt{3xy}\sqrt{6xy^3} = \sqrt{18x^2y^4}$
              $= 2x^2\sqrt{2x}$ $= \sqrt{9x^2y^4}\sqrt{2}$
                                                             $= 3xy^2\sqrt{2}$

33. $\sqrt{x^7}$ 34. $\sqrt{9x^5}$ 35. $\sqrt[3]{-27x^4}$
36. $\sqrt[3]{2x^3y^4}$ 37. $\sqrt{x}\sqrt{xy}$ 38. $\sqrt[3]{xy^2}\sqrt[3]{x^2y}$
39. $\sqrt[3]{4x^2}\sqrt[3]{16x^{10}}$ 40. $\sqrt{5xy^3}\sqrt{20x}$ 41. $\dfrac{\sqrt{xy}\sqrt{x}}{\sqrt{y}}$
42. $\dfrac{\sqrt[3]{xy}\sqrt[3]{x^4y}}{\sqrt[3]{x^2y^5}}$ 43. $\dfrac{\sqrt[3]{2x^4}\sqrt[3]{16y}}{\sqrt[3]{4x}}$ 44. $\dfrac{\sqrt{3y}\sqrt{6x}}{\sqrt{2x^3y^3}}$

Rationalize the denominator of each expression.

Examples a. $\dfrac{1}{\sqrt{3}}$ b. $\dfrac{x}{\sqrt{y}}$ c. $\dfrac{1}{\sqrt[3]{x}}$

3.3 Radical Expressions

Solutions

a. $\dfrac{1}{\sqrt{3}} = \dfrac{1\sqrt{3}}{\sqrt{3}\sqrt{3}}$

$= \dfrac{\sqrt{3}}{3}$

b. $\dfrac{x}{\sqrt{y}} = \dfrac{x\sqrt{y}}{\sqrt{y}\sqrt{y}}$

$= \dfrac{x\sqrt{y}}{y}$

c. $\dfrac{1}{\sqrt[3]{x}} = \dfrac{1\sqrt[3]{x^2}}{\sqrt[3]{x}\sqrt[3]{x^2}}$

$= \dfrac{\sqrt[3]{x^2}}{\sqrt[3]{x^3}}$

$= \dfrac{\sqrt[3]{x^2}}{x}$

45. $\dfrac{1}{\sqrt{2}}$ **46.** $\dfrac{6}{\sqrt{3}}$ **47.** $\dfrac{4}{\sqrt{2x}}$ **48.** $\dfrac{\sqrt{x}}{\sqrt{2y}}$

49. $\dfrac{1}{\sqrt[3]{4}}$ **50.** $\dfrac{3}{\sqrt[3]{9}}$ **51.** $\dfrac{1}{\sqrt[3]{y^2}}$ **52.** $\dfrac{1}{\sqrt[3]{xy^2}}$

Rationalize the numerator of each expression.

Examples

a. $\dfrac{\sqrt{x}}{2}$

b. $\dfrac{\sqrt[3]{y}}{y}$

Solutions

a. $\dfrac{\sqrt{x}}{2} = \dfrac{\sqrt{x}\sqrt{x}}{2\sqrt{x}} = \dfrac{x}{2\sqrt{x}}$

b. $\dfrac{\sqrt[3]{y}}{y} = \dfrac{\sqrt[3]{y}\sqrt[3]{y^2}}{y\sqrt[3]{y^2}} = \dfrac{1}{\sqrt[3]{y^2}}$

53. $\dfrac{\sqrt{2}}{2}$ **54.** $\dfrac{\sqrt[3]{2}}{2}$ **55.** $\dfrac{\sqrt[3]{x}}{y}$ **56.** $\dfrac{\sqrt{xy}}{x}$

Reduce the order of each radical.

Examples

a. $\sqrt[4]{5^2}$ **b.** $\sqrt[12]{81}$ **c.** $\sqrt[4]{x^2y^2}$

Solutions

a. $\sqrt[4]{5^2}$

$= \sqrt[2\cdot 2]{5^{1\cdot 2}}$

$= \sqrt{5}$

b. $\sqrt[12]{81}$

$= \sqrt[3\cdot 4]{3^{1\cdot 4}}$

$= \sqrt[3]{3}$

c. $\sqrt[4]{x^2y^2}$

$= \sqrt[2\cdot 2]{x^{1\cdot 2}y^{1\cdot 2}}$

$= \sqrt{xy}$

57. $\sqrt[6]{81}$ **58.** $\sqrt[10]{32}$ **59.** $\sqrt[4]{16x^2}$ **60.** $\sqrt[9]{8a^3}$

61. $\sqrt[6]{8x^3}$ **62.** $\sqrt[6]{125z^3}$ **63.** $\sqrt{(x-1)^4}$ **64.** $\sqrt[12]{(x-2y)^4}$

B *Write in standard form.*

65. $\sqrt[4]{\dfrac{(x^2-1)^2}{(x+1)^3}}$

66. $\sqrt{\dfrac{(x^4-1)^6}{x^2}}$

67. $\dfrac{\sqrt[3]{x^2(y+1)^2}}{\sqrt{x^4}}$ 68. $\dfrac{\sqrt[5]{x^2(x+1)^2}}{\sqrt[3]{(x-1)^2}}$

69. Is $\sqrt{a^2 + b^2} = a + b$ for all a and b with $a \geq 0$, $b \geq 0$? Why or why not?

70. Is there a natural number $n \geq 2$ for which $\sqrt[n]{a^n + b^n} = a + b$ for all $a \geq 0$ and $b \geq 0$? If so, what is it? If not, why not?

3.4 Operations on Radical Expressions

We have defined radical expressions so that they represent real numbers, and therefore the properties of the real numbers can be applied. For example, the distributive law permits us to express certain sums as products.

Examples

a. $2\sqrt{5} + 4\sqrt{5}$
$= (2 + 4)\sqrt{5}$
$= 6\sqrt{5}$

b. $5\sqrt{2} - 9\sqrt{2}$
$= (5 - 9)\sqrt{2}$
$= -4\sqrt{2}$

c. $5\sqrt{2} + \sqrt{75}$
$= 5\sqrt{2} + 5\sqrt{3}$
$= 5(\sqrt{2} + \sqrt{3})$

The distributive law in the form $c(a + b) = ca + cb$ permits us to write certain products as sums or differences.

Examples

a. $2(\sqrt{2} - 7) = 2\sqrt{2} - 14$

b. $(\sqrt{x} + 2)(\sqrt{x} - 1) = \sqrt{x}(\sqrt{x} - 1) + 2(\sqrt{x} - 1)$
$= x - \sqrt{x} + 2\sqrt{x} - 2$
$= x + \sqrt{x} - 2$

Binomial denominators

The distributive law also provides us with a means of rationalizing denominators of fractions in which radicals occur in one or both of two terms. To accomplish this, we first observe that

$$(a - \sqrt{b})(a + \sqrt{b}) = a^2 - (\sqrt{b})^2 = a^2 - b,$$

where the expression in the right-hand member contains no radical term. Each of the two factors of a product exhibiting this property is said to be the **conjugate** of the other. Now consider a fraction of the form

$$\frac{a}{b + \sqrt{c}},$$

where c is positive and $b \neq -\sqrt{c}$. If we multiply the numerator and denominator of this fraction by the conjugate of the denominator, then the denominator of the

3.4 Operations on Radical Expressions

resulting fraction will contain no term involving $\sqrt{c}$ and hence will be free of radicals. That is,

$$\frac{a}{b + \sqrt{c}} = \frac{a(b - \sqrt{c})}{(b + \sqrt{c})(b - \sqrt{c})} = \frac{ab - a\sqrt{c}}{b^2 - c},$$

where the denominator has been rationalized. This process is equally applicable to radical fractions of the form

$$\frac{a}{\sqrt{b} + \sqrt{c}},$$

since

$$\frac{a}{\sqrt{b} + \sqrt{c}} = \frac{a(\sqrt{b} - \sqrt{c})}{(\sqrt{b} + \sqrt{c})(\sqrt{b} - \sqrt{c})} = \frac{a\sqrt{b} - a\sqrt{c}}{b - c}.$$

Binomial Numerators

As noted on page 73, it is sometimes preferable, in working with fractions, to have their numerators rationalized. Thus, for example,

$$\frac{\sqrt{b} + \sqrt{c}}{a} = \frac{(\sqrt{b} + \sqrt{c})(\sqrt{b} - \sqrt{c})}{a(\sqrt{b} - \sqrt{c})} = \frac{b - c}{a(\sqrt{b} - \sqrt{c})}.$$

Exercise 3.4

Write each sum as a product.

Examples a. $2\sqrt{3} + 4\sqrt{3} - \sqrt{3}$ b. $3\sqrt{2} + \sqrt{8}$

Solutions a. $2\sqrt{3} + 4\sqrt{3} - \sqrt{3} = (2 + 4 - 1)\sqrt{3}$ b. $3\sqrt{2} + \sqrt{8} = 3\sqrt{2} + 2\sqrt{2}$
$\qquad\qquad\qquad\qquad\qquad\qquad = 5\sqrt{3} \qquad\qquad\qquad\qquad\qquad\qquad = 5\sqrt{2}$

1. $3\sqrt{5} + 4\sqrt{5}$
2. $\sqrt{6} - 2\sqrt{6}$
3. $4\sqrt{3} + 3\sqrt{3} - 2\sqrt{3}$
4. $4\sqrt{2} - \sqrt{2} + 3\sqrt{2}$
5. $5\sqrt{3} - 3\sqrt{12}$
6. $\sqrt{18} + \sqrt{2}$
7. $\sqrt{8} + \sqrt{98} - \sqrt{2}$
8. $\sqrt{75} - \sqrt{27} + \sqrt{3}$
9. $\sqrt[3]{2} + 3\sqrt[3]{16}$
10. $2\sqrt[3]{54} - \sqrt[3]{250}$
11. $\sqrt[4]{32} - 3\sqrt[4]{2}$
12. $6\sqrt[4]{3} + \sqrt[4]{48}$

Multiply factors and write all radicals in the result in simplest form. Assume all variables represent positive real numbers and that all radicands are positive.

Examples   a. $\sqrt{2x}(\sqrt{x} + \sqrt{2x})$ b. $(\sqrt{x} + \sqrt{y})(\sqrt{x} + 2\sqrt{y})$

Solutions   a. $\sqrt{2x}(\sqrt{x} + \sqrt{2x})$ b. $(\sqrt{x} + \sqrt{y})(\sqrt{x} + 2\sqrt{y})$
$= \sqrt{2x \cdot x} + \sqrt{2x \cdot 2x}$ $= (\sqrt{x})^2 + 3\sqrt{x}\sqrt{y} + 2(\sqrt{y})^2$
$= x\sqrt{2} + 2x$ $= x + 2y + 3\sqrt{xy}$

13. $3(2 - \sqrt{3})$
14. $\sqrt{5}(3 + 3\sqrt{5})$
15. $(\sqrt{3} - 1)(\sqrt{3} + 1)$
16. $(\sqrt{2} + \sqrt{3})(2\sqrt{2} - \sqrt{3})$
17. $(\sqrt{5} + 2)(1 - \sqrt{5})$
18. $(3 - \sqrt{2})(3 + \sqrt{2})$
19. $\sqrt{x}(\sqrt{2x} + \sqrt{8x})$
20. $\sqrt{2x}(\sqrt{3x} + \sqrt{x})$
21. $(\sqrt{x} - 1)(2\sqrt{x} + 3)$
22. $(\sqrt{2} + \sqrt{x})(1 - \sqrt{x})$
23. $(\sqrt{3} - \sqrt{x})(\sqrt{3} + \sqrt{x})$
24. $(2\sqrt{3} + \sqrt{x})(\sqrt{x} - \sqrt{3})$

Rationalize the denominators.

Examples   a. $\dfrac{1}{\sqrt{2} + 1}$ b. $\dfrac{\sqrt{x}}{\sqrt{x} + \sqrt{x - 2}}$

Solutions   a. $\dfrac{1}{\sqrt{2} + 1} = \dfrac{1(\sqrt{2} - 1)}{(\sqrt{2} + 1)(\sqrt{2} - 1)}$ b. $\dfrac{\sqrt{x}}{\sqrt{x} + \sqrt{x - 2}}$
$= \dfrac{\sqrt{2} - 1}{2 - 1}$ $= \dfrac{\sqrt{x}(\sqrt{x} - \sqrt{x - 2})}{(\sqrt{x} + \sqrt{x - 2})(\sqrt{x} - \sqrt{x - 2})}$
$= \sqrt{2} - 1$ $= \dfrac{x - \sqrt{x(x - 2)}}{x - (x - 2)}$
$= \dfrac{x - \sqrt{x^2 - 2x}}{2}$

25. $\dfrac{-1}{1 - \sqrt{3}}$ 26. $\dfrac{2}{\sqrt{2} + 2}$ 27. $\dfrac{x}{1 - \sqrt{x}}$

28. $\dfrac{x}{\sqrt{x} + 1}$ 29. $\dfrac{3x}{\sqrt{x} - 2}$ 30. $\dfrac{\sqrt{x}}{\sqrt{2x} + 3}$

Rationalize the numerators.

Examples   a. $\dfrac{1 - \sqrt{2}}{2}$ b. $\dfrac{1 - \sqrt{x + 1}}{\sqrt{x + 1}}$

Solutions

a. $\dfrac{1-\sqrt{2}}{2}$

$=\dfrac{(1-\sqrt{2})(1+\sqrt{2})}{2(1+\sqrt{2})}$

$=\dfrac{1-2}{2+2\sqrt{2}}$

$=\dfrac{-1}{2+2\sqrt{2}}$

b. $\dfrac{1-\sqrt{x+1}}{\sqrt{x+1}}$

$=\dfrac{(1-\sqrt{x+1})(1+\sqrt{x+1})}{\sqrt{x+1}(1+\sqrt{x+1})}$

$=\dfrac{1-(x+1)}{\sqrt{x+1}+x+1}$

$=\dfrac{-x}{\sqrt{x+1}+x+1}$

31. $\dfrac{\sqrt{3}+1}{2}$

32. $\dfrac{\sqrt{3}+2}{3}$

33. $\dfrac{1-\sqrt{x}}{x^2}$

34. $\dfrac{3-\sqrt{xy}}{xy}$

35. $\dfrac{2+\sqrt{2x-1}}{\sqrt{2x-1}}$

36. $\dfrac{\sqrt{x+4}-1}{\sqrt{x}}$

B Rationalize the denominators.

37. $\dfrac{\sqrt{x}+\sqrt{y}}{\sqrt{x}-\sqrt{y}}$

38. $\dfrac{4\sqrt{x}-\sqrt{3x+1}}{\sqrt{x}+\sqrt{3x+1}}$

39. $\dfrac{\sqrt{x+1}+\sqrt{x}}{\sqrt{x+1}-\sqrt{x}}$

40. $\dfrac{\sqrt{2x-1}}{\sqrt{x}+\sqrt{2x-1}}$

41. $\dfrac{x}{\sqrt{x}+\sqrt{y}+1}$

42. $\dfrac{y}{\sqrt{x}-\sqrt{y}+1}$

Rationalize the numerators.

43. $\dfrac{\sqrt{x-1}-1}{\sqrt{x-1}}$

44. $\dfrac{\sqrt{x+1}}{\sqrt{x+1}}$

45. $\dfrac{\sqrt{x}+\sqrt{y}}{\sqrt{x}-\sqrt{y}}$

46. $\dfrac{\sqrt{x+1}-\sqrt{x}}{\sqrt{x+1}+\sqrt{x}}$

47. $\dfrac{\sqrt{x}+\sqrt{y}-1}{x}$

48. $\dfrac{\sqrt{x}-\sqrt{y}+1}{y}$

3.5 Complex Numbers

In Section 3.4 we considered $\sqrt{b}$, $b > 0$, to be the positive number such that $(\sqrt{b})^2 = b$. For $b > 0$ there are no real numbers whose square is $-b$. In this section, we wish to consider a set of numbers containing numbers whose squares are negative and also containing a subset that can be identified with the set of real numbers. We shall see that this new set C of numbers, called the **complex numbers**, will provide solutions for all polynomial equations with real coefficients. To construct the set of complex numbers from the real numbers, we introduce a new symbol i to represent a specific complex number and then make the following definitions.

Definition 3.6 *An expression of the form $a + bi$, where $a, b \in R$, is a **complex number**.*

Definition 3.7 *The complex numbers $a + bi$ and $c + di$ are **equal** if and only if $a = c$ and $b = d$.*

Examples

a. The expressions
$$\frac{1}{3} + 5i, \quad 1 + \sqrt{2}i, \quad 0 + 1i, \quad \frac{1}{2} + 0i, \quad 5 + (-1)i$$
are all complex numbers.

b. The equality $a + bi = 1 + 7i$ holds if and only if $a = 1$ and $b = 7$.

Addition and multiplication

Complex numbers are added and multiplied according to the following rules. These definitions may appear arbitrary and unusual to start with (particularly for products) but we will see that they relate to the corresponding operations in R in a direct and useful way.

Definition 3.8 *If $z_1 = a + bi$ and $z_2 = c + di$, then*

$$\text{I} \quad z_1 + z_2 = (a + c) + (b + d)i,$$
$$\text{II} \quad z_1 \cdot z_2 = (ac - bd) + (ad + bc)i.$$

Examples

a. $(2 + 3i) + (6 + 1i)$
$= (2 + 6) + (3 + 1)i$
$= 8 + 4i$

b. $(2 + 3i)(6 + 1i)$
$= (2 \cdot 6 - 3 \cdot 1) + (2 \cdot 1 + 3 \cdot 6)i$
$= 9 + 20i$

For convenience, we write

a for $a + 0i$, $\quad bi$ for $0 + bi$, $\quad a - bi$ for $a + (-b)i$, $\quad i$ for $0 + 1i$.

With a identified with $a + 0i$, every real number is contained in C. Moreover, addition and multiplication of real numbers in C are no different than they are in R. For instance,

$$3 + 5 = (3 + 0i) + (5 + 0i)$$
$$= (3 + 5) + (0 + 0)i$$
$$= 8 + 0i = 8,$$

and

3.5 Complex Numbers

$$3 \cdot 5 = (3 + 0i) \cdot (5 + 0i)$$
$$= (3 \cdot 5 - 0 \cdot 0) + (3 \cdot 0 + 0 \cdot 5)i$$
$$= 15 + 0i = 15.$$

The complex numbers $a + bi$ with $b \neq 0$ are called **imaginary numbers**; the imaginary numbers bi with $a = 0$ are called **pure imaginary numbers**. Thus, our number systems are related as shown in Figure 3.1.

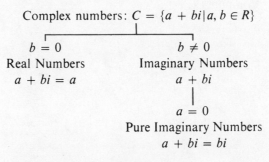

Figure 3.1

Special role of i

With $0 + 1i$ written as i, the special role of this specific complex number becomes apparent. We have

$$i^2 = (0 + 1i) \cdot (0 + 1i)$$
$$= (0 \cdot 0 - 1 \cdot 1) + (0 \cdot 1 + 1 \cdot 0)i$$
$$= -1 + 0i,$$

from which,

$$i^2 = -1.$$

Also, for positive real numbers b,

$$(\sqrt{b}i)^2 = b(-1) = -b.$$

Thus negative real numbers $-b$ have square roots in our enlarged number system C. In fact you can check that $(-\sqrt{b}i)^2$ is also equal to $-b$, so each nonzero real number has two square roots in C.

Because $i^2 = -1$, i is often written as $\sqrt{-1}$; similarly for real $b > 0$,

$$\sqrt{-b} = \sqrt{b}i = i\sqrt{b}.$$

The symbol $\sqrt{-b}$ must be used with care, however, since certain relationships involving square roots which are valid for real numbers are not valid for complex numbers. For example,

$$\sqrt{-2} \cdot \sqrt{-3} = (i\sqrt{2})(i\sqrt{3})$$
$$= i^2\sqrt{6} = -\sqrt{6}.$$

However, note that

$$\sqrt{-2} \cdot \sqrt{-3} \neq \sqrt{(-2)(-3)} = \sqrt{6}.$$

For this reason, complex numbers in the form $a + \sqrt{-b}$, $b > 0$, should be written in the form $a + i\sqrt{b}$ before making computations.

Making use of the fact that $i^2 = -1$, we can multiply complex numbers just as we multiply real polynomial expressions.

Examples Compute each product.

  a. $(2 + 3i)(6 + 1i)$ b. $(4 - \sqrt{-4})(4 + \sqrt{-4})$

Solutions

  a. $(2 + 3i)(6 + 1i)$
   $= 12 + 18i + 2i + 3i^2$
   $= 12 + 20i - 3$
   $= 9 + 20i$

  b. $(4 - \sqrt{-4})(4 + \sqrt{-4})$
   $= (4 - 2i)(4 + 2i)$
   $= 16 - 8i + 8i - 4i^2$
   $= 16 + 0i + 4$
   $= 20$

Field properties

The complex numbers, with addition and multiplication defined as in Definition 3.8, satisfy all the field axioms of the set R on page 8. Hence the complex numbers are a field.

In particular, since for any complex number $a + bi$,

$$(a + bi) + (0 + 0i) = (0 + 0i) + (a + bi) = a + bi,$$

the additive-identity in C is $0 + 0i = 0$. Also, since for any complex number $a + bi$

$$(a + bi)(1 + 0i) = (1 + 0i)(a + bi) = a + bi,$$

the multiplicative identity in C is $1 + 0i = 1$. Further, the additive inverse of $z = a + bi$ in C is

$$-z = -a - bi$$

because

$$(a + bi) + (-a - bi) = (-a - bi) + (a + bi) = 0.$$

Similarly, the multiplicative inverse of $z = a + bi$, $z \neq 0$, in C is

$$\frac{1}{z} = \frac{a}{a^2 + b^2} - \frac{b}{a^2 + b^2} i$$

since

$$(a + bi)\left(\frac{a}{a^2 + b^2} - \frac{b}{a^2 + b^2} i\right) = \left(\frac{a}{a^2 + b^2} - \frac{b}{a^2 + b^2} i\right)(a + bi) = 1.$$

Since Theorems 1.1 to 1.11 can be proven using only the field postulates, these theorems are also true for the complex numbers. The complex numbers

3.5 Complex Numbers

cannot be ordered, however, in such a way as to satisfy the order properties O-1 and O-2 on page 16.

Subtraction and division

The operations of subtraction and division for the field C can be introduced just as they were for R:

$$z_1 - z_2 = z_1 + (-z_2), \quad \text{and} \quad \frac{z_1}{z_2} = z_1 \cdot \frac{1}{z_2} \quad (z_2 \neq 0).$$

In the usual notation,

$$(a + bi) - (c + di) = (a - c) + (b - d)i,$$

and

$$\frac{a + bi}{c + di} = \frac{ac + bd}{c^2 + d^2} + \frac{bc - ad}{c^2 + d^2} i.$$

Examples

a. $(2 + 3i) - (5 + 6i)$
$= (2 - 5) + (3 - 6)i$
$= -3 - 3i$

b. $2 - (3 + 4i)$
$= (2 - 3) + (0 - 4)i$
$= -1 - 4i$

c. $\dfrac{4 + i}{2 + 3i} = \dfrac{4(2) + 1(3)}{2^2 + 3^2} + \dfrac{1(2) - 4(3)}{2^2 + 3^2} i$

$= \dfrac{11}{13} - \dfrac{10}{13} i$

d. $\dfrac{5}{-3 - i} = \dfrac{5(-3) + 0(-1)}{(-3)^2 + (-1)^2} + \dfrac{0(-3) - 5(-1)}{(-3)^2 + (-2)^2} i$

$= -\dfrac{15}{10} + \dfrac{5}{10} i$

$= -\dfrac{3}{2} + \dfrac{1}{2} i$

The quotient of two complex numbers can be rewritten by using the following theorem, which is analogous to Theorem 1.10 (fundamental principle of fractions) in the set of real numbers.

Theorem 3.5   If z_1, z_2, and z_3 are elements of C, and z_2 and z_3 are not zero, then

$$\frac{z_1}{z_2} = \frac{z_1 z_3}{z_2 z_3}.$$

The notion of the *conjugate* of a complex number can now be used to rewrite quotients with imaginary denominators as $a + bi$, $a, b \in R$.

84   3   Rational Exponents—Radicals

Definition 3.9   The **conjugate** of $z = a + bi$, denoted by $\bar{z}$, is $a - bi$.

Examples   **a.** The conjugate of $2 + 3i$ is $2 - 3i$.
b. The conjugate of $-3 - i$ is $-3 + i$.

To write the quotient $(a + bi)/(c + di)$ in standard form by using Theorem 3.5, you can multiply the numerator and the denominator by $c - di$, the conjugate of the denominator.

Examples   **a.** $\dfrac{4 + i}{2 + 3i} = \dfrac{(4 + i)(2 - 3i)}{(2 + 3i)(2 - 3i)}$ **b.** $\dfrac{5}{-3 - i} = \dfrac{5(-3 + i)}{(-3 - i)(-3 + i)}$

$= \dfrac{8 - 10i - 3i^2}{4 - 9i^2}$ $= \dfrac{-15 + 5i}{9 - i^2}$

$= \dfrac{8 - 10i + 3}{4 + 9} = \dfrac{11}{13} - \dfrac{10}{13}i$ $= \dfrac{-15 + 5i}{9 + 1} = -\dfrac{3}{2} + \dfrac{1}{2}i$

Exercise 3.5

A   Write each expression in the form $a + bi$.

Examples   **a.** $(2 + i) + (3 - 2i)$ **b.** $3i - (4 + i)$

Solutions   **a.** $(2 + i) + (3 - 2i)$ **b.** $3i - (4 + i)$
$= (2 + 3) + (1 - 2)i$ $= (0 + 3i) - (4 + 1i)$
$= 5 - i$ $= (0 - 4) + (3 - 1)i$
                             $= -4 + 2i$

1. $(5 + i) + (-4 - 2i)$ **2.** $(1 + i) + i$

3. $(3 - i) + 6$ **4.** $13 + (3 - 10i)$

5. $\left(4 - \dfrac{1}{2}i\right) + 3i$ **6.** $(6 + i) + (1 + 6i)$

7. $(5 - i) - (5 + i)$ **8.** $(6 + i) - (1 + 6i)$

9. $\left(\dfrac{1}{3} + 2i\right) - (1 - i)$ **10.** $\left(\dfrac{1}{2} - i\right) - (2 + i)$

11. $(3 + 4i) - (5 - 2i)$ **12.** $(3 - i) - (7 - 2i)$

Examples   **a.** $(1 + 3i)(2 - i)$ **b.** $\dfrac{1 + 3i}{2 - i}$

3.5 Complex Numbers

Solutions

a. $(1 + 3i)(2 - i)$
$= 2 + 5i - 3i^2$
$= 2 + 5i + 3 = 5 + 5i$

b. $\dfrac{1 + 3i}{2 - i} = \dfrac{(1 + 3i)(2 + i)}{(2 - i)(2 + i)}$
$= \dfrac{2 + 7i + 3i^2}{4 - i^2}$
$= \dfrac{-1 + 7i}{5} = -\dfrac{1}{5} + \dfrac{7}{5}i$

13. $(2 + i)(3 - i)$
14. $(1 - i)(5 + i)$
15. $\dfrac{(2 + i)}{(3 - i)}$
16. $\dfrac{(1 - i)}{(5 + i)}$
17. $(4 - i)(1 + 2i)$
18. $(3 + 4i)(1 + i)$
19. $\dfrac{(4 - i)}{(1 + 2i)}$
20. $\dfrac{(3 + 4i)}{(1 + i)}$
21. $i(3 + i)$
22. $3(2 + i)$
23. $\dfrac{i}{3 + i}$
24. $\dfrac{3}{(2 + i)}$

Examples

a. $1 + \sqrt{-2} + \sqrt{-4}$
b. $(1 + \sqrt{-2})(2 + \sqrt{-3})$

Solutions

a. $1 + \sqrt{-2} + \sqrt{-4}$
$= 1 + \sqrt{2}i + \sqrt{4}i$
$= 1 + (\sqrt{2} + 2)i$

b. $(1 + \sqrt{-2})(2 + \sqrt{-3})$
$= (1 + \sqrt{2}i)(2 + \sqrt{3}i)$
$= 2 + (2\sqrt{2} + \sqrt{3})i + \sqrt{6}i^2$
$= (2 - \sqrt{6}) + (2\sqrt{2} + \sqrt{3})i$

25. $(1 + \sqrt{-9}) + (3 - \sqrt{-2})$
26. $(2 - \sqrt{-4}) + (1 + \sqrt{-3})$
27. $(1 + \sqrt{-9})(3 - \sqrt{-2})$
28. $(2 - \sqrt{-4})(1 + \sqrt{-3})$
29. $\dfrac{1 + \sqrt{-9}}{3 - \sqrt{-2}}$
30. $\dfrac{2 - \sqrt{-4}}{1 + \sqrt{-3}}$
31. $\sqrt{-16} + \sqrt{-12}$
32. $\sqrt{-18} + \sqrt{-25}$

Write each expression in the form $a + bi$, a, or bi.

Examples

a. i^3
b. $\dfrac{i^2}{i^4}$

Solutions

a. $i^3 = i^2 \cdot i = -1 \cdot i = -i$
b. $\dfrac{i^2}{i^4} = \dfrac{1}{i^2} = \dfrac{1}{-1} = -1$

33. i^4
34. i^5
35. i^7
36. i^6
37. $\dfrac{i}{i^5}$
38. $\dfrac{i}{i^4}$
39. $\dfrac{i}{i^6}$
40. $\dfrac{i}{i^7}$

B Write each expression in the form $a + bi$.

41. $(1 + i)^3$
42. $(2 - i)^3$
43. $\dfrac{(1 + i)^2}{1 - i}$
44. $\dfrac{(-1 - 2i)^2}{2 - i}$
45. $\dfrac{1 - 2i}{(2 + i)^3}$
46. $\dfrac{3 - i}{(2 - 3i)^3}$

Prove each of the following statements.

Example If $z \in R$ then $\bar{z} = z$.

Solution Since $z \in R$, $z = a + 0i$ where $a \in R$. Thus, $\bar{z} = a - 0i = a + 0i = z$.

47. If $z_1, z_2 \in C$ then $\overline{z_1 \pm z_2} = \bar{z}_1 \pm \bar{z}_2$.
48. If $z_1, z_2 \in C$ then $\overline{z_1 \cdot z_2} = \bar{z}_1 \bar{z}_2$.
49. If $z \in C$ then $\overline{z^n} = \bar{z}^n$.
50. If $z \in C$ then $\bar{\bar{z}} = z$.

Chapter Review

[3.1] Write each expression as a product or quotient in which each variable occurs at most once in the expression and all exponents are positive.

1. $(x^3 y^2)^4$
2. $\dfrac{x^{-1} y^2}{x^2 y^{-1}}$
3. $\dfrac{(xy)^{-2}}{x^0 y^{-3}}$
4. $\dfrac{(x^{-1} y^0)^{-1}}{x^{-2}}$

Represent each expression as a single fraction involving positive exponents only.

5. $x^{-1} + x^{-3}$
6. $\dfrac{x^{-1} + y^{-2}}{(xy)^{-2}}$

Write each number in scientific notation.

7. 35100
8. 0.00018

Write each number in decimal form.

9. 3.14×10^8
10. 6.75×10^{-6}

Review Exercises

[3.2] *Write each expression as a product or quotient of powers in which each variable occurs only once and all exponents are positive. Assume that all variable bases are positive.*

11. $x^{1/2} \cdot x^{2/3}$ 12. $\left(\dfrac{x^3}{y^6}\right)^{-1/6}$ 13. $(x^8 y^{16})^{1/2}$ 14. $\left(\dfrac{y^8}{x^4 y^{12}}\right)^{1/4}$

Apply the distributive law to write each product as a sum.

15. $y^{1/3}(y + y^{2/3})$ 16. $(2y + y^{1/2})(2y - y^{1/2})$

Factor as indicated.

17. $x^{-1/4} + x^{1/2} = x^{-1/4}(?)$ 18. $x^{2/3} - x^{-2/3} = x^{-2/3}(?)$

[3.3] *Write in simplest form.*

19. $\sqrt{4x^5 y^3}$ 20. $\sqrt{3x} \cdot \sqrt{12xy}$

21. $\dfrac{\sqrt{2x}\sqrt{3xy}}{\sqrt{2y}}$ 22. $\dfrac{\sqrt[3]{6x}\sqrt[3]{9x^2 y}}{\sqrt[3]{2y}}$

23. Rationalize the denominator: $\sqrt[3]{\dfrac{x}{y}}$ 24. Rationalize the numerator: $\dfrac{\sqrt{3y}}{6y}$

Reduce the order of each radical.

25. $\sqrt[4]{25}$ 26. $\sqrt[6]{27x^3 y^3}$

[3.4] *Write each sum as a product.*

27. $3\sqrt{8} - 2\sqrt{50} + \sqrt{2}$ 28. $3\sqrt[3]{40} + 2\sqrt[3]{5}$

Multiply factors and write all radicals in the result in simplest form.

29. $(\sqrt{3} - \sqrt{2})(\sqrt{3} + \sqrt{2})$ 30. $(2\sqrt{x} - 5)(\sqrt{x} + 1)$

Rationalize the denominators.

31. $\dfrac{1}{3 - \sqrt{2}}$ 32. $\dfrac{\sqrt{x} + 1}{2\sqrt{x} - 1}$

Rationalize the numerators.

33. $\dfrac{\sqrt{3} + 1}{2}$

34. $\dfrac{\sqrt{x} - 1}{\sqrt{x} + 2}$

[3.5] *Write each expression in the form $a + bi$, a, or bi.*

35. $(3 + i) + (1 - 3i)$

36. $(3 + i) - (1 - 3i)$

37. $(3 + i)(1 - 3i)$

38. $(3 + i)(1 - 3i)$

39. $(1 + \sqrt{-1})^2$

40. $1 - \sqrt{-16}$

41. i^9

42. $\dfrac{1}{i^7}$

Review → sec. 1, 2, 3

4 Equations and Inequalities in One Variable

(4.1) Even 2-50
(4.2) Evens 2-42
(4.3) Evens 2-40

4.1 Equivalent Equations; First-Degree Equations

If we replace the variable x in $P(x) = Q(x)$ with an element from its replacement set, and if the resulting statement is true, the element is a **solution** of the equation and is said to **satisfy** the equation. Thus, 2 is a solution of $x + 3 = 5$, because $2 + 3 = 5$ is a true statement. On the other hand, 3 is not a solution of the equation, because $3 + 3 = 5$ is false. The set of all solutions of an equation is said to be the **solution set** of the equation. In the example we have been using here, $x + 3 = 5$, the solution set is $\{2\}$.

Equations that have the same solution set are called **equivalent equations**. For example, the equations

$$x + 3 = -3 \quad \text{and} \quad x = -6$$

are equivalent, because the solution set of each is $\{-6\}$.

Elementary transformations

To solve an equation, we usually either determine the members of the solution set by inspection or else generate a sequence of equivalent equations until we arrive at one with an obvious solution set. The following theorem, which is a direct consequence of the addition and multiplication laws for real and complex numbers, is frequently used in generating equivalent equations over the set of real and complex numbers.

Theorem 4.1   If $P(x)$, $Q(x)$, and $R(x)$ are expressions, then for all values of x for which $P(x)$, $Q(x)$, and $R(x)$ are real or complex numbers,

$$P(x) = Q(x)$$

Continued on overleaf

is equivalent to each of the following:

$$\text{I} \quad P(x) + R(x) = Q(x) + R(x),$$
$$\text{II} \quad P(x) - R(x) = Q(x) - R(x),\dagger$$
$$\left.\begin{array}{l} \text{III} \quad P(x) \cdot R(x) = Q(x) \cdot R(x) \\ \text{IV} \quad \dfrac{P(x)}{R(x)} = \dfrac{Q(x)\dagger}{R(x)} \end{array}\right\} \quad \text{for} \quad R(x) \neq 0.$$

Theorem 4.1 allows us to rewrite an equation $P(x) = Q(x)$ by adding, subtracting, multiplying or dividing by an expression $R(x)$ and, as long as we do not multiply or divide by zero, the result will be an equivalent equation. The goal is to obtain an equivalent equation in which the variable is isolated.

Example Solve $\dfrac{y+8}{y} = 5$.

Solution We use parts of Theorem 4.1 to generate the following sequence of equivalent equations.

$$(y)\frac{y+8}{y} = 5(y).$$
$$y + 8 = 5y \quad (y \neq 0)$$
$$8 = 5y - y$$
$$8 = 4y$$
$$2 = y$$

The equation $2 = y$ is equivalent to the original for $y \neq 0$. Further, 0 is not a solution of the original equation. Therefore the solution set is $\{2\}$.

Any application of any part of Theorem 4.1 is called an **elementary transformation**. An elementary transformation *always* produces an equivalent equation. Care must be exercised in the application of Parts III and IV, however, for we have specifically excluded multiplication or division by zero. For example, to solve the equation

$$\frac{x}{x-3} = \frac{3}{x-3} + 2 \tag{1}$$

we might first multiply each member by $(x - 3)$ to find an equation that is free of fractions. We have

$$(x-3)\frac{x}{x-3} = (x-3)\frac{3}{x-3} + (x-3)2,$$

---
† Parts II and IV can be considered special cases of Parts I and III, respectively.

4.1 Equivalent Equations; First-Degree Equations

or
$$x = 3 + 2x - 6, \qquad (2)$$

from which
$$x = 3.$$

Thus 3 is a solution of Equation (2). But, upon substituting 3 for x in Equation (1), we have

$$\frac{3}{3-3} = \frac{3}{3-3} + 2 \quad \text{or} \quad \frac{3}{0} = \frac{3}{0} + 2,$$

and neither member is defined. In obtaining Equation (2), each member of Equation (1) was multiplied by $(x - 3)$; but if x is 3, then $(x - 3)$ is zero, and Theorem 4.1-III is not applicable. Equation (2) is *not* equivalent to Equation (1), and in fact Equation (1) has no solution.

We can always decide whether what we think is a solution of an equation is such in reality by substituting the suggested solution in the original equation and determining whether or not the resulting statement is true. If each equation in a sequence is obtained by means of an elementary transformation, the sole purpose for such checking is to detect arithmetic errors. We shall dispense with checking solution sets in the examples that follow unless we apply what may be a non-elementary transformation—that is, unless we *multiply or divide by an expression that equals zero for some value or values of the variable.*

Solution of a linear equation

The equation
$$ax + b = 0, \qquad (3)$$

where $a \neq 0$, is a **first-degree**, or **linear, equation**. Any equation that can be reduced to this form by elementary transformations, therefore, is equivalent to a first-degree equation. We can show that such an equation always has one and only one solution. By Theorem 4.1-I,

$$ax = -b \qquad (4)$$

is equivalent to Equation (3); further, by Theorem 4.1-III, Equation (4) is equivalent to

$$x = -\frac{b}{a}, \qquad (5)$$

which, of course, has the unique solution $-b/a$. Since Equations (3), (4), and (5) are equivalent, Equation (3) has the unique solution $-b/a$.

An equation containing more than one variable, or containing symbols such as a, b, and c, representing constants, can often be solved for one of the symbols in terms of the remaining symbols by applying elementary transformations until the desired symbol is obtained by itself as one member of an equation.

Example Solve $ay = b + y$, for y.

Solution We generate the following sequence of equivalent equations.
$$ay - y = b$$
$$y(a - 1) = b$$
$$y = \frac{b}{a - 1} \quad (a \neq 1)$$

Exercise 4.1

A Solve. Consider R to be the replacement set of the variable.

Examples
a. $6y - 2(2y + 5) = 6(5 + y)$
b. $3x^2 + 2x - 1 = 3x(x - 4)$

Solutions
a. $6y - 4y - 10 = 30 + 6y$
$2y - 10 = 30 + 6y$
$2y - 40 = 6y$
$-40 = 4y$
$-10 = y$
The solution set is $\{-10\}$.

b. $3x^2 + 2x - 1 = 3x^2 - 12x$
$2x - 1 = -12x$
$-1 = -14x$
$\frac{1}{14} = x$
The solution set is $\left\{\frac{1}{14}\right\}$.

1. $3x + 2(8 - x) = 3(x - 2)$
2. $5(x + 1) = 4(x + 2) - 3$
3. $-2[x - (x + 3)] = x + 1$
4. $-[2x - (3 - x)] = 4x - 11$
5. $x^2 - 3 = 1 + (x + 1)(x - 2)$
6. $2 - x - x^2 = 1 - (x - 1)^2$
7. $2[3x - (4x + 1)] = 2x - 3$
8. $4x - (2 + 3x) = x - [2x - (3x - 2)]$
9. $[2x + (3 - x)] - [x - (2 + x)] = x$
10. $x - [(x - 1) - (2x - 2)] = -2x + 1$
11. $\{3x - [x - (2x + 1)] + 1\} = -2x$
12. $x - \{2x - [x + (x - 4)]\} = 2x$

Examples
a. $\dfrac{3x + 4}{5} = \dfrac{7x + 6}{10}$
b. $\dfrac{2}{x + 4} + \dfrac{x}{x + 4} = 2$

Solutions
a. $10\left(\dfrac{3x + 4}{5}\right) = 10\left(\dfrac{7x + 6}{10}\right)$
$2(3x + 4) = 7x + 6$
$6x + 8 = 7x + 6$
$2 = x$

The solution set is $\{2\}$.

b. $\dfrac{x + 2}{x + 4} = 2$
$x + 2 = 2(x + 4), \quad x \neq -4$
$x + 2 = 2x + 8,$
$-6 = x$

Check: $\dfrac{2}{-6 + 4} + \dfrac{-6}{-6 + 4} = 2$

The solution set is $\{-6\}$.

4.1 Equivalent Equations; First-Degree Equations

13. $\dfrac{x-5}{4} = 1 + \dfrac{x-9}{12}$

14. $\dfrac{2x+3}{6} = \dfrac{x+1}{9}$

15. $\dfrac{x}{3} - 1 = \dfrac{2x-3}{3}$

16. $\dfrac{x}{3} + \dfrac{x+1}{2} = 3$

17. $\dfrac{2}{x+1} = \dfrac{x}{x+1} + 1$

18. $\dfrac{3}{x-2} = \dfrac{1}{2} + \dfrac{2x-7}{2x-4}$

19. $\dfrac{2x+1}{x-4} - \dfrac{4x-3}{2(x-4)} = 0$

20. $\dfrac{3x}{x+3} + \dfrac{-9x-1}{3(x+3)} = \dfrac{4}{x+3}$

21. $\dfrac{3x}{x-4} = \dfrac{x}{2x-8} + \dfrac{1}{x-4}$

22. $\dfrac{6x}{3x+2} - \dfrac{4x}{3x+2} = \dfrac{1}{6x+4}$

Example $\dfrac{4}{2x+3} + \dfrac{3x}{4x^2-9} = \dfrac{1}{2x+3}$

Solution $\left(\dfrac{4}{2x+3} + \dfrac{3x}{4x^2-9}\right)(4x^2-9) = \dfrac{1}{2x+3}(4x^2-9)$

$$4(2x-3) + 3x = 2x - 3$$

$$9x = 9$$

$$x = 1$$

Check: $\dfrac{4}{2(1)+3} + \dfrac{3(1)}{4(1)^2-9} = \dfrac{1}{2(1)+3}$; $\dfrac{4}{5} - \dfrac{3}{5} = \dfrac{1}{5}$

The solution set is $\{1\}$.

23. $\dfrac{1}{x-1} + \dfrac{2}{x+1} = \dfrac{3x-1}{x^2-1}$

24. $\dfrac{4}{x+2} - \dfrac{1}{x} = \dfrac{2x-1}{x^2+2x}$

25. $\dfrac{3}{x+1} + \dfrac{x-4}{x^2-x-2} = \dfrac{-10}{x-2}$

26. $\dfrac{1}{x} + \dfrac{1}{x^2+x} = \dfrac{3}{x+1}$

27. $\dfrac{2}{x-1} + \dfrac{3x}{x^2-4x+3} = \dfrac{5}{x-3}$

28. $\dfrac{2x-1}{x-4} + \dfrac{x}{x^2-x-12} = \dfrac{2x-1}{x+3}$

29. $\dfrac{x-3}{x+2} + \dfrac{11x-8}{x^2-2x-8} = \dfrac{x+2}{x-4}$

30. $\dfrac{2x+1}{2x-3} + \dfrac{-2x+1}{2x^2-5x+3} = \dfrac{x}{x-1}$

Solve for the indicated variable. Leave the results in the form of an equation equivalent to the given equation. Indicate any restrictions on the variables.

Example Solve $l = a + (n-1)d$, for d.

Solution on overleaf

Solution   Add $-a$ to each member.

$$l - a = (n-1)d$$

Multiply each member by $\dfrac{1}{n-1}$ $(n \neq 1)$.

$$\dfrac{l-a}{n-1} = d$$

$$d = \dfrac{l-a}{n-1} \quad (n \neq 1)$$

31. $S = 2\pi rh$, for r
32. $S = 2\pi rh$, for h
33. $v = k + gt$, for k
34. $v = k + gt$, for t
35. $A = \dfrac{h}{2}(b+c)$, for c
36. $S = \dfrac{a}{1-r}$, for r
37. $l = a + (n-1)d$, for n
38. $\dfrac{1}{r} = \dfrac{1}{r_1} + \dfrac{1}{r_2}$, for r
39. $V = \pi R^2 h - \pi r^2 h$, for r^2
40. $V = \dfrac{4}{3}\pi R^3 - \dfrac{4}{3}\pi r^3$, for R^3

Example   Solve $x'(x - 3) + 4 = 5(x + x') - 9$, for x.

Solution

$x'(x - 3) + 4 = 5(x + x') - 9$

$x'x - 3x' + 4 = 5x + 5x' - 9$

$\quad\quad x'x - 5x = 5x' - 9 + 3x' - 4$

$\quad\quad\quad x(x' - 5) = 8x' - 13$

$$x = \dfrac{8x' - 13}{x' - 5} \quad (x' \neq 5)$$

41. $2y'y + 2x = 0$, for y'
42. $2x + 2y + 2xy' + 2yy' = 0$, for y'
43. $\dfrac{2xy - 2x^2 yy'}{y^4} = 0$, for y'
44. $\dfrac{(2x + 2yy')2y - 2y(x^2 + y^2)}{4y^2} = 1$, for y'
45. $x^2 y' - 3x - 2y^3 y' = 1$, for y'
46. $2xy' - 3y' + x^2 = 0$, for y'
47. $x_1 x_2 - 2x_1 x_3 = x_4$, for x_1
48. $3x_1 x_3 + x_1 x_2 = x_4$, for x_1
49. $\dfrac{y - y_1}{x - x_1} = 6$, for y
50. $\dfrac{y - y_1}{x - x_1} = 2$, for x

B 51. Find a value of k so that $2x - 3 = \dfrac{4 + x}{k}$ is equivalent to the equation $2x - 1 = -3$.

52. Find a value of k so that $3x - 1 = k$ is equivalent to the equation $2x + 5 = 1$.

53. For what value of k does $\dfrac{2x^2 - kx + 1}{x + 3} = 2x - 3$ have $\{-1\}$ as its solution set?

54. For what value of k does $\dfrac{2kx^3 - kx^2 + x - 1}{x + 2} = x$ have $\{1\}$ as its solution set?

55. For what value of k does $\dfrac{x - 3}{x + 2} + \dfrac{kx - 8}{x^2 - 2x - 8} = \dfrac{x + 2}{x - 4}$ have $\{x \mid x \neq -2, 4\}$ as its solution set?

56. For what value of k does $\dfrac{kx}{x + 1} + \dfrac{1}{2x + 2} = \dfrac{4}{x + 1}$ have the empty set as its solution set?

4.2 Second-Degree Equations

The equation
$$ax^2 + bx + c = 0 \quad (a \neq 0)$$
is a **second-degree**, or **quadratic**, **equation**. Any equation that can be reduced to this form by elementary transformations is therefore equivalent to a quadratic equation. We shall designate the form shown above as the **standard form** for such equations.

Solution by factoring

The following theorem, which is a consequence of Theorems 1.6 and 1.7, will prove useful to us in solving quadratic equations.

Theorem 4.2 If $a, b \in C$, then $ab = 0$ if and only if
$$a = 0 \quad \text{or} \quad b = 0 \quad \text{or both.}$$

Example Solve $x^2 + 2x - 15 = 0$.

Solution The equation
$$x^2 + 2x - 15 = 0$$
is equivalent to
$$(x + 5)(x - 3) = 0.$$
Since $(x + 5)(x - 3) = 0$ is true if and only if
$$x + 5 = 0 \quad \text{or} \quad x - 3 = 0,$$
we can see by inspection that the only values of x that satisfy the original equation are -5 and 3. Hence the solution set is $\{-5, 3\}$.

Number of solutions

The solution set of a quadratic equation over the real numbers can contain one or two elements. The equation in the forgoing example has two real solutions. However, consider the equation

$$x^2 - 2x + 1 = 0.$$

Since $x^2 - 2x + 1 = 0$ is equivalent to

$$(x - 1)^2 = 0,$$

and since the only value of x for which $(x - 1)^2 = 0$ is 1, this is the only member of the solution set of $x^2 - 2x + 1 = 0$. For reasons of convenience, we wish to consider every quadratic equation to have two roots. Accordingly, for any quadratic equation with one solution, we say that the solution is of **multiplicity two**; that is, we count it twice as a solution.

If we cannot readily factor a quadratic equation, we must use other methods of solution. Let us first consider the special case of the quadratic equation,

$$x^2 - q = 0.$$

Since $x^2 - q = 0$ is equivalent to $x^2 = q$, and since $x^2 = q$ implies that x must be a square root of q, we have as the solution set $\{\sqrt{q}, -\sqrt{q}\}$, where the members are real if $q \geq 0$ and imaginary if $q < 0$. This method of solving a quadratic equation is sometimes called **extraction of roots**.

General solution

Quadratic equations of the form

$$(x - p)^2 = q,$$

can be solved by observing that $x - p$ must be one of the square roots of q. That is, either

$$x - p = \sqrt{q} \quad \text{or} \quad x - p = -\sqrt{q},$$

and conversely. Thus the solution set of $(x - p)^2 = q$ is $\{p + \sqrt{q}, p - \sqrt{q}\}$.

Example

Solve $(x - 3)^2 = -4$.

Solution

By extraction of roots, we have

$$x - 3 = \sqrt{-4} \quad \text{or} \quad x - 3 = -\sqrt{-4},$$

from which

$$x - 3 = 2i \quad \text{or} \quad x - 3 = -2i.$$

Then, $x = 3 + 2i$ or $x = 3 - 2i$, and the solution set is $\{3 + 2i, 3 - 2i\}$.

Being able to find solution sets for quadratic equations of the form $(x - p)^2 = q$ enables us to find the solution set of any quadratic equation. Let us first consider the general quadratic equation in standard form

$$ax^2 + bx + c = 0,$$

4.2 Second-Degree Equations

for the special case in which $a = 1$, that is,
$$x^2 + bx + c = 0. \tag{1}$$
We can write this equation in the equivalent form
$$(x - p)^2 = q,$$
which we can solve as above. We begin the process by adding $-c$ to each member of Equation (1), which yields
$$x^2 + bx = -c. \tag{2}$$
If we then add $(b/2)^2$ to each member of Equation (2), we obtain
$$x^2 + bx + \left(\frac{b}{2}\right)^2 = -c + \left(\frac{b}{2}\right)^2, \tag{3}$$
in which the left-hand member is equal to $(x + b/2)^2$, and we have
$$\left(x + \frac{b}{2}\right)^2 = -c + \frac{b^2}{4}. \tag{4}$$

Since we have performed only elementary transformations, Equation (4) is equivalent to Equation (1), and we can solve (4) by the method discussed above.

The technique used to obtain Equations (3) and (4) is called **completing the square**. We can determine the term necessary to complete the square in (2) by dividing the coefficient b of the first-degree term by the number 2 and squaring the result. The expression obtained, $x^2 + bx + (b/2)^2$, is a perfect square and may be written in the form $(x + b/2)^2$. The process of completing the square has other applications in addition to solving quadratic equations.

Example Solve $x^2 - 2x - 5 = 0$ by completing the square.

Solution To complete the square, we first add 5 to both members of the equation to obtain
$$x^2 - 2x = 5.$$
Next we add the square of half of the coefficient of x to both members, which yields
$$x^2 - 2x + \left(\frac{-2}{2}\right)^2 = 5 + \left(\frac{-2}{2}\right)^2.$$
The left-hand member can now be written as a perfect square and we obtain,
$$(x - 1)^2 = 6,$$
from which
$$x - 1 = \sqrt{6}, \qquad x - 1 = -\sqrt{6},$$
$$x = 1 + \sqrt{6}; \qquad x = 1 - \sqrt{6}.$$
The solution set is $\{1 + \sqrt{6}, 1 - \sqrt{6}\}$.

Exercise 4.2

A *Solve by factoring.*

Example $x^2 + x = 12$

Solution Write an equivalent equation in standard form and then factor the left-hand member.

$$x^2 + x - 12 = 0$$
$$(x + 4)(x - 3) = 0$$

Determine solutions by inspection, or set each factor equal to zero and solve.

$$x + 4 = 0 \qquad x - 3 = 0$$
$$x = -4 \qquad x = 3$$

The solution set is $\{-4, 3\}$.

1. $x^2 + x = 2$
2. $x^2 = 6 - x$
3. $x^2 - 5x = 0$
4. $x^2 - 5x - 10 = 4$
5. $x^2 - x = -1 + x$
6. $x(2x - 1) = 1$
7. $6x^2 + 5x + 1 = 0$
8. $3x^2 = 16x - 5$
9. $2x^2 - 13x + 6 = 0$
10. $12x^2 - 13x + 3 = 0$
11. $6x^2 + 7x + 2 = 0$
12. $6x^2 + 13x - 2 = 0$
13. $2x^2 = 6 - x$
14. $3x^2 - 5 = 2x$
15. $x(2x + 9) = -9$
16. $(x - 5)(x + 1) = -8$
17. $5 = \dfrac{6}{x^2} - \dfrac{7}{x}$
18. $-3 = \dfrac{-10}{x + 2} + \dfrac{10}{x + 5}$

Solve by the extraction of roots.

Example $(x - 2)^2 = 5$

Solution Set $x - 2$ equal to each square root of 5.

$$x - 2 = \sqrt{5} \qquad x - 2 = -\sqrt{5}$$
$$x = 2 + \sqrt{5} \qquad x = 2 - \sqrt{5}$$

The solution set is $\{2 + \sqrt{5}, 2 - \sqrt{5}\}$.

19. $x^2 = 4$
20. $x^2 = 9$
21. $2x^2 = 6$
22. $5x^2 = 8$
23. $x^2 = -4$
24. $x^2 = -12$
25. $(x + 1)^2 = 4$
26. $(2x - 3)^2 = 16$
27. $x^2 = -9$
28. $(x + 2)^2 = 7$
29. $(x - 1)^2 = -1$
30. $(x - 3)^2 = -5$

Solve by completing the square.

Example $2x^2 - 6x - 3 = 0$

Solution Write an equivalent equation with the coefficient of x^2 equal to 1 and the constant term as the right-hand member.

$$x^2 - 3x = \frac{3}{2}$$

Add the square of half of the coefficient of the first degree term to each member.

$$x^2 - 3x + \left(-\frac{3}{2}\right)^2 = \frac{3}{2} + \left(-\frac{3}{2}\right)^2$$

Rewrite the left-hand member as the square of an expression.

$$\left(x - \frac{3}{2}\right)^2 = \frac{15}{4}$$

Then:

$$x - \frac{3}{2} = \frac{\sqrt{15}}{2} \qquad x - \frac{3}{2} = -\frac{\sqrt{15}}{2}$$

The solution set is $\left\{\dfrac{3 + \sqrt{15}}{2}, \dfrac{3 - \sqrt{15}}{2}\right\}$.

31. $x^2 + 8x - 9 = 0$ 32. $x^2 - 4x + 4 = 0$ 33. $x^2 + 11x + 30 = 0$
34. $4x^2 - 3x - 1 = 0$ 35. $3x^2 - 5x - 2 = 0$ 36. $5x^2 + 8x = 4$
37. $x^2 - x - 4 = 0$ 38. $x^2 - 2x - 2 = 0$ 39. $x^2 - 3x + 3 = 0$
40. $3x^2 + x + 7 = 0$ 41. $x^2 + 6x + 13 = 0$ 42. $9x^2 - 12x + 5 = 0$

4.3 The Quadratic Formula

Because the general quadratic equation

$$ax^2 + bx + c = 0 \quad (a \neq 0)$$

can be written equivalently in the form

$$x^2 + \frac{b}{a}x + \frac{c}{a} = 0,$$

the process of completing the square that we considered in Section 4.2 can be applied to obtain the following theorem which gives the solutions, or **roots**, of the general quadratic equation in terms of the coefficients. The proof is left as an exercise.

Theorem 4.3 If a, b, and $c \in R$ then the solutions to $ax^2 + bx + c = 0$ are

$$x = \frac{-b \pm \sqrt{b^2 - 4ac}}{2a}. \tag{1}$$

Equation (1) is called the **quadratic formula**. The symbol $\pm$ is used to condense the two equations

$$x = \frac{-b + \sqrt{b^2 - 4ac}}{2a} \quad \text{or} \quad x = \frac{-b - \sqrt{b^2 - 4ac}}{2a}$$

into a single equation. We need only substitute the coefficients a, b, and c of a given quadratic equation in the formula to find the solution set for the equation.

Determination of number of real solutions

An examination of the quadratic formula,

$$x = \frac{-b \pm \sqrt{b^2 - 4ac}}{2a},$$

shows that if $ax^2 + bx + c = 0$, $a, b, c \in R$, is to have a nonempty solution set in the set of real numbers, then $\sqrt{b^2 - 4ac}$ must be real. This, in turn, implies that only those quadratic equations for which $b^2 - 4ac \geq 0$ have real solutions. The number represented by $b^2 - 4ac$ is called the **discriminant** of the quadratic equation $ax^2 + bx + c = 0$. It yields the following information about the nature of the solution set of the equation for $a, b, c \in R$.

1. If $b^2 - 4ac = 0$, then there is precisely one real solution (multiplicity two).
2. If $b^2 - 4ac > 0$, then there are two real solutions.
3. If $b^2 - 4ac < 0$, then there are two imaginary solutions.

Exercise 4.3

A Solve for x by using the quadratic formula.

Example $x^2 - 5x + 5 = 0$

Solution Substitute 1 for a, -5 for b, and 5 for c in the quadratic formula and simplify.

$$x = \frac{5 \pm \sqrt{25 - 4 \cdot 1 \cdot 5}}{2 \cdot 1} = \frac{5 \pm \sqrt{5}}{2}.$$

The solution set is $\left\{ \dfrac{5 + \sqrt{5}}{2}, \dfrac{5 - \sqrt{5}}{2} \right\}$.

1. $x^2 - 7x + 12 = 0$
2. $x^2 - 9x + 20 = 0$
3. $4x^2 + 5x - 6 = 0$
4. $6x^2 + 7x - 3 = 0$

4.3 The Quadratic Formula

5. $6x^2 + 9x + 3 = 0$
6. $3x^2 + 11x - 4 = 0$
7. $x^2 + x - 3 = 0$
8. $x^2 + 3x - 1 = 0$
9. $2x^2 + 3x + 1 = 0$
10. $x^2 - x - 1 = 0$

Example $x^2 = 2x - 2$

Solution First write the equation in the standard form
$$x^2 - 2x + 2 = 0.$$

Substitute 1 for a, -2 for b, and 2 for c in the quadratic formula and simplify.
$$x = \frac{2 \pm \sqrt{4 - 4 \cdot 1 \cdot 2}}{2 \cdot 1} = \frac{2 \pm \sqrt{-4}}{2}$$
$$= \frac{2 \pm 2i}{2} = 1 \pm i.$$

The solution set is $\{1 + i, 1 - i\}$.

11. $x^2 + 8x + 17 = 0$
12. $x^2 + 6x + 13 = 0$
13. $2x^2 + 6x + 5 = 0$
14. $2x^2 + 10x + 17 = 0$
15. $x^2 + 3x + 5 = 0$
16. $x^2 + x - 1 = 0$
17. $-2x^2 + x = 1$
18. $x^2 = 10x - 28$
19. $\dfrac{x}{x^2 + 2x - 2} = 1$
20. $\dfrac{3x + 1}{x^2 + x + 5} = 1$
21. $x^2 + (x - 1)^2 = 0$
22. $2x^2 + (2x - 1)^2 = 4$
23. $(x + 4)^2 + (x - 2)^2 = 1$
24. $(2x + 3)^2 + (x - 4)^2 = 2$
25. $\dfrac{2x}{x - 1} - \dfrac{x + 1}{2} = 0$
26. $\dfrac{3}{2x + 1} - \dfrac{2x - 3}{x} = 0$
27. $\dfrac{2x}{x - 1} + \dfrac{3}{x^2 - x} = \dfrac{x + 1}{x}$
28. $\dfrac{3x - 1}{3x + 1} + \dfrac{2x}{2x + 1} = 1$

Use the discriminant to determine the nature of the solutions of each equation.

Example $3x^2 + 2x - 4 = 0$

Solution This equation is in standard form with $a = 3$, $b = 2$ and $c = -4$. The discriminant is
$$(2)^2 - 4(3)(-4) = 52 > 0.$$

Therefore the equation has two real solutions.

29. $x^2 + 2x - 8 = 0$

30. $2x^2 + 3x - 2 = 0$

31. $x^2 + 2 = -2x$

32. $x^2 = x + \dfrac{5}{4}$

33. $9x^2 + 1 = -6x$

34. $4x^2 = -12x - 9$

B *Solve for the indicated variable in terms of x.*

35. $k^2x - 1 = 0$, for k

36. $y^2x^2 - x = 0$, for y

37. $-k^2x + k = 0$, for k

38. $-xy^2 + 2y = 1$, for y

39. $k^2 + 2kx + x^2 = 0$, for k

40. $2y^2 + xy - 3x^2 = 0$, for y

41. Show that if $a, b, c \in R$, then the equation $ax^2 + bx + c = 0$ $(a \neq 0)$ can be written equivalently as

$$x = \frac{-b \pm \sqrt{b^2 - 4ac}}{2a}.$$

42. Show that if r_1 and r_2 are roots of the quadratic equation $ax^2 + bx + c = 0$, then $r_1 + r_2 = -b/a$ and $r_1 r_2 = c/a$.

evens thru 32

4.4 Equations Involving Radicals

Equality of like powers

In order to find solution sets for equations containing radical expressions, we shall need the following result.

Theorem 4.4 *If $U(x)$ and $V(x)$ are expressions in x, then the solution set of $U(x) = V(x)$ is a subset of the solution set of $[U(x)]^n = [V(x)]^n$, for each natural number n.*

This theorem, which follows simply from the fact that products of equal numbers are equal numbers, permits us to raise both members of an equation to the same natural-number power with the assurance that we do not lose any solutions of the original equation in the process. On the other hand, it does *not* assert that the resulting equation will be equivalent to the original equation, and indeed it will not always be so. The equation

$$[U(x)]^n = [V(x)]^n$$

may have additional solutions (called **extraneous solutions**) that are not solutions of $U(x) = V(x)$. For example, the solution set of $x = 10$ is $\{10\}$ whereas the solution set of $x^2 = 10^2 = 100$ is $\{10, -10\}$. Thus the equations $x = 10$ and $x^2 = 10^2$ are not equivalent.

4.4 Equations Involving Radicals

Necessity of checking solutions

Because raising both members of an equation to a natural-number power does not always produce an equivalent equation, each solution obtained using this process *must* be substituted for the variable in the original equation to check its validity. An application of Theorem 4.4 is not an elementary transformation.

Example Find the solution set of $\sqrt[3]{x-1} = -1$.

Solution If we raise each member of $\sqrt[3]{x-1} = -1$ to the third power, we obtain

$$(\sqrt[3]{x-1})^3 = (-1)^3,$$
$$x - 1 = -1,$$

which is equivalent to

$$x = 0.$$

Since $\sqrt[3]{0-1} = -1$, a solution of the original equation is 0. Moreover, 0 is the only real solution, since Theorem 4.4 guarantees that the solution set of the equation $\sqrt[3]{x-1} = -1$ is a subset of the solution set of $x = 0$.

Example Find the solution set of $\sqrt{x+2} + 4 = x$.

Solution We first write the equivalent equation

$$\sqrt{x+2} = x - 4$$

and then apply Theorem 4.4. We obtain

$$(\sqrt{x+2})^2 = (x-4)^2,$$
$$x + 2 = x^2 - 8x + 16.$$

This last equation is equivalent to

$$x^2 - 9x + 14 = 0,$$
$$(x - 2)(x - 7) = 0,$$

which clearly has solutions 2 and 7. Upon replacing x with 2 in the original equation, however, we obtain

$$\sqrt{2+2} + 4 = 2,$$
$$6 = 2,$$

which is false. Hence, 2 is not a solution of the original equation; it is an extraneous root. On the other hand, 7 does satisfy the original equation, so the solution set is {7}.

It is sometimes necessary to apply Theorem 4.4 more than once in solving certain equations.

Example Find the solution set of $\sqrt{x+4} + \sqrt{9-x} = 5$.

Solution It is helpful if this equation is first transformed so that each member of the equivalent equation contains only one of the radical expressions. Thus, by adding $-\sqrt{9-x}$ to each member, we obtain

$$\sqrt{x+4} = 5 - \sqrt{9-x}.$$

An application of Theorem 4.4 leads to

$$(\sqrt{x+4})^2 = (5 - \sqrt{9-x})^2,$$
$$x + 4 = 25 - 10\sqrt{9-x} + 9 - x,$$
$$2x - 30 = -10\sqrt{9-x},$$
$$x - 15 = -5\sqrt{9-x}.$$

Applying Theorem 4.4 again, we obtain

$$(x - 15)^2 = (-5\sqrt{9-x})^2,$$
$$x^2 - 30x + 225 = 25(9 - x),$$
$$x^2 - 5x = 0,$$
$$x(x - 5) = 0.$$

It is clear that this last equation has 0 and 5 as solutions. Both of these satisfy the original equation. Hence, the solution set is {0, 5}.

Exercise 4.4

A Solve and check. If there is no solution, so state.

Examples   **a.** $\sqrt{x} = 4$ **b.** $\sqrt{x+1} = 9$

Solutions   **a.** $(\sqrt{x})^2 = 4^2 = 16$ **b.** $(\sqrt{x+1})^2 = 9^2 = 81$

$x = 16$ $x + 1 = 81$

Check: $\sqrt{16} = 4$ $x = 80$

The solution set is {16}.   Check: $\sqrt{80+1} = 9$

The solution set is {80}.

1. $\sqrt{x} = 8$
2. $4\sqrt{x} = 9$
3. $\sqrt{y+8} = 1$
4. $\sqrt{y+3} = 2$
5. $x - 1 = \sqrt{2x+1}$
6. $2y + 3 = \sqrt{y+2}$
7. $\sqrt[3]{2-y} = 3$
8. $\sqrt[5]{7-x} = 2$
9. $\sqrt{x}\sqrt{x+9} = 20$
10. $\sqrt{x+3}\sqrt{x-9} = 8$

4.5 Substitution in Solving Equations

11. $\sqrt{y^2 - y} = \sqrt{y^2 + 2y - 3}$
12. $x - 2 = \sqrt{2x^2 - 3x + 2}$
13. $\sqrt{y + 4} = \sqrt{y + 20} - 2$
14. $\sqrt{x} + \sqrt{2} = \sqrt{x + 2}$
15. $\sqrt{x + 4} + \sqrt{x + 2} = 2$
16. $\sqrt{x - 1} + \sqrt{x + 2} = 1$
17. $(5 + x)^{1/2} + x^{1/2} = 5$
18. $(y + 7)^{1/2} + (y + 4)^{1/2} = 3$
19. $(y^2 - 3y + 5)^{1/2} - (y + 2)^{1/2} = 0$
20. $(z - 3)^{1/2} + (z + 5)^{1/2} = 4$

Solve for the indicated variable. Leave the results in the form of an equation. Assume that denominators are not zero.

21. $r = \sqrt{\dfrac{A}{\pi}}$, for A
22. $t = \sqrt{\dfrac{2v}{g}}$, for g
23. $x\sqrt{xy} = 1$, for y
24. $P = \pi\sqrt{\dfrac{l}{g}}$, for g
25. $x = \sqrt{a^2 - y^2}$, for y
26. $y = \dfrac{1}{\sqrt{1 - x}}$, for x
27. $R = \sqrt{\dfrac{V}{\pi h} + r^2}$, for V
28. $R = \sqrt[3]{\dfrac{3V}{4\pi} + r^3}$, for V

B Find all of the real solutions of each equation. If there are none, so state.

29. $\sqrt{5 + \sqrt{x}} = \sqrt{x} - 1$
30. $\sqrt{13 + \sqrt{x}} = \sqrt{x} + 1$
31. $\sqrt{x + \sqrt{x}} = 1 - \sqrt{x}$
32. $\sqrt{\sqrt{x} - x} = \sqrt{x} + 3$

4.5 Substitution in Solving Equations

Some equations that are not polynomial equations can nevertheless be solved by means of related polynomial equations. For example, though $y + 2\sqrt{y} - 8 = 0$ is not a polynomial equation, if the variable p is substituted for the radical expression $\sqrt{y}$, we then have $p^2 + 2p - 8 = 0$, which is a polynomial equation in p.

Example   Find the solution set of $y + 2\sqrt{y} - 8 = 0$, $y \geq 0$.

Solution   If we set $p = \sqrt{y}$ and substitute in the given equation, we have
$$p^2 + 2p - 8 = 0,$$
$$(p + 4)(p - 2) = 0,$$
which has 2 and -4 as solutions. The value $p = 2$ leads to $\sqrt{y} = 2$ or $y = 4$. The value $p = -4$ leads to $\sqrt{y} = -4$, which has no solution since $\sqrt{y}$ is always nonnegative when $y \geq 0$. The solution set is therefore $\{4\}$.

The technique of substituting one variable for another—or, more generally, a variable for an expression—is not limited to cases involving radicals, but is useful in any situation in which an equation is polynomial in form. For example, in the equation

$$\left(x + \frac{1}{x}\right)^{-2} + 6\left(x + \frac{1}{x}\right)^{-1} + 8 = 0,$$

we would set

$$p = \left(x + \frac{1}{x}\right)^{-1}.$$

Similarly, in the equation

$$(y + 3)^{1/2} - 4(y + 3)^{1/4} + 4 = 0,$$

we would set

$$p = (y + 3)^{1/4}.$$

Factoring

For some equations of the form $P(x) = 0$ the technique of substitution is simply a step which makes factoring $P(x)$ easier.

Example

Solve $x^4 - 3x^2 - 18 = 0$.

Solution 1

If we set $p = x^2$ and substitute in the given equation, we have

$$p^2 - 3p - 18 = 0,$$
$$(p - 6)(p + 3) = 0, \qquad (1)$$

which has 6 and -3 as solutions. The value $p = 6$ leads to $x^2 = 6$, or

$$x = \sqrt{6} \quad \text{or} \quad x = -\sqrt{6}.$$

The value $p = -3$ leads to $x^2 = -3$, or

$$x = \sqrt{3}i \quad \text{or} \quad x = -\sqrt{3}i.$$

Thus, the solution set is $\{\sqrt{6}, -\sqrt{6}, \sqrt{3}i, -\sqrt{3}i\}$.

Solution 2

Factoring the left-hand member of the given equation, we have

$$(x^2 - 6)(x^2 + 3) = 0, \qquad (2)$$

from which we obtain the solution set $\{\sqrt{6}, -\sqrt{6}, \sqrt{3}i, -\sqrt{3}i\}$.

Exercise 4.5

A Solve.

Example

$x^4 - 6x^2 + 5 = 0$

4.6 Solution of Linear Inequalities

Solution Set $x^2 = p$, so that $x^4 = p^2$. Rewrite and solve for p.
$$p^2 - 6p + 5 = 0$$
$$(p - 5)(p - 1) = 0$$
$$p = 1 \quad \text{or} \quad p = 5$$

For each value of p, solve $x^2 = p$.

$$x^2 = 1 \qquad x^2 = 5$$
$$x = 1, -1 \qquad x = \sqrt{5}, -\sqrt{5}$$

The solution set is $\{1, -1, \sqrt{5}, -\sqrt{5}\}$.

1. $x^4 - 10x^2 + 9 = 0$
2. $y^4 - 4y^2 - 77 = 0$
3. $z^4 - 7z^2 + 12 = 0$
4. $x^4 - 8x^2 + 7 = 0$
5. $(y + 1)^2 + 5(y + 1) - 24 = 0$
6. $(z - 1)^2 + 11(z - 1) + 30 = 0$
7. $x + 6\sqrt{x} - 7 = 0$
8. $y - 12y^{1/2} + 35 = 0$
9. $z^{2/3} - 3z^{1/3} - 18 = 0$
10. $x^{2/3} - x^{1/3} - 6 = 0$
11. $\dfrac{1}{y^2} - \dfrac{7}{y} - 18 = 0$
12. $z^{-2} - 5z^{-1} - 14 = 0$
13. $\dfrac{x^2}{(x + 1)^2} + \dfrac{x}{x + 1} = 30$
14. $\left(1 + \dfrac{1}{y}\right)^2 + 3\left(1 + \dfrac{1}{y}\right) = 40$
15. $\sqrt{x} - 6\sqrt[4]{x} + 8 = 0$
16. $\sqrt{x} - 4 = 0$
17. $\sqrt{x - 6} + 3\sqrt[4]{x - 6} - 18 = 0$
18. $\sqrt[3]{x^2} - 12\sqrt[3]{x} + 20 = 0$
19. $(x + 3)^{-2} + 4(x + 3)^{-1} = 32$
20. $z^{-4} - 8z^{-2} + 15 = 0$
21. $(x^2 + 2x + 1)^2 - 3(x + 1)^2 + 2 = 0$
22. $(4x^2 + 12x + 9)^2 - 2(2x + 3)^2 = 15$

B
23. $x - x^{2/3} = x^{1/3} - 1$
24. $x^{4/3} + x^{2/3} = -x^2 - 1$
25. $x^{1/2} + x^{1/3} = 4x^{1/6} + 4$
26. $x^6 + 9 = x^4 + 9x^2$
27. $8x + 4(2x + 1)^{2/3} = 9(2x + 1)^{1/3} + 5$
28. $2x + (x + 1)^{2/3} = 2(x + 1)^{1/3} - 1$

4.6 Solution of Linear Inequalities

Sentences such as
$$x + 3 \geq 10 \qquad (1)$$
and
$$\dfrac{-2y - 3}{3} < 5 \qquad (2)$$

are called **inequalities**. For appropriate values of the variable, one member of an inequality represents a real number that is less than ($<$), less than or equal to ($\leq$), greater than or equal to ($\geq$), or greater than ($>$) the real number represented by the other member.

For inequalities, we shall consider only real-number replacements for the variables. Any element of the replacement set of the variable for which an inequality is valid is called a **solution**, and the set of all solutions of an inequality is called the **solution set** of the inequality.

Elementary transformations

As in the case with equations, we shall solve a given inequality by generating a series of **equivalent inequalities** (inequalities having the same solution set) until we arrive at one for which the solution set is obvious. To do this, we shall need the following theorem applicable to inequalities. The proof of this theorem follows directly from the properties of order for the set of real numbers.

Theorem 4.5 *If $P(x)$, $Q(x)$, and $R(x)$ are expressions, then for all values of x for which $P(x)$, $Q(x)$, and $R(x)$ are real numbers, the sentence*

$$P(x) < Q(x)$$

is equivalent to each of the following:

I $P(x) + R(x) < Q(x) + R(x)$,

II $P(x) - R(x) < Q(x) - R(x)$,†

III $P(x) \cdot R(x) < Q(x) \cdot R(x)$

IV $\dfrac{P(x)}{R(x)} < \dfrac{Q(x)}{R(x)}$† $\Bigg\}$ for $R(x) > 0$,

V $P(x) \cdot R(x) > Q(x) \cdot R(x)$

VI $\dfrac{P(x)}{R(x)} > \dfrac{Q(x)}{R(x)}$† $\Bigg\}$ for $R(x) < 0$.

Similarly, the sentence

$$P(x) \leq Q(x)$$

is equivalent to sentences of the form I-VI, with $<$ (or $>$) replaced by $\leq$ (or $\geq$) under the same conditions as above.

The theorem can be interpreted as follows:

I and II. The addition or subtraction of the same expression to or from each member of an inequality produces an equivalent inequality in the same sense.

---

† Parts II, IV, and VI can be considered special cases of Parts I, III, and V, respectively.

4.6 Solution of Linear Inequalities

Example   By Theorem 4.5-II, the inequality
$$3x + 4 < 2x + 5$$
is equivalent to
$$3x + 4 - 2x - 4 < 2x + 5 - 2x - 4,$$
or
$$x < 1.$$

III and IV. If each member of an inequality is multiplied or divided by the same expression representing a *positive* number, the result is an equivalent inequality in the same sense.

Example   By Theorem 4.5-IV, the inequality
$$x^2 < x^3 \quad (x \neq 0)$$
is equivalent to
$$\frac{x^2}{x^2} < \frac{x^3}{x^2} \quad (x \neq 0),$$
$$1 < x.$$

V and VI. If each member of an inequality is multiplied or divided by the same expression representing a *negative* number, and the sense of the inequality is reversed, the result is an equivalent inequality.

Example   By Theorem 4.5-VI, the inequality
$$x^2 < -x^3 \quad (x \neq 0),$$
is equivalent to
$$\frac{x^2}{-x^2} > \frac{-x^3}{-x^2} \quad (x \neq 0),$$
$$-1 > x.$$

Note that Theorem 4.5 does not permit multiplying or dividing by zero, and variables in multipliers and divisors are restricted from values for which the expression vanishes. The result of applying any part of this theorem is an elementary transformation.

Solution of a linear inequality   Several applications of Theorem 4.5 can be applied to solve inequalities in the same way that Theorem 4.1 is applied to solve equations.

Example Find the solution set of $\dfrac{-x+3}{4} > -\dfrac{2}{3}$.

Solution Multiplying each member by 12 yields

$$3(-x+3) > -8,$$
$$-3x + 9 > -8.$$

Adding -9 to each member, we have

$$-3x > -17.$$

Finally, dividing each member by -3, we obtain

$$x < \frac{17}{3},$$

and the solution set is written

$$S = \left\{x \mid x < \frac{17}{3}\right\} = \left(-\infty, \frac{17}{3}\right).$$

Graphical representation The solution set in the foregoing example can be pictured on a line graph as shown in Figure 4.1. The colored line indicates points with coordinates in the solution set. Note that the open dot on the right-hand endpoint indicates that 17/3 *is not* a member of the solution set.

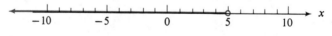

Figure 4.1

Inequalities sometimes appear in a form such as

$$-6 < 3x \le 15, \qquad (3)$$

where an expression is bracketed between two inequality symbols. As observed in Section 1.4, this means $-6 < 3x$ *and* $3x \le 15$. The solution set of such an inequality is obtained in the same manner as the solution set of any other inequality. In (3) above, each expression may be divided by 3 to obtain

$$-2 < x \le 5.$$

The solution set,

$$S = \{x \mid -2 < x \le 5\} = (-2, 5],$$

is shown on a line graph in Figure 4.2. The open dot at the left-hand endpoint of the interval indicates that -2 *is not* a member of the solution set, whereas the solid dot at the other end shows that 5 *is* a member of the solution set.

4.6 Solution of Linear Inequalities

Figure 4.2

Even 2-24

Exercise 4.6

A Solve each inequality. Write the solution set in interval notation and represent it on a line graph.

Examples

a. $\dfrac{3x+1}{5} < x - 2$

b. $-1 < 2x + 3 \leq 8$

Solutions Generate the following sequences of equivalent inequalities.

a. $3x + 1 < 5x - 10$

$11 < 2x$

$\dfrac{11}{2} < x$

$\left(\dfrac{11}{2}, +\infty\right)$

b. $-4 < 2x \leq 5$

$-2 < x \leq \dfrac{5}{2}$

$\left(-2, \dfrac{5}{2}\right]$

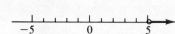 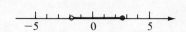

1. $x + 11 > 5$
2. $x - 3 < -2$
3. $3x + 1 \leq 10$
4. $5x - 4 \leq 11$
5. $2x + 7 \geq -5$
6. $3 - x \leq -1$
7. $x + 6 \leq 3x + 1$
8. $2x + 7 \leq x + 4$
9. $1 - 4x > x + 6$
10. $4x \geq -5x - 1$
11. $-(x + 3) \leq -1$
12. $-(2x + 4) \geq 3$
13. $-2 \leq \dfrac{3x + 1}{4}$
14. $1 \leq \dfrac{-2x - 1}{3}$
15. $\dfrac{3x + 4}{2} < 6$
16. $\dfrac{-3x + 2}{4} < 5$
17. $\dfrac{2x - 3}{3} \leq \dfrac{3x}{2}$
18. $\dfrac{3x - 4}{2} \leq \dfrac{-2x}{5}$
19. $-1 \leq x + 5 \leq 6$
20. $0 \leq 2x - 3 < 1$
21. $-4 < \dfrac{3x + 2}{5} \leq -2$
22. $4 \leq \dfrac{3x - 1}{2} \leq 10$
23. $5 \leq \dfrac{x + 1}{-3} \leq 8$
24. $-2 \leq \dfrac{x + 5}{2} \leq 0$

B In Exercises 25–30, assume that ε (epsilon) is a positive real number. Solve each inequality. Write the solution set in interval notation.

Examples   a. $-\varepsilon < 2x + 1$ b. $2x + 1 < \varepsilon$

Solutions   a.
$$-\varepsilon < 2x + 1$$
$$-1 - \varepsilon < 2x$$
$$-\frac{1+\varepsilon}{2} < x$$

The solution set $\left(-\frac{1+\varepsilon}{2}, \infty\right)$.

b.
$$2x + 1 < \varepsilon$$
$$2x < \varepsilon - 1$$
$$x < \frac{\varepsilon - 1}{2}$$

The solution set is $\left(-\infty, \frac{\varepsilon - 1}{2}\right)$.

25. $-\varepsilon < 4x - 5$
26. $4x - 5 < \varepsilon$
27. $-\varepsilon < 3x + 2 < \varepsilon$
28. $-\varepsilon < 2x + 4 < \varepsilon$
29. $-\varepsilon < -3x + 7 < \varepsilon$
30. $-\varepsilon < -2x + 1 < \varepsilon$

4.7 Solution of Nonlinear Inequalities

As in the case with linear inequalities, we can generate equivalent nonlinear inequalities by applying any part of Theorem 4.5. Additional procedures are necessary, however, to obtain the solution sets of such inequalities. For example, consider the inequality

$$x^2 + 4x < 5.$$

To determine values of x for which this condition holds, we might first rewrite the inequality equivalently as

$$x^2 + 4x - 5 < 0,$$

and then as

$$(x + 5)(x - 1) < 0.$$

It is clear here that only those values of x for which the factors $x + 5$ and $x - 1$ are opposite in sign will be in the solution set. These can be determined analytically by noting that $(x + 5)(x - 1) < 0$ implies either

     Case I: $x + 5 < 0$ and $x - 1 > 0$

or else

     Case II: $x + 5 > 0$ and $x - 1 < 0.$

Each of these two cases can be considered separately.

4.7 Solution of Nonlinear Inequalities

First, for Case I,

$$x + 5 < 0 \quad \text{and} \quad x - 1 > 0$$

imply

$$x < -5 \quad \text{and} \quad x > 1,$$

a condition which is not satisfied by any values of x. The solution set of Case I is $\emptyset$. The inequalities for Case II,

$$x + 5 > 0 \quad \text{and} \quad x - 1 < 0$$

imply

$$x > -5 \quad \text{and} \quad x < 1,$$

which lead to the solution set

$$\{x \mid -5 < x < 1\}.$$

Thus the complete solution set is

$$S = \{x \mid -5 < x < 1\} \cup \emptyset = \{x \mid -5 < x < 1\} = (-5, 1).$$

Critical numbers

Let us now consider an alternative method of solving quadratic inequalities. Notice in the foregoing solution process that the numbers -5 and 1 separate the set of real numbers into the three intervals shown in Figure 4.3.

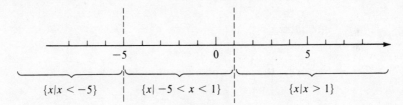

Figure 4.3

Each of these intervals either is or is not a part of the solution set of the inequality $x^2 + 4x < 5$. To determine which, we need only substitute an arbitrarily selected number from each interval and test it in the inequality. Let us use $-6, 0,$ and 2.

$$(-6)^2 + 4(-6) \stackrel{?}{<} 5 \qquad 0^2 + 4(0) \stackrel{?}{<} 5 \qquad 2^2 + 4(2) \stackrel{?}{<} 5$$
$$36 - 24 \stackrel{?}{<} 5 \qquad 0 + 0 \stackrel{?}{<} 5 \qquad 4 + 8 \stackrel{?}{<} 5$$
$$12 \stackrel{?}{<} 5 \qquad 0 \stackrel{?}{<} 5 \qquad 12 \stackrel{?}{<} 5$$

$$\text{No.} \qquad\qquad \text{Yes.} \qquad\qquad \text{No.}$$

Clearly, the only interval involved that is in the solution set is $\{x \mid -5 < x < 1\}$, and hence this interval is the solution set. The graph is shown in Figure 4.4 (on page 114). If the inequality in this example were $x^2 + 4x \leq 5$, the endpoints on the graph would be shown as closed dots.

Figure 4.4

In an inequality of the form $Q(x) < 0$ or $Q(x) > 0$, numbers for which either $Q(x) = 0$ or else $Q(x)$ is undefined are called **critical numbers**. For example, critical numbers in the preceding example are -5 and 1.

Examples   Find the critical numbers for each inequality.

     a. $2x^2 - x - 1 > 0$ b. $\dfrac{x}{x^2 - 4} < 0$

Solutions   a. Solving $2x^2 - x - 1 = 0$ we obtain $(2x + 1)(x - 1) = 0$, from which $x = -1/2$ or $x = 1$. Thus, the critical numbers are $-1/2$ and 1.

     b. $\dfrac{x}{x^2 - 4} = 0$ for $x = 0$. The expression $\dfrac{x}{x^2 - 4}$ is undefined when $x = \pm 2$. Thus, the critical numbers are 0, 2, and -2.

The critical numbers of an inequality separate the set of real numbers into disjoint intervals. Each of these intervals is either contained in or disjoint from the solution set of the inequality.

Example   Solve $x^2 - 3x - 4 \geq 0$.

Solution   Factoring the left-hand element yields

$$(x + 1)(x - 4) \geq 0,$$

and we observe that -1 and 4 are critical numbers because $(x + 1)(x - 4) = 0$ for these values. Hence, we wish to check the intervals shown on the number line.

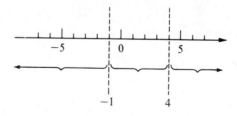

We now substitute selected arbitrary values in each interval, say -2, 0, and 5, for the variable in the original inequality.

     $(-2)^2 - 3(-2) - 4 \overset{?}{\geq} 0$ $(0)^2 - 3(0) - 4 \overset{?}{\geq} 0$ $(5)^2 - 3(5) - 4 \overset{?}{\geq} 0$

           Yes.   No.   Yes.

4.7 Solution of Nonlinear Inequalities

Hence, the solution set is

$$\{x|x \leq -1\} \cup \{x|x \geq 4\} = (-\infty, -1] \cup [4, \infty).$$

The graph is shown in the figure below.

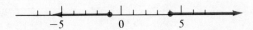

Variables in the denominator

Inequalities involving fractions have to be approached with care if any fraction contains a variable in the denominator. If each member of such an inequality is multiplied by an expression containing the variable, we have to be careful either to distinguish between those values of the variable for which the expression denotes a positive and negative number, respectively, or to make sure that the expression by which we multiply is always positive. However, we can use the notion of critical numbers to avoid these complications.

Example

Solve the inequality $\dfrac{x}{x-2} \geq 5$.

Solution

We first write the given inequality equivalently as

$$\frac{x}{x-2} - 5 \geq 0,$$

from which

$$\frac{x - 5(x-2)}{x-2} \geq 0,$$

$$\frac{-4x + 10}{x-2} \geq 0. \tag{1}$$

In this case, the critical numbers are $5/2$ and 2, because $\dfrac{-4x+10}{x-2}$ equals zero for $x = 5/2$, and $\dfrac{-4x+10}{x-2}$ is undefined for $x = 2$. Thus we want to check the intervals shown on the number line.

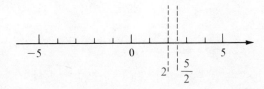

Substituting arbitrary values from each of the three intervals, say 0, 9/4, and 3, for the variable in the original inequality or in (1), we can identify the solution set,

$$\left\{x \Big| 2 < x \leq \frac{5}{2}\right\} = \left(2, \frac{5}{2}\right],$$

as we did in the preceding example. The graph is shown in the figure below. Note that the left-hand endpoint is an open dot (2 is not a member of the solution set) because the left-hand member of the original inequality is undefined for $x = 2$.

Exercise 4.7

A Solve each inequality. Write the solution set in interval notation.

Example $(x + 4)(x - 3) \leq 0$

Solution The critical numbers are -4 and 3 because $(x + 4)(x - 3) = 0$ for these values. Check the intervals shown on the number line in the figure by substituting an arbitrary value from each interval for the variable in the inequality. Here we use $-5, 0$ and 5.

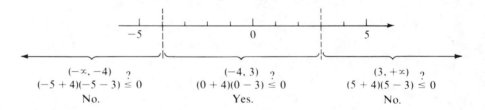

Next check the critical numbers themselves; here both are solutions. Therefore the solution set is $[-4, 3]$.

1. $(x + 1)(x - 5) > 0$
2. $(x + 6)(x + 2) < 0$
3. $(2x + 1)(3x - 8) \leq 0$
4. $(3x - 1)(3x - 10) \geq 0$
5. $x^2 + x - 6 > 0$
6. $x^2 + 3x - 4 \leq 0$
7. $x^2 - 2x - 8 > 0$
8. $x^2 + x - 12 \geq 0$
9. $x^2 + 4x > 3x + 20$
10. $8x + 8 \geq -x^2 + 2x$
11. $2x^2 - 2 \leq -x^2 + 5x$
12. $3x^2 + 2x < -3x^2 + x + 2$
13. $6x^2 + 6x \geq -2x^2 + 3$
14. $6x^2 - 6 > 5x - 2$
15. $x^2 < 0$
16. $x^2 + 1 \geq 0$

Example $x^2 - 2x - 1 \leq 0$

4.7 Solution of Nonlinear Inequalities

Solution Find the critical numbers by the quadratic formula:

$$x = \frac{2 \pm \sqrt{4+4}}{2} = \frac{2 \pm 2\sqrt{2}}{2} = 1 \pm \sqrt{2}$$

```
    |----|----|----|----|----|----|----|----|----|---->
   -5              0                   5
        (-∞, 1-√2)   (1-√2, 1+√2)     (1+√2, +∞)
```

Check each interval by testing an arbitrary point in it, say -4, 1, and 4.

$$(-4)^2 - 2(-4) - 1 \overset{?}{\leq} 0 \qquad 1^2 - 2(1) - 1 \overset{?}{\leq} 0 \qquad 4^2 - 2(4) - 1 \overset{?}{\leq} 0$$
$$\text{No.} \qquad\qquad \text{Yes.} \qquad\qquad \text{No.}$$

Test the critical numbers themselves; since they make the expression 0, they are solutions. The solution set is therefore $[1 - \sqrt{2}, 1 + \sqrt{2}]$.

17. $x^2 - 4x + 1 \geq 0$
18. $x^2 + 4x + 2 \leq 0$
19. $x^2 - x + 4 < 0$
20. $-x^2 + x + 1 > 0$

Example $\dfrac{x+1}{x} \geq 2$

Solution Write the equivalent inequalities

$$\frac{x+1}{x} - 2 \geq 0,$$

$$\frac{x+1-2x}{x} \geq 0,$$

$$\frac{1-x}{x} \geq 0.$$

The critical numbers are 0 and 1 since $\dfrac{1-x}{x}$ is zero when $x = 1$ and undefined for $x = 0$.

```
    |----|----|----|----|----|----|----|----|----|---->
   -5              0                   5
            (-∞, 0)     (0, 1)      (1, ∞)
```

Check each interval by testing an arbitrary point in it, say $-1, 1/2, 2$.

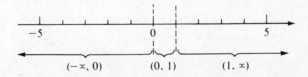

No. Yes. No.

Test the critical numbers; 1 is a solution but 0 is not. Therefore the solution set is $(0, 1]$.

21. $\dfrac{x-1}{x} \le 3$ 22. $\dfrac{x}{x-1} \ge 5$ 23. $\dfrac{x+1}{x-1} < 1$

24. $\dfrac{1}{x-1} \le 1$ 25. $\dfrac{2}{2+x} > 1$ 26. $\dfrac{x}{3-x} - 3 \le 6$

B Solve each inequality. Write the solution set in interval notation.

Example

$\dfrac{(x-1)(x+2)}{(x-3)} \le 0$

Solution

The critical numbers are 1, −2, and 3.

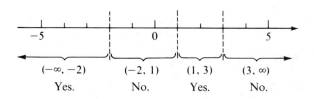

Check each interval by testing an arbitrary point in it, say −3, 0, 2, and 4. Test the critical numbers themselves; 1 and −2 are solutions but 3 is not. Therefore the solution set is $(-\infty, -2] \cup [1, 3)$.

27. $\dfrac{(x+1)}{(x-1)(x-3)} > 0$ 28. $\dfrac{(x-1)(x+1)}{(x-5)} < 0$

29. $\dfrac{(x-2)(x+1)}{x} < 0$ 30. $\dfrac{x}{(x+2)(x-4)} \ge 0$

31. $(x-1)(x+1)(x+3) \ge 0$ 32. $(x+1)(x-4)x < 0$

33. $\dfrac{(x+2)(x-2)}{x(x+4)} > 0$ 34. $x(x+2)(x-2)(x-4) \ge 0$

Assume ε is a positive real number and solve each inequality. Write the solution set in interval notation.

Example $x^2 - 2 < \varepsilon$

Solution

The critical values for this inequality are $-\sqrt{2+\varepsilon}$ and $\sqrt{2+\varepsilon}$. These values separate the number line into the disjoint intervals $(-\infty, -\sqrt{2+\varepsilon})$, $(-\sqrt{2+\varepsilon}, \sqrt{2+\varepsilon})$, and $(\sqrt{2+\varepsilon}, \infty)$. Check each interval by testing values from each of these intervals, say $-2\sqrt{2+\varepsilon}$, 0, and $2\sqrt{2+\varepsilon}$. Check the critical values $-\sqrt{2+\varepsilon}$ and $\sqrt{2+\varepsilon}$. The solution set is $(-\sqrt{2+\varepsilon}, \sqrt{2+\varepsilon})$.

35. $x^2 < \varepsilon$

36. $4x^2 < \varepsilon$

37. $x^2 - 1 < \varepsilon$

38. $x^2 - 4 < \varepsilon$

4.8 Equations and Inequalities Involving Absolute Value

In Section 1.4 the absolute value of a real number was defined by

$$|x| = \begin{cases} x & \text{if } x \geq 0, \\ -x & \text{if } x < 0. \end{cases}$$

For example, $|7| = 7$ and $|-5| = 5$. Since it is also true that $|5| = 5$, the equation $|x| = 5$ has two solutions, 5 and -5. This in no way contradicts our earlier discussion concerning the number of solutions of a first-degree equation. Equations involving absolute values are not polynomial equations and hence are not assigned a degree. In fact, simple equations involving absolute values are contractions of two equations without absolute values. The same is true of inequalities. The relationships are indicated in Figure 4.5.

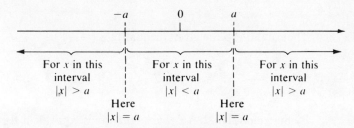

Figure 4.5

To summarize the relationships, note that for $a > 0$,

I $|x| = a$ is equivalent to $x = a$ or $x = -a$;
II $|x| < a$ is equivalent to $-a < x$ and $x < a$, i.e. $-a < x < a$;
III $|x| > a$ is equivalent to $x < -a$ or $x > a$.

To solve equations or inequalities involving absolute value, we translate into the two equivalent statements without absolute value and solve these two statements. We then recall that the everyday word "and" corresponds to set intersection and "or" (the nonexclusive "or," meaning "one or the other, or both") corresponds to set union. Our solution set will be the intersection of two sets for type II problems, the union of two sets for problems of type I or III.

Example $|x - 3| = 5$

Solution on overleaf

Solution   This equation is equivalent to

$$x - 3 = 5 \quad \text{or} \quad x - 3 = -5 \quad \text{by Statement I above.}$$

The solution to the first equality is $x = 8$, the solution to the second $x = -2$. The solution set is the union, $\{-2\} \cup \{8\} = \{-2, 8\}$.

Recall also from Section 1.4 that $|a - b| = c$ can be interpreted geometrically to mean that the distance between a and b is c. In the preceding example, we are given that the distance from x to 3 is 5. The values of x are then readily seen geometrically to be -2 and 8.

Example   Solve $|x - 3| < 5$.

Solution   By Statement II above, this inequality is equivalent to

$$-5 < x - 3 \quad \text{and} \quad x - 3 < 5,$$

or

$$-5 < x - 3 < 5,$$
$$-2 < x < 8.$$

The solution to the given inequality is $(-2, 8)$.

Example   Solve $|x - 3| > 5$.

Solution   By Statement III above, this inequality is equivalent to

$$x - 3 < -5 \quad \text{or} \quad x - 3 > 5.$$

The first inequality reduces to $x < -2$, so has solution set $(-\infty, -2)$. The second reduces to $x > 8$ and has solution set $(8, \infty)$. The solution to the given inequality is therefore the union $(-\infty, -2) \cup (8, \infty)$.

Inequalities involving absolute values and $\leq$ or $\geq$ are treated similarly. These relationships are apparent in Figure 4.5 on page 119.

IV $|x| \leq a$ is equivalent to $-a \leq x$ and $x \leq a$; i.e., $-a \leq x \leq a$;
V $|x| \geq a$ is equivalent to $x \leq -a$ or $x \geq a$.

Alternative method of solution   We can also solve absolute-value inequalities by using a method similar to the method of critical numbers used in the preceding section.

4.8 Equations and Inequalities Involving Absolute Value

Example Solve $|x^2 - 1| < 3$.

Solution The critical numbers are obtained by solving the equation
$$|x^2 - 1| = 3,$$
or, equivalently, the two equations
$$x^2 - 1 = 3, \quad \text{or} \quad x^2 - 1 = -3,$$
$$x^2 = 4 \quad \quad \quad x^2 = -2.$$

Thus, the critical numbers are the real numbers -2 and 2. These values separate the set of real numbers into three intervals as shown in the figure below.

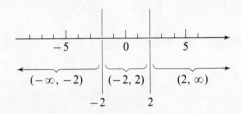

Substituting arbitrary values from each of the three intervals, say $-3, 0,$ and 3, for the variable in the original inequality we have

$$|(-3)^2 - 1| \stackrel{?}{<} 3 \quad\quad |(0)^2 - 1| \stackrel{?}{<} 3 \quad\quad |(3)^2 - 1| \stackrel{?}{<} 3$$
No. Yes. No.

Hence, the solution set is
$$\{x \mid -2 < x < 2\} = (-2, 2).$$

Exercise 4.8

A Solve. Write the solution set in interval notation.

Examples a. $|x - 2| \leq 5$ b. $|x + 3| \geq 5$

Solutions
a. Write the equivalent inequality from Statement IV on page 120.
$$-5 \leq x - 2 \leq 5,$$
$$-3 \leq x \leq 7.$$
The solution set is $[-3, 7]$.

b. From Statement V on page 120,
$$x + 3 \geq 5 \quad \text{or} \quad x + 3 \leq -5,$$
$$x \geq 2 \quad \text{or} \quad x \leq -8.$$
The solution set is $(-\infty, -8] \cup [2, \infty)$.

1. $|x + 2| = 3$
2. $|4 - x| = 1$
3. $|3 - x| = 4$
4. $|x + 5| = 1$
5. $|2x + 1| = 3$
6. $|3x - 2| = 4$

7. $|x - 2| < 5$
8. $|x + 3| < 2$
9. $\left|x - \frac{1}{2}\right| < \frac{2}{3}$
10. $\left|x + \frac{3}{4}\right| < \frac{1}{2}$
11. $|x - 4| > 1$
12. $|x + 2| > 2$
13. $|3 + x| > 7$
14. $|2x + 1| > 3$
15. $|x - 4| \leq 1$
16. $|x + 2| \leq 5$
17. $|2x - 3| \leq 3$
18. $|3x + 4| \leq \frac{1}{2}$
19. $\left|x - \frac{1}{2}\right| \geq 2$
20. $|x + 1| \geq \frac{1}{2}$
21. $|3x - 1| \geq 3$
22. $\left|\frac{1}{2}x - \frac{1}{4}\right| \geq \frac{1}{4}$
23. $|x| \leq 3$
24. $|x| \geq 1$

Rewrite each of the following, using absolute-value notation.

Examples a. $-1 < x < 1$ b. $-3 < x < 1$

Solutions a. $|x| < 1$ b. $-2 < x + 1 < 2$
 $|x + 1| < 2$

25. $-3 \leq x \leq 3$
26. $-4 \leq x \leq 4$
27. $-2 < x < 6$
28. $0 \leq x \leq 8$
29. $-8 \leq x \leq -2$
30. $-10 < x < 0$
31. $-\frac{4}{3} < x < \frac{2}{3}$
32. $\frac{1}{2} < x < \frac{9}{2}$

Solve. Write the solution set in interval notation.

33. $|4x^2 - 2| < 2$
34. $|2x^2 - 6| < 4$
35. $|x^2 - 4x| \geq 4$
36. $|x^2 + 5x| \geq 6$
37. $\left|\frac{x}{2x + 1}\right| \leq 3$
38. $\left|\frac{2x}{x - 1}\right| \leq 4$

4.9 Word Problems

Equations and inequalities can be used to express symbolically quantitative relations in word problems. The problem may be explicitly concerned with numbers, or it may be concerned with numerical measures of physical quantities. In either

4.9 Word Problems

event, we seek the set of numbers (the solution set) for which the stated relationship holds. The following suggestions are frequently helpful in expressing symbolically the conditions of the problem:

i. Determine the quantities asked for and represent them by symbols. Since at this time we are using one variable only, all relevant quantities should be represented in terms of this variable.
ii. Where applicable, draw a sketch and label all known quantities thereon; label the unknown quantities in terms of symbols.
iii. Find in the problem a quantity that can be represented in two different ways and write this representation as an equation. The equation may derive from:
   a. the problem itself, which may state a relationship explicitly; for example, "What number added to 4 gives 7?" produces the equation $x + 4 = 7$;
   b. formulas or relationships that are part of your general mathematical background; for example, $A = \pi r^2$, $d = rt$.
iv. Solve the resulting equation.
v. Use the solution to answer the question asked by the problem. Check your answer.

In some cases, the mathematical model we obtain for a physical situation is a quadratic equation that has two real solutions. It may be that one, but not both, of the solutions fits the physical situation.

Example   Find two consecutive natural numbers whose product is 72.

Solution
i. Let n denote the first of the two natural numbers, so that n and $n + 1$ are the two consecutive natural numbers.
ii. No figure would be helpful in this example.
iii. The product of n and $n + 1$ must be 72, thus we write the equation

$$n(n + 1) = 72$$

as our model.
iv. Solving this equation, we have

$$n^2 + n - 72 = 0$$
$$(n + 9)(n - 8) = 0$$

with solution set $\{8, -9\}$.
v. Since -9 is *not* a natural number, we reject it as a possible answer to the original question. The solution 8, however, leads to the consecutive natural numbers 8 and 9. Check: Does $8 \cdot 9 = 72$? Yes.

As additional examples, observe that we would not accept -6 feet as the height of a man, nor 27/4 for the number of persons in a room.

Exercise 4.9

A Solve the following word problems.

In Exercises 1–20, use a mathematical model in the form of a first-degree equation in one variable.

Example   A collection of coins consisting of dimes and quarters has a value of $12.75. If there are 33 more dimes than quarters, how many of each are in the collection?

Solution   We follow the suggestions i–v, omitting ii as not applicable.

    i. Represent the unknown quantities symbolically. Let x represent the number of quarters; then $x + 33$ represents the number of dimes.

    iii. Write an equation relating the values of quarters and dimes to the total value.

$$\begin{bmatrix} \text{value of} \\ \text{quarters} \\ \text{in cents} \end{bmatrix} + \begin{bmatrix} \text{value of} \\ \text{dimes} \\ \text{in cents} \end{bmatrix} = \begin{bmatrix} \text{total} \\ \text{value} \\ \text{in cents} \end{bmatrix}$$

$$25x + 10(x + 33) = 1275$$

    iv. Solve for x.

$$25x + 10x + 330 = 1275$$
$$35x = 945$$
$$x = 27$$

    v. Therefore $x + 33 = 60$ and there are 27 quarters and 60 dimes in the collection. Check: Do 27 quarters and 60 dimes have a value of $12.75? Yes.

1. A man has $2.35 in change consisting of two more nickels than dimes. How many dimes and how many nickels does he have?

2. The admission at a baseball game was $2.00 for adults and $1.25 for children. The receipts were $542.50 for 350 paid admissions. How many adults, and how many children, attended the game?

3. How many pounds of an alloy containing 32% silver must be melted with 25 pounds of an alloy containing 48% silver to obtain an alloy containing 42% silver?

4. How many pounds of an alloy containing 80% silver must be melted with 5 lbs. of an alloy containing 55% silver to obtain an alloy containing 75% silver?

5. How much water should be added to 50 gallons of a solution which is 50% acid to obtain a 12% solution?

6. How much water should be added to 6 gallons of pure acid to obtain a 15% solution?

4.9 Word Problems

A karat is a measure comprising 24 units used to specify the proportion of gold in an alloy, e.g. 12K (12 karat) gold is 12/24 or 50% gold.

7. How much 22K gold must be melted with 1 ounce of 14K gold to produce 18K gold?

8. How much 12K gold must be melted with two ounces of 20K gold to produce 18K gold?

Example   A man has an annual income of $6500 from two investments. He has $15,000 more invested at 6% than he has invested at 8%. How much does he have invested at each rate?

Solution
  i. Represent the amount invested at each rate symbolically. Let A represent the amount in dollars invested at 8%; then $A + 15,000$ represents the amounts invested at 6%.

  iii. Write an equation relating the interest from each investment and the total interest.

$$\begin{bmatrix} \text{interest from} \\ 8\% \text{ investment} \end{bmatrix} + \begin{bmatrix} \text{interest from} \\ 6\% \text{ investment} \end{bmatrix} = [\text{total interest}]$$

$$0.08A \quad + \quad 0.06(A + 15,000) \quad = \quad 6500$$

  iv. Solve for A.

$$8A + 6A + 90,000 = 650,000$$
$$14A = 560,000$$
$$A = 40,000; \text{ therefore } A + 15,000 = 55,000$$

  v. The man had $40,000 invested at 8% and $55,000 invested at 6%.
  Check: Does 8% of $40,000 ($3200) added to 6% of $55,000 ($3300) equal $6500? Yes.

9. A sum of $4000 is invested, part at 12% and the remainder at 14%. Find the amount invested at each rate if the yearly income from the two investments is $510.

10. A sum of $5000 is invested, part at 10% and the remainder at 15%. Find the total yearly interest if the interest on each investment is the same.

11. A man has three times as much money invested in 12% bonds as he has in stocks paying 8%. How much does he have invested in each if his yearly income from the investments is $4840?

12. A man has $1000 more invested at 10% than he has invested at 8%. If his annual income from the two investments is $2260, how much does he have invested at each rate?

Example   An express train travels 150 miles in the same time that a freight train travels 100 miles. If the express goes 20 miles per hour faster than the freight, find each rate.

Solution
  i. Represent the unknown quantities symbolically. Let r represent a rate for the freight train; then $r + 20$ represents the rate of the express train.

  iii. The fact that the times are equal is the significant equality in the problem.

$$(t \text{ of freight}) = (t \text{ of express})$$

Solution continued on overleaf

Express the time of each train in terms of r (time = distance/rate).

$$\frac{100}{r} = \frac{150}{r + 20}$$

iv. Solve for r.

$$(r + 20)100 = (r)150$$
$$100r + 2000 = 150r$$
$$-50r = -2000$$
$$r = 40; \quad \text{therefore} \quad r + 20 = 60$$

v. The freight train's rate is 40 miles per hour; the express train's rate is 60 miles per hour. Check: Does the time of the freight train (100/40) equal the time of the express train (150/60)? Yes.

13. An airplane travels 1260 miles in the same time that an automobile travels 420 miles. If the rate of the airplane is 120 miles per hour greater than the rate of the automobile, find the rate of each.

14. Two cars start together and travel in the same direction, one going twice as fast as the other. At the end of 3 hours they are 96 miles apart. How fast is each traveling?

15. A freight train leaves town A for town B, traveling at an average rate of 50 miles per hour. Three hours later a passenger train also leaves town A for town B, on a parallel track, traveling at an average rate of 80 miles per hour. How far from town A does the passenger train pass the freight train?

16. A boy walked to his friend's house at the rate of 4 miles per hour and he ran back home at the rate of 6 miles per hour. How far apart are the two houses if the round trip took 20 minutes?

If two resistors with resistances R_1 ohms and R_2 ohms, respectively, are connected in series, the equivalent resistance is $R = R_1 + R_2$. If the same two resistors are connected in parallel, the equivalent resistance R satisfies $1/R = 1/R_1 + 1/R_2$.

Example

Two resistors connected in parallel have an equivalent resistance of 10 ohms. One of them has a resistance of 20 ohms. What is the resistance of the other?

Solution

i. Represent the unknown resistance symbolically. Let R_1 represent the unknown resistance.
iii. The equation is

$$\frac{1}{R_1} + \frac{1}{20} = \frac{1}{10}.$$

iv. Solve for R_1.

$$20 + R_1 = 2R_1$$
$$R_1 = 20.$$

v. Use the solution to answer the original question. The unknown resistance is 20 ohms. Check: Is $1/20 + 1/20 = 1/10$? Yes.

4.9 Word Problems

17. Two equal resistors are connected in series and the equivalent resistance is 10 ohms. Find the resistance of each resistor.

18. Two equal resistors are connected in parallel and the equivalent resistance is 20 ohms. Find the resistance of each resistor.

19. Two equal resistors are connected in series with each other and the combination is connected in parallel with a 10 ohm resistor. If the equivalent resistance is 20/3 ohms, find the resistance of each of the two resistors with unknown resistance.

20. A resistor of unknown resistance is connected in series with a 5 ohm resistor and the combination is connected in parallel with a 20 ohm resistor. If the equivalent resistance is 100/9 ohms, find the value of the unknown resistance.

In Exercises 21–30, use a mathematical model in the form of a second-degree equation in one variable.

Example

Find two numbers whose sum is 27 and whose product is 180.

Solution

i. Let x represent one of the numbers so that $27 - x$ represents the other.

iii. The equation is

$$x(27 - x) = 180.$$

iv. Solve for x.

$$27x - x^2 = 180$$
$$x^2 - 27x + 180 = 0$$
$$x = \frac{27 \pm \sqrt{729 - 720}}{2} = \frac{27 \pm 3}{2}$$
$$x = 15 \quad \text{or} \quad x = 12$$

When $x = 15$, $27 - x = 12$; when $x = 12$, $27 - x = 15$. The solutions are 12 and 15.

v. Check: $12 + 15 = 27?$; $12 \cdot 15 = 180?$ Yes.

21. Find two numbers whose sum is 25 and whose product is 154.

22. Find two numbers whose sum is 5 and whose product is -24.

23. Find two consecutive integers whose product is 132.

24. Find two consecutive natural numbers such that the sum of their squares is 365.

25. Two airplanes with lines of flight at right angles to each other pass each other (at slightly different altitudes) at noon. One is flying at 140 miles per hour and the other is flying at 180 miles per hour. How far apart are they at 12:30 PM? *Hint*: Use the Pythagorean Theorem.

26. A box without a top is to be made from a square piece of tin by cutting a two-centimeter square from each corner and folding up the sides. If the box is to hold 128 cubic centimeters, what should be the length of each side of the original square?

27. A ball thrown vertically upward reaches a height h in feet given by the equation $h = 56t - 16t^2$, where t is the time in seconds after the throw. How long will it take the ball to reach a height of 24 feet on its way up? How long after the throw will the ball return to the height from which it was thrown?

28. The distance s a body falls in a vacuum is given by $s = v_0 t + \frac{1}{2}gt^2$, where s is measured in feet, t is measured in seconds, v_0 is the initial velocity in feet per second, and g is the constant of acceleration due to gravity (approximately 32 ft/sec/sec). How long will it take a body to fall 150 feet if v_0 is 20 feet per second? How long will it take if the body starts from rest?

29. A man and his son working together can paint their house in four days. The man can do the job alone in six days less than the son can do it. How long would it take each of them to paint the house alone? *Hint*: What part of the job could each of them do in one day?

30. A theater that is rectangular in shape seats 720 people. The number of rows needed to seat the people would be four fewer if each row held six more seats. How many seats would then be in each row?

31. A resistor with resistance R is connected in series with a 10-ohm resistor, and the combination is connected in parallel with another resistor of resistance R. If the equivalent resistance is 15/4 ohms, find the value of R. *Hint*: See formulas on page 126.

32. A resistor with resistance R is connected in series with a 5-ohm resistor, and the combination is connected in parallel with another resistor of resistance R. If the equivalent resistance is 100/9 ohms, find the value of R.

33. A resistor with resistance R is connected in parallel with a resistor of resistance $R + 5$. If the equivalent resistance is 5/3 ohms, find the value of R.

34. A resistor with resistance $R + 5$ is connected in parallel with a resistor of resistance $R + 10$. If the equivalent resistance is 60/7 ohms, find the value of R.

In Exercises 35–40, use a mathematical model in the form of an inequality in one variable.

Example

A student must have an average of 80% to 90%, inclusive, on five tests in a course to receive a *B*. His grades on the first four tests were 98%, 76%, 86%, and 92%. What grade on the fifth test would qualify him for a *B* in the course?

Solution

i. Let x represent a grade (in percent) on the last test.

iii. Write an inequality expressing the word sentence.

$$80 \leq \frac{98 + 76 + 86 + 92 + x}{5} \leq 90$$

iv. Solve for x.

$$400 \leq 352 + x \leq 450$$

$$48 \leq x \leq 98$$

v. Any grade equal to or greater than 48 and less than or equal to 98 will qualify the student for a *B*.

Review Exercises

35. In the preceding example, what grade on the fifth test would qualify the student for a B if his grades on the first four tests were 78%, 64%, 88%, and 76%?

36. In the example above, what grade on the fifth test would qualify him for a B if his first three test scores were 78%, 96%, and 65% and he missed the fourth test but the instructor agreed to substitute the grade he received on the fifth test for his fourth test score?

37. The Fahrenheit and centigrade temperatures are related by $C = \frac{5}{9}(F - 32)$. Within what range must the temperature be in Fahrenheit degrees for the temperature in centigrade degrees to lie between $-10°$ and $20°$?

38. Within what range must the temperature be in Fahrenheit degrees for the temperature in centigrade degrees to be between $0°$ and $100°$?

39. A ball thrown vertically reaches a height h in feet given by the equation $h = 56t - 16t^2$, where t is time measured in seconds. During what period(s) of time is the ball between 40 feet and 48 feet high?

40. The volume of a sphere is given by the formula $V = 4\pi r^3/3$, where r is the radius. In what range must the radius be if the volume is to be between 2 cubic feet and 4 cubic feet?

Chapter Review

[4.1] Solve each equation.

1. $3 + \dfrac{x}{5} = \dfrac{7}{10}$

2. $\dfrac{3x}{4} - \dfrac{5x - 1}{8} = \dfrac{1}{2}$

3. $\dfrac{x}{x + 1} + \dfrac{1}{3} = 2$

4. $1 - \dfrac{y + 1}{y - 1} = \dfrac{3}{y}$

5. Solve $\dfrac{x + y}{5} = \dfrac{x - y}{3}$ for y in terms of x.

6. Solve $\dfrac{x + y}{5} = \dfrac{x - y}{3}$ for x in terms of y.

[4.2] Solve by factoring.

7. $(x - 2)(x + 1) = 4$

8. $x(x + 4) = 21$

Solve by extraction of roots.

9. $6x^2 = 30$

10. $(x - 5)^2 = 2$

Solve by completing the square.

11. $x^2 + 5x - 2 = 0$

12. $2x^2 + x - 1 = 0$

[4.3] Solve for x by using the quadratic formula.

13. $2x^2 - x + 2 = 0$
14. $x^2 = 4 - 2x$

Solve for x in terms of k by using the quadratic formula.

15. $kx^2 - 3x + 1 = 0$
16. $x^2 + kx - 4 = 0$

Solve each equation or inequality in Exercises 17–26.

[4.4] 17. $x - 5\sqrt{x} + 6 = 0$
18. $\sqrt{x + 1} + \sqrt{x + 8} = 7$

[4.5] 19. $y^4 - 3y^2 + 2 = 0$
20. $y^{-2} - y^{-1} - 42 = 0$

[4.6] 21. $\dfrac{x - 3}{5} \geq 7$
22. $2(x + 1) < \dfrac{1}{3}x$

[4.7] 23. $x^2 + 3x - 10 < 0$
24. $\dfrac{2}{1 - x} > 3$

[4.8] 25. $|3x + 1| = 5$
26. $\left|x - \dfrac{2}{3}\right| = \dfrac{5}{3}$

27. Solve $|x - 3| > 4$ and graph the solution set on a number line.
28. Write $-5 \leq x \leq 3$ using a single inequality involving an absolute-value symbol.

[4.9] 29. In a recent election, the winning candidate received 150 votes more than his opponent. How many votes did each candidate receive if there were 4376 votes cast?

30. When the length of each side of a square is increased by five centimeters, the area is increased by 85 square centimeters. Find the length of a side of the original square.

31. A man sailed a boat across a lake and back in two and a half hours. If his rate returning was two miles per hour less than his rate going, and if the distance each way was six miles, find his rate each way.

32. A salesperson has to average at least $1000 per month in volume over a six month period to qualify for a bonus. If the volumes for the first five months for a person were $850, $1050, $1200, $950 and $1000, how much volume does the salesperson have to do in the sixth month to qualify for the bonus?

Supplemental Exercises for Chapters 1–4

Each of the exercises in this set is more difficult than those at the end of each section and requires the use of several of the techniques learned in the preceding chapters as well as a deeper understanding of the concepts than that required by the regular exercises.

Simplify each expression.

1. $2x\{x + (x + 2)[(x - 1)(x + 2) - 3] + 4\} - (3x^2 + 2x - 1)$

2. $-(x + 2)\{(3x + 1)[x(x - 4) + x^2(2x + 2)] + (x - 1)[(x^2 - 1) + x(x - 1)]\}$

3. $\dfrac{A}{2\sqrt{Ax - B^2}} - \left[\dfrac{B}{1 + [(Ax - B^2)/B^2]} \cdot \dfrac{A}{2B\sqrt{Ax - B^2}}\right]$

4. $\dfrac{1}{2}[w^3(9w + 1)^5]^{-1/2}[w^3(5)(9w + 1)^4(9) + (9w + 1)^5(3w^2)]$

Write each given rational expression as a rational expression in lowest terms. Assume no denominator is zero.

5. $\dfrac{\dfrac{4(x + h)}{1 + (x + h)^2} - \dfrac{4x}{(1 + x^2)}}{h}$

6. $\dfrac{1}{h}\left[\dfrac{2(x + h) + 3}{(x + h) - 1} - \dfrac{2x + 3}{x - 1}\right]$

7. $(x^3 + 2)(-3)(x^2 + 1)^{-4}(2x) + (x^2 + 1)^{-3}(2)(x^3 + 2)(3x^2)$

8. $\dfrac{(x^3 - 1)^2(3)(x^2 + 1)^2(2x) - (x^2 + 1)^3(2)(x^3 - 1)(3x^2)}{(x^3 - 1)^4}$

Write each expression as an equivalent fraction with all exponents positive.

9. $\dfrac{1}{9}\left[\dfrac{(9-x^2)^{1/2} - \frac{1}{2}x(9-x^2)^{-1/2}(-2x)}{9-x^2}\right]$

10. $(x^2+1)^{1/3}(\frac{1}{2})(x^3-1)^{-1/2}(3x^2) + (x^3-1)^{1/2}(\frac{1}{3})(x^2+1)^{-2/3}(2x)$

11. $\dfrac{(3x+2)^{1/2}(\frac{1}{3})(2x+3)^{-2/3}(2) - (2x+3)^{1/3}(\frac{1}{2})(3x+2)^{-1/2}(3)}{3x+2}$.

12. Assume $x^n + y^n = a^n$ and simplify the expression

$$\dfrac{-(n-1)x^{n-2}y^{n-1} + (n-1)x^{n-1}y^{n-2}(-x^{n-1}/y^{n-1})}{y^{2n-2}}.$$

Find y'' in terms of x and y only and simplify the expression.

13. $y'' = \dfrac{(2y-3x)(3y'-2) - (3y-2x)(2y'-3)}{(2y-3x)^2}$; $y' = \dfrac{3y-2x}{2y-3x}$

14. $y'' = [(y-2x)(1+2y') - (x+2y)(y'-2)](y-2x)^{-2}$; $y' = \dfrac{2x+4y}{2y-4x}$

Find the real solutions of each equation.

15. $(x-2)^3(4)(x+1)^3 + (x+1)^4(3)(x-2)^2 = 0$

16. $2x(x+1)^3(x+2) + 3x^2(x+1)^2(x+2) + x^2(x+1)^3 = 0$

17. $(6x-7)^3(2)(8x^2+9)(16x) + (8x^2+9)^2(3)(6x-7)^2(6) = 0$

18. $\dfrac{m^2}{(3m+1)\sqrt{m^2+1}} + \dfrac{-\sqrt{m^2+1}}{(3m+1)^2} = 0$

19. Solve $\pi R^3\left[\dfrac{2y(R-y)(R+y) - (R+y)^2(R-2y)}{(R-y)^2 y^2}\right] = 0$ for y in terms of R.

20. Solve $\dfrac{-2kA}{x^3} + \dfrac{2kB}{(L-x)^3} = 0$ for x in terms of k, L, A, and B. Assume $A \neq -B$.

21. Solve $\dfrac{2(V_0 - Cx)Cx - C(V_0 - Cx)^2}{C^2 x^2} = 0$ for x in terms of V_0 and C. Assume $C \neq 0$, $C \neq -2$.

22. Solve $\dfrac{2(Ax-B)A(Bx-A)^2 - 2(Bx-A)B(Ax-B)^2}{(Bx-A)^4} = 0$ for x in terms of A and B.
Assume $A \neq 0$, $B \neq 0$, and $B \neq A$.

Solve. Represent the solution set in interval notation and graphically.

23. $\dfrac{4}{3}(x^{1/3} + x^{-2/3}) > 0$

24. $\dfrac{x^{3/2} - (\frac{3}{2})(x-4)x^{1/2}}{x^3} > 0$

25. $\dfrac{5x - 15\sqrt{x^2 + 4}}{75\sqrt{x^2 + 4}} < 0$

26. $x\left(\dfrac{-2x}{2\sqrt{4 - x^2}}\right) + \sqrt{4 - x^2} > 0$

27. $-\dfrac{2}{9}(x^2 + 4)^{-4/3}(4x^2) + (\tfrac{4}{3})(x^2 + 4)^{-1/3} > 0$

28. $2(x^2 + 1)^{2/3} + (\tfrac{2}{3})(x^2 + 1)^{-2/3}(4x^2) > 0$

5 Functions and Graphs

In Chapter 4 the problems with which we dealt involved only one variable. In Chapter 5 we begin to concern ourselves with problems that involve two variables.

5.1 Pairings of Real Numbers

Ordered pairs

The replacement sets of expressions involving one variable are sets of numbers. The replacement sets for expressions involving two variables must contain members which are pairs of numbers, one number to replace each variable.

When the numbers of a number pair are to be considered in a specified order, the pair is called an **ordered pair**, and the pair is denoted by a symbol such as (3, 2), (2, 3), (−1, 5), or (0, 3). Each of the two numbers in an ordered pair is called a **component** of the ordered pair; in the ordered pair (x, y), x is called the **first component** and y is called the **second component**.

$R \times R$, or R^2, and the geometric plane

The most important set of ordered pairs with which we shall be concerned is the set of all ordered pairs of real numbers. This set is called the Cartesian product of R and R and is denoted by $R \times R$ (read "R cross R") or R^2. The fact that each member of R^2 corresponds to a point in the geometric plane, and that the coordinates of each point in the geometric plane are the components of a member of R^2, is the basis for all plane graphing. As you probably recall from your earlier study of algebra, the correspondence between points in the plane and ordered pairs of

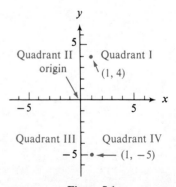

Figure 5.1

5.1 Pairings of Real Numbers

real numbers is usually established through a **Cartesian** (or **rectangular**) **coordinate system**, as shown in Figure 5.1. For this correspondence, the first component of an ordered pair is called the **abscissa** and the second component the **ordinate**.

Solutions of an equation in two variables

Equations in two variables, such as

$$3x + 2y = 12, \quad x^2y + 3x = y^5, \quad \text{and} \quad xy = y^2 - 5,$$

where x and y represent real numbers, have ordered pairs of numbers as solutions. For example, if the components of (2, 3) are substituted for the variables x and y, in that order, in the equation

$$3x + 2y = 12, \tag{1}$$

the result is

$$3(2) + 2(3) = 12,$$

which is true. Hence, (2, 3) is a solution of Equation (1). Since many equations (and inequalities) in two variables have an infinite number of solutions, we shall sometimes use the set-builder notation

$$\{(x, y) | \text{condition on } x \text{ and } y\}$$

to represent the set of all solutions. Thus the solution set of Equation (1) above can be denoted by

$$\{(x, y) | 3x + 2y = 12, \quad x, y \in R\}$$

(read "the set of all ordered pairs (x, y) such that $3x + 2y = 12$ and x and y are real numbers").

Relations

In mathematics and applications of mathematics, we are often concerned with relationships that involve the pairing of numbers. Such relationships as Fahrenheit and Celsius temperatures, standard height and weight tables, and square-root tables are common examples. The following definition gives us a basis for considering such relationships mathematically.

Definition 5.1   A **relation** is a set of ordered pairs.

Relations determined by sets of ordered pairs

It follows from Definition 5.1 that any subset of $R \times R$ is a relation. The set of all first components of the pairs in a relation is called the **domain** of the relation, and the set of all second components is called the **range** of the relation. For example,

$$\{(1, 5), (2, 10), (3, 15)\} \quad \begin{array}{l}\text{elements in the domain} \\ \text{elements in the range}\end{array}$$

is a relation with domain $\{1, 2, 3\}$ and range $\{5, 10, 15\}$.

Relations determined by rules

Relations are often defined by equations or inequalities. Such an equation or inequality can be thought of as a "rule" for finding the value or values in the range of a relation paired with a given value in the domain. For brevity, we shall simply refer to equations or inequalities such as

$$y = \frac{1}{x-2}, \quad \text{or} \quad y = \sqrt{x-2}, \quad \text{or} \quad y > 2x+1$$

as relations.

If a relation is defined by an equation with variables x and y, and the domain (set of values for x) is not specified, we shall understand that *the domain is the set of all real numbers for which a real number exists in the range.*

Examples

Find the domain of each of the following relations.

a. $y = \dfrac{1}{x-2}$ \qquad **b.** $y = \sqrt{x-2}$

Solutions

a. Since $1/(x-2)$ is a real number for every value x except 2, the domain of the relation is $\{x \mid x \neq 2\}$.

b. Since $\sqrt{x-2}$ is a real number only when $x - 2 \geq 0$, the domain of the relation is $\{x \mid x \geq 2\}$.

Functions

From the definition of a relation, each element in the domain is said to be paired with one or more elements in the range. A relation may be a pairing in which one or more elements in the domain are associated with *more than one element in the range*, as shown in Figure 5.2. A relation may also be a pairing in which each element in the domain is associated with *only one element in the range*, as illustrated in Figures 5.3-a and 5.3-b. Such a relation is called a **function**.

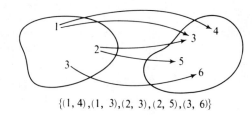

$\{(1, 4), (1, 3), (2, 3), (2, 5), (3, 6)\}$

Figure 5.2

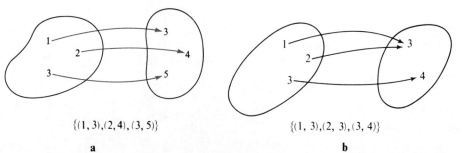

$\{(1, 3), (2, 4), (3, 5)\}$ \qquad $\{(1, 3), (2, 3), (3, 4)\}$

a \qquad b

Figure 5.3

5.1 Pairings of Real Numbers

Definition 5.2 A *function* is a relation in which each element of the domain is paired with one and only one element of the range.

It follows from Definition 5.2 that no two ordered pairs belonging to a function can have the same first components and different second components. For example, both

$$\{(3, 4), (4, 5), (6, 8)\} \qquad (2)$$

and

$$\{(3, 4), (3, 5), (6, 8)\} \qquad (3)$$

are relations. However, only set (2) is a function. Set (3) is not a function because two elements, 4 and 5, in the range are paired with one element, 3, in the domain.

Graphical characterization of a function

A function associates each element in its domain with one and only one element in its range. In a graphical sense, this implies that no two of the ordered pairs in a function have graphs which lie on the same vertical line.

As you should recall from your earlier study of algebra, graphs in $R \times R$ are often continuous lines and curves. Figure 5.4 shows three such graphs. Imagine a

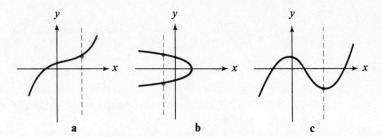

Figure 5.4

vertical line moving across each of these from left to right. If the line at any position cuts the graph of the relation in more than one point, then the relation is not a function. For example, Figures 5.4-a and 5.4-c show the graphs of relations that are functions; Figure 5.4-b shows the graph of a relation that is not a function, because the vertical line shown in the figure meets the graph in two places. This means that for the particular value of x involved, the relation associates two distinct values for y.

Function notation

In general, functions are denoted by single symbols; for example, f, g, h, and F might designate functions. Furthermore, a symbol such as $f(x)$, read "f of x" or "the value of f at x", represents the value in the range of a function f associated with the value x in the domain. This is the same use we made of $P(x)$ in Section 2.1 when we discussed algebraic expressions and, in particular, polynomial expressions.

Examples If $f(x) = x^2 + 3$, find the indicated elements in the range.

a. $f(2)$ **b.** $f(3)$

Solutions Substituting the given elements in the domain for x, we have

a. $f(2) = (2)^2 + 3 = 7$ **b.** $f(3) = (3)^2 + 3 = 12$

For a given function f it is often useful to know what change in the range value is produced by a given change h in the domain value. As the domain value changes from a to $a + h$, the range value changes from $f(a)$ to $f(a + h)$ and the change in the range value is

$$f(a + h) - f(a).$$

Example Find $f(a + h) - f(a)$ for $f(x) = 2x + 1$.

Solution We first compute

$$f(a + h) = 2(a + h) + 1$$
$$= 2a + 2h + 1.$$

We then subtract $f(a)$ to obtain

$$f(a + h) - f(a) = (2a + 2h + 1) - (2a + 1)$$
$$= 2h.$$

In some applications for specific domain values a and b, it is useful to know the difference $f(b) - f(a)$ between two specific range values. We introduce the notation "$f(x)|_a^b$" for that difference. Thus,

$$f(x)|_a^b = f(b) - f(a).$$

Examples For the given function f and values a and b, find $f(x)|_a^b$.

a. $f(x) = x^2$; $a = 1, b = 2$ **b.** $f(x) = 2x - 3$; $a = 0, b = 1$

Solutions **a.** $x^2|_1^2 = (2)^2 - (1)^2$ **b.** $(2x - 3)|_0^1 = (2(1) - 3) - (2(0) - 3)$
$\qquad\qquad = 4 - 1$ $\qquad\qquad\qquad\qquad\qquad = -1 - (-3)$
$\qquad\qquad = 3$ $\qquad\qquad\qquad\qquad\qquad = 2$

Note that $f(x)$ represents an element in the range of a function and hence plays the same role as y in equations involving the variables x and y. For example,

$$y = x + 3 \quad \text{and} \quad f(x) = x + 3$$

define the same function.

Exercise 5.1

A Supply the missing components so that the ordered pairs are solutions of the given equations.

(a) (0,) (b) (1,) (c) (2,) (d) (−3,) (e) $\left(\frac{2}{3}, \right)$

1. $2x + y = 6$
2. $y = 9 - x^2$
3. $y = \dfrac{3x}{x^2 - 2}$
4. $y = 0$
5. $y = \sqrt{3x + 11}$
6. $y = |x - 1|$

Specify the domain of each relation. State whether or not each relation is a function.

Example {(3, 5), (4, 8), (4, 9), (5, 10)}

Solution
(a) Domain (the set of first components): {3, 4, 5}
(b) The relation is not a function, because two ordered pairs, (4, 8) and (4, 9), have the same first components.

7. {(2, 3), (5, 7), (7, 8)}
8. {(−1, 6), (0, 2), (3, 3)}
9. {(2, −1), (3, 4), (3, 6)}
10. {(−4, 7), (−4, 8), (3, 2)}
11. {(5, 5), (6, 6), (7, 7)}
12. {(0, 0), (2, 4), (4, 2)}

Specify the domain of the relation.

Examples
a. $y = \sqrt{16 - x^2}$
b. $y = \dfrac{1}{x(x + 2)}$

Solutions
a. For what values of x is $16 - x^2 \geq 0$?
 The domain is $\{x \mid -4 \leq x \leq 4\}$.
b. For what values of x is $x(x + 2) \neq 0$?
 The domain is $\{x \mid x \neq 0, -2\}$.

13. $y = x + 7$
14. $y = 2x - 3$
15. $y = x^2$
16. $y = \dfrac{1}{x}$
17. $y = \dfrac{1}{x - 2}$
18. $y = \dfrac{1}{x^2 + 1}$
19. $y = \sqrt{x}$
20. $y = \sqrt{4 - x}$
21. $y = \sqrt{4 - x^2}$
22. $y = \sqrt{x^2 - 9}$
23. $y = \dfrac{4}{x(x - 1)}$
24. $y = \dfrac{x}{(x - 1)(x + 2)}$

State whether or not the given equation defines a function.

Examples   **a.** $x^2 y = 3$ **b.** $x^2 + y^2 = 36$

Solutions Solve explicitly for y.

a. $y = \dfrac{3}{x^2}$

Yes. There is only one value of y associated with each value of x ($x \neq 0$).

b. $y^2 = 36 - x^2$, or

$y = \pm\sqrt{36 - x^2}$

No. There are two values of y associated with values of x satisfying $|x| < 6$.

25. $x + y = 3$
26. $y = -x^2$
27. $y = \sqrt{x^2 - 5}$
28. $y = \sqrt{16 - x^2}$
29. $x^2 + y^2 = 16$
30. $y = \pm\sqrt{x^2}$
31. $y^2 = x^3$
32. $y^4 = x^2$

Find the indicated range value for the given function.

Examples   **a.** $f(x) = 3x + 4$, $f(-1)$ **b.** $g(x) = x^2 - 1$, $g(a - 1)$

Solutions   **a.** Substitute -1 for x.

$f(-1) = 3(-1) + 4 = 1.$

b. Substitute $a - 1$ for x.

$g(a - 1) = (a - 1)^2 - 1 = a^2 - 2a.$

33. $f(x) = x + 3$, $f(2)$
34. $f(x) = x - 2$, $f(-1)$
35. $f(x) = -2x + 4$, $f(a + 1)$
36. $f(x) = -\dfrac{3}{2}x + 1$, $f(a - 2)$
37. $g(x) = x^2 + 2x$, $g(-2)$
38. $g(x) = -3x^2 + x + 1$, $f(1)$
39. $g(x) = x^2 + 1$, $g(1/a)$
40. $g(x) = -2x^2 + x - 4$, $g(a + 1)$
41. $s(x) = x^3 - 2x^2 + x - 1$, $s(1)$
42. $s(x) = 3x^3 - 2x^2 - 2x + 4$, $s(-2)$
43. $h(t) = t^4 - 2t^2 + 1$, $h(a - 2)$
44. $h(t) = -3t^4 + 2t^3 + t$, $h(1/a)$
45. $h(x) = \dfrac{1}{x^2 - 4}$, $h(-1)$
46. $h(x) = \dfrac{2x}{x - 3}$, $h(2)$
47. $h(x) = \sqrt{4 - x^2}$, $h(-1)$
48. $h(x) = \dfrac{1}{\sqrt{x^2 - 1}}$, $h(2)$

Find $f(a + h) - f(a)$ for the given function.

Example $f(x) = x^2 + 1$

5.1 Pairings of Real Numbers

Solution First compute

$$f(a + h) = (a + h)^2 + 1$$
$$= a^2 + 2ah + h^2 + 1.$$

Next subtract $f(a)$ and simplify to obtain

$$f(a + h) - f(a) = (a^2 + 2ah + h^2 + 1) - (a^2 + 1)$$
$$= 2ah + h^2.$$

49. $f(x) = 2x - 1$ 50. $f(x) = 4x - 2$ 51. $f(x) = -3x + 2$
52. $f(x) = -5x + 4$ 53. $f(x) = 2 - x^2$ 54. $f(x) = x^2 + 2x$

Compute $f(x)|_a^b$ for the given function and values of a and b.

Examples
a. $f(x) = x^3$; $a = 0, b = 1$ b. $f(x) = \sqrt{x}$; $a = 1, b = 4$

Solutions
a. $x^3|_0^1 = (1)^3 - (0)^3 = 1$ b. $\sqrt{x}|_1^4 = \sqrt{4} - \sqrt{1} = 1$

55. $f(x) = x^2 - 4$; $a = -1, b = 2$ 56. $f(x) = x^4$; $a = -1, b = 1$
57. $f(x) = \sqrt{x}$; $a = 4, b = 9$ 58. $f(x) = \sqrt{x + 1}$; $a = 15, b = 24$

B 59. If $f(x) = x^2 - x + 1$, find and simplify $\dfrac{f(a + h) - f(a)}{h}$, $h \neq 0$.

60. If $f(x) = 1/x$, find and simplify $\dfrac{f(a + h) - f(a)}{h}$, $h \neq 0$.

For each pair of functions find the indicated range values.

Example $f(x) = 2x + 3$, $g(x) = x + 1$ a. $f[g(2)]$ b. $g[f(2)]$

Solution
a. First compute $g(2) = 2 + 1 = 3$, then compute
$$f[g(2)] = f(3) = 2(3) + 3 = 9.$$
b. First compute $f(2) = 2(2) + 3 = 7$, then compute
$$g[f(2)] = g(7) = 7 + 1 = 8.$$

61. $f(x) = x + 4$, $g(x) = x^2$ a. $f[g(3)]$ b. $g[f(3)]$
62. $f(x) = x^2 + 1$, $g(x) = \dfrac{1}{x}$ a. $f[g(1)]$ b. $g[f(1)]$
63. $f(x) = x$, $g(x) = \dfrac{1}{x}$ a. $f[g(a)]$ b. $g[f(a)]$
64. $f(x) = \dfrac{1}{x}$, $g(x) = \dfrac{1}{x^2}$ a. $f[g(a + 4)]$ b. $g[f(a + 4)]$

5.2 Linear Equations in Two Variables

A **first-degree equation**, or **linear equation**, in x and y is an equation that can be written equivalently in the form

$$Ax + By + C = 0 \quad (A \text{ and } B \text{ not both } 0). \tag{1}$$

Graphs of first-degree equations

We shall call (1) the **standard form** for a linear equation. The graph of any such equation (technically, of its solution set) in R^2 is a straight line, although we do not prove this here.

Since two distinct points determine a straight line, it is evident that we need find only two solutions of such an equation to determine its graph—that is, the graph of the solution set of the equation. In practice, the two solutions easiest to find are usually those with first and second components, respectively, equal to zero—that is, the solutions $(0, y_1)$ and $(x_1, 0)$.

The numbers x_1 and y_1 are called the **x- and y-intercepts** of the graph. As an example, consider the function defined by the equation

$$3x + 4y = 12. \tag{2}$$

If $y = 0$, we have $x = 4$, and the x-intercept is 4. If $x = 0$, then $y = 3$, and the y-intercept is 3. Thus the graph of (2) appears as in Figure 5.5.

If the graph intersects both axes at the origin, then the intercepts do not represent two separate points. It is then necessary to plot at least one other point at a distance far enough removed from the origin to establish the line with pictorial accuracy.

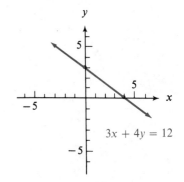

Figure 5.5

Example

Graph $8x + 8y = 0$.

Solution

If $y = 0$, we have $x = 0$, and the x-intercept is 0. Thus the graph intersects both axes at the origin. Hence it is necessary to plot another point on the line. Letting $x = 2$ and solving for y, we obtain $y = -2$. Graphing the points corresponding to $(0, 0)$ and $(2, -2)$ and drawing the line through them we obtain the figure shown.

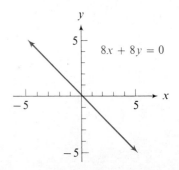

5.2 Linear Equations in Two Variables

Equations of horizontal lines

Two special types of linear equations are worth noting. First, an equation such as

$$y = 4$$

may be considered an equation in two variables:

$$0x + y = 4$$

For each x, this equation assigns $y = 4$. That is, any ordered pair of the form $(x, 4)$ is a solution of the equation. For instance,

$$(1, 4), \quad (2, 4), \quad (3, 4), \quad \text{and} \quad (4, 4)$$

are all solutions of the equation. If we graph these points and connect them with a straight line, we have the graph shown in Figure 5.6-a.

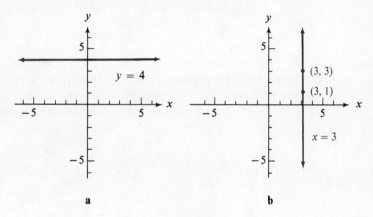

Figure 5.6

Since the equation

$$y = 4$$

assigns to each x the same value for y, the function defined by this equation is called a **constant function**.

Equations of vertical lines

The other special case of the linear equation is of the type

$$x = 3,$$

which may be looked upon as

$$x + 0y = 3.$$

Here, only one value is permissible for x, namely 3, whereas any value may be assigned to y. That is, any ordered pair of the form (, y) is a solution of this equation. If we choose two solutions, say (3, 1) and (3, 3), we can draw the graph shown in Figure 5.6-b. It is clear that this equation does *not* define a function.

Linear functions

If $B = 0$, $A \neq 0$ in Equation (1) as in the example at the bottom of the preceding page, then

$$Ax + C = 0$$

or equivalently

$$x = -\frac{C}{A}.$$

We have noted above that equations of this type do not define functions of x. If, however, $B \neq 0$ in Equation (1), then this equation defines a function. Such a function is called a **linear function**.

Any two distinct points in a plane can be looked upon as the endpoints of a line segment. Two fundamental properties of a line segment are its **length** and its **inclination** with respect to the x-axis.

Directed distance

Let P_1, with coordinates (x_1, y_1), and P_2, with coordinates (x_2, y_2), be endpoints of a line segment. If we construct through P_2 a line parallel to the y-axis, and through P_1 a line parallel to the x-axis, the lines will meet at a point P_3, as shown in Figure 5.7. The x-coordinate of P_3 is evidently the same as the x-coordinate of P_2, and the

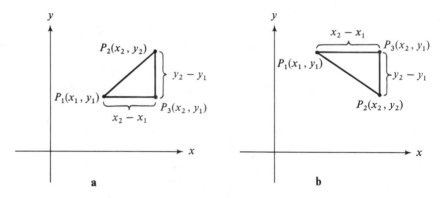

Figure 5.7

y-coordinate of P_3 is the same as that of P_1; hence the coordinates of P_3 are (x_2, y_1). By inspection, we observe that the distance from P_2 to P_3 is simply the absolute value of the difference in the y-coordinates of the two points, $|y_2 - y_1|$, and the distance between P_1 and P_3 is the absolute value of the difference of the x-coordinates of these points, $|x_2 - x_1|$.

In general, since $y_2 - y_1$ is positive or negative as P_2 is above or below P_3, respectively, and $x_2 - x_1$ is positive or negative as P_3 is to the right or left of P_1, respectively, it is also convenient to consider the distances from P_1 to P_3, and from P_3 to P_2 as positive or negative according as $x_2 - x_1$ and $y_2 - y_1$ are positive or negative, respectively. The values $x_2 - x_1$ and $y_2 - y_1$ are called the **directed distances** from P_1 to P_3 and from P_3 to P_2, respectively.

5.2 Linear Equations in Two Variables

Distance between two points

The Pythagorean theorem can be used to find the length, d, of the line segment joining P_1 and P_2. This theorem asserts that the square on the hypotenuse of any right triangle is equal to the sum of the squares on the legs. Thus, we have

$$d^2 = (x_2 - x_1)^2 + (y_2 - y_1)^2,$$

and by considering only the positive (or nonnegative) square root of the right-hand member, we obtain the **distance formula**

$$d = \sqrt{(x_2 - x_1)^2 + (y_2 - y_1)^2}. \tag{3}$$

Since the distances $x_2 - x_1$ and $y_2 - y_1$ are squared, it makes no difference here whether they are positive or negative—the result is the same. Equation (3) is a formula for the distance between any two points in the plane in terms of the coordinates of the points. The distance is always positive—or 0 if the points coincide.

If the points P_1 and P_2 lie on the same horizontal line, then the distance between them is

$$\sqrt{(x_2 - x_1)^2 + (y_1 - y_1)^2} = \sqrt{(x_2 - x_1)^2}$$
$$= |x_2 - x_1|;$$

and if they lie on the same vertical line, then the distance is

$$\sqrt{(x_1 - x_1)^2 + (y_2 - y_1)^2} = \sqrt{(y_2 - y_1)^2}$$
$$= |y_2 - y_1|.$$

Slope of a line segment

The second useful property of the line segment joining two points, its inclination, can be measured by comparing the *rise* of the segment with its *run*, as shown in Figure 5.8.

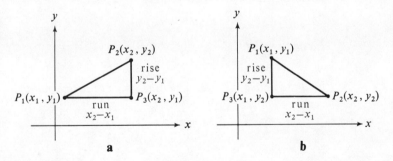

Figure 5.8

The ratio of *rise* to *run* is called the **slope** of the line segment and is designated by the letter m. Since the rise is simply $y_2 - y_1$ and the run is $x_2 - x_1$, the slope of the line segment joining P_1 and P_2 is given by the slope formula

$$m = \frac{y_2 - y_1}{x_2 - x_1} \quad (x_2 \neq x_1).$$

If P_2 is to the right of P_1, $x_2 - x_1$ will necessarily be positive, and the slope will be positive or negative as $y_2 - y_1$ is positive or negative. Thus positive slope indicates that a line rises to the right; negative slope indicates that it falls to the right. Since

$$\frac{y_2 - y_1}{x_2 - x_1} = \frac{-(y_1 - y_2)}{-(x_1 - x_2)} = \frac{y_1 - y_2}{x_1 - x_2},$$

the restriction that P_2 be to the right of P_1 is not necessary, and the order in which the points are considered is immaterial in determining slope.

If a line segment is parallel to the x-axis, then $y_2 - y_1 = 0$, and the line has slope 0; but if it is parallel to the y-axis, then $x_2 - x_1 = 0$, and its slope is not defined. These two special cases are shown in Figure 5.9.

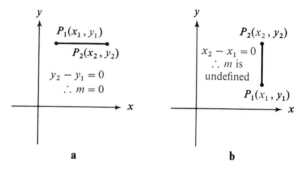

Figure 5.9

In the next section, we shall see how the slope concept is applied in discussing linear functions and their graphs.

Exercise 5.2

A Find the x- and y-intercepts of the graph of each equation.

Example $2x + 6y = 4$

Solution   For $y = 0$ we have $2x = 4$, or $x = 2$. Thus, the x-intercept is 2.
For $x = 0$ we have $6y = 4$, or $y = 2/3$. Thus, the y-intercept is 2/3.

1. $3x + 2y = 6$
2. $4x + y = 4$
3. $-2x + y = 1$
4. $-3x - 2y = 4$
5. $2x - 3y = 0$
6. $-3x + 6y = 0$
7. $2x = 4y$
8. $-5x = 12y$
9. $x = 3$
10. $-2x = 4$
11. $y = -3$
12. $-4y = 12$

5.2 Linear Equations in Two Variables

Example

Graph.

$3x + 4y = 24$

Solution

Determine the intercepts.

If $x = 0$, then $y = 6$;

if $y = 0$, then $x = 8$.

Sketch the line through (0, 6) and (8, 0), as shown.

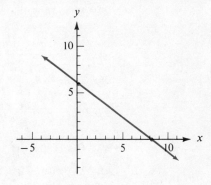

13. $y = 3x + 1$
14. $y = x - 5$
15. $3x - y = -2$
16. $2x + y = 3$
17. $y = -2x$
18. $3x = 2y$
19. $2x + 3y = 6$
20. $3x - 2y = 8$
21. $2x + 5y = 10$
22. $2x + 3y = 9$
23. $x = -2$
24. $y = -3$

Example

Graph $f(x) = x - 1$. Represent $f(5)$ and $f(3)$ by drawing line segments from (5, 0) to (5, $f(5)$) and from (3, 0) to (3, $f(3)$).

Solution

$f(5) = 5 - 1 = 4$

The ordinate at $x = 5$ is 4.

$f(3) = 3 - 1 = 2$

The ordinate at $x = 3$ is 2.

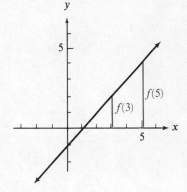

25. Plot the graph of $f(x) = 2x + 4$. Represent $f(0)$ and $f(4)$ by drawing line segments from (0, 0) to (0, $f(0)$) and from (4, 0) to (4, $f(4)$).

26. Plot the graph of $f(x) = 2x + 1$. Represent $f(3)$ and $f(-2)$ by drawing line segments from (3, 0) to (3, $f(3)$) and from (-2, 0) to (-2, $f(-2)$).

Find the distance between each of the given pairs of points, and find the slope of the line segment joining them.

Example

(3, -5), (2, 4)

Solution

Consider (3, -5) as P_1 and (2, 4) as P_2.

$$d = \sqrt{(x_2 - x_1)^2 + (y_2 - y_1)^2}$$

$$= \sqrt{[2 - 3]^2 + [4 - (-5)]^2}$$

$$= \sqrt{1 + 81}$$

$$m = \frac{y_2 - y_1}{x_2 - x_1}$$

$$= \frac{4 - (-5)}{2 - 3}$$

$$= \frac{9}{-1}$$

Distance, $\sqrt{82}$; slope -9

27. $(1, 1), (4, 5)$	28. $(-1, 1), (5, 9)$	29. $(-3, 2), (2, 14)$
30. $(-4, -3), (1, 9)$	31. $(2, 1), (1, 0)$	32. $(-3, 2), (0, 0)$
33. $(5, 4), (-1, 1)$	34. $(2, -3), (-2, -1)$	35. $(3, 5), (-2, 5)$
36. $(2, 0), (-2, 0)$	37. $(0, 5), (0, -5)$	38. $(-2, -5), (-2, 3)$

Find the lengths of the sides of the triangle having vertices as given.

39. $(10, 1), (3, 1), (5, 9)$	40. $(0, 6), (9, -6), (-3, 0)$
41. $(5, 6), (11, -2), (-10, -2)$	42. $(-1, 5), (8, -7), (4, 1)$

B 43. Show that the triangle described in Exercise 40 is a right triangle. *Hint*: Use the converse of the Pythagorean theorem.

44. The two line segments with endpoints at $(0, -7), (8, -5)$ and $(5, 7), (8, -5)$ are perpendicular. Find the slope of each line segment. Compare the slopes. Do the same for the perpendicular line segments with endpoints at $(8, 0), (6, 6)$ and $(-3, 3), (6, 6)$. Can you make a conjecture about the slopes of perpendicular line segments?

45. The graph of a linear function contains the points $(2, -3)$ and $(6, -1)$. Find an equation that defines the function.

46. Determine algebraically whether or not the points lie on the same line.

    a. $(2, 7), (-2, -5), (0, 1)$ **b.** $(9, 5), (-3, -1), (0, 1)$

47. Show by similar triangles that the coordinates of the midpoint of the line segment joining the points $P_1(x_1, y_1)$ and $P_2(x_2, y_2)$ are given by

$$x = \frac{x_1 + x_2}{2} \quad \text{and} \quad y = \frac{y_1 + y_2}{2}.$$

48. Using the results of Exercise 47, find the coordinates of the midpoint of the line segment joining:

    a. $(2, 4)$ and $(6, 8)$ **b.** $(-4, 6)$ and $(6, -10)$

49. Suppose the function defined by $y = 2x + 1$ has as domain the set of real numbers between and including -1 and 1. Plot the graph of the function on a rectangular coordinate system. What is the range of the function?

50. Graph $x + y = 6$ and $5x - y = 0$ on the same set of axes. Estimate the coordinates of the point of intersection. What can you say about the coordinates of this point in relation to the two linear equations?

5.3   Forms for Linear Equations

Point-slope form

Assuming that the slope of the line segment joining any two points on a line does not depend on the points (see Exercise 41), consider a line in the plane with given slope m that passes through a given point (x_1, y_1), as shown in Figure 5.10. If

5.3 Forms for Linear Equations

we choose any other point on the line and assign to it the coordinates (x, y), it is evident that the slope of the line is given by

$$\frac{y - y_1}{x - x_1} = m,$$

from which

$$y - y_1 = m(x - x_1). \qquad (1)$$

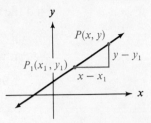

Figure 5.10

Note that (1) is satisfied also by $(x, y) = (x_1, y_1)$. Since now x and y are the coordinates of *any* point on the line, (1) is an equation of the line passing through (x_1, y_1) with slope m. This is called the **point-slope form** for a linear equation. Any linear equation in point-slope form can be written equivalently in standard form.

Example Find an equation in standard form of the line with slope 3/4 passing through the point $(-4, 1)$.

Solution Substituting 3/4, -4, and 1 for m, x_1, and y_1, respectively, in Equation (1), we have

$$y - 1 = \frac{3}{4}(x - (-4)),$$

$$y - 1 = \frac{3}{4}x + 3,$$

from which

$$3x - 4y + 16 = 0.$$

Slope-intercept form

Now consider the equation of the line with slope m passing through a given point on the y-axis having coordinates $(0, b)$, as shown in Figure 5.11. Substituting the components of $(0, b)$ in the point-slope form of a linear equation,

$$y - y_1 = m(x - x_1),$$

we obtain

$$y - b = m(x - 0),$$

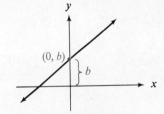

Figure 5.11

from which

$$y = mx + b. \qquad (2)$$

Equation (2) is called the **slope-intercept form** for a linear equation. Any linear equation in standard form can be written equivalently in the slope-intercept form by solving for y in terms of x if $B \neq 0$ in Equation (1) of Section 5.2.

Example Find the slope and *y*-intercept of the line with equation $2x + 3y - 6 = 0$.

Solution We can write $2x + 3y - 6 = 0$ equivalently as

$$y = -\frac{2}{3}x + 2.$$

Hence, we can now read the slope of the line, $-2/3$, and the *y*-intercept, 2, directly from the last form of the equation.

Intercept form If the *x*- and *y*- intercepts of the graph of

$$y = mx + b \qquad (3)$$

are *a* and *b* ($a, b \neq 0$), respectively, as shown in Figure 5.12, then the slope *m* is clearly equal to $-b/a$. Replacing *m* in Equation (3) with $-b/a$, we have

$$y = -\frac{b}{a}x + b,$$

$$ay = -bx + ab,$$

$$bx + ay = ab,$$

and multiplying each member by $\dfrac{1}{ab}$ produces

$$\frac{x}{a} + \frac{y}{b} = 1. \qquad (4)$$

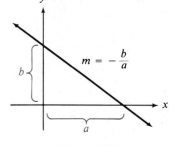

Figure 5.12

Equation (4) is called the **intercept form** for a linear equation.

Example Find an equation in standard form of the line with *x*-intercept 3 and *y*-intercept -2.

Solution We substitute 3 and -2 for *a* and *b*, respectively, in $x/a + y/b = 1$, to obtain

$$\frac{x}{3} + \frac{y}{-2} = 1,$$

$$-2x + 3y + 6 = 0,$$

or

$$2x - 3y - 6 = 0.$$

Any of the four forms discussed in this section may be used in working with linear functions and their graphs.

5.3 Forms for Linear Equations

Exercise 5.3

A *Find the equation, in standard form, of the line passing through each of the given points and having the given slope.*

Example $(3, -5)$, $m = -2$

Solution Substitute given values in the point-slope form of the linear equation.
$$y - y_1 = m(x - x_1)$$
$$y - (-5) = -2(x - 3)$$
$$y + 5 = -2x + 6$$
$$2x + y - 1 = 0$$

1. $(2, 1)$, $m = 4$
2. $(-2, 3)$, $m = 5$
3. $(5, 5)$, $m = -1$
4. $(-3, -2)$, $m = \dfrac{1}{2}$
5. $(0, 0)$, $m = 3$
6. $(-1, 0)$, $m = 1$
7. $(0, -1)$, $m = -\dfrac{1}{2}$
8. $(2, -1)$, $m = \dfrac{3}{4}$
9. $(-2, -2)$, $m = -\dfrac{3}{4}$
10. $(2, -3)$, $m = 0$
11. $(-4, 2)$, $m = 0$
12. $(-1, -2)$, parallel to y-axis

Write each equation in slope-intercept form; specify the slope of the line and the y-intercept.

Example $2x - 3y = 5$

Solution Solve explicitly for y.
$$-3y = 5 - 2x$$
$$3y = 2x - 5$$
$$y = \dfrac{2}{3}x - \dfrac{5}{3}$$

Compare with the general slope-intercept form $y = mx + b$.
Slope, $2/3$; y-intercept, $-5/3$.

13. $x + y = 3$
14. $2x + y = -1$
15. $3x + 2y = 1$
16. $3x - y = 7$
17. $x - 3y = 2$
18. $2x - 3y = 0$
19. $4x - 3y = 7$
20. $-2x + 6y = 8$
21. $y - 2 = 4$
22. $3y - 4 = 6$
23. $x = 5$
24. $2x - 5 = 0$

Find the equation, in standard form, of the line with the given intercepts. The x-intercept is given first.

Example $\quad$ 3; $-1/2$

Solution $\quad$ Substitute 3 and $-1/2$ for a and b, respectively, in the intercept form $\dfrac{x}{a} + \dfrac{y}{b} = 1$.

$$\frac{x}{3} + \frac{y}{-1/2} = 1$$

$$x - 6y - 3 = 0$$

25. 2; 3 $\qquad$ 26. 4; -1 $\qquad$ 27. -2; -5

28. -1; 7 $\qquad$ 29. $-\dfrac{1}{2}; \dfrac{3}{2}$ $\qquad$ 30. $\dfrac{2}{3}; -\dfrac{3}{4}$

Write the equation, in standard form, of the line passing through the given point and parallel to the graph of the given equation. Hint: Parallel lines have the same slope.

31. $(2, 1)$; $\quad y = 3x + 4$ $\qquad$ 32. $(-3, 2)$; $\quad y = -4x + 9$

33. $(-2, 5)$; $\quad 2x - 3y = 6$ $\qquad$ 34. $(-5, -1)$; $\quad 5x - 4y = 7$

35. $(2, 3)$; $\quad x = 6$ $\qquad$ 36. $(5, -4)$; $\quad y = 2$

B $\quad$ 37. Show that, for $x_2 \neq x_1$,

$$y - y_1 = \left(\frac{y_2 - y_1}{x_2 - x_1}\right)(x - x_1)$$

is an equation of the line joining the points (x_1, y_1) and (x_2, y_2). This is the **two-point form** of the linear equation.

In Exercises 38 and 39 use the form given in Exercise 37 to find the equation of the line through the given points.

38. $(2, 1)$ and $(-1, 3)$ $\qquad$ 39. $(3, 0)$ and $(5, 0)$

40. Consider the linear function $y = F(x)$. Assuming that $(2, 3)$ and $(-1, 4)$ are known to be in F, find $F(x)$ in terms of x.

41. Show that the slope of the line segment joining any two points on a line does not depend on the points.

5.4 Inverse Functions

In Section 5.1 we considered rules that can be used to find the elements in the range of a function that are paired with given elements of the domain. It is sometimes useful to be able to reverse this procedure. In this section, for a given function, we discuss how to find a rule that gives us the element or elements in the domain that are paired with a given element in the range. This rule will determine the inverse of the given function.

Inverse functions

By Definition 5.2, a function is a set of ordered pairs (x, y) such that no two have the same first components and different second components. When the components of every ordered pair in a function f are interchanged, the resulting relation may or may not be a function. For example, if

$$(1, 5), \quad (2, 5) \quad \text{and} \quad (3, 6)$$

are ordered pairs in f, then

$$(5, 1), \quad (5, 2), \quad \text{and} \quad (6, 3)$$

are members of the relation formed by interchanging the components of these ordered pairs. Clearly these latter pairs cannot be members of a function, since two of them, $(5, 1)$ and $(5, 2)$, have the same first components and different second components. If, however, no two ordered pairs of a function f have the same second component, the function is called **one-to-one** and the function obtained by interchanging the first and second components of every pair in f is also a function. This new function is called the *inverse function* of f. The notation f^{-1} (read "f inverse") is frequently used to denote the inverse function of f.

Definition 5.3 *If a function is a one-to-one function, then the set of ordered pairs obtained by interchanging the first and second components of each ordered pair in f is called the **inverse function of f** and is denoted f^{-1}.*

Example If $f = \{(2, 1), (3, 2), (4, 3)\}$, find f^{-1}.

Solution First note that f is a one-to-one function. Interchange the components of each ordered pair of f:

$$f^{-1} = \{(1, 2), (2, 3), (3, 4)\}.$$

It is evident from Definition 5.3 and the example above that the domain and range of f^{-1} are just the range and domain, respectively, of f. Thus, if $y = f(x)$ defines a function f, and if f is one-to-one, then the inverse function f^{-1} is obtained as follows:

1. Interchange x and y in $y = f(x)$ to write $x = f(y)$.
2. Solve $x = f(y)$ for y to obtain $y = f^{-1}(x)$.

Example Find the inverse of the function defined by $y = 4x - 3$.

Solution Interchanging the roles of x and y, we have

$$x = 4y - 3.$$

Solving for y in terms of x, we have

$$y = f^{-1}(x) = \frac{1}{4}(x + 3).$$

Graphs of inverse functions

The graphs of inverse functions are related in an interesting way. To see this, we first observe in Figure 5.13 that the line $y = x$ is the perpendicular bisector of the line segment joining the graphs of the ordered pairs (a, b) and (b, a). Therefore, because for every ordered pair (a, b) in f the ordered pair (b, a) is in f^{-1}, the graphs of $y = f^{-1}(x)$ and $y = f(x)$ are reflections of each other about the graph of $y = x$.

Figure 5.14 shows the graphs of the linear function

$$y = f(x) = 4x - 3$$

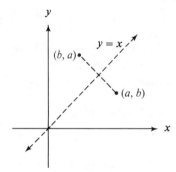

Figure 5.13

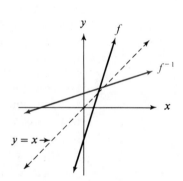

Figure 5.14

and its inverse

$$y = f^{-1}(x) = \frac{1}{4}(x + 3),$$

together with the graph of $y = x$.

$f^{-1}[f(x)] = x$
and
$f[f^{-1}(x)] = x$

Since an element in the domain of the inverse f^{-1} of a function f is the range element in the corresponding ordered pair of f and vice versa, it follows that for every x in the domain of f,

$$f^{-1}[f(x)] = x$$

(read "f inverse of f of x is equal to x"), and for every x in the domain of f^{-1},

$$f[f^{-1}(x)] = x$$

5.4 Inverse Functions

(read "f of f inverse of x is equal to x"). Using Equation (3) above as an example, we note that if f is the linear function defined by

$$f(x) = 4x - 3,$$

then f is a one-to-one function. Now f^{-1} is defined by

$$f^{-1}(x) = \frac{1}{4}(x + 3),$$

and we see that

$$f^{-1}[f(x)] = \frac{1}{4}[(4x - 3) + 3] = x$$

and

$$f[f^{-1}(x)] = 4\left[\frac{1}{4}(x + 3)\right] - 3 = x.$$

Exercise 5.4

A Graph each given function and its inverse function, using the same set of axes.

Example $\{(4, 1), (-2, 3), (1, 2)\}$

Solution Interchange the components of each ordered pair in f to obtain the inverse $\{(1, 4), (3, -2), (2, 1)\}$.

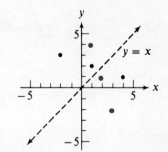

1. $\{(-2, 3), (4, 7), (5, 9)\}$
2. $\{(-3, 1), (2, -1), (3, 4)\}$
3. $\{(-1, -1), (2, 2), (3, 3)\}$
4. $\{(-4, 4), (0, 0), (4, -4)\}$

Example $x + 3y = 6$

Solution Interchange the variables in the defining equation to obtain the equation defining the inverse.

$$y + 3x = 6$$

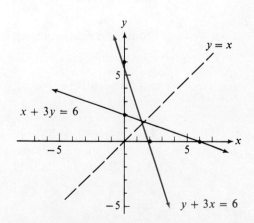

5. $y = 2x + 6$
6. $y = 3x - 6$
7. $y = 4 - 2x$

8. $y = 6 + 3x$ 9. $3x - 4y = 12$ 10. $x - 6y = 6$

11. $4x + y = 4$ 12. $2x - 3y = 12$

B In Exercise 13–20, each equation defines a one-to-one function f. Find an equation defining f^{-1} and show that $f[f^{-1}(x)] = f^{-1}[f(x)] = x$. Hint: Solve explicitly for y and let $y = f(x)$.

13. $y = x$ 14. $y = -x$ 15. $2x + y = 4$ 16. $x - 2y = 4$

17. $3x - 4y = 12$ 18. $3x + 4y = 12$ 19. $yx = 1$ 20. $x^3 - y = 1$

21. If $f^{-1}(4) = 2$, find $f(2)$.   22. If $f^{-1}(0) = 1$, find $f(1)$.

23. If $f(2) = 4$, find $f^{-1}(4)$.   24. If $f(0) = 0$, find $f^{-1}(0)$.

25. Show that the line $y = x$ is the perpendicular bisector of the line segment between the points (a, b) and (b, a) for every value of a and b. Notice that this proves that the graphs of $y = f(x)$ and $y = f^{-1}(x)$ are reflections of one another in the line $y = x$. Hint: If the products of the slopes of two lines is -1 then the lines are perpendicular.

26. Show that the graph of the inverse of a linear function is a straight line.

5.5   Special Functions

Absolute-value functions

Functions involving the absolute value of one or both of the variables not only are useful in more advanced courses in mathematics; they also offer interesting properties in their own right. We recall that

$$|x| = \begin{cases} x, & \text{if } x \geq 0, \\ -x, & \text{if } x < 0. \end{cases}$$

Now consider the function defined by

$$y = |x|. \tag{1}$$

From the definition of $|x|$, for nonnegative x we have

$$y = x, \quad x \geq 0, \tag{2}$$

and for negative x,

$$y = -x, \quad x < 0. \tag{3}$$

If we graph (2) and (3) on the same set of axes, we have the graph of $y = |x|$ shown in Figure 5.15 on page 157. In the case of any equation involving $|x|$ or $|f(x)|$, we can always plot individual points to deduce the graph.

For instance, if

$$y = |x| + 1, \tag{4}$$

5.5 Special Functions

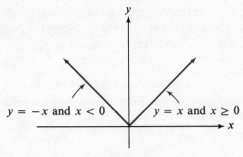

Figure 5.15

we can find solutions by assigning values to x and computing values for y. Some solutions of (4) are

$$(-2, 3), (-1, 2), (0, 1), (1, 2), \text{ and } (2, 3),$$

which can be graphed as in Figure 5.16-a. The graph of $y = |x| + 1$ where $x \in R$, appears in Figure 5.16-b.

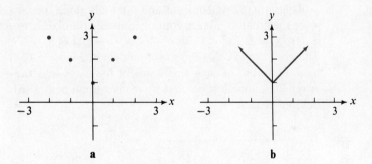

Figure 5.16

As an alternative approach, the definition of $|x|$ implies that

$$y = |x| + 1$$

is equivalent to

$$y = x + 1 \quad \text{for} \quad x \geq 0,$$

and

$$y = -x + 1 \quad \text{for} \quad x < 0,$$

and these can be graphed separately over the specified domains.

Related graphs It is usually advisable, where possible, to avoid graphing large numbers of points. For instance, comparing Equations (1) and (4), that is,

$$y = |x| \tag{1}$$

and

$$y = |x| + 1, \tag{4}$$

we observe that for each x the ordinate in (4) is one unit greater than that in (1); consequently, the graph of (4) is simply the graph of (1) with each ordinate increased by one unit. Whenever we can, we should use such considerations to help us graph equations.

Bracket function

Another interesting function (sometimes called the **bracket function**) is defined by the equation

$$f(x) = [x], \qquad (5)$$

where the brackets denote "the greatest integer contained in," or "the greatest integer not greater than." Thus,

$$[2] = 2, \quad \left[\frac{7}{4}\right] = 1, \quad [-2] = -2,$$

$$\left[\frac{-3}{2}\right] = -2, \quad \text{and} \quad \left[-\frac{5}{2}\right] = -3.$$

To graph (5), we consider unit intervals along the x-axis. If $0 \leq x < 1$, $[x]$ is 0, since the greatest integer contained in any number between 0 and 1 is 0. Similarly, if $1 \leq x < 2$, $[x]$ is 1; for $-2 \leq x < -1$, $[x]$ is -2; etc., as shown in Table 5.1. The graph of (5), therefore, is as shown in Figure 5.17. The heavy dot on the left-hand endpoint of each line segment indicates that the endpoint is a part of the graph. The function defined by (5) is sometimes called a "step function," for an obvious reason.

$x \in$	$[x]$
$[-2, -1)$	-2
$[-1, 0)$	-1
$[0, 1)$	0
$[1, 2)$	1
$[2, 3)$	2

Table 5.1

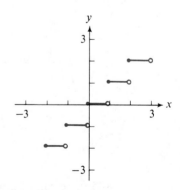

Figure 5.17

Functions defined piecewise

The graph of a function of the following type is usually obtained by graphing it separately over each subset of the domain determined by the definition:

$$y = f(x) = \begin{cases} x & \text{if } x \leq 1 \\ -x + 2 & \text{if } x > 1. \end{cases} \qquad (6)$$

5.5 Special Functions

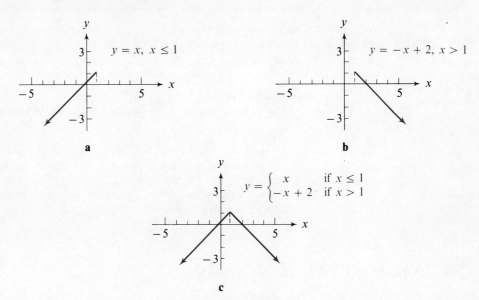

Figure 5.18

Thus, to graph (6) we first graph $y = f(x) = x$ for $x \leq 1$ to obtain Figure 5.18-a. We then graph $y = f(x) = -x + 2$ over $x > 1$ to obtain Figure 5.18-b. We then sketch the graphs in Figures 5.18-a and 5.18-b on the same set of coordinate axes to obtain Figure 5.18-c, which shows the completed graph of (6).

Exercise 5.5

A Graph the function defined by the given equation. For Exercises 9–12 graph the function over the interval $-5 \leq x \leq 5$.

1. $y = |x| + 2$
2. $y = -|x| + 3$
3. $f(x) = |x + 1|$
4. $f(x) = |x - 2|$
5. $y = -|2x - 1|$
6. $y = |3x + 2|$
7. $g(x) = |2x| - 3$
8. $y = |3x| + 2$
9. $y = 2[x]$
10. $y = [2x]$
11. $y = [x + 1]$
12. $f(x) = [x] + 1$
13. $f(x) = \begin{cases} -x & \text{if } x \leq -1 \\ 3x + 4 & \text{if } x > -1 \end{cases}$
14. $f(x) = \begin{cases} -2x + 1 & \text{if } x \leq 2 \\ -3 & \text{if } x > 2 \end{cases}$
15. $f(x) = \begin{cases} x - 2 & \text{if } x \leq 1 \\ x/2 - 3/2 & \text{if } x > 1 \end{cases}$
16. $f(x) = \begin{cases} -x + 1 & \text{if } x \leq 0 \\ x - 2 & \text{if } x > 0 \end{cases}$

B
17. $y = 3|x| - x$
18. $y = -2|x| + x$
19. $y = |x + 1| - |x|$
20. $y = |x + 1| + |x|$
21. $y = ||x| - 1|$
22. $y = ||2x| - 4|$

23. $f(x) = \begin{cases} x & \text{if } x \leq 1 \\ 1 & \text{if } 1 < x \leq 2 \\ -x + 3 & \text{if } x > 2 \end{cases}$ 24. $f(x) = \begin{cases} 2x + 3 & \text{if } x \leq 0 \\ 2x & \text{if } 0 < x \leq 1 \\ 2 & \text{if } x > 1 \end{cases}$

25. The postage on a letter sent by first-class mail is c cents per ounce or fraction thereof. Write an equation relating the cost (C) of mailing a letter and the weight of the letter in ounces (x).

26. A travel agency is offering a charter flight from Los Angeles to Paris. The round trip fare is $1,000 per person if more than 9 but fewer than 26 take the charter. The fare per person decreases $2.50 for every person after the 25th and up to the 75th, and it decreases $3.00 for every person after the 75th. If the capacity of the plane is 150 persons, graph the fare per person as a function of the number of persons taking the charter flight.

5.6 Graphs of First-Degree Inequalities

An inequality of the form

$$Ax + By + C \leq 0 \quad \text{or} \quad Ax + By + C < 0,$$

A and B not both 0, defines a relation.

Graph of an inequality

The graphs of such relations will be a *region* of the plane. For example, consider the relation

$$2x + y - 3 < 0. \tag{1}$$

When the inequality is rewritten in the equivalent form

$$y < -2x + 3, \tag{2}$$

we see that solutions (x, y) are such that for each x, y is less than $-2x + 3$. The graph of the equation

$$y = -2x + 3 \tag{3}$$

is simply a straight line, as illustrated in Figure 5.19-a. To graph inequality (1), we need only observe that any point below this line has a y-coordinate that satisfies (2), and consequently the solution set of (2) corresponds to the entire region below the line. The region is indicated on the graph with shading. That the line itself is not in the graph is shown by means of a broken line, as in Figure 5.19-b. Had the inequality been

$$2x + y - 3 \leq 0,$$

the line would be a part of the graph and would be shown as a solid line. In general, the graphs of the inequalities

$$Ax + By + C < 0 \quad \text{or} \quad Ax + By + C > 0$$

5.6 Graphs of First-Degree Inequalities

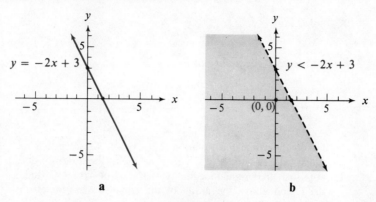

Figure 5.19

are the points in a half-plane on one side of the graph of the associated equation

$$Ax + By + C = 0,$$

depending on the constants and inequality symbols involved.

To determine which half-plane to shade in constructing graphs of first-degree inequalities, one can select any point in either half-plane and test its coordinates in the given inequality to see whether or not the selected point lies in the graph. If the coordinates satisfy the inequality, then the half-plane containing the selected point is shaded; if not, the opposite half-plane is shaded. A very convenient point to use in this process is the origin, provided the origin is not contained in the graph of the associated equation.

Example   Graph $3x + 4y \geq 12$.

Solution   Graph the associated equation

$$3x + 4y = 12.$$

If $x = 0$, $y = 3$; if $y = 0$, $x = 4$. The line is the edge of the half-plane that is the graph of the inequality. Observe that the origin is *not* part of the graph of the inequality because $(0, 0)$ does not satisfy the original inequality.

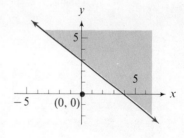

$$3(0) + 4(0) \not\geq 12$$

Hence, the region above the line is shaded. Note that the line is included in the graph.

Inequalities do not ordinarily define functions, according to our definition in Section 5.1, because it usually is not true that each element of the domain is associated with a unique element in the range.

Exercise 5.6

A Graph each inequality in R^2.

Example $2x + y \geq 4$

Solution Graph the associated equation $2x + y = 4$. Observe that the origin is not part of the graph of the inequality, because $(0, 0)$ does not satisfy the original inequality; that is, $2(0) + (0) \not\geq 4$. Hence the region above the graph of the equation $2x + y = 4$ is shaded. The line is included in the graph.

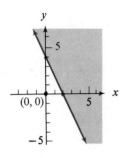

1. $y < x$
2. $y > x$
3. $y \leq x + 2$
4. $y \geq x - 2$
5. $x + y < 5$
6. $2x + y < 2$
7. $x - y < 3$
8. $x - 2y < 5$
9. $x \leq 2y - 4$
10. $2x \leq y + 1$
11. $3 \geq 2x - 2y$
12. $0 \geq x + y$

Example $x > 2$

Solution Graph the associated equation $x = 2$.

Shade region to the right of the graph of $x = 2$, that is, the set of all points such that $x > 2$.

The line is excluded from the graph.

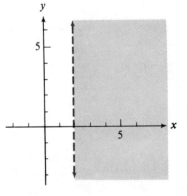

13. $x > 0$
14. $y < 0$
15. $x < 0$
16. $x < -2$
17. $-1 < x < 5$
18. $0 \leq y \leq 1$
19. $|x| < 3$
20. $|y| > 1$

B Graph each inequality. Hint: Consider the graphs in each quadrant separately: $x, y \geq 0$; $x \leq 0, y \geq 0$; $x, y \leq 0$; and $x \geq 0, y \leq 0$.

21. $|x| + |y| \leq 1$
22. $|x| + |y| \geq 1$
23. $|x| - |y| \leq 1$
24. $|x| - |y| \geq 1$
25. $|x| - |y| \geq 0$
26. $|x| - |y| \leq 0$
27. $|x + y| \geq 1$
28. $|x + y| \leq 1$

Chapter Review

[5.1] *Specify the domain of each relation.*

1. $\{(4, -1), (2, -4), (3, -5)\}$
2. $\{(1, 2), (1, 3), (1, 6)\}$
3. $y = \dfrac{1}{x + 4}$
4. $y = \sqrt{x - 6}$

Let $f(x) = x - 3$ and $g(x) = x^2 + 4$. Find each of the following.

5. $g(3)$
6. $f(-2)$
7. $f(x - h)$
8. $g(x + h)$

Find the indicated range value for the given function.

9. $f(x) = \dfrac{1}{2x - 1}$, $f(1)$
10. $f(x) = \dfrac{x}{2x + 1}$, $f(0)$
11. $f(x) = \sqrt{4 - x^2}$, $f(0)$
12. $f(x) = \dfrac{1}{\sqrt{x + 1}}$, $f(4)$

Evaluate $f(x)|_2^4$ for each of the following.

13. $f(x) = x^2 + 2$
14. $f(x) = \dfrac{1}{\sqrt{x - 1}}$

[5.2] *Graph.*

15. $3x - 4y = 4$
16. $2x - y = 1$
17. $x - y = 0$
18. $x + y = 0$
19. $x - 2y = 1$
20. $3x + y = 0$

Find the distance between each of the given pairs of points, and find the slope of the line segment joining them.

21. $(2, 0)$ and $(-3, 4)$
22. $(-6, 1)$ and $(-8, 2)$

[5.3] *Find the equation, in standard form, of the line passing through each of the given points and having the given slope.*

23. $(2, -7)$, $m = 4$
24. $(-6, 3)$, $m = \dfrac{1}{2}$

Write each equation in slope-intercept form; specify the slope and the y-intercept of the line.

25. $4x + y = 6$

26. $3x - 2y = 16$

Find the equation, in standard form, of the line with the given intercepts.

27. $x = -3;\ y = 2$

28. $x = \dfrac{1}{3};\ y = -4$

Find an equation of the line which passes through the given point and is parallel to the graph of the given equation.

29. $4x - 3y = 12;\ (-2, 1)$

30. $x + 6y = 0;\ (3, -4)$

[5.4] Graph each function and its inverse using the same set of axes.

31. $\{(-3, 1), (-1, 3), (2, 4)\}$

32. $\{(1, 1), (2, 4), (3, 6)\}$

33. $2x - y = 6$

34. $3x - 4y = 8$

[5.5] Graph each function.

35. $y = \left|\dfrac{x}{5}\right|$

36. $y = -\left|x - \dfrac{1}{4}\right|$

37. $f(x) = \begin{cases} x - 1 & \text{if } x \le 2 \\ \dfrac{1}{2}x & \text{if } x > 2 \end{cases}$

38. $f(x) = \begin{cases} x + 4 & \text{if } x \le -1 \\ -3x & \text{if } x > -1 \end{cases}$

[5.6] Graph each inequality in R^2.

39. $2x - y < 6$

40. $2y + x > 0$

41. $-2 \le y < 3$

42. $3 < x \le 5$

6 Nonlinear Relations and Functions

In Chapter 5 we studied linear, or first-degree, relations and functions. In this chapter we shall continue our study with a consideration of relations and functions whose graphs are curves rather than simple straight lines.

6.1 Quadratic Functions

We begin by considering equations of degree two.

Graph of a quadratic function

Consider the quadratic equation in two variables,

$$y = x^2 - 4. \tag{1}$$

As with linear equations in two variables, solutions of this equation are ordered pairs (x, y). We need replacements for both x and y in order to obtain a statement we may adjudge to be true or false. As before, such ordered pairs can be found by arbitrarily assigning values to x and computing related values for y. For instance, assigning the value -3 to x in Equation (1), we obtain

$$y = (-3)^2 - 4,$$
$$y = 5,$$

and $(-3, 5)$ is a solution. Similarly, we find that

$$(-2, 0), \quad (-1, -3), \quad (0, -4), \quad (1, -3), \quad (2, 0), \quad \text{and} \quad (3, 5)$$

are also solutions of (1). Locating the corresponding points on the plane, we have the graph in Figure 6.1-a on page 166. Clearly, these points do not lie on a straight line, and we might reasonably inquire whether the graph of the solution set of

$$y = x^2 - 4$$

165

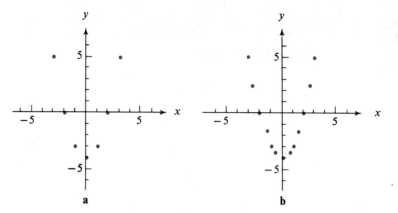

Figure 6.1

forms any kind of a meaningful pattern on the plane. By graphing additional solutions of (1)—solutions with x-components between those already found—we may be able to obtain a clearer picture. Accordingly, we find the solutions

$$\left(-\frac{5}{2}, \frac{9}{4}\right), \left(-\frac{3}{2}, -\frac{7}{4}\right), \left(-\frac{1}{2}, -\frac{15}{4}\right), \left(\frac{1}{2}, -\frac{15}{4}\right), \left(\frac{3}{2}, -\frac{7}{4}\right), \left(\frac{5}{2}, \frac{9}{4}\right),$$

and by graphing these points in addition to those found earlier, we have the graph in Figure 6.1-b. It now appears reasonable to connect these points in sequence, say from left to right, by a smooth curve as in Figure 6.2, and to assume that the resulting curve is a good approximation to the graph of (1). This curve is an example of a parabola.

More generally, the graph of the solution set of any second-degree equation of the form

$$y = ax^2 + bx + c, \qquad (2)$$

where a, b, and c are real and $a \neq 0$, is a parabola. Since for each x an equation of the form (2) will determine only one y, such an equation defines a function, called a **quadratic function**, having as domain the entire set of real numbers and as range some subset of the reals. For example, we observe from the graph in Figure 6.2 that the range of the function defined by (1) is the set of real numbers

$$\{y | y \geq -4\}.$$

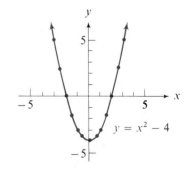

Figure 6.2

Axis of symmetry

The parabola that is the graph of an equation of the form (2) will have a lowest (minimum) point (will open upward) or a highest (maximum) point (will open downward), depending on whether $a > 0$ or $a < 0$, respectively. Such a point is called the **vertex** of the parabola. The line through the vertex and parallel to the

6.1 Quadratic Functions

y-axis is called the **axis of symmetry**, or simply the **axis**, of the parabola; it separates the parabola into two parts, each the mirror image of the other.

If we observe that the graphs of

$$y = ax^2 + bx + c \tag{3}$$

and

$$y = ax^2 + bx \tag{4}$$

have the same axis (Figure 6.3), we can find an equation for the axis of (3) by inspecting Equation (4). Factoring the right-hand member of $y = ax^2 + bx$ yields

$$y = x(ax + b),$$

and thus we can see that 0 and $-b/a$ are the x-intercepts of the graph. Since the axis of symmetry bisects the segment with these endpoints, an equation for the axis of symmetry is

$$x = -\frac{b}{2a}.$$

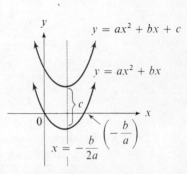

Figure 6.3

Example

Find an equation for the axis of symmetry of the graph of

$$y = 2x^2 - 5x + 7.$$

Solution

By comparing the given equation with $y = ax^2 + bx + c$, we see that $a = 2$ and $b = -5$. Hence, we have

$$x = -\frac{b}{2a} = -\frac{-5}{2(2)} = \frac{5}{4},$$

and $x = 5/4$ is an equation for the axis of symmetry.

Since the vertex of a parabola lies on its axis, obtaining an equation for this axis will give us the x-coordinate of the vertex. The y-coordinate can then be easily obtained by substitution in the equation for the parabola.

Sketching graphs of quadratic functions

When graphing a quadratic function, it is desirable first to select components for ordered pairs that ensure that the more significant parts of the graph are displayed. For a parabola, these parts include the x- and y-intercepts, if they exist, and the maximum or minimum point on the curve. Thus, when graphing a quadratic equation we perform the following four steps.

1. Find and plot the intercepts if possible.
2. Find and plot the axis of symmetry and the highest (or lowest) point.
3. If there are too few points obtained in steps 1 and 2 to get an accurate picture of the graph, then plot a few more selected points that will be helpful.
4. Sketch the curve through the points obtained above.

Example Graph $y = x^2 - 3x - 4$.

Solution Setting $x = 0$, we find that the y-intercept is -4. Setting $y = 0$, we have
$$0 = x^2 - 3x - 4 = (x - 4)(x + 1),$$
and the x-intercepts are 4 and -1. Since $a > 0$, the curve opens upward, and the x-coordinate of the minimum point is

$$x = -\frac{b}{2a} = -\frac{-3}{2(1)} = \frac{3}{2}.$$

By substituting $3/2$ for x in $y = x^2 - 3x - 4$, we obtain

$$y = \left(\frac{3}{2}\right)^2 - 3\left(\frac{3}{2}\right) - 4 = \frac{9}{4} - \frac{9}{2} - 4 = -\frac{25}{4}.$$

Graphing the intercepts and the coordinates of the minimum point and then sketching the curve produce the graph shown.

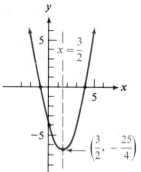

In the above example, we plotted four distinct points in steps 1 and 2; thus step 3 was unnecessary. The following example illustrates the use of step 3.

Example Graph $y = x^2 + x + 3$.

Solution Setting $x = 0$, we find that the y-intercept is 3, and by using the quadratic formula we find that $x^2 + x + 3 = 0$ has no real solutions; thus, there are no x-intercepts. Since the coefficient of x^2 (equal to 1) is greater than 0, the curve opens upward and the x-coordinate of the minimum point is

$$x = -\frac{b}{2a} = -\frac{1}{2(1)} = -\frac{1}{2}.$$

By substituting $-1/2$ for x in $y = x^2 + x + 3$, we obtain $y = 11/4$. Thus, the lowest point is at $(-1/2, 11/4)$. To obtain a more accurate picture of the graph, we can plot two more points. It is useful to plot several points on the same side of the axis of symmetry and then use the symmetry to sketch the curve. So we let $x = 1/2$ and find $y = 15/4$, and $x = 1$ and find $y = 5$. Now we sketch the curve to obtain the graph shown.

Solutions, zeros, and x-intercepts

Consider the graph of the function
$$f(x) = ax^2 + bx + c, \qquad (5)$$
for $a \neq 0$, and the solution set of the equation
$$ax^2 + bx + c = 0. \qquad (6)$$

6.1 Quadratic Functions

Any value of x for which $f(x) = 0$ in (5) will be a solution of (6). Since any point on the x-axis has y-coordinate zero (that is, $f(x) = 0$), the x-intercepts of the graph of (5) are the real solutions of (6). Values of x for which $f(x) = 0$ are called the **zeros of the function**. Thus we have three different names for a single idea:

1. the *real solutions* of the equation $ax^2 + bx + c = 0$,
2. the *zeros of the function* $f(x) = ax^2 + bx + c$,
3. the *x-intercepts* of the graph of $y = ax^2 + bx + c$.

We recall from Chapter 4 that a quadratic equation may have no real solution, one real solution, or two real solutions. If the equation has no real solution, we find that the graph of the related quadratic equation in two variables does not touch the x-axis; if there is one real solution, the graph is tangent to the x-axis; if there are two real solutions, the graph crosses the x-axis in two distinct points. These cases are illustrated in Figure 6.4.

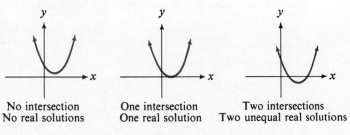

Figure 6.4

Inverse function

In Section 5.4 we observed that the graph of the inverse of a linear function is also a straight line, and hence the inverse is also a linear function if its graph is not vertical. Now note the graphs of

$$y = x^2$$

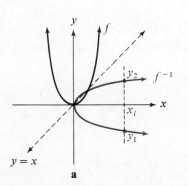
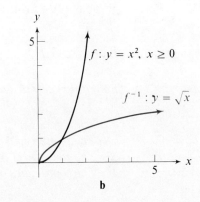

Figure 6.5

and

$$x = y^2 \quad \text{or} \quad y = \pm\sqrt{x},$$

shown in Figure 6.5-a. We observe in the latter that two different y's are associated with each x for all but one value in its domain. The equation $x = y^2$, or $y = \pm\sqrt{x}$, does not define a function. However, if the domain of f is restricted so that $x \geq 0$, the function has an inverse f^{-1} with range $y \geq 0$, as shown in Figure 6.5-b.

Quadratic inequalities

Relations of the form

$$y < ax^2 + bx + c \tag{7}$$

or

$$y > ax^2 + bx + c \tag{8}$$

can be graphed in the same manner in which we graphed relations defined by *linear* inequalities in two variables in Section 5.6. We first graph the associated equation and then shade an appropriate region as required.

Example

Graph $y < x^2 + 2.$ (9)

Solution

We first graph

$$y = x^2 + 2. \tag{10}$$

Now the coordinates of a selected point not on the graph of the associated equation can be used to determine which of the two resulting regions of the plane is the graph of the inequality. Upon substitution of 0 for x and 0 for y in $y < x^2 + 2$, we have $(0) < (0)^2 + 2$, a true statement. Hence the region containing the origin is shaded. Since the graph of (10) is not part of the graph of (9), a broken curve is used.

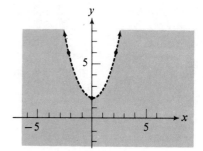

Exercise 6.1

A For each quadratic function find:

a. The x-intercepts, if they exist.
b. The y-intercept.
c. The equation of the axis of symmetry.
d. The coordinates of the maximum (or minimum) point.

Example

$y = x^2 - 2x - 3$

6.1 Quadratic Functions

Solution

a. Setting $y = 0$ and solving $x^2 - 2x - 3 = 0$, we have $x = 3$ and $x = -1$. The x-intercepts are 3, and -1.

b. Setting $x = 0$ we obtain $y = -3$. The y-intercept is -3.

c. Substituting -2 and 1 for b and a respectively in the equation $x = -b/2a$, we obtain $x = 1$, the equation for the axis of symmetry.

d. Substituting 1 (from part c) for x in the equation $y = x^2 - 2x - 3$, we obtain $y = -4$. Since $a > 0$, the graph opens upward and the point $(1, -4)$ is a minimum point.

1. $y = x^2 - 3x + 2$
2. $y = x^2 + x - 12$
3. $y = -2x^2 - 5x - 2$
4. $y = -6x^2 - x + 1$
5. $y = x^2 - 2x + 1$
6. $y = x^2 - 6x + 9$
7. $y = -4x^2 + 4x - 1$
8. $y = -9x^2 + 12x - 4$
9. $y = x^2 + 1$
10. $y = x^2 + 4$
11. $y = -2x^2 - 9$
12. $y = -3x^2 - 4$

Example Graph.

Solution $y = -x^2 + 7x - 6$

Setting $x = 0$, we find that the y-intercept is -6. Since the solutions of

$$-x^2 + 7x - 6 = -(x - 1)(x - 6) = 0$$

are 1 and 6, these are the x-intercepts. The axis of symmetry has equation

$$x = -\frac{b}{2a} = \frac{7}{2}.$$

Setting $x = 7/2$ in $y = -x^2 + 7x - 6$, we obtain

$$y = -\left(\frac{7}{2}\right)^2 + 7\left(\frac{7}{2}\right) - 6 = \frac{25}{4}.$$

Since the curve opens downward, the maximum point is $(7/2, 25/4)$. Using these points, we sketch the graph as shown.

13. $y = x^2 - 5x + 4$
14. $y = x^2 - 3x + 2$
15. $y = x^2 - 6x - 7$
16. $y = x^2 - 3x - 4$
17. $y = 2x^2 - 3x - 2$
18. $y = 6x^2 + x - 1$
19. $f(x) = -x^2 + 5x - 4$
20. $f(x) = -x^2 - 8x + 9$
21. $f(x) = -x^2 + 3x - 2$
22. $f(x) = -x^2 - x + 12$
23. $g(x) = \frac{1}{2}x^2 + 2$
24. $g(x) = -\frac{1}{2}x^2 - 2$
25. $y = x^2 + 9$
26. $y = 2x^2 + 8$

27. Graph $f(x) = x^2 + 1$. Represent $f(0)$ and $f(4)$ by drawing line segments from $(0, 0)$ to $(0, f(0))$ and from $(4, 0)$ to $(4, f(4))$.

28. Graph $g(x) = x^2 + 1$. Represent $g(-3)$ and $g(2)$ by drawing line segments from $(-3, 0)$ to $(-3, g(-3))$ and from $(2, 0)$ to $(2, g(2))$.

29. Graph $f(x) = x^2 + 1$. Draw the line segment joining the points $(4, f(4))$ and $(0, f(0))$. Compute $\dfrac{f(4) - f(0)}{4 - 0}$ and interpret this quantity geometrically.

30. Graph $g(x) = x^2 + 1$. Draw the line segment joining the points $(2, g(2))$ and $(-3, g(-3))$. Compute $\dfrac{g(2) - g(-3)}{2 - (-3)}$ and interpret this quantity geometrically.

For each given function f, graph f and f^{-1} using the same set of axes.

31. $f(x) = x^2 - 4, \quad x \geq 0$
32. $f(x) = x^2 + 4, \quad x \leq 0$
33. $f(x) = x^2 - 4x + 4, \quad x \geq 2$
34. $f(x) = x^2 - 2x - 3, \quad x \leq 1$

Graph.

Example $\quad y \geq x^2 + 2x$

Solution First graph $y = x^2 + 2x$. Use $(-1, 0)$ as a test point. Since $0 \geq (-1)^2 + 2(-1)$, the point with coordinates $(-1, 0)$ is in the graph. Shade the portion of the plane above the curve, and show the curve as a continuous parabola since the points on it satisfy the inequality.

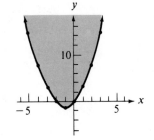

35. $y > x^2$
36. $y < x^2$
37. $y \geq x^2 + 3$
38. $y \leq x^2 + 3$
39. $y < 3x^2 + 2x$
40. $y > 3x^2 + 2x$
41. $y \leq x^2 + 3x + 2$
42. $y \geq x^2 + 3x + 2$

43. On a single set of axes, sketch the family of four curves that are the graphs of
$$y = x^2 + k \quad (k = -2, 0, 2, 4).$$
What effect does varying k have on the graph?

44. On a single set of axes, sketch the family of six curves that are the graphs of
$$y = kx^2 \quad \left(k = \tfrac{1}{2}, 1, 2, -\tfrac{1}{2}, -1, -2\right).$$
What effect does varying k have on the graph?

45. On a single set of axes, sketch the family of three curves that are the graphs of
$$y = x^2 + kx \quad (k = -1, 0, 1).$$
What effect does varying k have on the graph?

46. On a single set of axes sketch the family of three curves that are the graphs of
$$y = -x^2 + kx \quad (k = -1, 0, 1).$$
What effect does varying k have on the graph?

47. Graph the set of points whose coordinates satisfy both
$$y \leq 4 - x^2 \quad \text{and} \quad y \geq x^2 - 4.$$

48. Graph the set of points whose coordinates satisfy both
$$x \geq y^2 - 4 \quad \text{and} \quad y \geq x^2 - 4.$$

B 49. Find two numbers having sum 8 and product as great as possible.

50. Find the maximum possible area of a rectangle with perimeter 100 inches.

51. For $f(x) = 2x^2 + x + 1$, compute $\dfrac{f(x) - f(a)}{x - a}$ and interpret this quantity geometrically.

52. For $f(x) = -x^2 + 3x + 2$, compute $\dfrac{f(x) - f(a)}{x - a}$ and interpret this quantity geometrically.

6.2 Conic Sections

In addition to equations typified by $y = ax^2 + bx + c$, there are three other types of second-degree equations in two variables with graphs that are of particular interest. We shall discuss each of them separately.

First, consider the relation
$$x^2 + y^2 = 25. \tag{1}$$

Solving Equation (1) explicitly for y, we have
$$y = \pm\sqrt{25 - x^2}. \tag{2}$$

Assigning values to x, we find the following ordered pairs in the relation:
$$(-5, 0), (-4, 3), (-3, 4), (0, 5), (3, 4), (4, 3), (5, 0),$$
$$(-4, -3), (-3, -4), (0, -5), (3, -4), (4, -3).$$

Plotting these points on the plane, we have the graph shown in Figure 6.6-a on page 174. Connecting these points with a smooth curve, we obtain the graph in Figure 6.6-b, a **circle** with radius 5 and center at the origin.

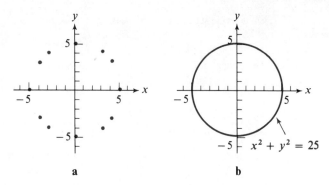

Figure 6.6

Equation of a circle

Since the number 25 in the right-hand member of Equation (1) clearly determines the length of the radius of the circle, we can generalize and observe that any equation of the form

$$x^2 + y^2 = r^2 \quad (r > 0)$$

has as its graph a circle with radius r and with center at the origin. (See Exercises 30–32, Exercise 6.2, for a more general approach.)

Note that in the preceding example it is not necessary to assign any values to x for which $|x| > 5$, because then y^2 would have to be negative. Since, except for -5 and $+5$, each permissible value for x is associated with two values for y—one positive and one negative—the relation (1) is not a function. We could represent the relationship specified by (2) by using two equations,

$$y = f(x) = \sqrt{25 - x^2} \tag{3a}$$

and

$$y = g(x) = -\sqrt{25 - x^2}, \tag{3b}$$

whose graphs would appear as in Figure 6.7-a and 6.7-b on page 175, respectively. Each of these equations does define a function. The domain of each is

$$\{x \mid |x| \leq 5\},$$

while the ranges differ. The range of the function defined by (3a) is

$$\{y \mid 0 \leq y \leq 5\},$$

and that of the function defined by (3b) is

$$\{y \mid -5 \leq y \leq 0\}.$$

6.2 Conic Sections

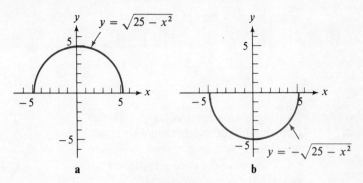

Figure 6.7

The second kind of quadratic (second-degree) relation of interest, which actually has the first as a limiting case, is that typified by

$$4x^2 + 9y^2 = 36. \tag{4}$$

We obtain solutions of (4) by first solving the equation explicitly for y,

$$y = \pm \frac{2}{3}\sqrt{9 - x^2},$$

and then assigning values to x satisfying $-3 \leq x \leq 3$. (Why these values only?) We obtain, for example,

$$(-3, 0), \left(-2, \frac{2}{3}\sqrt{5}\right), \left(-1, \frac{4}{3}\sqrt{2}\right), (0, 2), \left(1, \frac{4}{3}\sqrt{2}\right), \left(2, \frac{2}{3}\sqrt{5}\right), (3, 0),$$

$$\left(-2, -\frac{2}{3}\sqrt{5}\right), \left(-1, -\frac{4}{3}\sqrt{2}\right), (0, -2), \left(1, -\frac{4}{3}\sqrt{2}\right), \left(2, -\frac{2}{3}\sqrt{5}\right).$$

Equation of an ellipse

Locating the corresponding points on the plane and connecting them with a smooth curve, we have the graph shown in Figure 6.8. This curve is called an **ellipse**. In general, the graph of the equation

$$Ax^2 + By^2 = C; \quad A, B, C > 0; \quad A \neq B$$

is an ellipse with center at the origin.

The third kind of quadratic relation with which we are presently concerned is typified by

$$x^2 - y^2 = 9.$$

Solving the equation for y, we obtain

$$y = \pm\sqrt{x^2 - 9},$$

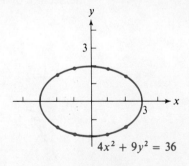

Figure 6.8

which has as a part of its solution set the ordered pairs

$$(-5, 4), (-4, \sqrt{7}), (-3, 0), (3, 0), (4, \sqrt{7}), (5, 4),$$
$$(-5, -4), (-4, -\sqrt{7}), (4, -\sqrt{7}), (5, -4).$$

Equation of a hyperbola

Plotting the corresponding points and connecting them with a smooth curve, we obtain the curve shown in Figure 6.9. This curve is called a **hyperbola**. In general, the equation

$$Ax^2 - By^2 = C; \quad A, B, C > 0$$

graphs into a hyperbola with center at the origin which opens to the left and right; and the relation

$$By^2 - Ax^2 = C; \quad A, B, C > 0$$

graphs into a hyperbola with center at the origin which opens upward and downward.

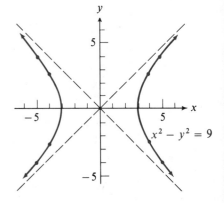

Figure 6.9

Asymptotes of a hyperbola

The dashed lines shown in Figure 6.9 are called **asymptotes** of the graph. They comprise the graph of the equation $x^2 - y^2 = 0$. While we shall not discuss the notion in detail here, it is true that, in general, the graph of $Ax^2 - By^2 = C$ "approaches" the two straight lines in the graph of $Ax^2 - By^2 = 0$ for each A, B, $C > 0$ as $|x|$ increases. (See Exercise 35.)

The graphs of the foregoing relations, together with the parabola of the preceding section, are called **conic sections**, or **conics**, because such curves result from the intersection of a plane and a right circular cone, as shown in Figure 6.10.

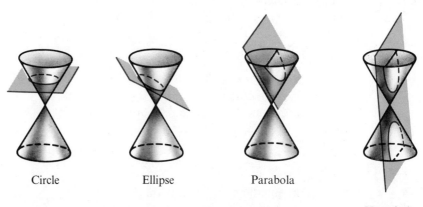

Circle   Ellipse   Parabola

Hyperbola

Figure 6.10

Use of the form of an equation in graphing

You should make use of the form of a quadratic equation as an aid in graphing the equation. The ideas developed in this and the preceding section may be summarized as follows:

1. A quadratic equation of the form

$$y = ax^2 + bx + c, \quad a \neq 0, \quad (5)$$

has a graph that is a parabola, opening upward if $a > 0$ and downward if $a < 0$. An equation of the form

$$x = ay^2 + by + c, \quad a \neq 0,$$

is the inverse of the relation defined by Equation (5). Thus it has a graph that is a parabola, opening to the right if $a > 0$ and to the left if $a < 0$.

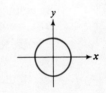

2. A quadratic equation of the form

$$Ax^2 + By^2 = C, \quad A^2 + B^2 \neq 0,$$

has a graph that is

(a) a **circle** if $A = B$ and A, B, and C have like signs;

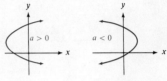

(b) an **ellipse** if $A \neq B$ and A, B, and C have like signs;

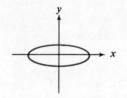

$A < B, C > 0 \qquad A > B, C > 0$

(c) a **hyperbola** if A and B are opposite in sign and $C \neq 0$;

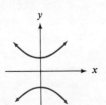

$A < 0, B, C > 0 \qquad A, C > 0, B < 0$

(d) **two distinct lines** through the origin if A and B are opposite in sign and $C = 0$ (see Exercise 25);

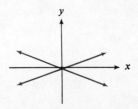

(e) **two distinct parallel lines** if one of A and B equals 0 and the other has the same sign as C (see Exercise 26);

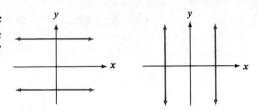

Sketching conic sections

After you recognize the general form of the curve, the graph of a few points should suffice to sketch the complete graph. The intercepts, for instance, are always easy to locate.

Example

Graph $x^2 + 4y^2 = 8$.

Solution

By comparing the equation with the above case 2(b), we note immediately that its graph is an ellipse. If $y = 0$, then $x = \pm\sqrt{8}$; and if $x = 0$, then $y = \pm\sqrt{2}$. We can accordingly sketch the graph.

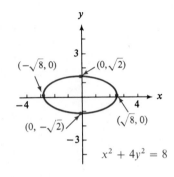

Example

Graph $x^2 - y^2 = 3$. (6)

Solution

By comparing the equation with case 2(c), we see that its graph is a hyperbola. If $y = 0$, then $x = \pm\sqrt{3}$; and if $x = 0$, then y^2 would have to be negative, an impossibility in the field of real numbers. Thus the graph does not cross the y-axis. By assigning a few other arbitrary values to one of the variables, say x, e.g., $(4,\ )$ and $(-4,\ )$, we can find additional ordered pairs

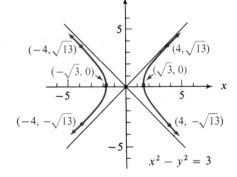

$(4, \sqrt{13}), (4, -\sqrt{13}),$

$(-4, \sqrt{13}), (-4, -\sqrt{13})$

which satisfy (6). The graph can then be sketched as shown in the figure. Observe that the asymptotes, which are the graphs of $x^2 - y^2 = 0$, are given by $y = x$ and $y = -x$. They can be helpful in sketching the hyperbola.

Exercise 6.2

A Name and sketch the graph of each of the following relations. Specify the vertices (or intercepts), and for the hyperbolas give equations of the asymptotes.

Example $4x^2 - 36 = -9y^2$

Solution   Rewrite the defining equation equivalently in standard form,
$$4x^2 + 9y^2 = 36.$$
By inspection, the graph is an ellipse; x-intercepts are -3 and 3; y-intercepts are -2 and 2. Sketch the graph.

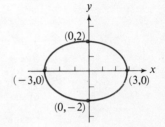

1. $y = 4x^2$
2. $y = 9x^2$
3. $y = -8x^2$
4. $y = -3x^2$
5. $y^2 = 16x$
6. $y^2 = 4x$
7. $y^2 = -x$
8. $y^2 = -4x$
9. $x^2 + y^2 = 49$
10. $x^2 + y^2 = 64$
11. $4x^2 + 25y^2 = 100$
12. $x^2 + 2y^2 = 8$
13. $4x^2 = 4y^2$
14. $x^2 - 9y^2 = 0$
15. $x^2 = 9 + y^2$
16. $x^2 = 2y^2 + 8$
17. $4x^2 + 4y^2 = 1$
18. $9x^2 + 9y^2 = 2$
19. $3x^2 - 12 = -4y^2$
20. $12 - 3y^2 = 4x^2$
21. $4x^2 - y^2 = 0$
22. $4x^2 + y^2 = 0$
23. $y^2 = 9$
24. $x^2 = 4$

B 25. Discuss the graph of any equation of the form $Ax^2 - By^2 = 0$, $A, B > 0$.

26. Discuss the graph of any equation of the form $Ax^2 = C$, $A, C > 0$.

27. Discuss the graph of any equation of the form $Ax^2 = 0$, $A \neq 0$.

28. Explain why the graph of $x^2 + y^2 = -1$ is the null set. Generalize from the result and discuss the graph of any equation of the form
$$Ax^2 + By^2 = C, \ A^2 + B^2 \neq 0, \ A, B \geq 0, \ C < 0.$$

29. Discuss the graph of any equation of the form $Ax^2 + By^2 = 0$, $A, B > 0$.

30. Use the distance formula to show that the graph of the equation $x^2 + y^2 = r^2$, $r \geq 0$ is the set of all points located a distance r from the origin.

31. Show that the graph of $(x - 2)^2 + (y + 3)^2 = 16$ is a circle with center at $(2, -3)$ and radius 4. *Hint*: Use distance formula.

32. Show that the graph of $(x - h)^2 + (y - k)^2 = r^2$, $r > 0$, is a circle with center at (h, k) and radius r.

33. Show that the general form of an equation for an ellipse, $Ax^2 + By^2 = C$, can be written in the intercept form $\dfrac{x^2}{a^2} + \dfrac{y^2}{b^2} = 1$, where a and b are x- and y-intercepts, respectively.

34. Show that the general form of an equation for a hyperbola, $Ax^2 - By^2 = C$, can be written in the intercept form $\dfrac{x^2}{a^2} - \dfrac{y^2}{b^2} = 1$, where a is an x-intercept.

35. By solving $Ax^2 - By^2 = C$ $(A, B, C > 0)$ for y, obtain the expression

$$y = \pm \sqrt{\frac{A}{B}} |x| \left(\sqrt{1 - \frac{C}{Ax^2}} \right),$$

and argue that the graph of $Ax^2 - By^2 = C$ approaches the straight-line graphs of $y = \pm \sqrt{A/B}\, |x|$ as $|x|$ increases.

36. Graph $x^2 + y^2 = 25$ and $4x^2 + y^2 = 36$ on the same set of axes. What is the significance of the coordinates of the points of intersection?

37. Graph the inequality $x^2 + y^2 \leq 25$ by observing that the graph of the equation $x^2 + y^2 = a^2$ is a circle of radius a and then examining what $a \leq 5$ implies.

38. Graph the inequality $4x^2 + 9y^2 \leq 36$.

6.3 Variation as a Functional Relationship

Direct variation

There are two types of functional relationships, widely used in the sciences, to which custom has assigned special names. First, the variable y is said to **vary directly** as the variable x if

$$y = kx \quad (k \text{ a positive constant}). \tag{1}$$

Note that since Equation (1) associates one and only one y with each x, a direct variation defines a function.

Examples

a. The circumference of a circle varies directly as the radius since

$$c = 2\pi r.$$

b. The area of a circle varies directly as the square of the radius since

$$A = \pi r^2.$$

Inverse variation

The second important type of variation arises from the equation

$$xy = k \quad (k \text{ a positive constant}), \tag{2}$$

in which x and y are said to **vary inversely**. When (2) is written in the form

$$y = \frac{k}{x}, \tag{3}$$

y is said to vary inversely as x.

6.3 Variation as a Functional Relationship

Examples

a. For an ideal gas at constant absolute temperature (T), the volume (V) and pressure (P) vary inversely since

$$VP = kT \quad (k, T \text{ constants}),$$

or

$$V = \frac{kT}{P}.$$

b. For a right circular cylinder with constant volume (V), the height (h) and the square of the radius (r) vary inversely since

$$V = \pi r^2 h \quad (V \text{ a constant}),$$

or

$$h = \frac{V}{\pi r^2}.$$

Since Equation (3) associates only one y with each x ($x \neq 0$), an inverse variation defines a function, with domain $\{x \mid x \neq 0\}$.

The names "direct" and "inverse," as applied to variation, arise from the fact that in direct variation an assignment of increasing absolute values of x results in an association with increasing absolute values of y, whereas in inverse variation an assignment of increasing absolute values of x results in an association with decreasing absolute values of y.

Constant of variation

The constant involved in an equation defining a direct or inverse variation is called the **constant of variation**. If we know that one variable varies directly or inversely as another, and if we have one set of associated values for the variables, we can find the constant of variation involved. For example, suppose we know that y varies directly as x^2, and that $y = 4$ when $x = 7$. We express the fact that y varies directly as x^2 by writing

$$y = kx^2, \tag{4}$$

and then substitute 7 for x and 4 for y in (4) to obtain

$$4 = k \cdot 7^2 = k \cdot 49,$$

from which

$$k = \frac{4}{49}.$$

The equation specifically expressing the direct variation is

$$y = \frac{4}{49} x^2.$$

Now we can use this equation to find the value of one of the variables given the value of the other.

Also, direct and inverse variation may take place concurrently. Thus, y may vary directly as x and inversely as z, giving rise to the equation

$$y = k\frac{x}{z}.$$

Joint variation

In the event that one variable varies as the product of two or more other variables, we refer to the relationship as **joint variation**. Thus, if y varies jointly as u, v, and w, we have

$$y = kuvw. \tag{5}$$

It should be pointed out that the way in which the word "variation" is used herein is a technical one, and when the ideas of direct, inverse, or joint variation are encountered, you should always think of equations of the form (1), (3), or (5). For instance, the equations

$$y = 2x + 1 \quad \text{and} \quad y = \frac{1}{2}x - 2$$

do not describe examples of variation within our meaning of the word.

Proportionality and variation

An alternative term frequently used to describe the variation relationship discussed in this section is the word "proportional." Thus, to say that "y is directly proportional to x" or "y is inversely proportional to x" is another way of describing direct and inverse variation.

The use of the word "proportion" arises from the fact that any two solutions of an equation expressing a direct variation satisfy a fractional equation of the form

$$\frac{a}{b} = \frac{c}{d},$$

which is commonly called a **proportion**. For example, consider the problem in which the volume of a gas varies directly as the absolute temperature and inversely as the pressure, and can accordingly be represented by a relation of the form

$$V = \frac{kT}{P}. \tag{6}$$

For any set of values T_1, P_1, and V_1,

$$k = \frac{V_1 P_1}{T_1}, \tag{6a}$$

and for any other set of values T_2, P_2, and V_2,

$$k = \frac{V_2 P_2}{T_2}. \tag{6b}$$

Equating the right-hand members of (6a) and (6b), we get

$$\frac{V_1 P_1}{T_1} = \frac{V_2 P_2}{T_2}, \tag{6c}$$

6.3 Variation as a Functional Relationship

from which any one of the six values can be determined if the other five values are known.

Example

If V varies directly as h and r^2, and $V = 4\pi$ when $h = 1$ and $r^2 = 4$, find h when $r^2 = 9$ and $V = 18\pi$.

Solution

Write an equation expressing the relationship between V, h, and r^2.

$$V = khr^2$$

Solve for k.

$$k = \frac{V}{hr^2}$$

Write a proportion relating the variables.

$$\frac{V_1}{h_1 r_1^2} = \frac{V_2}{h_2 r_2^2}$$

Substitute the known values for the variables.

$$\frac{4\pi}{(1)(4)} = \frac{18\pi}{h_2(9)}$$

Solve for h_2.

$$h_2 = \frac{(1)(4)(18\pi)}{4\pi(9)} = 2$$

Exercise 6.3

A Solve. In Exercises 1–8, first find the constant of variation.

Example

If V varies directly as T and inversely as P, and $V = 40$ when $T = 300$ and $P = 30$, find V when $T = 324$ and $P = 24$.

Solution

Write an equation expressing the relationship between the variables.

$$V = \frac{kT}{P} \qquad (1)$$

Substitute the initially known values for V, T, and P. Solve for k.

$$40 = \frac{k(300)}{30}$$

$$4 = k$$

Solution continued on overleaf

Rewrite Equation (1) with k replaced by 4.

$$V = \frac{4T}{P}$$

Substitute the second set of values for T and P and solve for V.

$$V = \frac{4(324)}{24} = 54$$

1. If y varies directly as x^2, and $y = 9$ when $x = 3$, find y when $x = 4$.
2. If r varies directly as s and inversely as t, and $r = 12$ when $s = 8$ and $t = 2$, find r when $s = 3$ and $t = 6$.
3. The distance a particle falls in a certain medium is directly proportional to the square of the length of time it falls. If the particle falls 16 feet in two seconds, how far will it fall in 10 seconds?
4. In Exercise 3, how far will the body fall *during* the seventh second?
5. The pressure exerted by a liquid at a given point varies directly as the depth of the point beneath the surface of the liquid. If a certain liquid exerts a pressure of 40 pounds per square foot at a depth of 10 feet, what is the pressure at 40 feet?
6. The volume (V) of a gas varies directly as its temperature (T) and inversely as its pressure (P). A gas occupies 20 cubic feet at a temperature of 300° A (absolute) and a pressure of 30 pounds per square inch. What will the volume be if the temperature is raised to 360° A and the pressure decreased to 20 pounds per square inch?
7. The maximum-safe uniformly distributed load (L) for a horizontal beam varies jointly as its breadth (b) and the square of the depth (d), and inversely as the length (l). An 8-foot beam with $b = 2$ feet and $d = 4$ feet will safely support a uniformly distributed load of up to 750 pounds. How many uniformly distributed pounds will an 8-foot beam support safely if $b = 2$ and $d = 6$?
8. The resistance (R) of a wire varies directly as its length (l) and inversely as the square of its diameter (d). Fifty feet of wire of diameter 0.012 inch has a resistance of 10 ohms. What is the resistance of 50 feet of the same type of wire if the diameter is increased to 0.015 inch?

Represent the relationship as a proportion by eliminating the constant of variation and then solve for the required variable.

Example

If V varies directly as T and varies inversely as P, and $V = 40$ when $T = 300$ and $P = 30$, find V when $T = 324$ and $P = 24$.

Solution

Write an equation expressing the relationship between the variables.

$$V = \frac{kT}{P}$$

6.3 Variation as a Functional Relationship

Solve for k.

$$k = \frac{VP}{T}$$

Write a proportion relating the variables for two different sets of conditions.

$$\frac{V_1 P_1}{T_1} = \frac{V_2 P_2}{T_2}$$

Substitute the known values of the variables.

$$\frac{(40)(30)}{300} = \frac{V_2(24)}{324}$$

Solve for V_2.

$$V_2 = \frac{(324)(40)(30)}{300(24)} = 54$$

9. Exercise 1 of this set.
10. Exercise 2 of this set.
11. Exercise 5 of this set.
12. Exercise 6 of this set.
13. Exercise 7 of this set.
14. Exercise 8 of this set.

B 15. Graph on the same set of axes the linear functions defined by $y = kx$, $x \geq 0$, when $k = 1, 2,$ and 3, respectively. Note that the constant of variation and the slope of the graph of the equation are the same.

16. Graph on the same set of axes the quadratic functions defined by $y = kx^2$, $x \geq 0$, when $k = 1, 2,$ and 3, respectively. What effect does a change in k have on the graph of $y = kx^2$?

17. Graph on the same set of axes the equations $y = kx$, $y = kx^2$, and $y = kx^3$, when $k = 2$ and $x \geq 0$. What effect does increasing n have on the graph of $y = kx^n$?

18. Graph on the same set of axes the functions defined by $xy = k$, for $k = 1$ and 2, and $x > 0$. What effect does a change in k have on the graph of $xy = k$?

19. From the formula for the circumference of a circle, $c = \pi d$, show that the ratio of the circumference of two circles equals the ratio of their respective diameters.

20. From the formula for the area of a circle, $A = \pi r^2$, show that the ratio of the areas of two circles equals the ratio of the squares of their respective radii.

21. Show that if y varies directly as x, and z varies directly as x, then $y + z$ varies directly as x.

22. Show that if y varies directly as x, and z varies directly as x, then yz varies directly as x^2. What is the constant of variation for yz varying directly as x^2?

23. Suppose y varies directly as x. Does it necessarily follow that x varies directly as y? If so, what is the constant of variation for x varying directly as y?

24. Suppose y varies directly as x, and x varies directly as z. Does it necessarily follow that y varies directly as z? If so, what is the constant of variation for y varying directly as z?

6.4 Zeros of a Polynomial Function

In our previous work we obtained x-intercepts to help us graph first- and second-degree equations in two variables. These intercepts are also helpful in graphing polynomial equations of degree greater than two. Since the x-intercepts of the graph of a polynomial function are the zeros of the function, we shall first consider some facts which will aid us in finding zeros of such functions.

Continuity

For real-number replacements for x, we know that

$$y = P(x) = a_n x^n + a_{n-1} x^{n-1} + \cdots + a_0 \quad (a_n \neq 0)$$

can be represented by a graph in the coordinate plane. Moreover, although we shall not prove it here, we have the following theorem on continuity.

Theorem 6.1 *If $P(x) = a_n x^n + a_{n-1} x^{n-1} + \cdots + a_0$ $(a_n \neq 0)$, and if k is between $P(x_1)$ and $P(x_2)$, then there exists at least one c between x_1 and x_2 such that $P(c) = k$.*

More intuitively, there are no breaks or jumps in the graph of $P(x)$, so we know that $P(x)$ must assume all values between any two of its values. Thus, the graph of P must be a continuous, unbroken curve (see Figure 6.11).

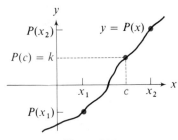

Figure 6.11

The location theorem

To assist us in the search for real zeros of a polynomial function, we have the following theorem that follows from Theorem 6.1.

Theorem 6.2 *Let $P(x)$ be a polynomial with real coefficients. If $x_1 < x_2$, and if $P(x_1)$ and $P(x_2)$ are opposite in sign, then there is at least one value c between x_1 and x_2 such that $P(c) = 0$.*

This theorem expresses the fact that if the points with coordinates $(x_1, P(x_1))$ and $(x_2, P(x_2))$ are on opposite sides of the x-axis, then the graph of $y = P(x)$ must cross the x-axis at some (at least one) point c between x_1 and x_2 (see Figure 6.12).

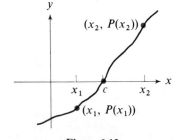

Figure 6.12

6.4 Zeros of a Polynomial Function

Example

Locate the x-intercepts of the graph of $y = P(x) = x^3 + x - 1$ within a unit interval, that is, within an interval between two consecutive integers.

Solution

To find approximate values for solutions to

$$P(x) = x^3 + x - 1 = 0,$$

we observe that $P(0) = -1$ while $P(1) = (1)^3 + (1) - 1 = 1$. Since $P(0)$ and $P(1)$ have opposite signs, Theorem 6.2 guarantees that there is a solution between 0 and 1. It is evident from the above equation that for increasing positive values of x, $P(x)$ becomes increasingly large. For increasing values of $|x|$, where $x < 0$, we note that $P(x) < 0$ and $|P(x)|$ takes on increasingly large values. Thus, all of the real-valued solutions of $P(x) = 0$ lie between 0 and 1.

Remainder theorem

The following two theorems will also be useful in obtaining information about the zeros of polynomial functions.

Theorem 6.3 If $P(x)$ is a real polynomial, then for every real number c there exists a unique real polynomial $Q(x)$ such that

$$P(x) = (x - c)Q(x) + P(c).$$

To see that this is true, recall that from Theorem 2.3 we know that there exists a real polynomial $Q(x)$ and a real number r such that

$$P(x) = (x - c)Q(x) + r.$$

Since this is true for all $x \in R$, then it is true for $x = c$, and we have

$$P(c) = (c - c)Q(c) + r,$$

$$P(c) = 0 \cdot Q(c) + r,$$

$$P(c) = r.$$

Thus, $P(x) = (x - c)Q(x) + P(c)$.

This theorem is called the **remainder theorem** because it asserts that the remainder, when $P(x)$ is divided by $(x - c)$, is the value of P at c, that is, $P(c)$. Since synthetic division (Section 2.5) offers a quick means of obtaining this remainder, we can usually find values $P(c)$ more rapidly by synthetic division than by direct substitution.

Example

Given $P(x) = x^3 - x^2 + 3$, find $P(3)$ by means of the remainder theorem.

Solution on overleaf

Solution   Synthetically dividing $x^3 - x^2 + 3$ by $x - $, we have

$$\underline{}\ \begin{array}{cccc} 1 & -1 & 0 & 3 \\ & 3 & 6 & 18 \\ \hline 1 & 2 & 6 & 21 \end{array}$$

and, by inspection, $r = P(3) = 21$.

Factor theorem

Observe that if $P(r) = 0$, then, by the remainder theorem,

$$P(x) = (x - r)Q(x) + P(r)$$
$$= (x - r)Q(x) + 0,$$

and, by the definition of a factor, $(x - r)$ is a factor of $P(x)$. This proves the following theorem, called the **factor theorem**.

Theorem 6.4   *If $P(x)$ is a polynomial with real-number coefficients, and $P(r) = 0$, then $(x - r)$ is a factor of $P(x)$.*

Converse of factor theorem

It might be noted, as a converse of the factor theorem, that if $(x - r)$ is a factor of $P(x)$, so that $P(x) = (x - r)Q(x)$, then $P(r) = 0$, since

$$P(r) = (r - r)Q(r) = 0 \cdot Q(r) = 0.$$

Although Theorems 6.3 and 6.4 were stated for polynomials with real-number coefficients, both the remainder theorem and the factor theorem are also true for polynomials with complex-number coefficients. However, in our work we shall be primarily concerned with polynomials with real-number coefficients.

Bounds on zeros

The following theorem is sometimes helpful in isolating real zeros of a polynomial function with real coefficients.

Theorem 6.5   *Let $P(x)$ be a polynomial with real-number coefficients.*

I   *If $r_1 \geq 0$, and the coefficients of the terms in $Q(x)$ and the term $P(r_1)$ are all of the same sign in the right-hand member of*

$$P(x) = (x - r_1)Q(x) + P(r_1),$$

then $P(x) = 0$ can have no solution greater than r_1.

II   *If $r_2 \leq 0$, and the coefficients of the terms in $Q(x)$ and the term $P(r_2)$ alternate in sign (zero suitably denoted by $+0$ or -0) in the right-hand member of*

$$P(x) = (x - r_2)Q(x) + P(r_2),$$

then $P(x) = 0$ can have no solution less than r_2.

We shall verify only the first part of the theorem here.

6.4 Zeros of a Polynomial Function

For all $x > r_1$, $(x - r_1) > 0$. Moreover, if all the coefficients in $Q(x)$ are of the same sign, say positive, then, since $x > r_1 \geq 0$, we have

$$Q(x) > 0 \quad \text{and} \quad (x - r_1)Q(x) > 0.$$

Since $P(r_1) \geq 0$ by hypothesis, we have

$$P(x) = (x - r_1)Q(x) + P(r_1) > 0,$$

and the first part is proved. The second part follows from the first by considering

$$P(-x) = (-x - r_2)Q(-x) + P(r_2).$$

This theorem permits us to place upper and lower bounds on the set of real zeros of the polynomial function

$$P(x) = a_n x^n + \cdots + a_0,$$

and consequently on the real members of the solution set of $P(x) = 0$.

Example Show that 2 and -2 are upper and lower bounds, respectively, for the location of the zeros of

$$P(x) = 18x^3 - 12x^2 - 11x + 10.$$

Solution Using synthetic division to divide $P(x)$ by $x - 2$, we have

$$\begin{array}{r|rrrr}
2 & 18 & -12 & -11 & 10 \\
  &   & 36 & 48 & 74 \\
\hline
  & 18 & 24 & 37 & 84.
\end{array}$$

Thus, $Q(x) = 18x^2 + 24x + 37$ and $P(2) = 84 > 0$. Hence, by Theorem 6.5-I, 2 is an upper bound for the zeros of P. Next, dividing $P(x)$ by $x + 2$, we have

$$\begin{array}{r|rrrr}
-2 & 18 & -12 & -11 & 10 \\
   &   & -36 & 96 & -170 \\
\hline
   & 18 & -48 & 85 & -160.
\end{array}$$

Here $Q(x) = 18x^2 - 48x + 85$ and $P(-2) = -160$. Therefore, by Theorem 6.5-II, -2 is a lower bound for the zeros of P.

Example Find the least nonnegative integer and the greatest nonpositive integer that are, by Theorem 6.5, upper and lower bounds, respectively, for the real zeros of

$$P(x) = x^4 - x^3 - 10x^2 - 2x + 12.$$

Solution on overleaf

Solution We shall first seek an upper bound by dividing $P(x)$ successively by $x - 1$, $x - 2$, and so on. Each row after the first in the following array is the bottom row in the respective synthetic division involved.

	1	−1	−10	−2	12
1	1	0	−10	−12	0
2	1	1	−8	−18	−24
3	1	2	−4	−14	−30
4	1	3	2	6	36

Since the numbers in the last row are all positive, 4 is an upper bound. Next, we divide by $x + 1$, $x + 2$, and so on, in search of a lower bound.

	1	−1	−10	−2	12
−1	1	−2	−8	6	6
−2	1	−3	−4	6	0
−3	1	−4	2	−8	36

Since the numbers in the row following -3 alternate from positive to negative, etc., -3 is a lower bound. (Had the numbers in the row been 1, 0, 2, -8, 36, then the sign "$-$" could arbitrarily have been assigned to 0 to give the desired pattern of alternating signs.)

Enlarging the replacement set for x to include complex values allows us to prove some very useful theorems concerning zeros of polynomial functions.

Conjugate complex zeros Recalling from Section 3.5 that the conjugate of the complex number $z = a + bi$ is $\bar{z} = a - bi$, where $a, b \in R$, we state an important property of polynomials with real coefficients. The proof is left as an exercise.

Theorem 6.6 *If $P(z)$ is a polynomial with real-number coefficients, and $P(z) = 0$ for some $z \in C$, then $P(\bar{z}) = 0$.*

This theorem guarantees that for a function defined by a polynomial with *real coefficients, the complex zeros always occur in conjugate pairs.*

Example Given that $2 - i$ is a zero of

$$P(x) = x^3 - 6x^2 + 13x - 10,$$

find all zeros of P.

6.4 Zeros of a Polynomial Function

Solution By Theorem 6.6, $\overline{2-i} = 2 + i$ is a zero of P. Thus, by the factor theorem, we have

$$P(x) = [x - (2-i)][x - (2+i)]Q(x)$$
$$= (x^2 - 4x + 5)Q(x).$$

Hence,

$$Q(x) = \frac{x^3 - 6x^2 + 13x - 10}{x^2 - 4x + 5} = x - 2.$$

Thus,

$$P(x) = [x - (2-i)][x - (2+i)](x - 2) = 0$$

and, by inspection, the solutions of this equation—and hence the zeros of P—are $2 - i, 2 + i$, and 2.

Fundamental theorem of algebra When Theorem 6.6 is coupled with the following theorem, which is called the **fundamental theorem of algebra**, a great deal of information relative to the zeros of polynomial functions becomes readily available.

Theorem 6.7 *Every polynomial function of degree $n \geq 1$ over the complex numbers has at least one real or complex zero.*

The proof of this theorem involves concepts beyond those available to us, and is omitted.

An nth-degree polynomial function has n zeros Repeated applications of the fundamental theorem of algebra and the factor theorem can be used to prove the following theorem. We leave the details as an exercise.

Theorem 6.8 *Every polynomial of degree $n \geq 1$ over the complex numbers can be expressed as a product of a constant and n linear factors of the form $x - x_j$, where $x_j \in C$.*

If a factor $(x - x_j)$ occurs k times in such a linear factorization of $P(x)$, then x_j is said to be a **zero of multiplicity k**. With this agreement, Theorem 6.8 shows that every polynomial function defined by a polynomial $P(x)$ of degree n with complex coefficients has exactly n zeros.

Note that any theorem stated in terms of zeros of polynomial functions applies to solutions of polynomial equations and vice versa; a *zero* of

$$P(x) = a_n x^n + a_{n-1} x^{n-1} + \cdots + a_0$$

is a *solution* of $P(x) = 0$.

Examples **a.** $P(x) = x^3 + 2x - 1$ has three zeros in the set of complex numbers. **b.** $x^5 - 2x^3 - x + 1 = 0$ has five solutions in the set of complex numbers.

Exercises 6.4

A In Exercises 1–6 use Theorem 6.2 to verify the given statement.

1. $f(x) = x^3 - 3x + 1$ has a zero between 0 and 1.
2. $f(x) = 2x^3 + 7x^2 + 2x - 6$ has a zero between -2 and -1.
3. $g(x) = x^4 - 2x^2 + 12x - 17$ has a zero between -3 and -2.
4. $g(x) = 2x^4 + 3x^3 - 14x^2 - 15x + 9$ has a zero between -2 and -1.
5. $P(x) = 2x^2 + 4x - 4$ has zeros between -3 and -2 and between 0 and 1.
6. $P(x) = x^3 - x^2 - 2x + 1$ has zeros between -2 and -1, between 0 and 1, and between 1 and 2.

Use synthetic division to find the specified values of each polynomial.

7. $P(x) = x^3 + 2x^2 + x - 1$; $P(1)$, $P(2)$, and $P(3)$
8. $P(x) = x^3 - 3x^2 - x + 3$; $P(1)$, $P(2)$, and $P(3)$
9. $P(x) = 2x^4 - 3x^3 + x + 2$; $P(-2)$, $P(2)$, and $P(4)$
10. $P(x) = 3x^4 + 3x^2 - x + 3$; $P(-2)$, $P(2)$, and $P(4)$
11. $P(x) = 3x^5 - x^3 + 2x^2 - 1$; $P(-3)$, $P(2)$, and $P(3)$
12. $P(x) = 2x^6 - x^4 + 3x^3 + 1$; $P(-3)$, $P(2)$, and $P(3)$

Find an upper bound and a lower bound for the real zeros of each polynomial function.

13. $P(x) = x^3 + 2x^2 - 7x - 8$
14. $P(x) = x^3 - 8x + 5$
15. $P(x) = x^4 - 2x^3 - 7x^2 + 10x + 10$
16. $P(x) = x^3 - 4x^2 - 4x + 12$
17. $P(x) = x^5 - 3x^3 + 24$
18. $P(x) = x^5 - 3x^4 - 1$
19. $P(x) = 2x^5 + x^4 - 2x - 1$
20. $P(x) = 2x^5 - 2x^2 + x - 2$

In Exercises 21–28, one or more zeros are given for each of the polynomial functions; find the other zeros.

21. $P(x) = x^2 + 4$; $2i$ is one zero.
22. $P(x) = 3x^2 + 27$; $-3i$ is one zero.
23. $P(x) = x^3 - 3x^2 + x - 3$; 3 and i are zeros.
24. $Q(x) = x^3 - 5x^2 + 7x + 13$; -1 and $3 - 2i$ are zeros.

25. $Q(x) = x^4 + 5x^2 + 4$; $-i$ and $2i$ are zeros.

26. $P(x) = x^4 + 11x^2 + 18$; $3i$ and $\sqrt{2}i$ are zeros.

27. $Q(x) = x^4 + 3x^3 + 4x^2 + 27x - 45$; $-3i$ is a zero.

28. $Q(x) = x^5 - 2x^4 + 8x^3 - 16x^2 + 16x - 32$; $2i$ (multiplicity 2).

29. A cubic equation with real coefficients has roots -2 and $1 + i$. What is the third root? Write the equation in the form $P(x) = 0$, given that the leading coefficient (the coefficient of the highest power of x) is 1.

30. A cubic equation with real coefficients has roots 4 and $2 - i$. What is the third root? Write the equation in the form $P(x) = 0$, given that the leading coefficient is 1.

31. One zero of $P(x) = 2x^3 - 11x^2 + 28x - 24$ is $2 - 2i$. Factor $P(x)$ over the complex numbers.

32. One zero of $Q(x) = 3x^3 - 10x^2 + 7x + 10$ is $2 + i$. Factor $Q(x)$ over the complex numbers.

B

33. One root of $x^4 - 10x^3 + 35x^2 - 50x + 34 = 0$ is $4 - i$. Find the remaining roots.

34. One root of $5x^4 + 34x^3 + 40x^2 - 78x + 51 = 0$ is $-4 - i$. Find the remaining roots.

35. Argue that every polynomial equation with real coefficients and of odd degree has at least one real root.

36. One zero of a polynomial P is i. Is $-i$ necessarily a zero of P? Why or why not? The number i is a zero of $P(x) = x^3 + i$. Determine by synthetic division whether or not $-i$ is a zero.

37. Determine the three cube roots of -1. *Hint*: Consider the roots of the equation $x^3 + 1 = 0$.

38. Show that there are four fourth roots of -1. *Hint*: Consider the roots of the equation $x^4 + 1 = 0$, that is, of $(x^2 + i)(x^2 - i) = 0$.

39. Prove Theorem 6.6. *Hint*: Show that $\overline{P(z)} = P(\bar{z})$.

40. Fill in the details of the proof of Theorem 6.8.

41. Show that if P and Q are two polynomials of degree n and $P(x) = Q(x)$ for more than n values of x, then $P(x) = Q(x)$ for every x.

6.5 Rational Zeros of Polynomial Functions

Prime and composite numbers

The results of this section depend on the notion of a prime number. If a is a natural number, and $a \neq 1$, then a is a **prime number** if and only if a has no factor in N other than itself and 1; otherwise, a is a **composite number**. For example, 2 and 3 are prime numbers, and 6 (equal to $2 \cdot 3$) is composite; but 1 is considered to be neither prime nor composite.

The **unique-factorization theorem**, or **fundamental theorem of arithmetic** (which we shall not prove), follows.

Theorem 6.9 *If a is a composite number, then a is the product of only one set of prime factors; that is, there are unique prime numbers $p_1, \ldots, p_n$ and natural numbers $k_1, \ldots, k_n$ so that*

$$a = p_1^{k_1} \cdot p_2^{k_2} \cdot \cdots \cdot p_n^{k_n}.$$

Examples

a. $24 = 2 \cdot 2 \cdot 2 \cdot 3 = 2^3 \cdot 3$
c. $30 = 2 \cdot 3 \cdot 5$
b. $36 = 2 \cdot 2 \cdot 3 \cdot 3 = 2^2 3^2$
d. $90 = 2 \cdot 3 \cdot 3 \cdot 5 = 2 \cdot 3^2 \cdot 5$

Two integers a and b are said to be **relatively prime** if and only if they have no prime factors in common. For example, -6 and 35 are relatively prime, since $-6 = -2 \cdot 3$ and $35 = 5 \cdot 7$; but 6 and 8 are not, since they have the prime factor 2 in common. The fraction a/b is said to express a rational number in **lowest terms** if and only if a and b are relatively prime.

Identifying possible rational zeros

If all the coefficients of the polynomial

$$P(x) = a_n x^n + a_{n-1} x^{n-1} + \cdots + a_0$$

are integers, then we can identify all possible rational zeros of P by means of the following theorem.

Theorem 6.10 *If the rational number p/q, in lowest terms, is a solution of*

$$P(x) = a_n x^n + a_{n-1} x^{n-1} + \cdots + a_0 = 0,$$

where $a_j \in J$, then p is an integral factor of a_0 and q is an integral factor of a_n.

To see that this is true, let p/q be a solution of $P(x) = 0$, which is reduced to lowest terms. Then we have

$$a_n \left(\frac{p}{q}\right)^n + a_{n-1} \left(\frac{p}{q}\right)^{n-1} + \cdots + a_0 = 0,$$

and we can multiply each member here by q^n to obtain

$$a_n p^n + a_{n-1} p^{n-1} q + \cdots + a_0 q^n = 0.$$

Adding $-a_0 q^n$ to each member and factoring p from each term in the left-hand member of the resulting equation, we have

$$p(a_n p^{n-1} + a_{n-1} p^{n-2} q + \cdots + a_1 q^{n-1}) = -a_0 q^n.$$

Since the set J of integers is closed with respect to addition and multiplication, the expression in parentheses in the left-hand member here represents an integer, say r, so that we have

$$pr = -a_0 q^n,$$

where pr is an integer having p as a factor. Hence, p is a factor of $-a_0 q^n$. But, by Theorem 6.9, p and q^n have no factor in common, because, by hypothesis, p/q is

6.5 Rational Zeros of Polynomial Functions

in lowest terms; hence p must be a factor of a_0. In a similar manner, by writing the equation

$$a_n p^n + a_{n-1} p^{n-1} q + \cdots + a_0 q^n = 0$$

in the form

$$-a_n p^n = a_{n-1} p^{n-1} q + \cdots + a_0 q^n,$$

we can factor q from each term in the right-hand member and show that q must be a factor of a_n.

Example

List all possible rational zeros of

$$P(x) = 2x^3 - 4x^2 + 3x + 9.$$

Solution

Rational zeros, p/q, must, by Theorem 6.10, be such that p is an integral factor of 9 and q is an integral factor of 2. Hence

$$p \in \{-9, -3, -1, 1, 3, 9\}, \quad q \in \{-2, -1, 1, 2\},$$

and the set of possible rational zeros of P is

$$\left\{ -9, -\frac{9}{2}, -3, -\frac{3}{2}, -1, -\frac{1}{2}, \frac{1}{2}, 1, \frac{3}{2}, 3, \frac{9}{2}, 9 \right\}.$$

Test for rational zeros

It is important to observe that Theorem 6.10 does not assure us that a polynomial function with integral coefficients indeed has a rational zero; it simply enables us to identify possibilities for rational zeros. These can then be checked by synthetic division.

As a special case of Theorem 6.10, it is evident that if a function P is defined by the equation

$$P(x) = x^n + a_{n-1} x^{n-1} + \cdots + a_0,$$

in which $a_i \in J$ and $a_n = 1$, then any rational zero of P must be an integer, and, moreover, must be an integral factor of a_0.

Example

Find all rational zeros of $P(x) = x^3 - 4x^2 + x + 6$.

Solution

The only possible rational zeros of P are $-6, -3, -2, -1, 1, 2, 3$, and 6. Using synthetic division, we set up the following array:

	1	−4	1	6
1	1	−3	−2	4
−1	1	−5	6	0

Solution continued on overleaf

The third row in this array is the bottom row in division of $x^3 - 4x^2 + x + 6$ by $x + 1$. Thus we can cease our trials with -1, because by the remainder theorem, -1 is a zero of P, and the remaining zeros can be obtained from the equation $x^2 - 5x + 6 = 0$, whose coefficients are taken from the last line in the array. We can write the equation as $(x - 2)(x - 3) = 0$ and observe that 2 and 3 are also zeros.

Exercise 6.5

A Find all integral zeros of each function.

1. $f(x) = 3x^3 - 13x^2 + 6x - 8$
2. $f(x) = x^4 - x^2 - 4x + 4$
3. $f(x) = x^4 + x^3 + 2x - 4$
4. $f(x) = 5x^3 + 11x^2 - 2x - 8$
5. $P(x) = 2x^4 - 3x^3 - 8x^2 - 5x - 3$
6. $P(x) = 3x^4 - 40x^3 + 130x^2 - 120x + 27$

Find all rational zeros of each function.

7. $f(x) = 2x^3 + 3x^2 - 14x - 21$
8. $f(x) = 3x^4 - 11x^3 + 9x^2 + 13x - 10$
9. $P(x) = 4x^4 - 13x^3 - 7x^2 + 41x - 14$
10. $P(x) = 2x^3 - 4x^2 + 3x + 9$
11. $Q(x) = 2x^3 - 7x^2 + 10x - 6$
12. $Q(x) = x^3 + 3x^2 - 4x - 12$

Find all zeros of each function. *Hint*: First find all rational zeros.

13. $P(x) = 3x^3 - 5x^2 - 14x - 4$
14. $P(x) = x^3 - 4x^2 - 5x + 14$
15. $P(x) = 2x^4 + 3x^3 + 2x^2 - 1$
16. $P(x) = 8x^4 - 22x^3 + 29x^2 - 66x + 15$
17. $P(x) = 12x^4 + 7x^3 + 7x - 12$
18. $P(x) = 6x^4 - 13x^3 + 2x^2 - 4x + 15$

B 19. Factor the polynomial $2x^3 + 3x^2 - 2x - 3$ over C.

20. Factor the polynomial $x^4 - 6x^3 - 3x^2 - 24x - 28$ over C.

21. Show that $\sqrt{3}$ is irrational. *Hint*: Consider the equation $x^2 - 3 = 0$.

22. Show that $\sqrt{2}$ is irrational.

23. Show that if p is a prime number, then $\sqrt{p}$ is irrational.

24. Show that if n is any natural number *not* of the form $n = k^2$, $k \in N$, then $\sqrt{n}$ is irrational.

6.6 Graphing Polynomial Functions

In Section 5.2 we graphed linear functions
$$f(x) = a_1 x + a_0 \quad (a_1 \neq 0),$$
and in Section 6.1 we graphed quadratic functions
$$f(x) = a_2 x^2 + a_1 x + a_0 \quad (a_2 \neq 0).$$
We can graph any real polynomial function
$$P(x) = a_n x^n + a_{n-1} x^{n-1} + \cdots + a_0 \quad (a_n \neq 0)$$
by similar methods—that is, by combining the plotting of points with a consideration of certain general properties of the defining equations. The set of zeros of a polynomial function is one such property. The zeros of a polynomial were considered in Sections 6.4 and 6.5. Some additional properties will be discussed in this section.

General form of a polynomial graph

To determine whether the graph of a polynomial ultimately goes up or down to the right, that is, for large values of x, we can examine the leading coefficient a_n. If $a_n > 0$, then the graph ultimately goes up to the right, as in Figure 6.13-a and 6.13-c. If $a_n < 0$, then the graph ultimately goes down to the right, as in Figures 6.13-b and 6.13-d. To determine whether the graph of a polynomial ultimately goes up or down to the left, that is, for large negative values of x, we must examine both the leading coefficient a_n and the degree n. If $a_n > 0$ and n is even, then the graph ultimately goes up to the left, as in Figure 6.13-c. If $a_n > 0$ and n is odd, then the graph ultimately goes down to the left, as in Figure 6.13-a. If $a_n < 0$, we obtain

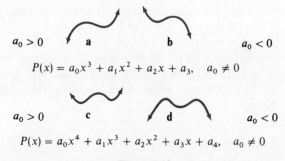

Figure 6.13

just the opposite results: If n is even, the graph ultimately goes down to the left; if n is odd, the graph ultimately goes up to the left. In each case, then, the leading term $a_n x^n$ governs the behavior of the graph of the polynomial function for large values of $|x|$.

Notice that for each graph in Figure 6.13 there are intervals on which the function values increase as x increases and intervals on which the function values decrease as x increases. A function is said to be **increasing** on an interval if the function values increase as x increases on the given interval. Similarly, a function is said to be **decreasing** on an interval if the function values decrease as x increases on the given interval.

Finding turning points

Another fact about polynomial functions involves *turning points*, or local maximum and minimum values of y. Thus, for example, in Figure 6.13-a there is one local maximum as well as one local minimum, or a total of two turning points. In Figure 6.13-c there is one local maximum and two local minima, or a total of three turning points. In general, we have the following result.

Theorem 6.11 *If $P(x) = a_n x^n + a_{n-1} x^{n-1} + \cdots + a_0 \ (a_n \neq 0)$ is a real polynomial of degree n, then its graph is a smooth curve that has at most $n - 1$ turning points.*

Since the proof of this theorem involves ideas from the calculus, it is omitted. As a consequence of this theorem, the graphs of third- and fourth-degree polynomial functions might appear as in Figure 6.13-a and 6.13-d, respectively.

To determine where the turning points of the graph of a polynomial lie, we use the following result. Its proof also involves ideas from the calculus and is also omitted.

Theorem 6.12 *The first components of the turning points of the graph of*
$$P(x) = a_n x^n + a_{n-1} x^{n-1} + \cdots + a_1 x + a_0$$
are solutions of the equation
$$P'(x) = n a_n x^{n-1} + (n-1) a_{n-1} x^{n-2} + \cdots + 2 a_2 x + a_1 = 0.$$

To find the turning points of the graph of a polynomial function $y = P(x)$, we first solve $P'(x) = 0$, where $P'(x)$ is as defined in Theorem 6.12. We then substitute the solutions that are real for x in $P(x)$ to obtain the second coordinates of the points. By plotting these points we can often determine by inspection whether each point is a local maximum or a local minimum. If inspection is not sufficient, then plotting a few more points on either side of the point will enable us to determine its nature.

Example

Find the turning points of
$$P(x) = x^3 - 3x^2 - 9x + 15.$$

6.6 Graphing Polynomial Functions

Solution

We first obtain
$$P'(x) = 3(1)x^{3-1} + 2(-3)x^{2-1} + (-9).$$
Then we solve
$$P'(x) = 3x^2 - 6x - 9$$
$$= 3(x - 3)(x + 1) = 0,$$
obtaining $x = 3$ and $x = -1$. Next we use the remainder theorem and synthetic division to compute $P(3)$ and $P(-1)$. We have

```
-1 | 1  -3  -9  15        3 | 1  -3  -9   15
       -1   4   5              3   0  -27
     1  -4  -5  20            1   0  -9  -12
```

and the possible turning points are $(-1, 20)$ and $(3, -12)$. Now we plot $(3, -12)$ and $(-1, 20)$. Since the curve ultimately goes down to the left and up to the right, we see by inspection of the figure that $(-1, 20)$ is a local maximum and $(3, -12)$ is a local minimum.

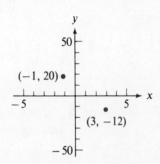

Graphing polynomial functions

When graphing a polynomial function, we determine the general form of the graph, locate and graph the turning points and intercepts, if necessary plot a few more points $(x, P(x))$, and then sketch the curve through these points.

Example

Graph $P(x) = 4x^3 + 3x^2 - 6x$.

Solution

Since P is defined by a cubic polynomial with positive leading coefficient, we expect a graph of form similar to that in Figure 6.13-a. To find the turning points of the graph, we solve
$$P'(x) = (3)4x^{3-1} + (2)3x^{2-1} + (-6)$$
$$= 12x^2 + 6x - 6 = 6(2x - 1)(x + 1) = 0,$$
obtaining $x = 1/2$ and $x = -1$. Using the remainder theorem and synthetic division to compute $P(-1)$ and $P(1/2)$, we have

```
-1 | 4   3  -6   0        1/2 | 4  3   -6    0
       -4   1   5                  2  5/2  -7/4
     4  -1  -5   5               4  5  -7/2  -7/4.
```

Thus, $(-1, 5)$ and $(1/2, -7/4)$ are on the graph.

Solution continued on overleaf

By inspection of the figure we see that $(-1, 5)$ is a local maximum and $(1/2, -7/4)$ is a local minimum. The y-intercept is $P(0) = 0$. Thus $(0, 0)$ is on the graph. To obtain the values of the x-intercepts, we solve $P(x) = 4x^3 + 3x^2 - 6x = 0$. Factoring, we obtain

$$4x^3 + 3x^2 - 6x = x(4x^2 + 3x - 6) = 0.$$

By inspection, one solution is $x = 0$; by using the quadratic formula to solve $4x^2 + 3x - 6 = 0$, we see that

$$x = \frac{-3 + \sqrt{105}}{8} \approx 0.91$$

and

$$x = \frac{-3 - \sqrt{105}}{8} \approx -1.66.$$

We approximate the points with these coordinates on the x-axis and sketch the curve, as shown in the figure.

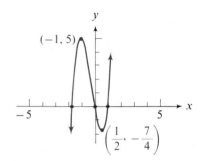

In the foregoing examples, all of the solutions of the equation $P'(x) = 0$ yielded turning points. This is *not* always the case, as the following example illustrates.

Example Find the turning points of $P(x) = x^3 - 3x^2 + 3x - 1$.

Solution We first obtain

$$P'(x) = 3(1)x^{3-1} + 2(-3)x^{2-1} + 3.$$

Solving

$$P'(x) = 3x^2 - 6x + 3 = 0,$$

we obtain

$$3x^2 - 6x + 3 = 3(x - 1)^2 = 0.$$

Hence, $x = 1$ and the *only possible* turning point (since 1 is the only solution to $P'(x) = 0$) is $(1, 0)$. The degree of $P(x)$ is 3 and the leading coefficient is positive; thus the graph ultimately goes down to the left and up to the right. If the point $(1, 0)$ were a turning point then there would have to be at least one more turning point for the graph to have this form. Hence, the graph has no turning points. By obtaining several additional ordered pairs we obtain the graph shown.

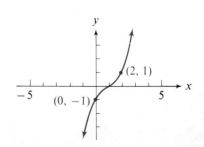

Exercise 6.6

A Find all of the possible turning points and state which are local maxima, which are local minima, and which are neither.

1. $P(x) = 2x^3 - 9x^2 + 12x + 10$
2. $P(x) = 2x^3 + 3x^2 - 72x + 50$
3. $P(x) = x^4 - 2x^2 + 1$
4. $P(x) = 3x^4 + 4x^3 - 12x^2 + 10$
5. $P(x) = 6x^4 - 8x^3 + 10$
6. $P(x) = 6x^5 - 15x^4 + 10x^3 + 20$

Graph. Include enough of the domain to show all turning points.

7. $P(x) = x^3 - 3x$
8. $P(x) = 2x^3 - 3x^2 - 12x + 18$
9. $P(x) = 4x^3 - 12x^2 + 9x$
10. $P(x) = 2x^3 - 5x^2 + 4x + 1$
11. $P(x) = \dfrac{x^3}{3} - \dfrac{x^2}{2} - 6x$
12. $P(x) = \dfrac{2x^3}{3} + \dfrac{5x^2}{2} - 3x$
13. $P(x) = x^3 - 4x^2 - 3x$
14. $P(x) = 3x^3 + 2x^2 - \dfrac{4}{3}x + 1$
15. $P(x) = -x^3 + 2x^2 - 1$
16. $P(x) = -10x^3 + 9x^2 + 12x$
17. $P(x) = -3x^3 + 9x^2 + 7x + 6$
18. $P(x) = \dfrac{-x^3}{3} + x$

B Graph. Include enough of the domain to show all turning points.

19. $P(x) = -x^4 + x^2$
20. $P(x) = -x^4 - 4x$
21. $P(x) = x^4 - 10x^2 + 9$
22. $P(x) = x^4 - 5x^2 + 4$
23. $P(x) = x^4 - 4x^3 + 4x^2 + 12$
24. $P(x) = x^4 - x^3 - 2x^2 + 3x - 3$
25. $P(x) = x^3$
26. $P(x) = -x^3 + 8$
27. Use Theorem 6.12 to find the coordinates of the vertex of the graph of $P(x) = ax^2 + bx + c$.
28. Find conditions on the real numbers a, b, and c which guarantee that the graph of $P(x) = ax^3 + bx^2 + cx + d$ has no turning points. Has one turning point. Has two turning points. If no such conditions exist, then so state.

6.7 Rational Functions

So far in this chapter we have discussed polynomial functions. In this section we discuss techniques for graphing rational functions.

A function defined by an equation of the form

$$y = \frac{P(x)}{Q(x)}, \tag{1}$$

Vertical asymptotes

where $P(x)$ and $Q(x)$ are polynomials in x and $Q(x)$ is not the zero polynomial, is called a **rational function**. Rational functions with real coefficients are of importance in the calculus and provide some interesting problems with respect to their graphs.

Since $P(x)/Q(x)$ is not defined for values of x for which $Q(x) = 0$, it is evident that we shall not be able to find points on the graph of (1) having such x-coordinates. We can, however, consider the graph for values of x as close as we please to a value, say x_0, for which $Q(x_0) = 0$, but still with $x \neq x_0$. This consideration is usually described by saying that x "approaches" x_0, "grows close" to x_0, etc., and correspondingly that $Q(x)$ approaches 0. Thus, the closer $Q(x)$ approaches 0, if $P(x)$ does not approach 0 at the same time, then the larger $|y|$ becomes in (1). For example, Figure 6.14-a shows the behavior of

$$y = \frac{2}{x-2}, \qquad (2)$$

as x approaches 2 from the right. In situations such as this, the vertical line that the curve approaches is called a **vertical asymptote**.

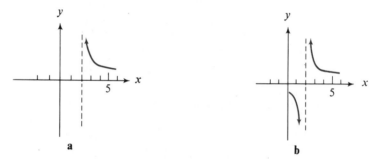

Figure 6.14

Theorem 6.13 *The graph of the function defined by $y = P(x)/Q(x)$ has a vertical asymptote $x = a$ for each value a at which $Q(x)$ vanishes and $P(x)$ does not vanish.*

Figure 6.14-a shows the behavior of the graph of Equation (2) as x approaches 2 from the right. We are equally interested in its behavior as x approaches 2 from the left. Figure 6.14-b illustrates this behavior. As long as $x > 2$, we have $x - 2 > 0$ and $2/(x-2) > 0$; but if $x < 2$, then we have $x - 2 < 0$ and $2/(x-2) < 0$.

Horizontal asymptotes

The graphs of some rational functions have **horizontal asymptotes**, which can, in general, be identified by using the following theorem.

Theorem 6.14 *The graph of the rational function defined by*

$$y = R(x) = \frac{a_n x^n + a_{n-1} x^{n-1} + \cdots + a_0}{b_m x^m + b_{m-1} x^{m-1} + \cdots + b_0}, \qquad (3)$$

6.7 Rational Functions

where a_n, $b_m \neq 0$, and n, m are nonnegative integers, has

    I   *a horizontal asymptote at* $y = 0$ *if* $n < m$,
    II   *a horizontal asymptote at* $y = a_n/b_m$ *if* $n = m$,
    III *no horizontal asymptotes if* $n > m$.

Though we shall not give a rigorous proof of this theorem, we can certainly make the results plausible. If $n < m$, we can divide the numerator and denominator of the right-hand member of (3) by x^m to obtain, for $x \neq 0$,

$$y = R(x) = \frac{\dfrac{a_n}{x^{m-n}} + \dfrac{a_{n-1}}{x^{m-n+1}} + \cdots + \dfrac{a_0}{x^m}}{b_m + \dfrac{b_{m-1}}{x} + \cdots + \dfrac{b_0}{x^m}}.$$

Now, as $|x|$ grows greater and greater, each term containing an x in its denominator grows closer and closer to 0, and we find the expression on the right approaching $0/b_0$, so that y approaches 0. But if, as $|x|$ increases without bound, y grows close to 0, then the graph of $y = R(x)$ approaches the line $y = 0$ asymptotically and actually must approach 0 from one of the directions shown in Figure 6.15-a. A

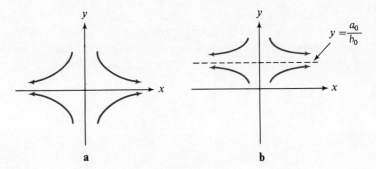

Figure 6.15

similar argument shows that if $n = m$, then, as $|x|$ increases without bound, the graph of $y = R(x)$ approaches the line $y = a_n/b_m$ from one of the directions shown in Figure 6.15-b. If $n > m$, then as $|x|$ becomes greater and greater, so does $|y|$. Thus the curve has no horizontal asymptote.

Example   Find all horizontal and vertical asymptotes of the graph of the function

$$y = \frac{x^2 - 4}{x - 1}.$$

Solution   We observe by Theorem 6.13 that $x = 1$ is a vertical asymptote and by Theorem 6.14 that there are no horizontal asymptotes.

Helpful items for graphing

Identifying asymptotes is one aid to graphing a rational function. Other helpful items are the following:

1. the zeros of the function, because these give us the *x*-intercepts,
2. the domain and range, because these let us know where we can expect to find parts of the graph and where we cannot, and
3. some specific points on the graph, because these give us guidelines in sketching.

Example

Graph $y = \dfrac{x-1}{x-2}$.

Solution

We can begin by observing that the numerator of the right-hand member will be equal to 0 when x is equal to 1. Therefore, when $x = 1$ we have $y = 0$, and 1 is an x-intercept. Also, when $x = 0$ we have $y = 1/2$, so that $1/2$ is a y-intercept. Thus we can begin our graph as shown in Figure a. By inspection, $y = (x - 1)/(x - 2)$ is defined for all x except $x = 2$, so that $\{x \mid x \neq 2\}$ is the domain. Similarly, if we solve the defining equation for x in terms of y, we have

$$x = \frac{2y-1}{y-1},$$

which is defined for all values of y except 1. Hence $\{y \mid y \neq 1\}$ is the range of the function. From the defining equation and Theorems 6.13 and 6.14, we see that there is a vertical asymptote at $x = 2$ and a horizontal asymptote at $y = 1$. We can then add this information to our graph, as indicated in Figure b. To determine the

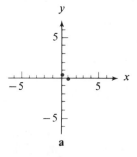
a

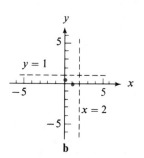
b

behavior of the function near the vertical asymptote, we consider one side of the asymptote at a time. On the left side x is just less than 2, say $2 - p$, $0 < p < 1/10$, so the denominator $x - 2$ in the expression for y is

$$2 - p - 2 = -p,$$

which is barely negative, whereas the numerator

$$x - 1 = 2 - p - 1 = 1 - p$$

is definitely positive; hence y is negative and $|y|$ large. Thus the graph goes down on the left side of the vertical asymptote. The graph appears in Figure c. To the right of the vertical asymptote, we consider $x = 2 + p$, $0 < p < 1/10$, to find that the graph goes up on the right of the vertical asymptote. This along with the knowledge that $y = 1$ is a horizontal asymptote leads us to the complete graph of the function $y = (x - 1)/(x - 2)$, as shown in Figure d.

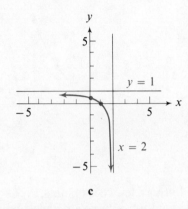

c

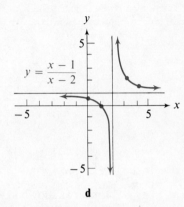

d

Oblique asymptotes

In Theorem 6.14 if $n = m + 1$, that is, if the degree of the numerator is one greater than that of the denominator, we can argue that though the graph has no horizontal asymptote, it does have an **oblique asymptote**. We shall illustrate a special case only, but the technique involved is quite general.

Example

Find all asymptotes for the graph of the function

$$y = \frac{x^2 - 4}{x - 1}.$$

Solution

This is the same function we investigated in the example on page 203, where we found the vertical asymptote $x = 1$ and no horizontal asymptote. If we rewrite $y = (x^2 - 4)/(x - 1)$ by dividing $x^2 - 4$ by $x - 1$, we obtain

$$y = x + 1 - \frac{3}{x - 1}.$$

Now, as x grows greater and greater, $3/(x - 1)$ becomes closer and closer to 0 and the graph of $y = (x^2 - 4)/(x - 1)$ approaches the graph of $y = x + 1$. Hence, the graph of $y = x + 1$, which is an oblique line, is an asymptote to the curve.

Example

Graph $y = \dfrac{x^2 - 4}{x - 1}$.

Solution

This is the same function we investigated in the preceding example, for which we found the vertical asymptote $x = 1$ and the oblique asymptote $y = x + 1$. If $x = 0$ then $y = 4$, and if $y = 0$ then $x = 2$ or $x = -2$, so that the y-intercept

Solution continued on overleaf

is 4 and the x-intercepts are 2 and -2. We therefore have the situation shown in Figure a. Without further information, we have reason to suspect that the graph will

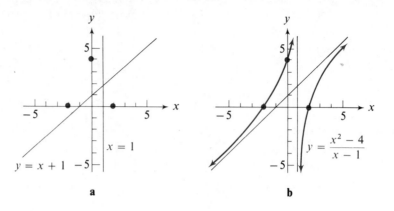

a b

appear as shown in Figure b. Plotting a few check points, say $(-1, 3/2)$, $(1/2, 15/2)$, $(3/2, -7/2)$, and $(3, 5/2)$, would tend to confirm our conjecture.

Exercise 6.7

A *Determine the vertical asymptotes of the graph of each function.*

Example $x^2 y - 4y = 1$

Solution Express y explicitly in terms of x.

$$y(x^2 - 4) = 1$$

$$y = \frac{1}{x^2 - 4} = \frac{1}{(x - 2)(x + 2)}$$

By Theorem 6.13 there are vertical asymptotes at $x = 2$ and $x = -2$.

1. $y = \dfrac{1}{x - 3}$
2. $y = \dfrac{1}{x + 4}$
3. $y = \dfrac{4}{(x + 2)(x - 3)}$
4. $y = \dfrac{8}{(x - 1)(x + 3)}$
5. $y = \dfrac{2x - 1}{x^2 + 5x + 4}$
6. $y = \dfrac{x + 3}{2x^2 - 5x - 3}$
7. $xy + y = 4$
8. $x^2 y + xy = 3$

6.7 Rational Functions

Graph.

9. $y = \dfrac{1}{x}$
10. $y = \dfrac{1}{x+4}$
11. $y = \dfrac{1}{x-3}$
12. $y = \dfrac{1}{x-6}$
13. $y = \dfrac{4}{(x+2)(x-3)}$
14. $y = \dfrac{8}{(x-1)(x+3)}$
15. $y = \dfrac{2}{(x-3)^2}$
16. $y = \dfrac{1}{(x+4)^2}$

Determine any vertical, horizontal, or oblique asymptotes of the graphs of each of the following functions.

Example

$$y = \frac{6x^2 + 1}{2x^2 + 5x - 3}$$

Solution

The equation can be written equivalently as

$$y = \frac{6x^2 + 1}{(2x - 1)(x + 3)}.$$

By Theorem 6.13 there are vertical asymptotes at $x = 1/2$ and $x = -3$. By Theorem 6.14 there is a horizontal asymptote at $y = 6/2 = 3$.

17. $y = \dfrac{x}{x^2 - 4}$
18. $y = \dfrac{3x - 6}{x^2 + 3x + 2}$
19. $y = \dfrac{x^2 - 9}{x - 4}$
20. $y = \dfrac{x^3 - 27}{x^2 - 1}$
21. $y = \dfrac{x^2 - 3x + 2}{x^2 - 3x - 4}$
22. $y = \dfrac{x^2}{x^2 - x - 6}$

Graph. Use information concerning the zeros of the function and concerning vertical, horizontal, and oblique asymptotes.

23. $y = \dfrac{x}{x - 2}$
24. $y = \dfrac{x - 1}{x + 3}$
25. $y = \dfrac{2x - 4}{x^2 - 9}$
26. $y = \dfrac{3x}{x^2 - 5x + 4}$
27. $y = \dfrac{x^2 - 4}{x^3}$
28. $y = \dfrac{x - 2}{x^2}$
29. $y = \dfrac{x + 1}{x(x^2 - 4)}$
30. $y = \dfrac{x^2 + x - 2}{x(x^2 - 9)}$
31. $y = \dfrac{x^2 - 4x + 4}{x - 1}$
32. $y = \dfrac{x^2 + 4}{x - 2}$
33. $y = \dfrac{x^3 + 1}{x^2}$
34. $y = \dfrac{x^3 - x^2 - 2x}{x^2 - 1}$

B Graph. Use information concerning the zeros of the function, and vertical, horizontal, and oblique asymptotes if they exist. Hint: Reduce the rational expression to lowest terms and keep in mind that the function is undefined for all values at which the original denominator vanishes.

35. $y = \dfrac{x^2 - 9}{x + 3}$

36. $y = \dfrac{x^2 - 1}{x + 1}$

37. $y = \dfrac{x^3 - 3x^2 + 2x}{x - 1}$

38. $y = \dfrac{-x^3 + 2x^2 - x}{x}$

39. $y = \dfrac{x - 1}{x^2 - x}$

40. $y = \dfrac{x}{x^2 - 4x}$

41. Describe the asymptotic behavior of $y = \dfrac{x^3 + x^2 + 2}{x + 1}$.

42. Sketch the graph of $y = \dfrac{x^3 + x^2 + 2}{x + 1}$.

43. Is it possible for the graph of a rational function to cross a vertical asymptote? Why or why not?

44. Find the point where the graph of $y = \dfrac{x^2 + 1}{x^2 - 2x - 1}$ crosses its horizontal asymptote.

45. Find the point where the graph of $y = \dfrac{2x^3 + x^2 + 3x}{x^2 + 1}$ crosses its oblique asymptote.

46. Give an example of a rational function with the property that its graph crosses its oblique asymptote more than once.

Chapter Review

[6.1] Find the x-intercepts, the axis of symmetry, and the maximum or minimum point of the graph of each function by analytic methods.

1. $f(x) = x^2 - x - 6$

2. $f(x) = -x^2 + 7x - 10$

3. Graph the function of Exercise 1. Draw a line segment from $(4, 0)$ to $(4, f(4))$.

4. Graph the function of Exercise 2. Draw a line segment from $(3, 0)$ to $(3, f(3))$.

Graph f and f^{-1} on the same set of axes.

5. $f(x) = x^2 - 9,\ x \geq 0$

6. $f(x) = x^2 - 4x - 5,\ x \leq 2$

Graph the given inequality.

7. $y < x^2 - 9$

8. $y \geq x^2 + 6x + 5$

Review Exercises

[6.2] *Name and sketch the graph of the given equation.*

9. $x^2 - 16 = 4y^2$
10. $2y^2 = 8 - 2x^2$
11. $x^2 = 36 - 4y^2$
12. $x^2 - y - 9 = 0$
13. $x^2 - 9y^2 = 0$
14. $-x^2 + 9y^2 = 0$

[6.3] 15. If y varies directly as x^2, and $y = 20$ when $x = 2$, find y when $x = 5$.

16. If r varies directly as x^2 and inversely as z^3, and $r = 4$ when $x = 3$ and $z = 2$, find r when $x = 2$ and $z = 3$.

17. The number of posts needed to string a telephone line over a given distance varies inversely as the distance between posts. If it takes 80 posts separated by 120 feet to string a wire between two points, how many posts would be required if the posts were 150 feet apart?

18. The speed of a gear varies directly as the number of teeth it contains. If a gear with 10 teeth rotates at 240 revolutions per minute (RPM), with what speed would a gear with 18 teeth revolve under the same conditions?

[6.4] 19. Use Theorem 6.2 to show that $P(x) = 2x^3 + x^2 - 4x - 2$ has a zero between 1 and 2.

20. Use Theorem 6.2 to show that $P(x) = x^4 - x$ has a zero between 1/2 and 3/2.

21. Use synthetic division to find $P(2)$, where $P(x) = x^3 - 3x^2 + 2x + 1$.

22. Use synthetic division to find $P(-3)$, where $P(x) = x^4 - 3x^2 - 1$.

Find an upper bound and a lower bound for the real zeros of the given polynomial function.

23. $P(x) = 2x^3 + x^2 - 4x - 2$
24. $P(x) = x^4 + 2x^3 - 7x^2 - 10x + 10$

Find all of the zeros of the given polynomial function.

25. $P(x) = x^3 - 2x^2 + 4x - 8$; $2i$ is one zero.
26. $Q(x) = 2x^3 - 11x^2 + 28x - 24$; $2 + 2i$ is one zero.

[6.5] 27. Find all integral zeros of $Q(x) = x^4 - 3x^3 - x^2 - 11x - 4$.

28. Find all rational zeros of $P(x) = 2x^3 - 11x^2 + 12x + 9$.

29. Find all zeros of $P(x) = x^3 + 2x^2 + 2x + 4$.

30. Find all zeros of $P(x) = 6x^4 - x^3 - 13x^2 + 2x - 2$.

[6.6] Find all possible turning points and state which are local maxima, which are local minima, and which are neither.

31. $P(x) = 2x^3 - 9x^2 + 12x$

32. $P(x) = x^3 - x^2$

Graph the function.

33. $P(x) = 2x^3 - 9x^2 + 12x$

34. $P(x) = 2x^3 + 3x^2$

35. $P(x) = x^3 - x^2$

36. $P(x) = x^4 - 2x^2 - 3$

[6.7] Determine all asymptotes and graph each function.

37. $y = \dfrac{3}{x + 2}$

38. $y = \dfrac{2x}{(x + 3)(x - 4)}$

39. $y = \dfrac{x + 2}{x - 3}$

40. $y = \dfrac{2x}{x^2 - 3x + 2}$

7 Exponential and Logarithmic Functions

The functions studied in Chapters 5 and 6 are rational functions as defined in Section 6.7. In this chapter we consider two classes of functions that are not rational.

7.1 Exponential Functions

Powers of the form b^x, where $b \in R$, $b > 0$, and x denotes a *rational number*, were discussed in Section 3.2. These can be used to define functions. Notice that the base b is restricted here to positive real numbers to ensure that b^x be real for all rational numbers x.

Since we now want to define a function over R using b^x, we must be able to interpret powers with *irrational exponents*, such as

$$b^{\sqrt{2}}, b^{-\sqrt{3}}, \text{ and } b^\pi,$$

to be real numbers. The following theorem, which is presented without proof, will be useful in doing this.

Theorem 7.1 Let $x, y \in Q$, and let $x > y$. Then
$$b^x > b^y \text{ if } b > 1, \quad b^x = b^y \text{ if } b = 1, \quad \text{and} \quad b^x < b^y \text{ if } 0 < b < 1.$$

Powers with real-number exponents

You should recall that irrational numbers can be approximated by rational numbers to as great a degree of accuracy as desired. For example, $\sqrt{2} \approx 1.4$, or $\sqrt{2} \approx 1.414$, etc. Since 2^x is defined for rational x, by Theorem 7.1 we can write the sequence of inequalities

$$2^1 < 2^2,$$
$$2^{1.4} < 2^{1.5},$$
$$2^{1.41} < 2^{1.42},$$
$$2^{1.414} < 2^{1.415},$$
$$\vdots$$

It seems plausible and is actually true, though we shall not prove it, that the difference between the number on the left and that on the right can be made as small as we please and that there is just one number, which we denote by $2^{\sqrt{2}}$, that lies between the number on the left and the number on the right, no matter how long this process of approximation is continued. Since we can produce the same type of argument for any irrational exponent x and any positive base b, we shall assume that b^x $(b > 0)$ is defined in this way for all real values of x and that the laws of exponents, appropriately reworded for real-number exponents, are valid for such powers.

Since for any given $b > 0$ and for each $x \in R$, there is only one value for b^x, the equation

$$f(x) = b^x \quad (b > 0) \tag{1}$$

defines a function. Because $1^x = 1$ for all $x \in R$, (1) defines a constant function if $b = 1$. If $b \neq 1$, we say that (1) defines an **exponential function**. Exponential functions can perhaps be visualized most clearly by considering their graphs. We illustrate two typical examples, in which $0 < b < 1$ and $b > 1$, respectively. Assigning values to x in the equations

$$f(x) = \left(\frac{1}{2}\right)^x \quad \text{and} \quad f(x) = 2^x,$$

we find some ordered pairs in each solution set and sketch the graphs shown in Figures 7.1-a and 7.1-b, respectively.

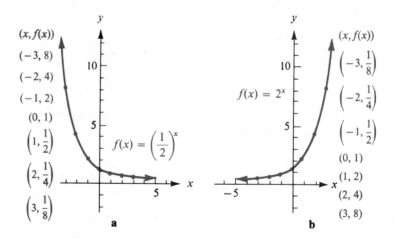

Figure 7.1

Increasing and decreasing functions

Notice that, in accordance with Theorem 7.1, the graph of the function determined by $f(x) = (1/2)^x$ goes *down* to the right, and the graph of the function determined by $f(x) = 2^x$ goes *up* to the right. For this reason, we say that the former function is a *decreasing* function and the latter is an *increasing* function. In either case, the domain is the set of real numbers, and the range is the set of positive real numbers.

7.1 Exponential Functions

Frequently used bases

There are two exponential functions that are of special interest because they are frequently used in mathematics and other sciences. The base for one of these functions is 10 and the base for the other is an irrational number which we shall denote by e. The value of e to eight decimal places is 2.71828183. Table I on page 413 shows values of e^x and e^{-x} for selected values of x; alternatively, a calculator can be used to find such powers.

Examples

a. $e^2 = 7.3891$

b. $e^{-2} = 0.1353$

Exercise 7.1

A *Find the second component of each of the ordered pairs that makes the pair a solution of the corresponding equation.*

1. $y = 3^x$; (0,), (1,), (2,)
2. $y = -2^x$; (0,), (1,), (2,)
3. $y = -5^x$; (0,), (1,), (2,)
4. $y = 4^x$; (0,), (1,), (2,)
5. $y = \left(\frac{1}{2}\right)^x$; (−3,), (0,), (3,)
6. $y = \left(\frac{1}{3}\right)^x$; (−3,), (0,), (3,)
7. $y = 10^x$; (−2,), (1,), (0,)
8. $y = 10^{-x}$; (0,), (1,), (2,)
9. $y = e^x$; (−1,), (0,), (1,)
10. $y = e^{-x}$; (−1,), (0,), (1,)

Graph the function.

11. $y = 4^x$
12. $y = 5^x$
13. $y = 10^x$
14. $y = 10^{-x}$
15. $y = 2^{-x}$
16. $y = 3^{-x}$
17. $y = \left(\frac{1}{3}\right)^x$
18. $y = \left(\frac{1}{4}\right)^x$
19. $y = e^x$
20. $y = e^{-x}$

21. Graph $f: y = 10^x$ and $f^{-1}: x = 10^y$ on the same set of axes.
22. Graph $f: y = 2^x$ and $f^{-1}: x = 2^y$, on the same set of axes.

Solve for x by inspection.

23. $10^x = \dfrac{1}{100}$
24. $\left(\dfrac{1}{2}\right)^x = 16$
25. $16^x = 8$
26. $8^x = 16$

B *Determine an integer n such that $n < x < n + 1$.*

27. $3^x = 16.2$
28. $4^x = 87.1$
29. $10^x = 0.016$
30. $2^x = 6$

31. Graph $y = b^x$, $b = 2, 3$ and 4, on the same set of coordinate axes. What effect does increasing b have on the curves if $b > 1$?

32. Graph $y = b^x$, $b = \frac{1}{4}, \frac{1}{3},$ and $\frac{1}{2}$, on the same set of coordinate axes. What effect does increasing b have on the curves if $0 < b < 1$?

7.2 Logarithmic Functions

Inverse of an exponential function

In the exponential function

$$y = b^x \quad (b > 0, b \neq 1), \tag{1}$$

illustrated in Figure 7.1 for $b = 1/2$ and $b = 2$, there is only one x associated with each y. Thus, by Definition 5.3 we have the inverse function

$$x = b^y \quad (b > 0, b \neq 1). \tag{2}$$

Observe that Equation (2) implies $x > 0$, because there is no real number y for which b^y is not positive.

The graphs of functions of this form can be illustrated by the example

$$x = 10^y.$$

We consider x the variable denoting an element in the domain and, in the defining equation, we assign specific values for x, say,

$$(0.01, \quad), (0.1, \quad), (1, \quad), (10, \quad), (100, \quad),$$

to obtain the ordered pairs

$$(0.01, -2), (0.1, -1), (1, 0), (10, 1), (100, 2).$$

These can be graphed and connected with a smooth curve as shown in Figure 7.2.

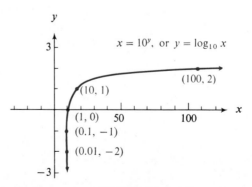

Figure 7.2

7.2 Logarithmic Functions

It is always useful to be able to express the variable denoting an element in the range explicitly in terms of the variable denoting an element in the domain. To do this for Equation (2), we use the notation

$$y = \log_b x \quad (x > 0, b > 0, b \neq 1). \tag{3}$$

Thus for $x > 0, b > 0, b \neq 1$, we have

$$y = \log_b x \quad \text{if and only if} \quad x = b^y.$$

Here, $\log_b x$ is read "the logarithm to the base b of x," or "the logarithm of x to the base b." The function defined by Equation (3) is called a **logarithmic function**.

Observe that $\log_b x$ is the *exponent* y such that the power b^y is equal to x; that is,

$$b^{\log_b x} = x.$$

Properties of logarithmic functions

From the graph in Figure 7.2, we generalize from $\log_{10} x$ to $\log_b x$, and observe that for $b > 1$, a logarithmic function has the following properties:

1. The domain is the set of positive real numbers, and the range is the set of all real numbers.
2. $\log_b x < 0$ for $0 < x < 1$, $\log_b x = 0$ for $x = 1$, and $\log_b x > 0$ for $x > 1$.

Logarithmic and exponential statements

It should be recognized that Equations (2) and (3) are two equations determining the same function, in the same way that $x = y + 4$ and $y = x - 4$ determine the same function, and we may use whichever equation suits our purpose. Thus, exponential statements may be written in logarithmic form, and logarithmic statements may be written in exponential form.

Examples

a. $5^2 = 25$ can be written as $\log_5 25 = 2$.
b. $8^{1/3} = 2$ can be written as $\log_8 2 = \dfrac{1}{3}$.
c. $3^{-3} = \dfrac{1}{27}$ can be written as $\log_3 \dfrac{1}{27} = -3$.

Examples

a. $\log_{10} 100 = 2$ can be written as $10^2 = 100$.
b. $\log_3 81 = 4$ can be written as $3^4 = 81$.
c. $\log_2 \dfrac{1}{2} = -1$ can be written as $2^{-1} = \dfrac{1}{2}$.

Notice that in each of the above examples the logarithm and the exponent is equal. Since a logarithm is an exponent, the following theorem follows directly from the properties of powers with real-number exponents.

Theorem 7.2 If $x_1, x_2 \in R$, $x_1, x_2, b > 0$, $b \neq 1$, $m \in R$, then

I $\quad \log_b(x_1 x_2) = \log_b x_1 + \log_b x_2,$

II $\quad \log_b \dfrac{x_2}{x_1} = \log_b x_2 - \log_b x_1,$

III $\quad \log_b(x_1)^m = m \log_b x_1.$

The validity of I can be shown as follows: Since
$$x_1 = b^{\log_b x_1} \quad \text{and} \quad x_2 = b^{\log_b x_2},$$
it follows that
$$x_1 x_2 = b^{\log_b x_1} \cdot b^{\log_b x_2} = b^{\log_b x_1 + \log_b x_2},$$
and, by the definition of a logarithm,
$$\log_b(x_1 x_2) = \log_b x_1 + \log_b x_2.$$

The validity of II and III can be established in a similar way and are left as exercises.

Examples

a. $\log_{10} 100 = \log_{10}(10 \cdot 10)$
$\phantom{\log_{10} 100} = \log_{10} 10 + \log_{10} 10$
$\phantom{\log_{10} 100} = 1 + 1 = 2$

b. $\log_2 \dfrac{1}{4} = \log_2 1 - \log_2 4$
$\phantom{\log_2 \dfrac{1}{4}} = 0 - 2 = -2$

c. $\log_3 81 = \log_3 3^4$
$ = 4 \log_3 3 = 4 \cdot 1 = 4$

d. $\log_2 \dfrac{1}{8} = \log_2 8^{-1}$
$\phantom{\log_2 \dfrac{1}{8}} = -1 \cdot \log_2 8 = -1 \cdot 3 = -3$

Logarithmic equations

An equation which involves logarithms can sometimes be solved by using Theorem 7.2 to rewrite the equation equivalently as an equation involving a single logarithm with coefficient 1 and then writing the equation in exponential notation.

Example

Solve $\log_{10}(x + 6) = 1 - \log_{10}(x - 3)$.

Solution

We first rewrite the given equation as
$$\log_{10}(x + 6) + \log_{10}(x - 3) = 1.$$
We then use Theorem 7.2-I to obtain
$$\log_{10}[(x + 6)(x - 3)] = 1.$$
Next we rewrite this equation in exponential notation to obtain
$$(x + 6)(x - 3) = 10^1.$$
Solving this equation for x, we obtain $x = -7$ and $x = 4$. Neither $\log_{10}(-7 + 6)$ nor $\log_{10}(-7 - 3)$ is defined. Thus -7 does not satisfy the original equation. Since 4 satisfies the equation, the solution set is $\{4\}$.

7.2 Logarithmic Functions

Exercise 7.2

A *Express in logarithmic notation.*

Examples
a. $2^3 = 8$
b. $\left(\dfrac{1}{2}\right)^{-1} = 2$
c. $4^{1/2} = 2$

Solutions
a. $\log_2 8 = 3$
b. $\log_{1/2} 2 = -1$
c. $\log_4 2 = \dfrac{1}{2}$

1. $4^2 = 16$
2. $5^3 = 125$
3. $3^3 = 27$
4. $8^2 = 64$
5. $\left(\dfrac{1}{2}\right)^2 = \dfrac{1}{4}$
6. $\left(\dfrac{1}{3}\right)^2 = \dfrac{1}{9}$
7. $8^{-1/3} = \dfrac{1}{2}$
8. $64^{-1/6} = \dfrac{1}{2}$
9. $10^2 = 100$
10. $10^0 = 1$
11. $10^{-1} = 0.1$
12. $10^{-2} = 0.01$

Express in exponential notation.

Examples
a. $\log_4 16 = 2$
b. $\log_{1/2} 8 = -3$
c. $\log_{64} 8 = \dfrac{1}{2}$

Solutions
a. $4^2 = 16$
b. $\left(\dfrac{1}{2}\right)^{-3} = 8$
c. $64^{1/2} = 8$

13. $\log_2 64 = 6$
14. $\log_5 25 = 2$
15. $\log_3 9 = 2$
16. $\log_{16} 256 = 2$
17. $\log_{1/3} 9 = -2$
18. $\log_{1/2} 8 = -3$
19. $\log_{10} 1000 = 3$
20. $\log_{10} 1 = 0$

Find the value of each expression.

21. $\log_7 49$
22. $\log_2 32$
23. $\log_4 64$
24. $\log_3 \dfrac{1}{3}$
25. $\log_5 \dfrac{1}{5}$
26. $\log_3 3$
27. $\log_2 2$
28. $\log_{10} 10$
29. $\log_{10} 100$
30. $\log_{10} 1$
31. $\log_{10} 0.1$
32. $\log_{10} 0.01$

Solve for x, y, or b.

Examples
a. $\log_2 x = 3$
b. $\log_b 2 = \dfrac{1}{2}$
c. $\log_{1/4} 16 = y$

Solution on overleaf

Determine the solution by inspection or by first writing the equation equivalently in exponential form.

Solutions

a. $2^3 = x$
   $x = 8$

b. $b^{1/2} = 2$
   $(b^{1/2})^2 = (2)^2$
   $b = 4$

c. $\left(\dfrac{1}{4}\right)^y = 16$
   $y = -2$

33. $\log_3 9 = y$
34. $\log_5 125 = y$
35. $\log_b 8 = 3$
36. $\log_b 625 = 4$
37. $\log_4 x = 3$
38. $\log_{1/2} x = -5$
39. $\log_2 \dfrac{1}{8} = y$
40. $\log_5 5 = y$
41. $\log_b 10 = \dfrac{1}{2}$
42. $\log_b 0.1 = -1$
43. $\log_2 x = 2$
44. $\log_{10} x = -3$

Express as the sum or difference of simpler logarithmic quantities.

Example

$$\log_b\left(\dfrac{xy}{z}\right)^{1/2}$$

Solution

By Theorem 7.2-III,

$$\log_b\left(\dfrac{xy}{z}\right)^{1/2} = \dfrac{1}{2}\log_b\left(\dfrac{xy}{z}\right).$$

By Theorem 7.2-I and 7.2-II,

$$\dfrac{1}{2}\log_b\left(\dfrac{xy}{z}\right) = \dfrac{1}{2}(\log_b x + \log_b y - \log_b z) = \dfrac{1}{2}\log_b x + \dfrac{1}{2}\log_b y - \dfrac{1}{2}\log_b z.$$

45. $\log_b(xy)$
46. $\log_b(xyz)$
47. $\log_b\left(\dfrac{x}{y}\right)$
48. $\log_b\left(\dfrac{xy}{z}\right)$
49. $\log_b x^5$
50. $\log_b x^{1/2}$
51. $\log_b \sqrt[3]{x}$
52. $\log_b \sqrt[3]{x^2}$
53. $\log_b \sqrt{\dfrac{x}{z}}$
54. $\log_b \sqrt{xy}$
55. $\log_{10} \sqrt[3]{\dfrac{xy^2}{z}}$
56. $\log_{10} \sqrt[5]{\dfrac{x^2 y}{z^3}}$

Express as a single logarithm with coefficient 1.

Example

$\dfrac{1}{2}(\log_b x - \log_b y)$

Solution

By Theorems 7.2-II and 7.2-III,

$$\dfrac{1}{2}(\log_b x - \log_b y) = \dfrac{1}{2}\log_b\left(\dfrac{x}{y}\right) = \log_b\left(\dfrac{x}{y}\right)^{1/2}.$$

7.3 Special Logarithms and Powers

57. $\log_b 2x + 3 \log_b y$
58. $3 \log_b x - \log_b 2y$
59. $\frac{1}{2} \log_b x + \frac{2}{3} \log_b y$
60. $\frac{1}{4} \log_b x - \frac{3}{4} \log_b y$
61. $3 \log_b x + \log_b y - 2 \log_b z$
62. $\frac{1}{3}(\log_b x + \log_b y - 2 \log_b z)$
63. $\log_{10}(x - 2) + \log_{10} x - 2 \log_{10} z$
64. $\frac{1}{2}(\log_{10} x - 3 \log_{10} y - 5 \log_{10} z)$

Solve for x.

Example $\log_{10}(x + 9) + \log_{10} x = 1$

Solution By Theorem 7.2-I, the given equation is equivalent to
$$\log_{10} x(x + 9) = 1$$
for $x + 9 > 0$ and $x > 0$. Write this equation in exponential form to obtain
$$x(x + 9) = 10^1,$$
$$x^2 + 9x - 10 = 0.$$
Solve for x to obtain $x = -10$ and $x = 1$. Since neither $\log_{10}(-10 + 9)$ nor $\log_{10}(-10)$ is defined, -10 does not satisfy the original equation. Since 1 satisfies the equation, the solution set is $\{1\}$.

65. $\log_{10} x + \log_{10} 2 = 3$
66. $\log_{10}(x - 1) - \log_{10} 4 = 2$
67. $\log_{10} x + \log_{10}(x + 21) = 2$
68. $\log_{10}(x + 3) + \log_{10} x = 1$
69. $\log_{10}(x + 2) + \log_{10}(x - 1) = 1$
70. $\log_{10}(x - 3) - \log_{10}(x + 1) = 1$

B 71. Show that $\log_b 1 = 0$ for $b > 0$, $b \neq 1$.
72. Show that $\log_b b = 1$ for $b > 0$, $b \neq 1$.
73. Prove Theorem 7.2-II.
74. Prove Theorem 7.2-III.

7.3 Special Logarithms and Powers

The two logarithmic functions
$$y = \log_{10} x \quad \text{and} \quad y = \log_e x,$$
where e is the irrational number introduced in Section 7.1, are of special interest. We shall first consider the logarithm and exponential functions with the base 10.

Determination of $\log_{10} x$

Values for $\log_{10} x$ are called **logarithms to the base 10**, or **common logarithms**. From the definition of $\log_{10} x$,

$$10^{\log_{10} x} = x \quad (x > 0); \tag{3}$$

that is, $\log_{10} x$ is the exponent that must be placed on 10 so that the resulting power is x. Thus, $\log_{10} x$ can easily be determined for all values of x that are *integral* powers of 10 by inspection, directly from the definition of a logarithm:

$$\log_{10} 10 = \log_{10} 10^1 = 1,$$
$$\log_{10} 100 = \log_{10} 10^2 = 2,$$

and so forth; similarly,

$$\log_{10} 1 = \log_{10} 10^0 = 0,$$
$$\log_{10} 0.1 = \log_{10} 10^{-1} = -1,$$
$$\log_{10} 0.01 = \log_{10} 10^{-2} = -2,$$
$$\log_{10} 0.001 = \log_{10} 10^{-3} = -3.$$

Using tables

Values for $\log_{10} x$ in case x is not an integral power of 10 can be obtained by using a table of logarithms (Table II on pages 415 and 416) and Theorem 7.2-I or by using a calculator. We shall first consider how these values are obtained by using the table.

For $1 < x < 10$, the table on pages 415 and 416 provides values of $\log_{10} x$, directly. Consider the excerpt from this table shown in Figure 7.3. Each number in

x	0	1	2	3	4	5	6	7	8	9
3.8	0.5798	0.5809	0.5821	0.5832	0.5843	0.5855	0.5866	0.5877	0.5888	0.5899
3.9	0.5911	0.5922	0.5933	0.5944	0.5955	0.5966	0.5977	0.5988	0.5999	0.6010
4.0	0.6021	0.6031	0.6042	0.6053	0.6064	0.6075	0.6085	0.6096	0.6107	0.6117
4.1	0.6128	0.6138	0.6149	0.6160	0.6170	0.6180	0.6191	0.6201	0.6212	0.6222
4.2	0.6232	0.6243	0.6253	0.6263	0.6274	0.6284	0.6294	0.6304	0.6314	0.6325
4.3	0.6335	0.6345	0.6355	0.6365	0.6375	0.6385	0.6395	0.6405	0.6415	0.6425
4.4	0.6435	0.6444	0.6454	0.6464	0.6474	0.6484	0.6493	0.6503	0.6513	0.6522
4.5	0.6532	0.6542	0.6551	0.6561	0.6571	0.6580	0.6590	0.6599	0.6609	0.6618
4.6	0.6628	0.6637	0.6646	0.6656	0.6665	0.6675	0.6684	0.6693	0.6702	0.6712

Figure 7.3

the column headed by x represents the first two digits of x, while each number in the row opposite x contains the third digit of x. The digits located at the intersection of a row and a column form the logarithm of x. For example, to find $\log_{10} 4.25$, we look at the intersection of the row opposite 4.2 under x and the column headed by 5. Thus,

$$\log_{10} 4.25 = 0.6284.$$

The equality sign is used here in an inexact sense; $\log_{10} 4.25 \approx 0.6284$ is more proper, because $\log_{10} 4.25$ is irrational and cannot be precisely represented by a rational number. We shall follow customary usage, however, writing = instead of $\approx$, and leave the intent to the context.

7.3 Special Logarithms and Powers

Now suppose we wish to find $\log_{10} x$ for values of x outside the range of the table—that is, for $0 < x < 1$ or $x > 10$ as shown in color in Figure 7.4. This can be done quite readily by first representing the number in scientific notation and then applying Theorem 7.2-I.

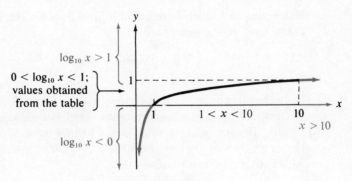

Figure 7.4

The following examples illustrate this method.

Examples

a. $\log_{10} 42.5 = \log_{10}(4.25 \times 10^1)$
$= \log_{10} 4.25 + \log_{10} 10^1$
$= 0.6284 + 1 = 1.6284,$

b. $\log_{10} 425 = \log_{10}(4.25 \times 10^2),$
$= \log_{10} 4.25 + \log_{10} 10^2$
$= 0.6284 + 2 = 2.6284.$

Characteristic and mantissa

Observe that in the preceding examples the decimal portion of the logarithm is always 0.6284, and the integral portion is just the exponent on 10 when the number is written in scientific notation. This process can be reduced to a mechanical one by considering $\log_{10} x$ to consist of two parts, an integral part (called the **characteristic**) and a nonnegative decimal fraction part (called the **mantissa**). Thus the table of values for $\log_{10} x$ for $1 < x < 10$ can be looked upon as a table of mantissas for $\log_{10} x$ for all $x > 0$.

Example

Find $\log_{10} 43{,}700$.

Solution

We first write

$$\log_{10} 43700 = \log_{10}(4.37 \times 10^4).$$

Upon examining the table of logarithms, we find that $\log_{10} 4.37 = 0.6405$, so that

$$\log_{10} 43700 = 0.6405 + 4 = 4.6405.$$

Now consider an example of the form $\log_{10} x$, for $0 < x < 1$.

Example Find $\log_{10} 0.00402$.

Solution We first write

$$\log_{10} 0.00402 = \log_{10}(4.02 \times 10^{-3}).$$

Examining the table, we find that $\log_{10} 4.02$ is 0.6042. Upon adding 0.6042 to the characteristic -3, we obtain

$$\log_{10} 0.00402 = 0.6042 - 3 = -2.3958.$$

In the preceding example, the decimal portion -0.3958 of -2.3958 is negative. Because the table of common logarithms has only positive entries, if we plan to locate this number in the table in any further work, it is more convenient to leave the logarithm in the form $0.6042 - 3$, in which the decimal part is positive. Alternatively, we can write

$$\begin{aligned}\log_{10} 0.00402 &= 0.6042 - 3 \\ &= 0.6042 + (7 - 10) \\ &= 7.6042 - 10,\end{aligned}$$

and again the decimal part is positive. The logarithm

$$6.6042 - 9$$

is an equally valid representation, but $7.6042 - 10$, in which a multiple of 10 is subtracted, is customary in most cases, though not necessary. In any case, *if we wish to use Table II then the decimal part of the logarithm must be positive.* The characteristic may be positive, negative, or zero.

In case x is between 1 and 10 but is not in the table, we can approximate $\log_{10} x$ with the value in Table II corresponding to the value closest to x, or with a value between the two function values corresponding to the two values closest to x.

Determination of 10^x Since $y = 10^x$ and $x = \log_{10} y$ are equivalent equations, if, for a given value of x, we want to find the value of y or 10^x, we reverse the process outlined above.

Example Find $10^{1.6395}$.

Solution We begin by locating the mantissa, 0.6395, of the exponent 1.6395 in the body of Table II and observing that it is paired with 4.36. Thus,

$$\begin{aligned}10^{1.6395} &= 10^{0.6395} \times 10^1 \\ &= 4.36 \times 10^1 = 43.6.\end{aligned}$$

If the decimal portion of the given value for x is negative, then in order to use the table to determine 10^x, we must first rewrite x as a positive decimal plus a negative integer.

7.3 Special Logarithms and Powers

Example Find $10^{-2.3605}$.

Solution We first add $(+3 - 3)$ to -2.3605 to obtain a positive decimal part. Thus,

$$10^{-2.3605} = 10^{(-2.3605+3)-3}$$
$$= 10^{0.6395-3} = 10^{0.6395} \times 10^{-3}.$$

We then use Table II and note that $10^{0.6395} = 4.36$. Hence,

$$10^{-2.3605} = 10^{0.6395} \times 10^{-3}$$
$$= 4.36 \times 10^{-3} = 0.00436.$$

In case the value of x cannot be found in the body of Table II, we can approximate 10^x with the power of 10 corresponding to the value in the table closest to x or with a value between the two powers of 10 corresponding to the two values in the body of the table closest to x.

Antilog$_{10}$ x Because the process for finding values of $\log_{10} x$ is reversed to find values of 10^x, we sometimes use the notation **antilog$_{10}$ x** for 10^x. Thus in the above example,

$$10^{-2.3605} = \text{antilog}_{10}(-2.3605) = 0.00436.$$

Determination of $\log_e x$ or $\ln x$ Values for $\log_e x$ are called **logarithms to the base e**, or **natural logarithms**. Frequently $\log_e x$ is denoted by **ln x**; we shall adopt that convention. Thus,

$$\ln x = \log_e x.$$

From the definition of $\log_b x$,

$$e^{\ln x} = x \quad (x > 0);$$

that is, $\ln x$ is the exponent that must be placed on e so that the resulting power is x.

Table III on page 416 includes values of $\ln x$, $0.1 \leq x \leq 190$. Values for $\ln x$ outside of this interval can be obtained in the same way that we obtained values for $\log_{10} x$, except that we no longer have integer characteristics.

Example Find $\ln 270$.

Solution We first observe that $270 = 2.70 \times 10^2$. Using Table III, we have

$$\ln 270 = \ln(2.70 \times 10^2) = \ln 2.70 + 2 \ln 10$$
$$= 0.9933 + 2(2.3026) = 0.9933 + 4.6052 = 5.5985$$

Example Find $\ln 0.27$.

Solution This time, let us observe that $0.27 = 2.70 \times 10^{-1}$. Hence,

$$\ln 0.27 = \ln(2.70 \times 10^{-1}) = \ln 2.70 + \ln 10^{-1}$$
$$= \ln 2.70 - \ln 10 = 0.9933 - 2.3026 = -1.3093.$$

As one would expect, the logarithm is negative. In this case we may leave the decimal part negative since we will be using a different table to compute powers of e.

Determination of e^x

The process used in the preceding examples can be reversed to find e^x. However, if a table for the exponential function $y = e^x$ is available (see Table I on page 413), e^x can be read directly from this table as we did in Section 7.1.

Examples

a. Find $e^{3.2}$.

b. Find $e^{-3.2}$.

Solutions

Reading directly from Table I, we obtain

a. $e^{3.2} = 24.533$

b. $e^{-3.2} = 0.0408$

As in the case with base 10, we sometimes use **antilog$_e$ x** to denote e^x. For example, $e^{3.2} = \text{antilog}_e 3.2 = 24.533$.

Determination of b^x ($b \neq 10$, and $b \neq e$)

Since tables of values of b^x for $b \neq 10$ and $b \neq e$ are not readily available, to compute the value of b^x for bases other than 10 and e we must use Theorem 7.2-III as the following example illustrates.

Example

Find $(1.01)^{14}$.

Solution

We first let $y = (1.01)^{14}$. Since $y = \log_{10} x$ defines a function we can equate $\log_{10}$ of each member of $y = (1.01)^{14}$ to obtain

$$\log_{10} y = \log_{10}(1.01)^{14}.$$

We now use Theorem 7.2-III to write

$$\log_{10} y = 14 \log_{10} 1.01 = 14(0.0043) = 0.0602.$$

Since the equations $\log_{10} y = 0.0602$ and $y = 10^{0.0602}$ are equivalent and $10^{0.0602} = 1.15$, we have

$$(1.01)^{14} = y = 10^{0.0602} = 1.15.$$

Using a calculator

It is a simple process to find values for $\log_{10} x$, $\ln x$, 10^x, e^x, and b^x† by using a calculator that has such function keys. Most calculators display these values in standard decimal form with more than four decimal place accuracy. To obtain values consistent with those obtained in the tables, we shall round off all results to an appropriate number of decimal places for the table involved.

Examples

a. $\log_{10} 0.234 = -0.63078414$
$= -0.6308$

b. $\ln 234 = 5.45532112$
$= 5.4553$

† The y^x function key can be used to find values b^x, for all $b > 0$.

7.3 Special Logarithms and Powers

c. $10^{-0.6308} = 0.23399146$
      $= 0.234$

d. $e^{-2.4} = 0.09071795$
      $= 0.0907$

Note that in Example (a) the decimal part of the logarithm is negative. Since we can obtain values for 10^x, $x < 0$, directly by using a calculator, as shown in Example (c), there is no need to change -0.6308 to a number with a positive decimal part.

Accuracy   It should be noted that for values of x that do not appear in the tables, the values for $\log_{10} x$, 10^x, $\ln x$, and e^x which are obtained on a calculator are more accurate than those obtained by using an approximation from the tables.

Exercise 7.3

A   Find the characteristic of the given logarithm.

Examples

a. $\log_{10} 248$

b. $\log_{10} 0.0057$

Solutions

Represent the number in scientific notation.

a. $\log_{10}(2.48 \times 10^2)$

b. $\log_{10}(5.7 \times 10^{-3})$

The exponent on the base 10 is the characteristic.

2

-3, or $7 - 10$

1. $\log_{10} 312$
2. $\log_{10} 0.02$
3. $\log_{10} 0.00851$
4. $\log_{10} 8.012$
5. $\log_{10} 0.00031$
6. $\log_{10} 0.0004$
7. $\log_{10}(15 \times 10^3)$
8. $\log_{10}(820 \times 10^4)$

Using Table II or a hand calculator, find the specified logarithm.

Examples

a. $\log_{10} 16.8$

b. $\log_{10} 0.043$

Solutions using the table

Represent the number in scientific notation.

a. $\log_{10}(1.68 \times 10^1)$

b. $\log_{10}(4.3 \times 10^{-2})$

Determine the mantissa from the table.

0.2253

0.6335

Add the characteristic as determined by the exponent on the base 10.

1.2253

$0.6335 - 2$ or $8.6335 - 10$

Solution continued on overleaf

226   7   Exponential and Logarithmic Functions

Calculator solutions

a. $\log_{10} 16.8 = 1.22530928$
$\phantom{\log_{10} 16.8} = 1.2253$

b. $\log_{10} 0.043 = -1.36653154$
$\phantom{\log_{10} 0.043} = -1.3665$

Note that in Example (b) $8.6335 - 10$ is equal to -1.3665.

9. $\log_{10} 6.73$ 10. $\log_{10} 891$ 11. $\log_{10} 0.813$
12. $\log_{10} 0.00214$ 13. $\log_{10} 0.08$ 14. $\log_{10} 0.000413$
15. $\log_{10}(2.48 \times 10^2)$ 16. $\log_{10}(5.39 \times 10^3)$ 17. $\log_{10}(3.41 \times 10^{-4})$
18. $\log_{10}(6.42 \times 10^{-6})$ 19. $\log_{10}(2.71 \times 10^2)$ 20. $\log_{10}(5.42 \times 10^4)$

Using Table II or a hand calculator, find the specified power.

Examples

a. $10^{2.7364}$ b. $10^{-1.1221}$

($10^{2.7364} = \text{antilog}_{10} 2.7364$ and $10^{-1.1221} = \text{antilog}_{10}(-1.221)$.)

Solutions using the table

a. Locate the mantissa, 0.7364, in the body of Table II and determine the associated power.
$$10^{2.7364} = 10^{0.7364+2} = 10^{0.7364} \times 10^2$$
$$= 5.45 \times 10^2 = 545$$

b. Add $(+2 - 2)$ to -1.1221 to obtain $0.8779 - 2$. Use Table II to find the associated power.
$$10^{-1.1221} = 10^{0.8779-2} = 10^{0.8779} \times 10^{-2}$$
$$= 7.55 \times 10^{-2} = 0.0755$$

Calculator solutions

a. $10^{2.7364} = 545.0043890 = 545$ b. $10^{-1.1221} = 0.07549184 = 0.0755$

21. $10^{0.6128}$ 22. $10^{0.2504}$ 23. $10^{0.5647}$
24. $10^{0.9258}$ 25. $10^{2.8470}$ 26. $10^{2.3874}$
27. $10^{3.3139}$ 28. $10^{2.8000}$ 29. $10^{-2.1851}$
30. $10^{-1.0991}$ 31. $10^{-2.3893}$ 32. $10^{-3.1543}$

Find the logarithm directly from Table III or by using a calculator.

Examples

a. $\ln 120$ b. $\ln 0.075$

Solutions using the table

a. $\ln 120 = \ln(1.2 + 10^2)$
$ = \ln 1.2 + \ln 10^2$
$ = \ln 1.2 + \ln 100$
$ = 0.1823 + 4.6052$
$ = 4.7875$

b. $\ln 0.075 = \ln(7.5 \times 10^{-2})$
$ = \ln 7.5 + \ln 10^{-2}$
$ = \ln 7.5 + (-2)\ln 10$
$ = 2.0149 + (-4.6052)$
$ = -2.5903$

Calculator solutions

a. $\ln 120 = 4.78749174$
$ = 4.7875$

b. $\ln 0.075 = -2.59026717$
$ = -2.5903$

7.3 Special Logarithms and Powers

33. ln 3	**34.** ln 8	**35.** ln 17	**36.** ln 98
37. ln 450	**38.** ln 650	**39.** ln 0.065	**40.** ln 0.035

Find the power directly from Table I or by using a calculator.

Examples

    a. $e^{0.85}$ **b.** $e^{-3.4}$

($e^{0.85} = \text{antilog}_e\, 0.85$ and $e^{-3.4} = \text{antilog}_3(-3.4)$.)

Solutions using the table

Obtain both values from Table I.

  a. $e^{0.85} = 2.3396$ **b.** $e^{-3.4} = 0.0334$

Calculator solutions

  a. $e^{0.85} = 2.33964685$ **b.** $e^{-3.4} = 0.03337327$
             $= 2.3396$ $= 0.0334$

41. $e^{0.50}$	**42.** $e^{1.5}$	**43.** $e^{3.4}$	**44.** $e^{4.5}$
45. $e^{0.23}$	**46.** $e^{1.4}$	**47.** $e^{-0.15}$	**48.** $e^{-0.95}$
49. $e^{-2.5}$	**50.** $e^{-4.2}$	**51.** $e^{-0.25}$	**52.** $e^{-3.6}$

Compute the following powers.

Example

    $(1.02)^8$

Solution using the table

Write $y = (1.02)^8$. Equate $\log_{10}$ of each member and use Theorem 7.2-III to write

$$\log_{10} y = \log_{10}(1.02)^8 = 8 \log_{10}(1.02)$$
$$= 8(0.0086) = 0.0688.$$

Use the fact that $\log_{10} y = 0.0688$ is equivalent to $y = 10^{0.0688}$ to write

$$(1.02)^8 = y = 10^{0.0688} = 1.17.$$

Calculator solution

Use the y^x key.

$$(1.02)^8 = 1.17165938 = 1.17$$

53. $(1.01)^{12}$	**54.** $(1.01)^7$	**55.** $(0.99)^8$
56. $(0.98)^7$	**57.** $(1.03)^8$	**58.** $(1.02)^{10}$

B 59. By definition, $\log_e x = N$ can be written in exponential form as $e^N = x$. Show that $N = \log_{10} x/\log_{10} e$, and hence that

$$\log_e x = \frac{\log_{10} x}{\log_{10} e}.$$

60. Use the results of Exercise 59 to show that

$$\log_e x = 2.3026 \log_{10} x.$$

The following exercises require the use of a calculator. Evaluate each of the following functions at $x = 0.1, 0.01, 0.001,$ *and* 0.0001. *Show four decimal place accuracy.*

61. $f(x) = 1 - x^x$
62. $f(x) = (1 + x)^{1/x} - e$
63. $f(x) = x \ln x$
64. $f(x) = x(\ln x)^2$

7.4 Solution of Exponential Equations; Applications

In this section, calculator values of powers and logarithms are used in the examples and exercises, with answers rounded off to four decimal places. If you use the tables, slightly different values may result.

When solving an exponential equation, $y = b^{f(x)}$, we are interested in finding the value for y, given a value of x, or in finding the value(s) for x, given a value of y.

Finding powers

The first case in which we want to find y or $b^{f(x)}$ for a given value of x is a direct application of the methods of Section 7.3, as illustrated by the following examples.

Example   Solve $y = ce^{kt}$ for y, given that $c = 25$, $k = -5$, and $t = 0.5$.

Solution   Substituting 25, -5, and 0.5 for c, k, and t, respectively, we have

$$y = 25e^{-5(0.5)} = 25e^{-2.5}$$
$$= 25(0.0821) = 2.0521.$$

Example   The amount y (measured in grams) of a radioactive element remaining at any time t is given by $y = 25e^{-0.5t}$, where t is in seconds. How much of the element is remaining after three seconds?

Solution   Substituting 3 for t, we have

$$y = 25e^{-0.5(3)} = 25e^{-1.5}$$
$$= 25(0.2231) = 5.5783.$$

There are approximately 5.5783 grams of material remaining after three seconds.

7.4 Solution of Exponential Equations; Applications

Example Solve $A = \left(1 + \dfrac{r}{n}\right)^{nt}$ for A, given that $r = 0.07$, $n = 4$, and $t = 3$.

Solution Substituting 0.07, 4, and 3 for r, n, and t, respectively, we have

$$A = \left(1 + \frac{0.07}{4}\right)^{4(3)} = (1.0175)^{12}.$$

Using the techniques of Section 7.3 or the y^x key on a calculator, $A = 1.2314$.

Finding exponents

For a given value of y, in order to solve an exponential equation $y = b^{f(x)}$ for x, it is necessary first to rewrite the equation so that the variable no longer appears in the exponent. In the cases when the base is either 10 or e, we simply rewrite the equation using logarithmic notation with the appropriate base. The following examples illustrate the procedure.

Example Find the solution of $10^{2x} = 6$.

Solution Rewriting the equation in logarithmic notation with base 10, we have

$$2x = \log_{10} 6 = 0.7782.$$

Thus,

$$x = 0.3891.$$

Example The amount A that results from a deposit of P dollars after t years when interest is compounded continuously at an annual rate of $r\%$ is given by the formula

$$A = Pe^{rt/100}.$$

If a savings and loan company compounds interest continuously, and if the annual rate is $5\frac{3}{4}\%$, how long will it take for a deposit of $\$5000$ to grow to $\$6000$?

Solution Substituting appropriate values in the given formula, we obtain the equation

$$6000 = 5000 e^{0.0575t}.$$

This is equivalent to

$$\frac{6}{5} = e^{0.0575t}.$$

Rewriting the equation in logarithmic notation with base e, we have

$$0.0575t = \ln \frac{6}{5} = 0.1823.$$

We then solve for t, obtaining $t = 3.1708$. Hence, it takes slightly more than three years for the value of the account to reach $\$6000$.

Example

If a particular component of a certain manufactured system shows no aging effect (in other words, it does not wear out with use), then the probability that the component will not fail within t years is

$$P = e^{-kt},$$

where $k > 0$ is a constant that depends on the component. If $k = 2$ for a given component, find t so that the probability that the component will not fail within t years is 1/2.

Solution

Since we want $P = 1/2$ and we know $k = 2$, we have

$$\frac{1}{2} = e^{-2t}.$$

Rewriting this equation in logarithmic notation with base e, we have

$$\ln(\tfrac{1}{2}) = -2t,$$
$$-0.6931 = -2t,$$
$$t = \frac{-0.6931}{-2} = 0.3466.$$

Hence the probability is 1/2 that the component will not fail within 0.3466 years.

The foregoing examples involved powers of 10 and e. When the base b is not 10 or e tables for $\log_b$ are not readily available and most calculators do not have such keys; thus we shall use $\log_{10}$ and Theorem 7.2-III to rewrite the equation so that the variable does not appear in the exponent. The method is illustrated in the next example.

Example

Find the solution of $5^x = 7$.

Solution

Equating $\log_{10}$ of each member we have

$$\log_{10} 5^x = \log_{10} 7;$$

from Theorem 7.2-III,

$$x \log_{10} 5 = \log_{10} 7.$$

Dividing each member by $\log_{10} 5$, we obtain

$$x = \frac{\log_{10} 7}{\log_{10} 5} = \frac{0.8451}{0.6990} \approx 1.2091.$$

Note that in seeking the decimal approximation, the logarithms are *divided*, not subtracted.

7.4 Solution of Exponential Equations; Applications

Note that the technique used in the preceding example is applicable in solving an exponential equation which involves any positive base b.

Exponential equations involving more than one variable can be solved for one of the variables in terms of the others, as the following examples illustrate.

Example Solve $y = Ce^{kt}$ for k.

Solution We first rewrite the equation as

$$\frac{y}{C} = e^{kt}.$$

We then rewrite this equation in logarithmic notation with the base e to obtain

$$\ln\left(\frac{y}{C}\right) = kt.$$

Solving for k, we obtain

$$k = \frac{1}{t} \ln\left(\frac{y}{C}\right).$$

Example Solve $y = (1 + r)^n$ for n in terms of y and r using logarithms to the base 10.

Solution We first equate $\log_{10}$ of each member to obtain

$$\log_{10} y = \log_{10}(1 + r)^n;$$

then from Theorem 7.2-III,

$$\log_{10} y = n \log_{10}(1 + r).$$

Thus,

$$n = \frac{\log_{10} y}{\log_{10}(1 + r)}.$$

Exercise 7.4

A *Solve for the indicated variable. Show the solution to two decimal places.*

1. $y = k10^t$, for y given $k = 5$, $t = 0.2455$.
2. $y = k10^t$, for y given $k = 5$, $t = 2.2455$.
3. $y = ke^{2t}$, for y given $k = 0.01$, $t = 5$.
4. $y = ke^{-4t}$, for y given $k = 10$, $t = 0.5$.
5. $A = (1 + r)^n$, for A given $r = 0.01$, $n = 8$.
6. $A = (1 + r)^n$, for A given $r = 0.0025$, $n = 20$.

Solve. Give the solution in logarithmic form using the base 10 and also as an approximation in decimal form. Show four decimal place accuracy.

Example $10^{2x+1} = 4$

Solution   Rewrite the equation in logarithmic notation with base 10; then solve for x.

$$2x + 1 = \log_{10} 4$$
$$2x = -1 + \log_{10} 4$$
$$x = -\frac{1}{2} + \frac{1}{2}\log_{10} 4$$

The solution is $-\frac{1}{2} + \frac{1}{2}\log_{10} 4 \approx -0.1990$.

7. $10^{3x} = 6$ 8. $10^{-2x} = 15$ 9. $10^{3x+1} = 9$
10. $10^{-2x+3} = 20$ 11. $10^{x^2} = 150$ 12. $10^{2x^2+1} = 200$

Solve. Give the solution in logarithmic form with base e and also as an approximation in decimal form. Show four decimal place accuracy.

Example $e^{3x+2} = 5$

Solution   Rewrite the equation in logarithmic notation with base e; then solve for x.

$$3x + 2 = \ln 5$$
$$3x = -2 + \ln 5$$
$$x = -\frac{2}{3} + \frac{1}{3}\ln 5$$

The solution is $-\frac{2}{3} + \frac{1}{3}\ln 5 \approx -0.1302$.

13. $e^{3x} = 5$ 14. $e^{-2x} = 10$ 15. $e^{2x+1} = 25$
16. $e^{-x+2} = 8$ 17. $e^{x^2/2} = 15$ 18. $e^{x^2+1} = 4$

Solve. Give the solution in logarithmic form with base 10 and also as an approximation in decimal form. Show four decimal place accuracy.

Example $3^{x-2} = 16$

Solution   Equate the $\log_{10}$ of both sides.

$$\log_{10} 3^{x-2} = \log_{10} 16$$

7.4 Solution of Exponential Equations; Applications

By Theorem 7.2-III,
$$(x - 2)\log_{10} 3 = \log_{10} 16,$$
from which
$$x - 2 = \frac{\log_{10} 16}{\log_{10} 3},$$
$$x = \frac{\log_{10} 16}{\log_{10} 3} + 2.$$

The solution is $\dfrac{\log_{10} 16}{\log_{10} 3} + 2 \approx 4.5237$.

19. $2^x = 7$ 20. $3^x = 4$ 21. $7^{2x-1} = 3$
22. $3^{x+2} = 10$ 23. $4^{x^2} = 16$ 24. $8^{x^2+1} = 64$

Solve. Leave the result in the form of an equation equivalent to the given equation.

25. $A = 10^{kt}$, for t 26. $A = B10^{-kt}$, for t 27. $y = e^{kt}$, for t
28. $y = Ce^{-kt}$, for t 29. $y = x^n$, for n 30. $y = Cx^{-n}$, for n

P dollars invested at an interest rate r compounded yearly yields an amount A after n years, given by $A = P(1 + r)^n$. *If the interest is compounded t times yearly, the amount is given by*
$$A = P\left(1 + \frac{r}{t}\right)^{tn}.$$

Example   One dollar compounded annually for 12 years yields $1.90. What is the rate of interest to the nearest $\frac{1}{2}\%$?

Solution   Use the given equation to obtain $(1 + r)^{12} = 1.90$. Equate $\log_{10}$ of each member and apply Theorem 7.2-III.
$$12 \log_{10}(1 + r) = \log_{10} 1.90 = 0.2788$$

Multiply each member by 1/12.
$$\log_{10}(1 + r) = \frac{1}{12}(0.2788) = 0.0232$$

Rewrite this equation in exponential notation.
$$1 + r = 10^{0.0232}$$

Determine $10^{0.0232}$ and solve for r.
$$1 + r = 10^{0.0232} = 1.0549$$
$$r = 0.0549$$

To the nearest $\frac{1}{2}\%$ the rate of interest is $5\frac{1}{2}\%$.

31. One dollar compounded annually for 10 years yields $1.95. What is the rate of interest to the nearest $\frac{1}{2}\%$?

32. How many years (nearest year) would it take for $1.00 to yield $2.19 if compounded annually at 9%?

33. Find the compounded amount of $5000 invested at $9\frac{1}{2}\%$ for 10 years when compounded annually; when compounded semiannually.

34. Two men, A and B, each invested $10,000 at $9\frac{1}{2}\%$ for 20 years with a bank that computed interest quarterly. A withdrew his interest at the end of each 3-month period, but B let his investment be compounded. How much more than A did B earn over the period of 20 years?

If S dollars is borrowed at an annual interest rate r (simple interest) and is to be repaid in equal monthly payments over a period of n months then each monthly payment p is

$$p = S\left[\frac{\frac{r}{12}}{1 - \left(1 + \frac{r}{12}\right)^{-n}}\right].$$

Example

A person borrows $64,000 at 12% annual simple interest for a period of 30 years. What are the monthly payments?

Solution

Using the given equation, we have

$$p = 64,000\left[\frac{0.01}{1 - (1.01)^{-360}}\right] = 658.31.$$

The monthly payment is $658.31.

In Exercises 35–38, assume simple interest.

35. A family buying a house costing $95,000 has $19,000 to use as a down payment. If they borrow the remaining amount at an annual rate of $11\frac{3}{4}\%$ for a 30-year period, what is their monthly payment?

36. A person wishes to borrow $100,000. One loan company will lend it at 11% annual interest for a period of 30 years. A second loan company will lend it at $11\frac{1}{2}\%$ annual interest but only for a period of 25 years. Which loan results in the lower monthly payment and how much difference is there in the monthly payments?

37. A family wishes to buy a house costing $125,000. They can obtain a loan at 12% annual interest. They want to pay the home off in 30 years. If they want to keep their monthly house payment at $1000, how much of the $125,000 do they have to have as a down payment?

38. Suppose in Exercise 37 the family had $50,000 to use as a down payment. How long would the term of the loan have to be (to the nearest month) to keep the monthly payments at $1000?

7.4 Solution of Exponential Equations; Applications

The present value of A dollars in t years when compounded continuously at $r\%$ annually is the amount that must be deposited in an account at the present so that it is worth A dollars in t years. The formula for present value (PV) is

$$PV = Ae^{-rt/100}.$$

39. What is the annual rate of interest of an account if when compounded continuously the present value of $10,000 in 10 years is $5000?

40. At an annual interest rate of 7.25% when compounded continuously, how long must $1000 be kept in an account to attain a value of $2000?

The chemist defines the pH (hydrogen potential) of a solution by

$$pH = \log_{10} \frac{1}{[H^+]} = \log_{10}[H^+]^{-1}$$

$$= -\log_{10}[H^+],$$

where $[H^+]$ is a numerical value for the concentration of hydrogen ions in aqueous solution in moles per liter.

Example

Calculate the pH of a solution whose hydrogen ion concentration is 3.7×10^{-6}.

Solution

Substitute 3.7×10^{-6} for $[H^+]$ in the relationship $pH = \log_{10} \frac{1}{[H^+]}$.

$$pH = \log_{10} \frac{1}{3.7 \times 10^{-6}} = \log_{10}\left(\frac{1}{3.7} \times 10^6\right)$$

$$= \log_{10} 1 - \log_{10} 3.7 + \log_{10} 10^6$$

$$= 0 - 0.5682 + 6 = 5.4318$$

41. Calculate the pH of a solution whose hydrogen ion concentration $[H^+]$ is 2.0×10^{-8}.

42. Calculate the pH of a solution whose hydrogen ion concentration is 6.3×10^{-7}.

43. Calculate the hydrogen ion concentration of a solution whose pH is 5.6.

44. Calculate the hydrogen ion concentration of a solution whose pH is 7.2.

45. The atmospheric pressure p, in inches of mercury, is given approximately by $p = 30.0(10)^{-0.09a}$, where a is the altitude in miles above sea level. What is the atmospheric pressure at sea level; at 3 miles above sea level?

46. Using the information in Exercise 45, find the atmospheric pressure 6 miles above sea level.

47. The amount of a radioactive element remaining at any time t is given by $y = y_0 e^{-0.4t}$, where t is in seconds and y_0 is the amount present initially. How much of the element would remain after 3 seconds if 40 grams were present initially?

48. The number of bacteria present in a culture is related to time by the formula $N = N_0 e^{0.04t}$, where N_0 is the amount of bacteria present at time $t = 0$, and t is time in hours. If 10,000 bacteria are present 10 hours after the beginning of an experiment, how many were present when $t = 0$?

49. If a component in a system shows no aging effect and the probability of failure after 10 units of time is 1/10, using the formula on page 230 find the value k associated with this component. The value $1/k$ is the average time until a failure occurs.

50. If a component in a system shows no aging effect and the probability of failure after 3 units of time is 0.001, find the average time until a failure occurs (see Exercise 49).

Chapter Review

[7.1]   1. Graph $y = 3^x$.   2. Graph $y = e^{5x}$.

[7.2]   *Express in logarithmic notation.*

3. $16^{-1/2} = \dfrac{1}{4}$ 4. $7^3 = 343$

Express in exponential notation.

5. $\log_2 8 = 3$ 6. $\log_{10} 0.0001 = -4$

Solve.

7. $\log_2 16 = y$ 8. $\log_{10} x = 3$

Express as the sum or difference of simpler logarithmic quantities.

9. $\log_{10} \sqrt[3]{xy^2}$ 10. $\log_{10} \dfrac{2R^3}{\sqrt{PQ}}$

Express as a single logarithm with coefficient 1.

11. $2 \log_b x - \dfrac{1}{3} \log_b y$ 12. $\dfrac{1}{3}(2 \log_{10} x + \log_{10} y) - 3 \log_{10} z$

Review Exercises

[7.3] *Find the logarithm. Show four decimal place accuracy.*

13. $\log_{10} 42$
14. $\log_{10} 0.00314$
15. $\log_{10} 682$
16. $\log_{10} 0.0414$

Find the power. Show three decimal place accuracy.

17. $10^{1.8287}$
18. $10^{0.4240}$
19. $10^{-1.3316}$
20. $10^{-0.1776}$

Find the logarithm. Show four decimal place accuracy.

21. $\ln 7$
22. $\ln 23$
23. $\ln 241$
24. $\ln 510$

Find the power. Show four decimal place accuracy.

25. $e^{6.0}$
26. $e^{1.8}$
27. $e^{-3.5}$
28. $e^{-0.42}$

Find the power. Show three decimal place accuracy.

29. $(1.01)^{36}$
30. $(0.99)^{36}$

[7.4] *Solve. Give the solution in logarithmic form using the base 10 and also as an approximation in decimal form. Show four decimal place accuracy.*

31. $10^{-2x} = 8$
32. $10^{-x-2} = 25$

Solve. Give the solution in logarithmic form with the base e and also as an approximation in decimal form. Show four decimal place accuracy.

33. $e^{4x} = 8$
34. $e^{-2x+2} = 16$

Solve. Give the solution in logarithmic form with the base 10 and also as an approximation in decimal form. Show four decimal place accuracy.

35. $3^x = 2$
36. $3^{x+1} = 80$

Solve. Leave the result in the form of an equation equivalent to the given equation.

37. $y = A + ke^{-t}$, for t
38. $y = A + ke^{-(t+c)}$, for t

Solve. Use the appropriate formula from pages 230 and 233–235.

39. One dollar compounded quarterly for 5 years yields $1.50. What is the annual rate of interest to the nearest $\frac{1}{4}\%$?

40. Calculate the hydrogen ion concentration of a solution if its pH is 6.2.

41. Find the interest rate at which $10,000 has a present value of $7000 in 5 years with continuously compounded interest.

42. Find the average time until failure of a component that shows no aging effect if the probability of failure after 3 units of time is 0.01.

43. A person borrows $25,000 at 18% annual simple interest. If the loan is to be repaid in 5 years, what are the monthly payments?

44. A person borrows $30,000 at 15% annual simple interest. If the monthly payments are $400, how long is the term of the loan?

Supplemental Exercises for Chapters 5–7

Each of the exercises in this set is more difficult than those at the end of each section and may require the use of several of the techniques learned in the preceding chapters and more insight into the concepts developed in those chapters.

Graph each of the following functions.

1. $f(x) = 2^{-x^2}$

2. $f(x) = 2^{x^2+1}$

3. $f(x) = \begin{cases} x^2 & \text{if } x \leq 1 \\ 2^x & \text{if } x > 1 \end{cases}$

4. $f(x) = \begin{cases} e^{-x} & \text{if } x \leq 0 \\ -x + 1 & \text{if } 0 < x \leq 1 \\ \ln x & \text{if } x > 1 \end{cases}$

5. $f(x) = \log_{10} x^2$

6. $f(x) = \log_{10}\left(\dfrac{1}{x^2}\right)$

7. $f(x) = \ln(1 - x^2)$

8. $f(x) = \ln(x^2 - 1)$

9. $P(x) = x^4 - \dfrac{8}{3}x^3 - 10x^2 + 24x + 20$

10. $P(x) = 3x^4 - 8x^3 + 6x^2$

11. $P(x) = \dfrac{x^5}{5} + \dfrac{x^4}{2} + \dfrac{x^3}{3}$

12. $f(x) = \dfrac{x^4}{x^3 - x}$

13. $f(x) = \dfrac{x^3 + x}{x^2 + x - 2}$

14. $R(x) = \dfrac{x^4 - x^3 + 1}{x - 1}$

15. $f(x) = \left| \,||x| - 1| - \dfrac{1}{2} \right|$

16. $f(x) = |\,||2x| - 4| - 2|$

Find the range of each of the following functions. Hint: Solve for x in terms of y and determine allowable y values, or graph the given function and determine the range from the graph.

239

17. $y = \dfrac{1 - x^2}{x^2}$

18. $y = \dfrac{1}{\sqrt{1 - x}}$

19. $y = 2^{-x^2}$

20. $y = \ln(1 - x^2)$

Solve the given equation or inequality.

21. $e^x - e^{-x} \geq 0$

22. $e^x - e^{-2x} \geq 0$

Find $f^{-1}(x)$ for the given function f.

23. $f(x) = e^x - e^{-x}$.

24. $f(x) = e^x + e^{-x}, x \geq 0$.

25. In the adjoining figure, L_1 is perpendicular to L_2. Use the fact that $\Delta P_1 Q P_2$ is similar to $\Delta P_3 A P_1$ to deduce $m_1 m_2 = -1$.

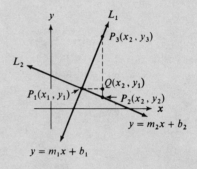

26. Use similar triangles to show that if the point $P(x, y)$ divides the segment from $P_1(x_1, y_1)$ to $P_2(x_2, y_2)$ in the ratio m_1/m_2, that is, if $P_1 P / P P_2 = m_2/m_1$, then

$$x = \dfrac{m_1}{m_1 + m_2} x_1 + \dfrac{m_2}{m_1 + m_2} x_2,$$

and

$$y = \dfrac{m_1}{m_1 + m_2} y_1 + \dfrac{m_2}{m_1 + m_2} y_2.$$

27. Assume a and b are known positive constants. Find $\log_a x$ in terms of $\log_b x$.

28. Show that for any positive constants a and b, with $a \neq 1$ and $b \neq 1$, $\log_a b = \dfrac{1}{\log_b a}$.

8 Systems of Equations and Inequalities

For a given collection of equations or inequalities in two variables, it is often necessary to determine the ordered pairs which satisfy all of the equations or inequalities in the collection. In such a case, the collection of equations or inequalities is called a **system in two variables**. Any ordered pair which is a solution of all of the equations or inequalities in a system is called a **solution of the system**, and the set of all solutions of a system is called the **solution set of the system**. Systems in more than two variables are defined similarly.

8.1 Systems of Linear Equations in Two Variables

We shall begin by considering the system

$$a_1 x + b_1 y + c_1 = 0 \quad (a_1, b_1 \text{ not both } 0) \tag{1}$$

$$a_2 x + b_2 y + c_2 = 0 \quad (a_2, b_2 \text{ not both } 0). \tag{2}$$

Linear dependence   The left-hand members of Equations (1) and (2) are said to be **linearly dependent** if one of them can be obtained from the other through multiplication by a constant. If the left-hand members of Equations (1) and (2) are *not* linearly dependent, then they are said to be **linearly independent**.

Examples   **a.** The expressions $2x + 4y - 8$ and $6x + 12y - 24$ are linearly dependent since

$$(2x + 4y - 8) = 6x + 12y - 24.$$

b. The expressions $2x + 4y - 8$ and $6x + 12y - 23$ are linearly independent.

Consistency   A system is said to be **consistent** if its solution set is nonempty. A system which is not consistent is **inconsistent**.

Examples

a. The system
$$x - 2 = 0$$
$$y + 3 = 0$$
is consistent. The solution set is $\{(2, -3)\}$.

b. The system
$$x + y = 0$$
$$x + y - 1 = 0$$
is inconsistent. The solution set is empty.

Geometric interpretation

In a geometric sense, because the graphs of both Equations (1) and (2) above are straight lines, we are confronted with three possibilities. One possibility is that the graphs of the equations are the same line, as illustrated in Figure 8.1. In this case, the solution set of the system is the same as the solution set of each equation. The left-hand members of the two equations in standard form are linearly dependent and the system is consistent.

A second possibility is that the graphs of the equations are parallel lines, as illustrated in Figure 8.2. Since the lines have no point in common, there is no ordered pair which satisfies both equations. Hence the solution set of the system is empty. In this case, the left-hand members are linearly independent and the system is inconsistent.

The last possibility is that the graphs of the equations intersect in a single point, as illustrated in Figure 8.3. Since the lines have exactly one point in common, there is exactly one ordered pair which satisfies both equations. Hence, the solution set of the system contains one ordered pair. In this case the left-hand members are linearly independent and the system is consistent.

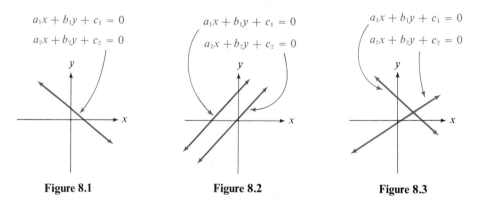

Figure 8.1   Figure 8.2   Figure 8.3

You will recall that equivalent equations have the same solution set. Systems may be said to be equivalent in a similar sense.

Equivalent systems

Definition 8.1   *If the solution set of one system is equal to (the same as) the solution set of another system, then the systems are* **equivalent**.

In seeking the solution set of a system of equations, our procedure will be to generate equivalent systems until we arrive at a system for which the solution set is obvious. One way to obtain an equivalent system is to replace one equation by a certain *linear combination* of the equations in the system. If $f(x, y)$ and $g(x, y)$ are

8.1 Systems of Linear Equations in Two Variables

polynomials, the polynomial $af(x, y) + bg(x, y)$, where a and b are not both zero, is said to be a linear combination of $f(x, y)$ and $g(x, y)$. We shall then refer to the equation

$$af(x, y) + bg(x, y) = 0$$

as a **linear combination** of the equations $f(x, y) = 0$ and $g(x, y) = 0$.

We now compare the two systems:

$$f(x, y) = 0$$
$$g(x, y) = 0 \qquad (3)$$

and

$$af(x, y) + bg(x, y) = 0$$
$$g(x, y) = 0 \qquad (4)$$

where $a \neq 0$. If (x_1, y_1) is a solution to (3), then $f(x_1, y_1) = 0$ and $g(x_1, y_1) = 0$, so

$$af(x_1, y_1) + bg(x_1, y_1) = 0,$$

and (x_1, y_1) is a solution to (4). On the other hand, suppose (x_2, y_2) is a solution to (4). Then $g(x_2, y_2) = 0$ and

$$af(x_2, y_2) + bg(x_2, y_2) = 0.$$

Thus, $af(x_2, y_2) = 0$, and since $a \neq 0$, $f(x_2\ y_2) = 0$, so that (x_2, y_2) is a solution to (3). Therefore, (3) and (4) have the same solution set. This proves the following theorem.

Theorem 8.1 *If either equation in the system*

$$f(x, y) = 0$$
$$g(x, y) = 0$$

is replaced by a linear combination of the two equations with nonzero coefficient for the replaced equation, the result is an equivalent system.

Theorem 8.1 is useful in solving linear systems of the form

$$a_1 x + b_1 y + c_1 = 0 \quad (a_1, b_1 \text{ not both 0}) \qquad (5)$$
$$a_2 x + b_2 y + c_2 = 0 \quad (a_2, b_2 \text{ not both 0}). \qquad (6)$$

By appropriate choice of multipliers a and b, the linear combination

$$a(a_1 x + b_1 y + c_1) + b(a_2 x + b_2 y + c_2) = 0 \qquad (7)$$

will be free of one variable; that is, the coefficient of one variable will be 0. We can then form an equivalent system by substituting Equation (7) for either Equation (5) or Equation (6).

Example

Solve

$$x - 3y + 5 = 0$$
$$2x + y - 4 = 0. \qquad (8)$$

Solution

We first form the linear combination

$$1(x - 3y + 5) + 3(2x + y - 4) = 0,$$

or

$$7x - 7 = 0,$$

from which

$$x - 1 = 0.$$

Replacing $x - 3y + 5 = 0$ with $x - 1 = 0$, we then have the equivalent system

$$x - 1 = 0$$
$$2x + y - 4 = 0. \qquad (9)$$

We can next replace the second equation in (9) with the linear combination

$$-2(x - 1) + 1(2x + y - 4) = 0,$$

or

$$y - 2 = 0,$$

to obtain the equivalent system

$$x - 1 = 0$$
$$y - 2 = 0, \qquad (10)$$

from which, by inspection, we can obtain the unique solution (1, 2). Hence the solution set of (8) is {(1, 2)}. Graphs of the systems (8), (9), and (10) appear in Figures a, b and c. The point of intersection in each case has coordinates (1, 2).

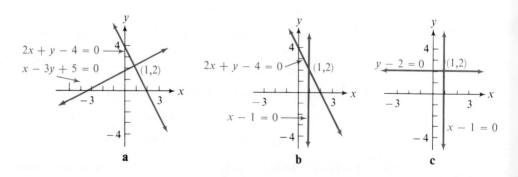

Solution of a system by substitution

Note that we could have proceeded somewhat differently from (9), as follows. A solution of system (9) must be of the form (1, y) (because any such ordered pair is a solution of $x - 1 = 0$); when x is replaced by 1 in $2x + y - 4 = 0$, we obtain $y = 2$, the only ordered pair that satisfies both of the equations

8.1 Systems of Linear Equations in Two Variables

in (9) is (1, 2). Hence, again the solution set of (8) is {(1, 2)}. In this latter way of proceeding from (9) to the solution of (8), we have used what is known as the **method of substitution**.

Choice of multipliers for linear combinations

Observe that the multipliers used in the foregoing example were first 1 and 3, and later -2 and 1. These were chosen because they produced coefficients that were additive inverses, first for the terms in y and later for the terms in x. In general, the linear combination

$$b_2(a_1 x + b_1 y + c_1) - b_1(a_2 x + b_2 y + c_2) = 0$$

will always be free of y, and

$$a_2(a_1 x + b_1 y + c_1) - a_1(a_2 x + b_2 y + c_2) = 0$$

will be free of x.

Examples

Form a linear combination of the equations in the system

$$2x + 6y + 7 = 0$$
$$5x + 4y + 3 = 0$$

to obtain an equation that is free of

a. the variable x. **b.** the variable y.

Solutions

a. $5(2x + 6y + 7) - 2(5x + 4y + 3) = 0$
$\qquad\qquad\qquad\qquad 22y + 29 = 0$

b. $4(2x + 6y + 7) - 6(5x + 4y + 3) = 0$
$\qquad\qquad\qquad\qquad -22x + 10 = 0$

Criteria for dependent or inconsistent equations

If the coefficients of the variables in one equation in a system are proportional to the corresponding coefficients in the other equation, then the equations might be either dependent or inconsistent. The equations in the system

$$a_1 x + b_1 y + c_1 = 0$$
$$a_2 x + b_2 y + c_2 = 0 \quad (a_2, b_2, c_2 \neq 0)$$

are dependent if

$$\frac{a_1}{a_2} = \frac{b_1}{b_2} = \frac{c_1}{c_2},$$

and inconsistent if

$$\frac{a_1}{a_2} = \frac{b_1}{b_2} \neq \frac{c_1}{c_2}.$$

(See Exercises 8.1-36 and 8.1-37.)

If
$$\frac{a_1}{a_2} \neq \frac{b_1}{b_2},$$
then the system has a unique solution.

Systems of linear equations are quite useful in expressing relationships in practical applications. By assigning separate variables to represent separate physical quantities, we can usually decrease the difficulty in symbolically representing these relationships.

Example   A company offers split-rail fence for sale in two prepackaged options. One option consists of 4 posts and 6 rails for $31; the other consists of 3 posts and 4 rails for $22. What are the individual values of posts and rails?

Solution   Let x represent the value of a post in dollars and y represent the value of a rail in dollars. Then
$$4x + 6y = 31$$
$$3x + 4y = 22.$$
This system is solved in the first example in the exercises.

Exercise 8.1

A   Solve each system in Exercises 1–18.

Example
$$4x + 6y = 31 \tag{1}$$
$$3x + 4y = 22 \tag{2}$$

Solution   Replace (1) by 3 times itself and (2) by 4 times itself to obtain
$$12x + 18y = 93 \tag{1'}$$
$$12x + 16y = 88. \tag{2'}$$
Replace (1') by itself plus -1 times (2').
$$2y = 5 \tag{1''}$$
$$12x + 16y = 88 \tag{2''}$$
From (1''), $y = 5/2$. Substituting this value into (2''), we obtain
$$12x + 40x = 88,$$
$$x = 4.$$
The solution set is $\{(4, 5/2)\}$.

8.1 Systems of Linear Equations in Two Variables

1. $x - y = 1$
   $x + y = 5$

2. $2x - 3y = 6$
   $x + 3y = 3$

3. $3x + y = 7$
   $2x - 5y = -1$

4. $2x - y = 7$
   $3x + 2y = 14$

5. $5x - y = -29$
   $2x + 3y = 2$

6. $6x + 4y = 12$
   $3x + 2y = 12$

7. $5x + 2y = 3$
   $x = 0$

8. $2x - y = 0$
   $x = -3$

9. $3x - 2y = 4$
   $y = -1$

10. $x + 2y = 6$
    $x = 2$

11. $\frac{1}{4}x - \frac{1}{3}y = -\frac{5}{12}$
    $\frac{1}{10}x + \frac{1}{5}y = \frac{1}{2}$

12. $\frac{2}{3}x - y = 4$
    $x - \frac{3}{4}y = 6$

13. $\frac{1}{7}x - \frac{3}{7}y = 1$
    $2x - y = -4$

14. $\frac{1}{3}x - \frac{2}{3}y = 2$
    $x - 2y = 6$

15. $6x + 4y = 12$
    $3x + 2y = 6$

16. $\frac{1}{3}x - \frac{2}{3}y = 2$
    $x - 2y = 6$

17. $3x - y = 4$
    $18x - 6y = -24$

18. $\frac{1}{2}x - \frac{3}{4}y = 1$
    $x - \frac{3}{2}y = 6$

19. Find a and b so that the graph of $ax + by + 3 = 0$ passes through the points $(-1, 2)$ and $(-3, 0)$.

20. Find a and b so that the graph of $ax + by - 4 = 0$ passes through the points $(0, 2)$ and $(2, -4)$.

21. Find a and b so that the solution set of the system
    $$ax + by = 2$$
    $$bx - ay = 2$$
    is $\{(1, 1)\}$.

22. Find a and b so that the solution set of the system
    $$ax + by = 4$$
    $$bx - ay = -3$$
    is $\{(1, 2)\}$.

23. Recall that the slope-intercept form of the equation of a straight line is $y = mx + b$. Find an equation of the line that passes through the points $(0, 2)$ and $(3, -8)$.

24. Find an equation of the line that passes through the points $(-6, 2)$ and $(4, 1)$.

25. Find a linear relationship between centigrade temperature, C, and Fahrenheit temperature, F, given that $F = 32°$ when $C = 0°$, and $F = 212°$ when $C = 100°$.

26. A man has \$1.80 in nickels and dimes, with three more dimes than nickels. How many dimes and nickels does he have?

27. How many pounds of an alloy containing 45% silver must be melted with an alloy containing 60% silver to obtain 40 pounds of a 48% silver alloy?

28. A man has three times as much money invested in 9% bonds as he has in stocks paying 8%. How much does he have invested in each if his yearly income from the investments is $1680?

29. A man has $1000 more invested at 9% than he has invested at 11%. If his annual income from the two investments together is $698, how much does he have invested at each rate?

30. An airplane travels 1260 miles in the same time that an automobile travels 420 miles. If the rate of the airplane is 120 miles per hour greater than the rate of the automobile, find the rate of each.

31. Two cars start together and travel in the same direction, one going twice as fast as the other. At the end of 3 hours, they are 96 miles apart. How fast is each traveling?

B Solve each system. Hint: Using substitutions, change the given system to one of the form

$$a_1 u + b_1 v = c_1$$
$$a_2 u + b_2 v = c_2.$$

32. $\dfrac{1}{x} + \dfrac{1}{y} = 2$

    $\dfrac{1}{x} - \dfrac{2}{y} = -1$

33. $\dfrac{4}{x} - \dfrac{3}{y} = -7$

    $\dfrac{-1}{x} - \dfrac{2}{y} = -1$

34. $\dfrac{1}{x+2} - \dfrac{2}{y+3} = 0$

    $\dfrac{2}{x+2} - \dfrac{1}{y+3} = \dfrac{3}{4}$

35. $\dfrac{3}{x+4} + \dfrac{1}{y-4} = \dfrac{1}{6}$

    $\dfrac{2}{x+4} + \dfrac{1}{y-4} = 0$

36. Show that the linear polynomials

$$a_1 x + b_1 y + c_1 \quad (a_1, b_1, c_1 \neq 0)$$
$$a_2 x + b_2 y + c_2 \quad (a_2, b_2, c_2 \neq 0)$$

are dependent if and only if

$$\frac{a_1}{a_2} = \frac{b_1}{b_2} = \frac{c_1}{c_2}.$$

Hint: Write the equations in slope-intercept form.

37. Show that the equations in the system

$$a_1 x + b_1 y + c_1 = 0$$
$$a_2 x + b_2 y + c_2 = 0 \quad (a_2, b_2, c_2 \neq 0)$$

are inconsistent if and only if

$$\frac{a_1}{a_2} = \frac{b_1}{b_2} \neq \frac{c_1}{c_2}.$$

Hint: Write the equations in slope-intercept form.

38. Find a and b so that the system

$$ax + by = -1$$
$$bx - ay = 3$$

is inconsistent.

8.2 Systems of Linear Equations in Three Variables

Solutions of an equation in three variables

A solution of an equation in three variables, such as

$$x + 2y - 3z + 4 = 0, \tag{1}$$

is an ordered triple of numbers (x, y, z), because all three of the variables must be replaced before we can decide whether or not the statement is true. Thus, $(0, -2, 0)$ and $(-1, 0, 1)$ are solutions of Equation (1), whereas $(1, 1, 1)$ is not. There are, of course, infinitely many members in the solution set of such an equation.

Solution of a system

The solution set of a system of linear (first-degree) equations in three variables is the intersection of the solution sets of the separate equations in the system. We are primarily interested in systems involving three equations, such as

$$x + 2y - 3z + 4 = 0$$
$$2x - y + z - 3 = 0$$
$$3x + 2y + z - 10 = 0.$$

We can find the members of this set by methods analogous to those used in the preceding section.

Theorem 8.2 *If any equation in the system*

$$f(x, y, z) = 0$$
$$g(x, y, z) = 0$$
$$h(x, y, z) = 0$$

is replaced by a linear combination, with nonzero coefficients, of itself and any one of the other equations in the system, then the result is an equivalent system.

The proof of this theorem is similar to that of Theorem 8.1 and is omitted here.

Example

Find the solution set of the system

$$x + 2y - 3z + 4 = 0 \tag{1}$$
$$2x - y + z - 3 = 0 \tag{2}$$
$$3x + 2y + z - 10 = 0. \tag{3}$$

Solution on overleaf

Solution We begin by replacing Equation (2) with the linear combination formed by multiplying Equation (1) by -2 and Equation (2) by 1, that is, with

$$-2(x + 2y - 3z + 4) + 1(2x - y + z - 3) = 0,$$

or

$$-5y + 7z - 11 = 0.$$

We obtain the equivalent system

$$x + 2y - 3z + 4 = 0 \tag{1}$$
$$-5y + 7z - 11 = 0 \tag{2'}$$
$$3x + 2y + z - 10 = 0, \tag{3}$$

where (2') is free of x. Next, if we replace (3) by the sum of -3 times (1) and 1 times (3), we have

$$x + 2y - 3z + 4 = 0 \tag{1}$$
$$-5y + 7z - 11 = 0 \tag{2'}$$
$$-4y + 10z - 22 = 0, \tag{3'}$$

where both (2') and (3') are free of x. If now (3') is replaced by the sum of 4 times (2') and -5 times (3'), we have

$$x + 2y - 3z + 4 = 0 \tag{1}$$
$$-5y + 7z - 11 = 0 \tag{2'}$$
$$-22z + 66 = 0, \tag{3''}$$

which is equivalent to the original (1), (2), and (3). But, from (3''), we see that for any solution of (1), (2'), and (3''), $z = 3$; that is, the solution will be of the form $(x, y, 3)$. If 3 is substituted for z in (2'), we have

$$-5y + 7(3) - 11 = 0, \quad \text{or} \quad y = 2,$$

and any solution of the system must be of the form $(x, 2, 3)$. Substituting 2 for y and 3 for z in (1) gives

$$x + 2(2) - 3(3) + 4 = 0, \quad \text{or} \quad x = 1,$$

so that the single member of the solution set of (1), (2'), and (3'') is $(1, 2, 3)$. Therefore, the solution set of the original system is $\{(1, 2, 3)\}$.

The foregoing process of solving a system of linear equations can be reduced to a series of mechanical procedures as shown in the following example.

Example Solve.

$$x + 2y - z + 1 = 0 \tag{1}$$
$$x - 3y + z - 2 = 0 \tag{2}$$
$$2x + y + 2z - 6 = 0 \tag{3}$$

8.2 Systems of Linear Equations in Three Variables

Solution We first multiply (1) by -1 and add the result to 1 times (2). We then multiply (1) by -2 and add the result to 1 times (3). In each linear combination the x-terms vanish.

$$-5y + 2z - 3 = 0 \tag{4}$$

$$-3y + 4z - 8 = 0 \tag{5}$$

Next we multiply (4) by -3 and add the result to 5 times (5). In this linear combination the y-terms vanish.

$$14z - 31 = 0 \tag{6}$$

Now (1), (4), and (6) constitute a set of equations equivalent to (1), (2), and (3).

$$x + 2y - z + 1 = 0 \tag{1}$$

$$-5y + 2z - 3 = 0 \tag{4}$$

$$14z - 31 = 0 \tag{6}$$

Solving for z in (6), we obtain

$$z = \frac{31}{14}.$$

We then substitute $31/14$ for z in (4) and solve for y to obtain

$$-5y + 2\left(\frac{31}{14}\right) - 3 = 0,$$

$$y = \frac{2}{7}.$$

Substituting $2/7$ for y and $31/14$ for z in either (1), (2), or (3), say (1), and solving for x, we obtain

$$x + 2\left(\frac{2}{7}\right) - \frac{31}{14} + 1 = 0,$$

$$x = \frac{9}{14}.$$

The solution set is $\left\{\left(\frac{9}{14}, \frac{2}{7}, \frac{31}{14}\right)\right\}$.

If at any step in the procedure used in the examples above, the resulting linear combination vanishes or yields a contradiction, the system contains linearly dependent left-hand members or else inconsistent equations, or both, and it either has an infinite number of solutions or else has no member in its solution set.

Graphs in three dimensions

By establishing a three-dimensional Cartesian coordinate system as shown in Figure 8.4, a one-to-one correspondence can be established between the points in a three-dimensional space and ordered triples of real numbers. If this is done, it can be shown that the graph of a linear equation in three variables is a plane. For example, the equation

$$x + y + z = 1$$

represents the plane through the points $(1, 0, 0)$, $(0, 1, 0)$, and $(0, 0, 1)$, as illustrated in Figure 8.4. Hence, the solution set of a system of three linear

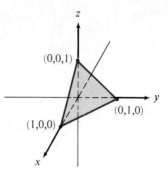

Figure 8.4

equations in three variables consists of the coordinates of the common intersection of three planes. Figure 8.5 shows the possibilities for the relative positions of the plane graphs of three linear equations in three variables. In case (a), the common

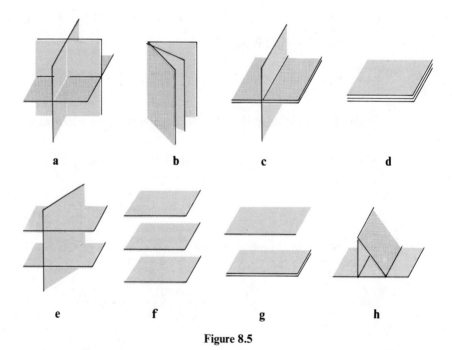

Figure 8.5

intersection consists of a single point, and hence the solution set of the corresponding system of three equations contains a single member. In cases (b), (c), and (d), the intersection is a line or a plane, and the solution of the corresponding system has infinitely many members. In cases (e), (f), (g) and (h), the three planes have no common intersection, and the solution set of the corresponding system is the null set. Linear equations corresponding to cases (a), (b), (c), and (d) are consistent, and the others inconsistent.

8.2 Systems of Linear Equations in Three Variables

The use of linear combinations to solve systems of linear equations can be extended to cover cases of n equations in n variables. Clearly, as n grows larger, the time and effort necessary to find the solution set of a system increase correspondingly. Fortunately, modern high-speed computers can handle such computations in stride for fairly large values of n. There are analytic methods other than those exhibited here which become preferable for very large systems.

Exercise 8.2

A Solve.

Example

$$x + 2y + z = 4 \quad (1)$$
$$2x + y - z = -1 \quad (2)$$
$$-x + y + z = 2 \quad (3)$$

Solution Replace (2) by itself plus -2 times (1), and (3) by itself plus (1).

$$x + 2y + z = 4 \quad (1')$$
$$-3y - 3z = -9 \quad (2')$$
$$3y + 2z = 6 \quad (3')$$

Replace (3') by itself plus (2').

$$x + 2y + z = 4 \quad (1'')$$
$$-3y - 3z = -9 \quad (2'')$$
$$-z = -3 \quad (3'')$$

From (3''), obtain $z = 3$. Substitute $z = 3$ into (2'') to obtain $y = 0$. Substitute $z = 3$ and $y = 0$ into (1'') to obtain $x = 1$. The solution set is therefore $\{(1, 0, 3)\}$.

1. $x + y + z = 2$
   $2x - y + z = -1$
   $x - y - z = 0$

2. $x + y + z = 1$
   $2x - y + 3z = 2$
   $2x - y - z = 2$

3. $x + y + 2z = 0$
   $2x - 2y + z = 8$
   $3x + 2y + z = 2$

4. $2x - 3y + z = 3$
   $x - y - 3z = -1$
   $-x + 2y - 3z = -4$

5. $x - 2y + z = -1$
   $2x + y - 3z = 3$
   $3x + 3y - 2z = 10$

6. $x - 2y + 4z = -3$
   $3x + y - 2z = 12$
   $2x + y - 3z = 11$

7. $4x - 2y + 3z = 4$
   $2x - y + z = 1$
   $3x - 3y + 4z = 5$

8. $x + 5y - z = 2$
   $3x - 9y + 3z = 6$
   $x - 3y + z = 4$

9. $x + z = 5$
   $y - z = -4$
   $x + y = 1$

10. $5y - 8z = -19$
    $5x - 8z = 6$
    $3x - 2y = 12$

11. $x - \frac{1}{2}y - \frac{1}{2}z = 4$
    $x - \frac{3}{2}y - 2z = 3$
    $\frac{1}{4}x + \frac{1}{4}y - \frac{1}{4}z = 0$

12. $x + 2y + \frac{1}{2}z = 0$
    $x + \frac{3}{5}y - \frac{2}{5}z = \frac{1}{5}$
    $4x - 7y - 7z = 6$

Use three variables in the solution of each of the following problems.

Example The parabola $y = ax^2 + bx + c$ passes through the points $(-1, 9)$, $(1, 3)$, and $(3, 5)$. Find the coefficients a, b, and c.

Solution Since $(-1, 9)$ is on the parabola,

$$9 = a(-1)^2 + b(-1) + c,$$

or

$$a - b + c = 9. \tag{1}$$

The other two points yield the equations

$$a + b + c = 3, \tag{2}$$

$$9a + 3b + c = 5. \tag{3}$$

Replace (2) by itself plus -1 times (1), and replace (3) by itself plus -9 times (1).

$$a - b + c = 9 \tag{1'}$$

$$2b = -6 \tag{2'}$$

$$12b - 8c = -76 \tag{3'}$$

From (2'), $b = -3$. Substitute $b = -3$ into (3') to obtain $c = 5$. Substitute $b = -3$ and $c = 5$ into (1') to obtain $a = 1$. Therefore the equation of the parabola is $y = x^2 - 3x + 5$.

13. The sum of three numbers is 15. The second equals two times the first and the third equals the second. Find the numbers.

14. The sum of three numbers is 2. The first number is equal to the sum of the other two, and the third number is the result of subtracting the first from the second. Find the numbers.

15. A box contains $6.25 in nickels, dimes, and quarters. There are 85 coins in all, with three times as many nickels as dimes. How many coins of each kind are there?

16. A man had $446 in ten-dollar, five-dollar, and one-dollar bills. There were 94 bills in all and 10 more five-dollar bills than ten-dollar bills. How many bills of each kind did he have?

17. The perimeter of a triangle is 155 centimeters. The side x is 20 centimeters shorter than the side y, and the side y is 5 centimeters longer than the side z. Find the lengths of the sides of the triangle.

18. The perimeter of a triangle is 120 centimeters. The side x is the same length as side y and 10 centimeters shorter than side z. Find the length of each side.

19. Find values for a, b, and c so that the graph of $x^2 + y^2 + ax + by + c = 0$ will contain the points $(0, 0)$, $(6, 0)$, and $(0, 8)$.

20. The equation for a circle can be written $x^2 + y^2 + ax + by + c = 0$. Find the equation of the circle with graph containing the points $(2, 3)$, $(3, 2)$, and $(-4, -5)$.

21. Find values for a, b and c so that the graph of $y = ax^2 + bx + c$ contains the points $(-1, 0)$, $(2, 12)$ and $(-2, 8)$.

8.3 Partial Fractions

22. Find values for a, b, and c so that the graph of $y = ax^2 + bx + c$ contains the points $(-1, 2)$, $(1, 6)$, and $(2, 11)$.

23. Three solutions of the equation $ax + by + cz = 1$ are $(0, 2, 1)$, $(6, -1, 2)$, and $(0, 2, 0)$. Find the coefficients a, b, and c.

24. Three solutions of the equation $ax + by + cz = 1$ are $(2, 1, 0)$, $(-1, 3, 2)$, and $(3, 0, 0)$. Find the coefficients a, b, and c.

B Solve each system. Hint: Using substitutions, change the given system to one of the form

$$a_1 u + b_1 v + c_1 w = d_1$$
$$a_2 u + b_2 v + c_2 w = d_2$$
$$a_3 u + b_3 v + c_3 w = d_3.$$

25.
$$\frac{1}{x} + \frac{1}{y} - \frac{1}{z} = 1$$
$$\frac{2}{x} - \frac{2}{y} + \frac{1}{z} = 1$$
$$\frac{-3}{x} + \frac{1}{y} - \frac{1}{z} = -3$$

26.
$$\frac{4}{x} - \frac{2}{y} + \frac{1}{z} = 4$$
$$\frac{3}{x} - \frac{1}{y} + \frac{2}{z} = 0$$
$$\frac{-1}{x} + \frac{3}{y} - \frac{2}{z} = 0$$

27.
$$\frac{2}{x-1} + \frac{1}{y+2} - \frac{1}{z-3} = \frac{-2}{3}$$
$$\frac{-3}{x-1} + \frac{1}{y+2} - \frac{2}{z-3} = \frac{16}{3}$$
$$\frac{2}{x-1} + \frac{3}{y+2} - \frac{1}{z-3} = 0$$

28.
$$\frac{2}{x+1} - \frac{1}{y-3} + \frac{1}{z} = -3$$
$$\frac{-3}{x+1} + \frac{2}{y-3} - \frac{2}{z} = \frac{13}{2}$$
$$\frac{-2}{x+1} + \frac{4}{y-3} - \frac{1}{z} = 9$$

29. Show that the system

$$x + y + 2z = 2$$
$$2x - y - z = 3$$

has an infinite number of members in its solution set. List two ordered triples that are solutions. Hint: Express x in terms of z alone, and express y in terms of z alone.

30. Give a geometric argument to show that any system of two consistent linear equations in three variables has an infinite number of solutions.

8.3 Partial Fractions

We can use the method of solving systems of equations that we considered in Sections 8.1 and 8.2 to help us rewrite certain kinds of fractions in what may be more useful forms. Recall from Section 2.7 that the sum $\dfrac{P(x)}{Q(x)} + \dfrac{R(x)}{S(x)}$ can be rewritten as the single fraction

$$\frac{P(x) \cdot S(x) + Q(x) \cdot R(x)}{Q(x) \cdot S(x)}.$$

It is frequently useful to be able to reverse this process in the case of certain kinds of fractions—in particular, for rational expressions in which the numerator is a polynomial of lesser degree than the denominator. Such rational expressions are customarily referred to as "proper" fractions.

Notice first that by the long-division algorithm, any rational expression can be written as a polynomial or the sum of a polynomial and a proper fraction. For example, by the long-division algorithm we find that

$$\frac{x^3 + 2x^2 + 2x + 3}{x^2 - 1} = x + 2 + \frac{3x + 5}{x^2 - 1}.$$

Distinct factors in the denominator

We shall now consider the problem of finding values c_1 and c_2 (if they exist) so that

$$\frac{ax + b}{(x - r_1)(x - r_2)} = \frac{c_1}{x - r_1} + \frac{c_2}{x - r_2}, \qquad (1)$$

where a, b, r_1, r_2 ($r_1 \neq r_2$) are known constants.

Suppose first that there are such constants $c_1, c_2 \in R$. Multiplying each member of Equation (1) by $(x - r_1)(x - r_2)$, we get

$$ax + b = c_1(x - r_2) + c_2(x - r_1).$$

Since we wish this equation to hold for *every value of* x other than r_1 and r_2, it must hold also for $x = r_1$ and $x = r_2$ (see Exercise 6.4-41). Substituting r_1 and r_2 for x in turn, we get the system of equations in c_1 and c_2,

$$ar_1 + b = c_1(r_1 - r_2) + c_2(r_1 - r_1)$$
$$ar_2 + b = c_1(r_2 - r_2) + c_2(r_2 - r_1),$$

from which, since $r_1 \neq r_2$,

$$c_1 = \frac{ar_1 + b}{r_1 - r_2} \quad \text{and} \quad c_2 = \frac{ar_2 + b}{r_2 - r_1}.$$

That these values satisfy the given fractional equation can be verified by direct substitution. Thus, not only do c_1 and c_2 exist, but they have the values shown. This proves the following theorem.

Theorem 8.3 *If $a, b, r_1, r_2 \in R$ are given, with $r_1 \neq r_2$, then there exist constants $c_1, c_2 \in R$ such that for $x \neq r_1, r_2$,*

$$\frac{ax + b}{(x - r_1)(x - r_2)} = \frac{c_1}{x - r_1} + \frac{c_2}{x - r_2}.$$

8.3 Partial Fractions

While proof of the foregoing theorem produces formulas for c_1 and c_2, it is preferable from the standpoint of efficiency to rely on the method used in the proof to obtain the numbers c_1 and c_2 in any particular problem. The technique is known as the method of **partial fractions**.

Example Apply the method of partial fractions to express $\dfrac{3x - 2}{x^2 - x}$ as a sum of fractions.

Solution We first observe that

$$\frac{3x - 2}{x^2 - x} = \frac{3x - 2}{x(x - 1)}. \tag{2}$$

Then by Theorem 8.3, we know there exist numbers $c_1, c_2 \in R$, such that

$$\frac{3x - 2}{x(x - 1)} = \frac{c_1}{x} + \frac{c_2}{x - 1}. \tag{3}$$

Multiplying each member of the equation by $x(x - 1)$, we obtain

$$3x - 2 = c_1(x - 1) + c_2(x). \tag{4}$$

We wish the members of Equation (3) to be equal for all values of x except 0 and 1. However, if this is to be the fact, then the members of (4) must be equal for all values of x, *including* 0 and 1. Hence, while we can arbitrarily select *any* values for x and solve for c_1 and c_2 in (4), let us take 0 and 1 because these values of x yield the values of c_1 and c_2 by inspection. Thus, if $x = 0$, we have from Equation (4)

$$3(0) - 2 = c_1(0 - 1) + c_2(0),$$

from which

$$c_1 = 2.$$

If $x = 1$, Equation (4) becomes

$$3(1) - 2 = c_1(1 - 1) + c_2(1),$$

$$c_2 = 1.$$

Therefore,

$$\frac{3x - 2}{x(x - 1)} = \frac{2}{x} + \frac{1}{x - 1}.$$

Theorem 8.3 generalizes to the following form, although the proof is omitted.

Theorem 8.4 If $P(x)$ and $Q(x)$ are real polynomials, with $P(x)$ of degree less than $Q(x)$, and if $Q(x) = (x - r_1)(x - r_2) \cdots (x - r_n)$, where no two factors are identical, then there exist constants $c_1, c_2, \ldots, c_n \in R$ such that

$$\frac{P(x)}{Q(x)} = \frac{c_1}{x - r_1} + \frac{c_2}{x - r_2} + \cdots + \frac{c_n}{x - r_n}.$$

Repeated factors in the denominator

If the denominator of a proper fraction can be factored into *equal* linear factors, then we need the following result, which is stated without proof.

Theorem 8.5 *If $P(x)$ and $Q(x)$ are real polynomials with $P(x)$ of degree less than $Q(x)$, and if $Q(x) = (x - r_1)^n$, then there exist constants $c_1, c_2, \ldots, c_n \in R$ such that*

$$\frac{P(x)}{Q(x)} = \frac{c_1}{(x - r_1)} + \frac{c_2}{(x - r_1)^2} + \cdots + \frac{c_n}{(x - r_1)^n}.$$

A further extension of Theorems 8.4 and 8.5 can be used to apply the method of partial fractions to rational expressions in which the denominator has one or more repeated factors. Each repeated factor contributes a sum such as the one in Theorem 8.5.

Example Express $\dfrac{5x^2 + 2}{x^3 - 2x^2 + x}$ as a sum of fractions.

Solution We first observe that $\dfrac{5x^2 + 2}{x^3 - 2x^2 + x} = \dfrac{5x^2 + 2}{x(x - 1)(x - 1)}$. Then we write

$$\frac{5x^2 + 2}{x(x - 1)(x - 1)} = \frac{c_1}{x} + \frac{c_2}{x - 1} + \frac{c_3}{(x - 1)^2}.$$

Multiplying each member of the equation by $x(x - 1)^2$, we obtain

$$5x^2 + 2 = c_1(x - 1)^2 + c_2 x(x - 1) + c_3 x.$$

Arbitrarily selecting any values of x, we can solve for c_1, c_2, and c_3. Let us use 0, 1, and 2, because these numbers yield simple computations. Substituting 0 for x gives

$$5(0)^2 + 2 = c_1(-1)^2 + 0 + 0,$$

$$c_1 = 2.$$

Substituting 1 for x gives

$$5(1)^2 + 2 = 0 + 0 + c_3(1),$$

$$c_3 = 7.$$

Substituting 2 for x gives

$$5(2)^2 + 2 = c_1(1)^2 + c_2(2)(1) + c_3(2).$$

8.3 Partial Fractions

Now, since $c_1 = 2$ and $c_3 = 7$, we have

$$22 = 2 + 2c_2 + 14,$$
$$c_2 = 3.$$

Hence,

$$\frac{5x^2 + 2}{x^3 - 2x^2 + x} = \frac{2}{x} + \frac{3}{x - 1} + \frac{7}{(x - 1)^2}.$$

Alternative method The constants c_i used to write a rational expression as a sum of fractions of the form $c_i/(x - r)^n$ can be obtained by equating the coefficients of two polynomials. This method is illustrated in the following example.

Example Express $\dfrac{2x + 1}{(x - 1)^3}$ as a sum of fractions.

Solution We first write

$$\frac{2x + 1}{(x - 1)^3} = \frac{c_1}{x - 1} + \frac{c_2}{(x - 1)^2} + \frac{c_3}{(x - 1)^3}.$$

Multiplying both members of this equation by $(x - 1)^3$, we obtain

$$2x + 1 = c_1(x - 1)^2 + c_2(x - 1) + c_3$$
$$= c_1 x^2 + (c_2 - 2c_1)x + (c_1 - c_2 + c_3).$$

Since both members of this equation are polynomials, and they are equal for all values of x except $x = 1$, their coefficients must be equal (Exercise 6.4-41). Thus, we have the system

$$c_1 = 0$$
$$-2c_1 + c_2 = 2$$
$$c_1 - c_2 + c_3 = 1.$$

Using the methods of Section 8.2, we solve this system to obtain

$$c_1 = 0, \quad c_2 = 2, \quad \text{and} \quad c_3 = 3.$$

Hence,

$$\frac{2x + 1}{(x - 1)^3} = \frac{2}{(x - 1)^2} + \frac{3}{(x - 1)^3}.$$

Note that Theorems 8.4 and 8.5 apply to fractions whose denominators can be factored into linear factors. Similar theorems apply to fractions whose denominators cannot be factored completely into linear factors. In particular, the numerator of each term of the partial-fraction expansion whose denominator is a nonfactorable quadratic expression will be a linear expression of the form $c_1 x + c_2$. Several such fractions are given in the exercises.

Exercise 8.3

A Use the method of partial fractions to express each fraction as a sum of fractions with powers of linear polynomials as denominators. The numerators may be rational numbers.

Example

$$\frac{2}{x^2 - x}$$

Solution

Factor $x^2 + x$ as $x(x + 1)$ and write

$$\frac{2}{x^2 + x} = \frac{c_1}{x} + \frac{c_2}{x + 1} = \frac{c_1(x + 1) + c_2 x}{x^2 + x}.$$

Thus, $2 = c_1(x + 1) + c_2 x$. Substitute $x = -1$ and $x = 0$ to obtain

$$2 = c_1(-1 + 1) + c_2(-1) = -c_2,$$

$$2 = c_1(0 + 1) + c_2 \cdot 0 = c_1.$$

Therefore, $c_1 = 2$, $c_2 = -2$, and

$$\frac{2}{x^2 + x} = \frac{2}{x} - \frac{2}{x + 1}.$$

1. $\dfrac{4}{x^2 + x}$
2. $\dfrac{x + 2}{x^2 + x}$
3. $\dfrac{x}{(x + 1)(x + 2)}$
4. $\dfrac{3}{x^2 + 4x + 3}$
5. $\dfrac{x - 2}{x^2 + 5x + 6}$
6. $\dfrac{2x - 1}{x^2 + 5x - 14}$

Example

$$\frac{x + 1}{x(x + 2)(x + 4)}$$

Solution

Write

$$\frac{x + 1}{x(x + 2)(x + 4)} = \frac{c_1}{x} + \frac{c_2}{x + 2} + \frac{c_3}{x + 4}$$

$$= \frac{c_1(x + 2)(x + 4) + c_2 x(x + 4) + c_3 x(x + 2)}{x(x + 2)(x + 4)}.$$

Thus, $x + 1 = c_1(x + 2)(x + 4) + c_2 x(x + 4) + c_3 x(x + 2)$. Substitute $x = 0, -2, -4$ to obtain

$$1 = c_1 \cdot 2 \cdot 4 + c_2 \cdot 0 + c_3 \cdot 0 = 8c_1$$

$$-1 = c_1 \cdot 0 \cdot 2 + c_2(-2) \cdot 2 + c_3(-2) \cdot 0 = -4c_2$$

$$-3 = c_1 \cdot (-2) \cdot 0 + c_2(-4) \cdot 0 + c_3(-4)(-2) = 8c_3.$$

8.3 Partial Fractions

Therefore, $c_1 = \frac{1}{8}$, $c_2 = \frac{1}{4}$, $c_3 = \frac{-3}{8}$, so that

$$\frac{x+1}{x(x+2)(x+4)} = \frac{\frac{1}{8}}{x} + \frac{\frac{1}{4}}{x+2} - \frac{\frac{3}{8}}{x+4}.$$

7. $\dfrac{2x+1}{x(x+1)(x+2)}$ 8. $\dfrac{x}{(x-1)(x+1)(x+3)}$

9. $\dfrac{x^2-x+1}{x(x+1)(x+2)}$ 10. $\dfrac{x^2+1}{x(x-1)(x+1)}$

11. $\dfrac{1}{x^3+3x^2+2x}$ 12. $\dfrac{x-1}{(x+1)(x^2-2x)}$

Example

$\dfrac{x+2}{x(x+1)^2}$

Solution

Write $\dfrac{x+2}{x(x+1)^2} = \dfrac{c_1}{x} + \dfrac{c_2}{x+1} + \dfrac{c_3}{(x+1)^2}$

$= \dfrac{c_1(x+1)^2 + c_2 x(x+1) + c_3 x}{x(x+1)^2}.$

Thus, $x + 2 = c_1(x+1)^2 + c_2 x(x+1) + c_3 x$. Substitute $x = 0, -1$ to obtain

$$2 = c_1 \cdot 1 + c_2 \cdot 0 + c_3 \cdot 0 = c_1$$

$$1 = c_1 \cdot 0 + c_2 \cdot 0 + c_3(-1) = -c_3.$$

Substitute any other value for x to find c_2; here we will use $x = 1$ to obtain

$$3 = c_1 \cdot 4 + c_2 \cdot 2 + c_3 = 2 \cdot 4 + 2 \cdot c_2 - 1,$$

so that $3 = 7 + 2c_2$, or $c_2 = -2$. Therefore,

$$\frac{x+2}{x(x+1)^2} = \frac{2}{x} - \frac{2}{x+1} - \frac{1}{(x+1)^2}.$$

13. $\dfrac{x+1}{x^2(x-2)}$ 14. $\dfrac{x+3}{(x+1)(x+2)^2}$ 15. $\dfrac{x^3}{(x-1)^2(x+1)^2}$

16. $\dfrac{x^2+1}{x^4-2x^2+1}$ 17. $\dfrac{1}{x^3-x^2}$ 18. $\dfrac{1}{x^4+x^2}$

Use long division and the method of partial fractions to express each fraction as the sum of a polynomial and fractions with powers of linear polynomials for denominators.

Example

$\dfrac{x^3 + 2x^2 + x + 2}{x^2 + x}$

Solution on overleaf

Solution Long division yields

$$\frac{x^3 + 2x^2 + x + 2}{x^2 + x} = x + 1 + \frac{2}{x^2 + x}.$$

The method of partial fractions yields

$$\frac{2}{x^2 + x} = \frac{2}{x} - \frac{2}{x + 1}.$$

Thus,

$$\frac{x^3 + 2x^2 + x + 2}{x^2 + x} = x + 1 + \frac{2}{x} - \frac{2}{x + 1}.$$

19. $\dfrac{x^4 + x^3 + x + 2}{x^2 + x}$

20. $\dfrac{2x^2 + 9x + 10}{x^2 + 5x + 6}$

21. $\dfrac{x^3}{(x - 1)^2}$

22. $\dfrac{x^3}{x^2 - 1}$

B 23. Find constants a, b, and c so that

$$\frac{x + 1}{x^3 + x^2 + x} = \frac{a}{x} + \frac{bx + c}{x^2 + x + 1}.$$

(*Hint:* Multiply each member of the equation by $x^3 + x^2 + x = x(x^2 + x + 1)$. Then simplify the right-hand member and equate the coefficients of like powers of x. Solve the resulting system of equations for a, b, and c).

24. Find constants a, b, and c so that

$$\frac{3x + 1}{x^3 - 8} = \frac{a}{x - 2} + \frac{bx + c}{x^2 + 2x + 4}.$$

25. Find constants a, b, and c so that

$$\frac{2x - 3}{x^3 - 2x^2 + 4x} = \frac{a}{x} + \frac{bx + c}{x^2 - 2x + 4}.$$

26. Find constants a, b, and c so that

$$\frac{3x - 2}{(x - 2)(x^2 - 3x + 3)} = \frac{a}{x - 2} + \frac{bx + c}{x^2 - 3x + 3}.$$

8.4 Systems of Nonlinear Equations

In Sections 8.1 and 8.2 we discussed systems of *linear* equations. In this section we introduce methods which can sometimes be used to solve systems containing nonlinear equations.

8.4 Systems of Nonlinear Equations

Use of graphs In solving a system containing nonlinear equations, it is frequently helpful to sketch the graphs of all of the equations in the system. By studying the graphs, we can approximate the intersection points of the curves. This usually tells us how many solutions with real components the system has, and it gives us approximate values for the solutions, which we can use as a rough check on any solutions we find algebraically.

Example Determine the number of solutions of the system

$$x^2 + y^2 = 25$$

$$x + y = 1$$

and approximate them.

Solution We sketch the graphs of $x^2 + y^2 = 25$ and $x + y = 1$ on the same set of coordinate axes. Observe from the figure that there are two points of intersection with coordinates approximately $(-3, 4)$ and $(4, -3)$. Thus, the system has two solutions with real components and they are approximately $(-3, 4)$ and $(4, -3)$.

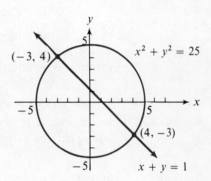

Substitution The substitution method is sometimes a convenient means of finding solution sets of systems in which nonlinear equations are present.

Example Solve.

$$x^2 + y^2 = 25 \qquad (1)$$

$$x + y = 1$$

Solution In the foregoing example, we found that the system has two solutions with real components. We now use the method of substitution to determine the solutions algebraically. Equation (2) can be written equivalently in the form

$$y = 1 - x, \qquad (3)$$

and we can replace y in (1) by $(1 - x)$ from (3). This produces

$$x^2 + (1 - x)^2 = 25, \qquad (4)$$

Solution continued on overleaf

which has as a solution set those values of x for which the ordered pair (x, y) is a common solution of (1) and (2). We can now find the solution set of (4):

$$x^2 + 1 - 2x + x^2 = 25,$$
$$2x^2 - 2x - 24 = 0,$$
$$2(x + 3)(x - 4) = 0,$$

which is satisfied if x is either -3 or 4. Now, by replacing x in the *first-degree equation* (3) by each of these numbers in turn, we obtain

$$y = 1 - (-3) = 4$$

and

$$y = 1 - 4 = -3,$$

respectively, so that the solution set of the system (1) and (2) is $\{(-3, 4), (4, -3)\}$.

If, in the preceding example, we had substituted the values -3 and 4 that we obtained for x in the *second-degree equation* (1), we would have obtained some extraneous solutions that do not satisfy Equation (2).

Linear combination

If both of the equations in a system of two equations in two variables are of the second degree in both variables, the use of linear combinations of the equations often provides a simpler means of solution than does substitution.

Example Solve.

$$3x^2 - 7y^2 + 15 = 0 \qquad (1)$$
$$3x^2 - 4y^2 - 12 = 0 \qquad (2)$$

Solution We graph the two equations on the same set of coordinate axes and observe that there are four solutions with real components. By forming the linear combination

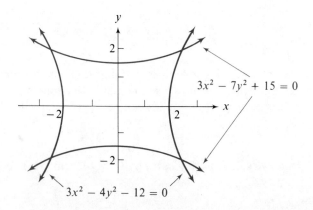

8.4 Systems of Nonlinear Equations

of -1 times (1) and 1 times (2), we obtain

$$3y^2 - 27 = 0$$
$$y^2 = 9,$$

from which

$$y = 3 \quad \text{or} \quad y = -3,$$

and we have the y-components of the members of the solution set of Equations (1) and (2). Substituting 3 for y in either (1) or (2), say (1), we have

$$3x^2 - 7(3)^2 + 15 = 0,$$
$$x^2 = 16,$$

from which

$$x = 4 \quad \text{or} \quad x = -4.$$

Thus, the ordered pairs $(4, 3)$ and $(-4, 3)$ are solutions of the system. Substituting -3 for y in (1) or (2) (this time we shall use (2)) gives us

$$3x^2 - 4(-3)^2 - 12 = 0,$$
$$x^2 = 16,$$

so that

$$x = 4 \quad \text{or} \quad x = -4.$$

Thus the ordered pairs $(4, -3)$ and $(-4, -3)$ are solutions of the system, and accordingly the complete solution set is $\{(4, 3), (4, -3), (-4, 3), (-4, -3)\}$.

Linear combination and substitution

The solution of some systems requires the application of both linear combinations *and* substitution.

Example

Solve.

$$x^2 + y^2 = 5 \quad (1)$$
$$x^2 - 2xy + y^2 = 1 \quad (2)$$

Solution

We graph the two equations on the same set of coordinate axes. Note that $x^2 - 2xy + y^2 = 1$ is equivalent to $(x - y)^2 = 1$, or $x - y = \pm 1$. Thus the graph of this equation consists of the two parallel lines $x - y = 1$ and $x - y = -1$. Observe that there are four solutions with real components.

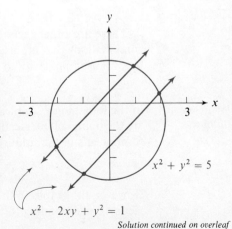

Solution continued on overleaf

By forming the linear combination of 1 times (1) and -1 times (2), we obtain
$$2xy = 4,$$
$$xy = 2. \qquad (3)$$

The system (1) and (2) is equivalent to the system (1) and (3). This latter system can be solved by substitution. From (3), we have
$$y = \frac{2}{x}.$$

Replacing y in (1) by $2/x$, we obtain
$$x^2 + \left(\frac{2}{x}\right)^2 = 5,$$
$$x^2 + \frac{4}{x^2} = 5, \qquad (4)$$
$$x^4 + 4 = 5x^2,$$
$$x^4 - 5x^2 + 4 = 0, \qquad (5)$$

which is a quadratic in x^2. The left-hand member of (5), when factored, yields
$$(x^2 - 1)(x^2 - 4) = 0,$$

from which we obtain
$$x^2 - 1 = 0 \quad \text{or} \quad x^2 - 4 = 0,$$

so that
$$x = 1, \quad x = -1, \quad x = 2, \quad x = -2.$$

Since the step from (4) to (5) was not an elementary transformation, we are careful to note that these values of x all satisfy (4). Substituting 1, -1, 2, and -2 in turn for x in (3), we obtain the corresponding values for y, and thus the solution set of the system (1) and (3) or, equivalently, the solution set of the system (1) and (2), is $\{(1, 2), (-1, -2), (2, 1), (-2, -1)\}$.

There are other techniques involving substitution in conjunction with linear combinations that are useful in handling systems of higher-degree equations, but they all bear similarity to those illustrated.

Imaginary solutions

In the foregoing examples, each solution is a member of $R \times R$ and their graphs are the points of intersection of the graphs of each equation. If one or more of the components of the solutions are complex numbers, we find these solutions in
$$C \times C = \{(x, y) \mid x \in C \text{ and } y \in C\}.$$

However, the graphs in the real plane of the equations do *not* have points of intersection corresponding to these solutions.

8.4 Systems of Nonlinear Equations

Example Solve.

$$x^2 + y^2 = 26 \qquad (1)$$
$$x + y = 8 \qquad (2)$$

Solution We graph the two equations on the same set of coordinate axes and observe that there are no solutions with real components. We use substitution to find the solutions.

Equation (2) can be written equivalently as

$$y = 8 - x. \qquad (3)$$

Substituting $8 - x$ for y in (1) and simplifying yields

$$x^2 + (8 - x)^2 = 26$$
$$x^2 + 64 - 16x + x^2 = 26$$
$$2x^2 - 16x + 38 = 0$$
$$x^2 - 8x + 19 = 0.$$

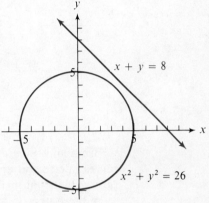

Using the quadratic formula to solve for x, we obtain

$$x = \frac{8 \pm \sqrt{64 - 76}}{2(1)} = \frac{8 \pm \sqrt{-12}}{2}$$
$$= \frac{8 \pm 2i\sqrt{3}}{2} = \frac{2(4 \pm i\sqrt{3})}{2} = 4 \pm i\sqrt{3}.$$

Then, substituting $4 + i\sqrt{3}$ for x in (3) gives $y = 4 - i\sqrt{3}$, and substituting $4 - i\sqrt{3}$ for x in (3) gives $y = 4 + i\sqrt{3}$. Hence, the solution set is

$$\{(4 + i\sqrt{3}, 4 - i\sqrt{3}), (4 - i\sqrt{3}, 4 + i\sqrt{3})\}.$$

Approximations to real solutions

When the degree of any equation in a set of equations is greater than two, or when the left-hand member of one of the equations of the form $f(x, y) = 0$ is not a polynomial, it may be very difficult or impossible to find common solutions analytically. In this case, we can at least obtain approximations to any real solutions by graphical methods.

Example Approximate the solutions to the system

$$y = 2^{-x}$$
$$y = \sqrt{0.01x}.$$

Solution on overleaf

Solution   We first graph these equations.

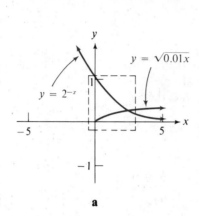

a

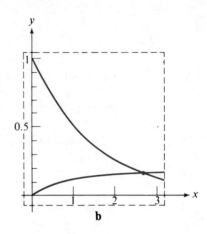

b

Examining Figure b, an enlargement of part of Figure a, we observe that the curves intersect at approximately (2.6, 0.16), and from geometric considerations we conclude that this ordered pair approximates the only member in the solution set of the system.

The preceding example shows, in particular, that the equation

$$2^{-x} - \sqrt{0.01x} = 0$$

has just one solution, approximately 2.6, and suggests a method of approximating solutions of similar equations.

Exercise 8.4

A   *Solve by the method of substitution. Check the solutions by sketching the graphs of the equations and estimating the coordinates of any points of intersection.*

Example

$y = x^2 + 2x + 1$ (1)

$y - x = 3$ (2)

Solution   Solve Equation (2) explicitly for y.

$y = x + 3$ (2')

Substitute $x + 3$ for y in (1).

$x + 3 = x^2 + 2x + 1$ (3)

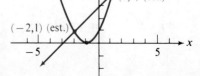

8.4 Systems of Nonlinear Equations

Solve for x.

$$x^2 + x - 2 = 0$$
$$(x + 2)(x - 1) = 0$$
$$x = -2, \quad x = 1$$

Substitute each of these values in turn in (2') to determine values for y.

If $x = -2$, then $y = 1$.

If $x = 1$, then $y = 4$.

The solution set is $\{(-2, 1), (1, 4)\}$.

1. $y = x^2 - 5$
   $y = 4x$
2. $y = x^2 - 2x + 1$
   $y + x = 3$
3. $x^2 + y^2 = 13$
   $x + y = 5$
4. $x^2 + 2y^2 = 12$
   $2x - y = 2$
5. $x + y = 1$
   $xy = -12$
6. $2x - y = 9$
   $xy = -4$

Solve using linear combinations.

Example

$$2x^2 + y^2 = 7 \qquad (1)$$
$$x^2 - y^2 = 2 \qquad (2)$$

Solution

Replace (1) by itself minus 2 times (2).

$$3y^2 = 3 \qquad (1')$$
$$x^2 - y^2 = 2 \qquad (2)$$

From (1') $y^2 = 1$; substitution in (2) yields $x^2 = 3$. Therefore

$$x = \pm\sqrt{3} \text{ and } y = \pm 1,$$

so the solution set is $\{(\sqrt{3}, 1), (\sqrt{3}, -1), (-\sqrt{3}, 1), (-\sqrt{3}, -1)\}$.

7. $x^2 + y^2 = 10$
   $9x^2 + y^2 = 18$
8. $x^2 + 4y^2 = 52$
   $x^2 + y^2 = 25$
9. $x^2 - y^2 = 7$
   $2x^2 + 3y^2 = 24$
10. $x^2 + 4y^2 = 25$
    $4x^2 + y^2 = 25$
11. $4x^2 - 9y^2 + 132 = 0$
    $x^2 + 4y^2 - 67 = 0$
12. $16y^2 + 5x^2 - 26 = 0$
    $25y^2 - 4x^2 - 17 = 0$

Solve.

Example

$$x^2 - 3xy + 2y^2 = 0 \qquad (1)$$
$$2x^2 - xy - 3y^2 = 12 \qquad (2)$$

Solution on overleaf

Solution   Factor the left-hand member of (1).

$$(x - y)(x - 2y) = 0$$

Since the product is 0, we must have

$$x - y = 0 \quad \text{or} \quad x - 2y = 0.$$

Thus,

$$x = y, \qquad (1')$$

or

$$x = 2y. \qquad (1'')$$

Substitute y for x in (2) to obtain

$$2y^2 - y^2 - 3y^2 = 12,$$
$$-2y^2 = 12,$$
$$y = i\sqrt{6} \quad \text{or} \quad y = -i\sqrt{6}.$$

Substitute each of these values in turn in (1') to determine values for x.

If $y = i\sqrt{6}$, then $x = i\sqrt{6}$.

If $y = -i\sqrt{6}$, then $x = -i\sqrt{6}$.

Thus $(i\sqrt{6}, i\sqrt{6})$ and $(-i\sqrt{6}, -i\sqrt{6})$ are solutions of the system.
We now substitute $2y$ for x in (2) and solve for y.

$$8y^2 - 2y^2 - 3y^2 = 12$$
$$3y^2 = 12$$
$$y = 2 \quad \text{or} \quad y = -2.$$

Substitute each of these values in turn in (1") to determine values for x.

If $y = 2$, then $x = 2(2) = 4$.

If $y = -2$, then $x = 2(-2) = -4$.

The solution set is $\{(4, 2), (-4, -2)(i\sqrt{6}, i\sqrt{6}), (-i\sqrt{6}, -i\sqrt{6})\}$.

13. $x^2 - xy + y^2 = 7$
    $x^2 + y^2 = 5$

14. $3x^2 - 2xy + 3y^2 = 34$
    $x^2 + y^2 = 17$

15. $3x^2 + 3xy - y^2 = 35$
    $x^2 - xy - 6y^2 = 0$

16. $x^2 - xy + y^2 = 21$
    $x^2 + 2xy - 8y^2 = 0$

17. $2x^2 - xy - 6y^2 = 0$
    $x^2 + 3xy + 2y^2 = 4$

18. $2x^2 + xy - y^2 = 0$
    $6x^2 + xy - y^2 = 1$

19. $x^2 + y^2 = 9$
    $y = 4$

20. $x^2 + 2y^2 = 6$
    $x + y = 10$

21. $2x^2 + xy - 4y^2 = 12$
    $x^2 - 2y^2 = 4$

22. $x^2 + 3xy - y^2 = -3$
    $x^2 - xy - y^2 = 1$

8.4 Systems of Nonlinear Equations

23. $y - 10^x = 0$
    $\log_{10} y - 2x = 1$

24. $y - 10^x = 0$
    $\log_{10} y + 3x = 4$

25. $10^y = \dfrac{1}{x}$
    $\log_{10} x = y^2$

26. $10^y = \sqrt{x}$
    $\log_{10} x^2 = y^2$

Solve by graphing. Approximate components of solutions to the nearest half unit.

27. $y = 10^x$
    $x + y = 2$

28. $y = 2^x$
    $y - x = 2$

29. $y = \log_{10} x$
    $y = x^2 - 2x + 1$

30. $y = 10^x$
    $y = x^2$

31. The sum of the squares of two positive numbers is 13. If twice the first number is added to the second, the sum is 7. Find the numbers.

32. The sum of two numbers is 6 and their product is 35/4. Find the numbers.

33. The annual income from an investment is $32. If the amount invested were $200 more and the rate 1/2% less, the annual income would be $35. What are the amount and rate of the investment?

34. At a constant temperature, the pressure P and volume V of a gas are related by the equation $PV = K$. The product of the pressure (in pounds per square inch) and the volume (in cubic inches) of a certain gas is 30 inch-pounds. If the temperature remains constant as the pressure is increased 4 pounds per square inch, the volume is decreased by 2 cubic inches. Find the original pressure and volume of the gas.

B 35. How many real solutions are possible for a simultaneous system of linearly independent equations that consists of:

  a. two linear equations in two variables?
  b. one linear equation and one quadratic equation in two variables?
  c. two quadratic equations in two variables?

  Support each of your answers with sketches.

36. What relationships must exist between the numbers a and b so that the solution set of the system

$$x^2 + y^2 = 25$$

$$y = ax + b$$

has two ordered pairs of real numbers? One ordered pair of real numbers? No ordered pairs of real numbers? *Hint*: Use substitution and consider the nature of the roots of the resulting quadratic equation.

37. Consider the system

$$x^2 + y^2 = 8 \quad (1)$$

$$xy = 4. \quad (2)$$

We can solve this system by substituting $4/x$ for y in (1) to obtain

$$x^2 + \frac{16}{x^2} = 8,$$

from which we obtain $x = 2$ or $x = -2$. Now if we obtain the y-components of the solution from (2), we find that for $x = 2$ we have $y = 2$, and that for $x = -2$ we have $y = -2$. But if we seek y-components from (1), for $x = 2$ we have $y = \pm 2$, and for $x = -2$, $y = \pm 2$. Discuss the fact that we seem to obtain two more solutions from (1) than from (2). What is the solution set of the system?

8.5 Systems of Inequalities

Graphs of systems of inequalities

In Section 5.6, we observed that the graph of the solution set of an inequality in two variables might be a region in the plane. The graph of the solution set of a system of inequalities in two variables, which consists of the intersection of the graphs of the inequalities in the system, might also be a plane region.

Example

Graph the solution set of the system

$$x + 2y \leq 6$$

$$2x - 3y \geq 12.$$

Solution

Graphing each inequality by the method discussed in Section 5.6, we obtain the figure shown, where the doubly shaded region is the graph of the solution set of the system.

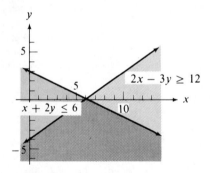

Example

Graph the solution set of the system

$$y \leq x + 2$$

$$y \geq x^2.$$

Solution

The graph of each inequality is shaded as shown in the figure. The doubly shaded region constitutes the graph of the solution set of the system.

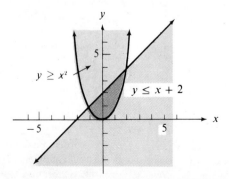

8.5 Systems of Inequalities

Example Graph the solution set of the system

$$y > 2$$
$$x > -2$$
$$x + y > 1.$$

Solution The triply shaded region in the figure constitutes the graph of the solution set of the system.

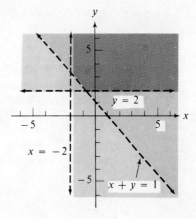

Exercise 8.5

A By double or triple shading, indicate the region in the plane representing the solution set of each system.

1. $y \geq x + 1$
   $y \geq 5 - x$

2. $y \geq 4$
   $x \geq 2$

3. $y - x \geq 0$
   $y + x \geq 0$

4. $2y - x \geq 1$
   $x < -3$

5. $y + 3x < 6$
   $y > 2$

6. $x > 3$
   $x + y \geq 5$

7. $x = 3$
   $x + y < 4$

8. $y = 2$
   $2y + x < 3$

9. $x - 3y < 6$
   $y + x = 1$

10. $2x + y \geq 4$
    $x - y = -2$

11. $y > x^2 + 1$
    $x + y > 4$

12. $y < x^2 + 4$
    $x - y \leq 4$

13. $x^2 + y^2 < 25$
    $y > 3$

14. $x^2 + y^2 \geq 25$
    $y > x^2$

15. $9x^2 + 16y^2 \geq 144$
    $y < x^2$

16. $y \geq 2$
    $x \geq 2$
    $y \geq x$

17. $y \leq 3$
    $x \leq 2$
    $y < x$

18. $x^2 + y^2 \leq 36$
    $x \geq 3$
    $y \geq 3$

B 19. $x^2 + y^2 \leq 25$
   $y \geq x^2 - 4$
   $y \leq -x^2 + 4$

20. $y \leq \log_{10} x$
    $y \geq x - 1$

21. $y \leq 10^x$
    $y \leq 2 - x^2$

22. $9 \leq x^2 + y^2 \leq 16$
    $-1 \leq x - y \leq 1$

23. $9 \leq x^2 + y^2 \leq 16$
    $x^2 + 1 \leq y \leq x^2 + 3$

24. $x^2 - 2 \leq y \leq 2 - x^2$
    $|x| \geq 1$

8.6 Convex Sets—Polygonal Regions

From the examples in Section 8.5 it is apparent that the graph of the solution set of a system of linear inequalities in two variables is simply the common intersection of a number of half-planes. Any such intersection is an example of a *convex set*.

Definition 8.2 *A set $\mathscr{S}$ of points is a **convex set** if and only if, for each two points P and Q in $\mathscr{S}$, the line segment $\overline{PQ}$ lies entirely in $\mathscr{S}$.*

Figure 8.6 shows three sets of points in a plane. Figures 8.6-a and 8.6-b show examples of convex sets, while the set in Figure 8.6-c is not convex because part of

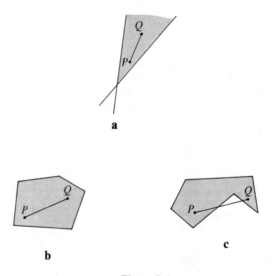

Figure 8.6

the line segment PQ does not lie in the set. If the boundary of a convex set is a (closed) polygon, then the boundary is called a **convex polygon**, and the region enclosed by the polygon (including the boundary) is called a **closed convex polygonal set**.

We have two immediate results from Definition 8.2 which are intuitively true and whose proofs are omitted.

Theorem 8.6 *Any half-plane is a convex set.*

Theorem 8.7 *The intersection of two convex sets is a convex set.*

8.6 Convex Sets—Polygonal Regions

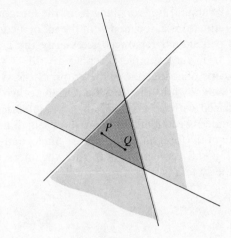

Figure 8.7

As suggested by Figure 8.7, this latter result extends to any number of convex sets. If the intersection of the graphs of a system of linear inequalities constitutes a (closed) polygonal set $\mathscr{P}$, then we can locate the set $\mathscr{P}$ in the plane by graphing the equations associated with the given inequalities and identifying the vertices of $\mathscr{P}$.

Example Shade the polygonal set specified by the system

$$x + y \leq 5 \qquad x \geq 0$$
$$x - y \leq 2 \qquad y \geq 0.$$
$$x - y \geq -2$$

Solution Graph the associated equations

$$x + y = 5$$
$$x - y = 2$$
$$x - y = -2$$
$$x = 0$$

and

$$y = 0,$$

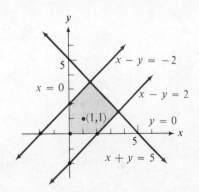

as shown. Note which half-plane is determined by each of the five inequalities, and shade the intersection of the five half-planes. The intersection is the pentagonal region shown. To check the result, note that (1, 1) is in the region and its coordinates satisfy each given inequality.

We should not conclude from the foregoing example that, simply because the five equations associated with the set of inequalities determine a convex pentagon, the pentagonal region is necessarily the graph of the system. If the inequality $y \geq 0$ in the example is replaced with $y \leq 0$, then the graph of the equations remains unchanged, but the graph of the system of inequalities is the triangular region shown in Figure 8.8-a. On the other hand, if the inequality $x + y \leq 5$ is replaced with $x + y \geq 5$, then the intersection of the graphs becomes the (unbounded) rectangular region shown in Figure 8.8-b. Finally we observe that if both of the foregoing changes are made, that is, if $x + y \geq 5$ replaces $x + y \leq 5$ and $y \leq 0$ replaced $y \geq 0$, then the intersection of the graphs of the inequalities in the system is $\emptyset$.

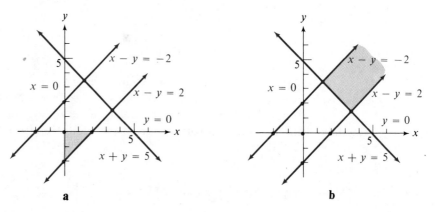

Figure 8.8

Exercise 8.6

Graph the convex polygonal set defined by the given system of inequalities.

1. $x - y \leq 2$
   $x - y \geq -2$
   $x + y \geq 2$
   $x + y \leq 6$

2. $0 \leq x \leq 5$
   $0 \leq y \leq 5$
   $x + y \leq 6$

3. $0 \leq x \leq 5$
   $0 \leq y \leq 4$
   $x + 2y \leq 10$
   $2x + y \leq 10$

4. $0 \leq x$
   $0 \leq y \leq 4$
   $x - 2y \leq 6$
   $2x + y \leq 12$
   $2x - y \leq 10$

5. $0 \leq x \leq 5$
   $y \geq x - 3$
   $x + y \leq 9$
   $3y \leq 2x + 12$

6. $0 \leq x$
   $0 \leq y \leq 6$
   $x + 2y \leq 13$
   $2x + y \leq 11$
   $3x + y \leq 15$

8.7 Linear Programming

7. Repeat Exercise 1 with $x - y \leq 2$ replaced with $x - y \geq 2$. Is the graph a closed polygonal set?

8. Repeat Exercise 3 with $x + 2y \leq 10$ replaced with $x + 2y \geq 10$. Is the graph a closed polygonal set?

9. Repeat Exercise 4 with $0 \leq y \leq 4$ replaced with $y \leq 4$. Is the graph a closed polygonal set?

10. Repeat Exercise 6 without the condition $x \geq 0$. Is the graph a closed polygonal set?

8.7 Linear Programming

For each ordered pair $(x, y) \in R^2$, if a, b are real numbers then the linear expression $ax + by$ has a value. In particular, if $\mathscr{P} \subset R^2$, and $\mathscr{P}$ is a closed polygonal subset of R^2, then $ax + by$ has a value for each $(x, y) \in \mathscr{P}$. It can be shown that the following result holds for the values of $ax + by$ over $\mathscr{P}$.

Theorem 8.8 *If $\mathscr{P} \subset R^2$ is a closed convex polygonal set, then the expression $ax + by$ takes on its maximum and minimum values over $\mathscr{P}$ at one of the vertices of $\mathscr{P}$.*

Thus, if $\mathscr{P}$ is the closed convex polygonal set with vertices $(1, 2), (3, 1), (5, 3), (4, 5)$, and $(3, 6)$, as pictured in Figure 8.9, then the linear expression $3x - 2y$ has a value for every ordered pair (x, y) in, or on the boundary of, $\mathscr{P}$. In particular, at the vertices we have the values shown in the table below.

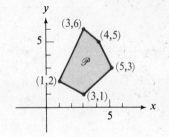

Figure 8.9

Vertex	(1, 2)	(3, 1)	(5, 3)	(4, 5)	(3, 6)
Value of $3x - 2y$	-1	7	9	2	-3

The maximum of these values is 9 and the minimum is -3. By Theorem 8.8, then, 9 is the greatest value $3x - 2y$ has over the entire set $\mathscr{P}$, and its least value over $\mathscr{P}$ is -3.

Theorem 8.8 has many important applications.

Example

A company manufactures two kinds of electric shavers, one using a cord and the other a cordless model. The company can make up to 500 cord models and 400 cordless models per day, but it can make a total of only 600 shavers per day. It takes 2 man-hours to manufacture the cord model and 3 man-hours to manufacture the cordless model, and the company has available at most 1400 man-hours per day. If there is a profit of $2.50 on each cord shaver and $3.50 on each cordless shaver, how many of each kind should the company make each day in order to realize the greatest profit?

Solution

Let x = number of cord shavers made in one day;
y = number of cordless shavers made in one day.

We have the following constraints on x and y:

$$0 \leq x \leq 500, \quad 0 \leq y \leq 400,$$
$$x + y \leq 600, \quad 2x + 3y \leq 1400.$$

Solving the associated equations for these inequalities in pairs yields the vertices of the polygonal region shown in the figure. We wish to maximize

$$2.5x + 3.5y.$$

We find the values for this expression at the vertices to be:

Vertex	Profit
(0, 0)	$ 0
(0, 400)	$1400
(100, 400)	$1650
(400, 200)	$1700
(500, 100)	$1601
(500, 0)	$1250

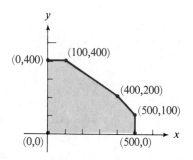

Clearly, the most profitable numbers of shavers are 400 cord models and 200 cordless models.

Exercise 8.7

Find the maximum and minimum values of the given expression over the given closed convex polygonal set of the specified set of Exercise 8.6, page 276.

1. $x + 3y$; the set in Exercise 1.
2. $3x + 5y$; the set in Exercise 2.
3. $3x + 5y$; the set in Exercise 3.
4. $x + 2y$; the set in Exercise 4.
5. $5x - 4y$; the set in Exercise 5.
6. $10x - 2y$; the set in Exercise 6.

8.7 Linear Programming

For each Exercise 7–10, use Tables A and B below, which show the number of units of labor, machinery, materials, and overhead necessary for a manufacturer to produce one unit of each of two products x and y. The tables also show the total available units of each of these items.

Table A

	Units needed for 1 unit of product x	Units needed for 1 unit of product y	Total units available
Labor	1	6	120
Machinery	1	3	66
Material	3	2	86
Overhead	10	3	250

Table B

	Units needed for 1 unit of product x	Units needed for 1 unit of product y	Total units available
Labor	2	7	63
Machinery	5	1	50
Material	1	2	21
Overhead	1	1	14

7. Using Table A, find the maximum profit if each unit of product x earns a profit of $50 and each unit of product y earns a profit of $60.

8. Using Table A, find the maximum profit if each unit of product x earns a profit of $60 and each unit of product y earns a profit of $50.

9. Using Table B, find the maximum profit if each unit of product x earns a profit of $50 and each unit of product y earns a profit of $60.

10. Using Table B, find the maximum profit if each unit of product x earns a profit of $60 and each unit of product y earns a profit of $50.

11. An electronics company makes two kinds of electronic ranges, a standard model that earns a $50 profit and a deluxe model that earns a $60 profit. The company has machinery capable of producing any number of deluxe ranges up to 400 per month and any number of standard ranges up to 500 per month, but it has enough man-hours available to produce

only 600 ranges of both kinds in a month. How many of each kind of range should the company produce to realize a maximum profit?

12. A pharmacy has 300 ounces of a drug it can use to make two different kinds of medicines. From each ounce of the drug, 15 bottles of medicine A or 25 bottles of medicine B can be produced. The pharmacy must keep all the bottles in its own storage room, which has a capacity of 6000 bottles in all. At least 600 bottles of medicine A and 1250 bottles of medicine B must be retained in inventory and the rest will be used. If medicine A sells for $2.75 per bottle and medicine B sells for $2.00 per bottle, how many ounces of the drug should be devoted to each medicine to maximize the total return?

Chapter Review

[8.1] Solve each system.

1. $x + 5y = 18$
   $x - y = -3$

2. $x + 5y = 11$
   $2x + 3y = 8$

3. $2x - 3y = 8$
   $3x + 2y = 7$

4. $3x - y = -7$
   $x - 2y = -4$

5. $2x - 3y = 1$
   $4x - 6y = 2$

6. $9x + 12y = 1$
   $6x + 8y = 0$

7. Find values for a and b so that the graph of $ax + by = 19$ passes through the points $(2, 3)$ and $(-3, 5)$.

8. Find a and b so that the solution set of

$$ax + by = 3$$
$$2ax - by = 1$$

is $\{(2, 1)\}$.

[8.2] Solve each system.

9. $x + 3y - z = 3$
   $2x - y + 3z = 1$
   $3x + 2y + z = 5$

10. $x + y + z = 2$
    $3x - y + z = 4$
    $2x + y + 2z = 3$

11. $2x + 3y - z = -2$
    $x - y + z = 6$
    $3x - y + z = 10$

12. Find value for a, b, and c so that the graph of $y = ax^2 + bx + c$ contains the points $(-1, 9)$, $(0, 4)$, and $(1, 3)$.

[8.3] Use the method of partial fractions to express each fraction as a sum of fractions with powers of linear polynomials for denominators.

13. $\dfrac{5}{2x^2 - x - 3}$

14. $\dfrac{x^2 + 1}{x^3 - x}$

Review Exercises

[8.4] Solve each system.

15. $x^2 + y = 3$
    $5x + y = 7$

16. $x^2 - xy + y^2 = 1$
    $2x^2 - xy + 2y^2 = 3$

17. $2x^2 + 5y^2 - 53 = 0$
    $4x^2 + 3y^2 - 43 = 0$

18. $x^2 + 3xy + 2y^2 = 0$
    $x^2 - 3xy + y^2 = 1$

19. $y - 10^{2x} = 0$
    $\log_{10} y^2 - x + 1 = 0$

20. $10^y - 2x = 0$
    $\log_{10} x + \log_{10} 2 = -1$

21. Approximate the solution of the system

$$y = 3^x$$
$$y = 3 - x$$

by graphical methods.

[8.5] By double shading, indicate the region representing the solution set of each system.

22. $y > x^2 - 4$
    $y < 2 - x$

23. $y + x^2 < 0$
    $y + 3 > 0$
    $y < x - 3$

24. $y - x^2 < 4$
    $y - 3 > 0$
    $y - x > 0$

[8.6] Graph the polygonal set defined by each system of inequalities.

25. $0 \leq x \leq 3$
    $0 \leq y \leq 4$
    $x + y \leq 5$

26. $0 \leq x \leq 3$
    $0 \leq y$
    $y \geq 2 + x$
    $x + y \leq 4$

[8.7] 27. Find the maximum and minimum values of $2x + y$ over the set in Exercise 25.

28. Find the maximum and minimum values of $3x + 2y$ over the set in Exercise 26.

29. Using Table A on page 279, find the maximum profit if each unit of product x earns a profit of $80 and each unit of product y earns a profit of $60.

30. Using Table B on page 279, find the maximum profit if each unit of product x earns a profit of $10 and each unit of product y earns a profit of $20.

9 Matrices and Determinants

Matrices, as introduced in this chapter, are today much used in mathematics and engineering, and also in the physical, social, and life sciences.

9.1 Definitions; Matrix Addition

A **matrix** is a rectangular array of numbers (or other suitable entities), which are called the **entries** or **elements** of the matrix. In this book, we shall consider only real numbers as entries. A matrix is customarily displayed in a pair of brackets or parentheses (we shall use brackets). Thus,

$$\begin{bmatrix} 1 & 2 & 3 \\ 4 & 5 & 6 \end{bmatrix} \text{ and } \begin{bmatrix} 2 \\ 1 \end{bmatrix}$$

are matrices. The **order**, or **dimension**, of a matrix is the ordered pair having as first component the number of (horizontal) **rows** and as second component the number of (vertical) **columns** in the matrix. Thus,

$$\begin{bmatrix} 1 & 2 & 3 \\ 4 & 5 & 6 \end{bmatrix}, \begin{bmatrix} 1 \\ 2 \\ 3 \end{bmatrix}, \text{ and } \begin{bmatrix} a_1 & a_2 & a_3 & a_4 \\ b_1 & b_2 & b_3 & b_4 \\ c_1 & c_2 & c_3 & c_4 \\ d_1 & d_2 & d_3 & d_4 \end{bmatrix}$$

are 2×3 (read "two by three"), 3×1 (read "three by one"), and 4×4 (read "four by four") matrices, respectively. Note that the number of *rows* is given first, and then the number of *columns*. A matrix consisting of a single row is called a **row matrix** or a **row vector**, whereas a matrix consisting of a single column is called a **column matrix** or a **column vector**.

9.1 Definitions; Matrix Addition

Matrices are frequently represented by capital letters. Thus, we might want to talk about the matrices A and B, where

$$A = \begin{bmatrix} a_1 & a_2 \\ b_1 & b_2 \end{bmatrix} \quad \text{and} \quad B = [b_1 \quad b_2].$$

To indicate that A is a 2×2 matrix, we can write $A_{2 \times 2}$. Similarly, $B_{1 \times 2}$ is a matrix with one row and two columns.

To represent the entries of a matrix, double-subscript notation can be used. A single letter, say a, is used to denote an entry in a matrix, and then *two* subscripts are appended, the first subscript telling in which *row* the entry occurs, and the second telling in which *column*. Thus, we write

$$A = \begin{bmatrix} a_{11} & a_{12} & a_{13} \\ a_{21} & a_{22} & a_{23} \\ a_{31} & a_{32} & a_{33} \end{bmatrix},$$

where a_{21} is the element in the second *row* and first *column*, a_{33} is the element in the third *row* and third *column*, and, in general, a_{ij} is the element in the *i*th *row* and *j*th *column*.

Definition 9.1 Two matrices, A and B, are **equal** if and only if both matrices are of the same order and $a_{ij} = b_{ij}$ for each i, j.

Examples a. $\begin{bmatrix} 2 & 1 \\ 3 & 0 \end{bmatrix} = \begin{bmatrix} \frac{4}{2} & 2-1 \\ \sqrt{9} & 0 \end{bmatrix}$ b. $\begin{bmatrix} 2 & 1 \\ 3 & 0 \end{bmatrix} \neq \begin{bmatrix} 2 & 3 \\ 1 & 0 \end{bmatrix}$ c. $\begin{bmatrix} 1 \\ 2 \end{bmatrix} \neq [1 \quad 2]$

Definition 9.2 The **transpose** of a matrix A, denoted by A^t, is the matrix in which the rows are the columns of A and the columns are the rows of A.

Thus, if a_{ij} and b_{ij} represent the entries in the *i*th row and *j*th column of A and A^t, respectively, then we have

$$b_{ij} = a_{ji}$$

for each i and j.

Examples a. $\begin{bmatrix} 2 & 1 \\ 3 & 0 \end{bmatrix}^t = \begin{bmatrix} 2 & 3 \\ 1 & 0 \end{bmatrix}$ b. $\begin{bmatrix} 1 & 2 & 3 \\ 4 & 5 & 6 \end{bmatrix}^t = \begin{bmatrix} 1 & 4 \\ 2 & 5 \\ 3 & 6 \end{bmatrix}$ c. $\begin{bmatrix} 1 \\ 2 \end{bmatrix}^t = [1 \quad 2]$

Definition 9.3 The **sum** of two matrices of the same order, $A_{m \times n}$ and $B_{m \times n}$, is the matrix $(A + B)_{m \times n}$, in which the entry in the *i*th row and *j*th column is $a_{ij} + b_{ij}$, for $i = 1, 2, 3, \ldots, m$ and $j = 1, 2, 3, \ldots, n$.

Examples

a. $\begin{bmatrix} 3 & 1 & 2 \\ 2 & 1 & 4 \end{bmatrix} + \begin{bmatrix} 1 & 0 & 2 \\ -1 & 3 & 0 \end{bmatrix} = \begin{bmatrix} 3+1 & 1+0 & 2+2 \\ 2+(-1) & 1+3 & 4+0 \end{bmatrix} = \begin{bmatrix} 4 & 1 & 4 \\ 1 & 4 & 4 \end{bmatrix}$

b. $\begin{bmatrix} 1 & 2 \\ 1 & -3 \end{bmatrix} + \begin{bmatrix} 1 & -2 \\ 2 & 3 \end{bmatrix} = \begin{bmatrix} 1+1 & 2+(-2) \\ 1+2 & -3+3 \end{bmatrix} = \begin{bmatrix} 2 & 0 \\ 3 & 0 \end{bmatrix}$

Thus, the sum of two matrices of the same order is obtained by adding the corresponding entries; the sum of two matrices of different orders is not defined.

Definition 9.4   A matrix with each entry equal to 0 is a **zero matrix**.

Zero matrices are generally denoted by the symbol **0**. This distinguishes the zero matrix from the real number 0. For example,

$$\mathbf{0}_{2 \times 4} = \begin{bmatrix} 0 & 0 & 0 & 0 \\ 0 & 0 & 0 & 0 \end{bmatrix}$$

is the 2 × 4 zero matrix.

Definition 9.5   The **negative of a matrix** $A_{m \times n}$, denoted by $-A_{m \times n}$, is formed by replacing each entry a_{ij} in the matrix $A_{m \times n}$ with $-a_{ij}$.

For example, if

$$A_{3 \times 2} = \begin{bmatrix} 3 & -1 \\ 2 & -2 \\ -4 & 5 \end{bmatrix}, \quad \text{then} \quad -A_{3 \times 2} = \begin{bmatrix} -3 & 1 \\ -2 & 2 \\ 4 & -5 \end{bmatrix}.$$

The sum $B_{m \times n} + (-A_{m \times n})$ is called the **difference** of $B_{m \times n}$ and $A_{m \times n}$ and is denoted $B_{m \times n} - A_{m \times n}$.

Properties of sums

At this point, we are able to establish the following facts concerning sums of matrices with real-number entries. We shall prove Parts III and IV; the rest are left as exercises.

Theorem 9.1   If A, B, and C are m × n matrices with real-number entries, then:

I $(A + B)_{m \times n}$ is a matrix with real-number entries.   Closure law for addition.

II $(A + B) + C = A + (B + C)$ Associative law for addition.

III   The matrix $\mathbf{0}_{m \times n}$ has the property that for every matrix $A_{m \times n}$,   Additive-identity law.

$A + \mathbf{0} = A \quad \text{and} \quad \mathbf{0} + A = A.$

9.1 *Definitions; Matrix Addition*

IV For every matrix $A_{m \times n}$, the matrix   *Additive-inverse law.*
    $-A_{m \times n}$ has the property that

$$A + (-A) = \mathbf{0} \quad \text{and} \quad (-A) + A = \mathbf{0}.$$

V $A + B = B + A$ *Commutative law for addition.*

To prove 9.1-III, observe that since each entry of the zero matrix is 0, it follows that the entries of $A_{m \times n} + \mathbf{0}_{m \times n}$ are $a_{ij} + 0 = a_{ij}$ and the entries of $\mathbf{0}_{m \times n} + A_{m \times n}$ are $0 + a_{ij} = a_{ij}$. Thus, $A + \mathbf{0} = \mathbf{0} + A = A$.

Examples

a. $\begin{bmatrix} a_{11} & a_{12} \\ a_{21} & a_{22} \end{bmatrix} + \begin{bmatrix} 0 & 0 \\ 0 & 0 \end{bmatrix} = \begin{bmatrix} a_{11} & a_{12} \\ a_{21} & a_{22} \end{bmatrix}$ **b.** $\begin{bmatrix} a_1 \\ a_2 \\ a_3 \end{bmatrix} + \begin{bmatrix} 0 \\ 0 \\ 0 \end{bmatrix} = \begin{bmatrix} a_1 \\ a_2 \\ a_3 \end{bmatrix}$

To see that 9.1-IV is true, let the entries of $A_{m \times n}$ and $-A_{m \times n}$ be a_{ij} and $-a_{ij}$, respectively. Since each entry of $A + (-A)$ is $a_{ij} - a_{ij}$, or 0, we have $A + (-A) = \mathbf{0}$. Similarly, $(-A) + A = \mathbf{0}$.

Example

If

$$A = \begin{bmatrix} 1 & -1 & 2 \\ 3 & -1 & 1 \end{bmatrix},$$

then

$$A + (-A) = \begin{bmatrix} 1 & -1 & 2 \\ 3 & -1 & 1 \end{bmatrix} + \begin{bmatrix} -1 & 1 & -2 \\ -3 & 1 & -1 \end{bmatrix} = \begin{bmatrix} 0 & 0 & 0 \\ 0 & 0 & 0 \end{bmatrix} = \mathbf{0}.$$

Example

As an example of Theorem 9.1-II, consider the matrices

$$A = \begin{bmatrix} 1 & 3 \\ -1 & 2 \end{bmatrix}, \quad B = \begin{bmatrix} 2 & 1 \\ -3 & 2 \end{bmatrix}, \quad \text{and} \quad C = \begin{bmatrix} -1 & 1 \\ 0 & 2 \end{bmatrix}.$$

Then

$$(A + B) + C = \left(\begin{bmatrix} 1 & 3 \\ -1 & 2 \end{bmatrix} + \begin{bmatrix} 2 & 1 \\ -3 & 2 \end{bmatrix} \right) + \begin{bmatrix} -1 & 1 \\ 0 & 2 \end{bmatrix}$$

$$= \begin{bmatrix} 3 & 4 \\ -4 & 4 \end{bmatrix} + \begin{bmatrix} -1 & 1 \\ 0 & 2 \end{bmatrix} = \begin{bmatrix} 2 & 5 \\ -4 & 6 \end{bmatrix},$$

and

$$A + (B + C) = \begin{bmatrix} 1 & 3 \\ -1 & 2 \end{bmatrix} + \left(\begin{bmatrix} 2 & 1 \\ -3 & 2 \end{bmatrix} + \begin{bmatrix} -1 & 1 \\ 0 & 2 \end{bmatrix} \right)$$

$$= \begin{bmatrix} 1 & 3 \\ -1 & 2 \end{bmatrix} + \begin{bmatrix} 1 & 2 \\ -3 & 4 \end{bmatrix} = \begin{bmatrix} 2 & 5 \\ -4 & 6 \end{bmatrix}.$$

Example continued on overleaf

Hence,
$$A + (B + C) = (A + B) + C.$$

Example As an example of Theorem 9.1-V, consider the matrices
$$A = \begin{bmatrix} 1 & -1 \\ 2 & 3 \end{bmatrix} \quad \text{and} \quad B = \begin{bmatrix} 3 & 4 \\ -2 & -4 \end{bmatrix}.$$

Then
$$A + B = \begin{bmatrix} 1 & -1 \\ 2 & 3 \end{bmatrix} + \begin{bmatrix} 3 & 4 \\ -2 & -4 \end{bmatrix} = \begin{bmatrix} 4 & 3 \\ 0 & -1 \end{bmatrix}$$

and
$$B + A = \begin{bmatrix} 3 & 4 \\ -2 & -4 \end{bmatrix} + \begin{bmatrix} 1 & -1 \\ 2 & 3 \end{bmatrix} = \begin{bmatrix} 4 & 3 \\ 0 & -1 \end{bmatrix}.$$

Hence,
$$A + B = B + A.$$

Solution of matrix equations Theorem 9.1 can be used to solve matrix equations of the form $X + A = B$, as follows:

$$(X + A) + (-A) = B + (-A)$$
$$X + (A + (-A)) = B + (-A) \quad \text{by Theorem 9.1-II}$$
$$X + 0 = B + (-A) \quad \text{by Theorem 9.1-IV}$$
$$X = B + (-A) \quad \text{by Theorem 9.1-III}$$

Example Solve $X + \begin{bmatrix} -3 & 1 \\ 2 & -1 \end{bmatrix} = \begin{bmatrix} 1 & -1 \\ 2 & -2 \end{bmatrix}.$

Solution
$$X = \begin{bmatrix} 1 & -1 \\ 2 & -2 \end{bmatrix} + \left(-\begin{bmatrix} -3 & 1 \\ 2 & -1 \end{bmatrix}\right)$$
$$= \begin{bmatrix} 1 & -1 \\ 2 & -2 \end{bmatrix} + \begin{bmatrix} 3 & -1 \\ -2 & 1 \end{bmatrix}$$
$$= \begin{bmatrix} 4 & -2 \\ 0 & -1 \end{bmatrix}$$

9.1 Definitions; Matrix Addition

Exercise 9.1

A State the order and find the transpose of each matrix.

Examples

a. $\begin{bmatrix} 2 & 4 \\ 1 & -3 \\ 6 & 0 \end{bmatrix}$

b. $\begin{bmatrix} 0 & 2 \\ -1 & 0 \end{bmatrix}$

Solutions

a. 3×2 matrix; $\begin{bmatrix} 2 & 4 \\ 1 & -3 \\ 6 & 0 \end{bmatrix}^t = \begin{bmatrix} 2 & 1 & 6 \\ 4 & -3 & 0 \end{bmatrix}$

b. 2×2 matrix; $\begin{bmatrix} 0 & 2 \\ -1 & 0 \end{bmatrix}^t = \begin{bmatrix} 0 & -1 \\ 2 & 0 \end{bmatrix}$

1. $\begin{bmatrix} 6 & -1 \\ 2 & 3 \end{bmatrix}$

2. $\begin{bmatrix} 4 & 1 \\ 0 & -2 \end{bmatrix}$

3. $\begin{bmatrix} 2 & -7 & 3 \\ 1 & 4 & 0 \end{bmatrix}$

4. $\begin{bmatrix} -3 & 1 \\ 6 & 0 \\ 0 & 2 \end{bmatrix}$

5. $\begin{bmatrix} 2 & 3 & -1 \\ 4 & 0 & 1 \\ -2 & 3 & 1 \end{bmatrix}$

6. $\begin{bmatrix} 4 & -1 & -2 \\ 3 & 0 & 0 \\ 2 & 1 & 1 \end{bmatrix}$

7. $\begin{bmatrix} 4 & -3 & -1 & 0 \\ 2 & 1 & 1 & 6 \end{bmatrix}$

8. $\begin{bmatrix} -2 & 1 & 3 & 2 \\ 4 & 0 & 0 & -2 \\ -1 & 3 & 2 & 4 \end{bmatrix}$

Write each sum or difference as a single matrix.

Example

$\begin{bmatrix} 2 & 1 & 4 \\ 3 & -1 & 0 \end{bmatrix} + \begin{bmatrix} 6 & 3 & 0 \\ -2 & 1 & 0 \end{bmatrix}$

Solution

$\begin{bmatrix} 2 & 1 & 4 \\ 3 & -1 & 0 \end{bmatrix} + \begin{bmatrix} 6 & 3 & 0 \\ -2 & 1 & 0 \end{bmatrix} = \begin{bmatrix} 2+6 & 1+3 & 4+0 \\ 3-2 & -1+1 & 0+0 \end{bmatrix} = \begin{bmatrix} 8 & 4 & 4 \\ 1 & 0 & 0 \end{bmatrix}$

9. $\begin{bmatrix} 2 & 3 \\ 1 & 6 \end{bmatrix} + \begin{bmatrix} 1 & -2 \\ 2 & 3 \end{bmatrix}$

10. $\begin{bmatrix} 4 & -1 & 3 \\ 2 & 1 & 0 \end{bmatrix} + \begin{bmatrix} 3 & -1 & 0 \\ 4 & 0 & -2 \end{bmatrix}$

11. $\begin{bmatrix} 3 & 0 & -1 \\ 2 & 1 & 2 \end{bmatrix} + \begin{bmatrix} 6 & -1 & 0 \\ 0 & 2 & 4 \end{bmatrix}$

12. $[1 \quad 3 \quad 5] + [0 \quad -2 \quad 1]$

13. $\begin{bmatrix} 4 & -3 \\ 2 & 1 \end{bmatrix} - \begin{bmatrix} 6 & 0 \\ -2 & 1 \end{bmatrix}$

14. $\begin{bmatrix} 4 & -1 & 2 \\ 3 & 1 & -4 \end{bmatrix} - \begin{bmatrix} -1 & -1 & 2 \\ 3 & 1 & 4 \end{bmatrix}$

15. $\begin{bmatrix} 10 & 3 & 2 \\ 5 & 1 & 7 \\ 6 & 1 & 9 \end{bmatrix} - \begin{bmatrix} 8 & 12 & 15 \\ -2 & 5 & 6 \\ -3 & 1 & 9 \end{bmatrix}$

16. $\begin{bmatrix} 3 & -1 & 2 \\ 4 & -2 & 1 \\ 6 & 3 & 2 \end{bmatrix} - \begin{bmatrix} 2 & -1 & 2 \\ 4 & -1 & 1 \\ 6 & 3 & 1 \end{bmatrix}$

17. $\begin{bmatrix} 4 \\ 3 \\ -1 \end{bmatrix} + \begin{bmatrix} 6 \\ 0 \\ -2 \end{bmatrix}$

18. $\begin{bmatrix} 2 & 3 \\ 1 & 0 \\ -1 & 2 \end{bmatrix} + \begin{bmatrix} -2 & 0 \\ -3 & 0 \\ 4 & -1 \end{bmatrix}$

19. $\begin{bmatrix} 2 & 3 & 4 \\ -1 & 6 & 2 \\ 1 & 0 & 3 \end{bmatrix} + \begin{bmatrix} 0 & 0 & 0 \\ 0 & 0 & 0 \\ 0 & 0 & 0 \end{bmatrix}$

20. $\begin{bmatrix} 2 & -3 \\ 4 & -1 \\ -2 & 1 \end{bmatrix} + \begin{bmatrix} -2 & 3 \\ -4 & 1 \\ 2 & -1 \end{bmatrix}$

Solve each of the following matrix equations.

Examples

a. $X + \begin{bmatrix} 2 & 3 \\ 1 & 7 \end{bmatrix} = \begin{bmatrix} 9 & -4 \\ 2 & 0 \end{bmatrix}$

b. $X - \begin{bmatrix} 2 & 1 \\ -3 & 0 \end{bmatrix} = \begin{bmatrix} -1 & 2 \\ 1 & 0 \end{bmatrix}$

Solutions

a. $X = \begin{bmatrix} 9 & -4 \\ 2 & 0 \end{bmatrix} - \begin{bmatrix} 2 & 3 \\ 1 & 7 \end{bmatrix}$

$= \begin{bmatrix} 7 & -7 \\ 1 & -7 \end{bmatrix}$

b. $X = \begin{bmatrix} -1 & 2 \\ 1 & 0 \end{bmatrix} + \begin{bmatrix} 2 & 1 \\ -3 & 0 \end{bmatrix}$

$= \begin{bmatrix} 1 & 3 \\ -2 & 0 \end{bmatrix}$

21. $X - \begin{bmatrix} -1 & 0 \\ 0 & 0 \end{bmatrix} = \begin{bmatrix} 3 & -1 \\ 2 & 1 \end{bmatrix}$

22. $X + \begin{bmatrix} 3 & -1 \\ 2 & 1 \end{bmatrix} = \begin{bmatrix} 5 & 1 \\ -3 & 5 \end{bmatrix}$

23. $\begin{bmatrix} 1 & 3 \\ -1 & 0 \end{bmatrix}^t = \begin{bmatrix} 2 & -2 \\ -1 & 3 \end{bmatrix}^t - X$

24. $X + \begin{bmatrix} 3 & 2 \\ -1 & 4 \end{bmatrix}^t = \begin{bmatrix} 2 & 6 \\ 1 & 5 \end{bmatrix}^t$

B 25. Show that $[A^t_{2 \times 2}]^t = A_{2 \times 2}$. Does an analogous result seem valid for $n \times n$ matrices? For $m \times n$ matrices?

26. Show that $[A_{2 \times 2} + B_{2 \times 2}]^t = A^t_{2 \times 2} + B^t_{2 \times 2}$. Does an analogous result seem valid for $n \times n$ matrices?

27. Prove Theorem 9.1-I.

28. Prove Theorem 9.1-II.

29. Prove Theorem 9.1-V.

9.2 Matrix Multiplication

We shall be interested in two kinds of products involving matrices: (1) the product of a matrix and a real number, and (2) the product of two matrices.

Definition 9.6 The **product** of a real number c and an $m \times n$ matrix A with entries a_{ij} is the matrix cA with corresponding entries ca_{ij}, where $i = 1, 2, 3, \ldots, m$ and $j = 1, 2, 3, \ldots, n$.

9.2 Matrix Multiplication

Examples

a. $3\begin{bmatrix} 2 & 1 \\ 0 & 5 \end{bmatrix} = \begin{bmatrix} 3(2) & 3(1) \\ 3(0) & 3(5) \end{bmatrix} = \begin{bmatrix} 6 & 3 \\ 0 & 15 \end{bmatrix}$
b. $-4\begin{bmatrix} 1 \\ 2 \\ -1 \end{bmatrix} = \begin{bmatrix} -4(1) \\ -4(2) \\ -4(-1) \end{bmatrix} = \begin{bmatrix} -4 \\ -8 \\ 4 \end{bmatrix}$

Properties of products of matrices and real numbers

The following theorem states some simple algebraic laws for the multiplication of matrices by real numbers.

Theorem 9.2 If A and B are $m \times n$ matrices, and $c, d \in R$, then

I cA is an $m \times n$ matrix,

II $c(dA) = (cd)A$,

III $(c + d)A = cA + dA$,

IV $c(A + B) = cA + cB$,

V $1A = A$,

VI $(-1)A = -A$,

VII $0A = \mathbf{0}$,

VIII $c\mathbf{0} = \mathbf{0}$.

We shall prove only Part IV, leaving the remaining parts as exercises. Since the elements of $A + B$ are of the form $a_{ij} + b_{ij}$, it follows, by definition, that the elements of $c(A + B)$ are of the form $c(a_{ij} + b_{ij})$. But, since a_{ij}, b_{ij}, and c denote real numbers,

$$c(a_{ij} + b_{ij}) = ca_{ij} + cb_{ij}.$$

Now, the elements of cA are of the form ca_{ij}, and those of cB are of the form cb_{ij}, so that the elements of $cA + cB$ are of the form $ca_{ij} + cb_{ij}$ and part IV is proved.

Products of matrices

Turning now to the *product of two matrices*, we first define the product of a row matrix and a column matrix.

Definition 9.7 For $A_{1 \times p} = [a_1 \cdots a_p]$ and $B_{p \times 1} = \begin{bmatrix} b_1 \\ \vdots \\ b_p \end{bmatrix}$, the product

$$A_{1 \times p} B_{p \times 1} = a_1 b_1 + a_2 b_2 + \cdots + a_p b_p.$$

Examples

a. $\begin{bmatrix} 1 & -2 \end{bmatrix} \begin{bmatrix} 3 \\ -4 \end{bmatrix} = 1(3) + (-2)(-4) = 11$

b. $\begin{bmatrix} 1 & 0 & -1 & 2 \end{bmatrix} \begin{bmatrix} 3 \\ 1 \\ -2 \\ 4 \end{bmatrix} = 1(3) + 0(1) + (-1)(-2) + 2(4) = 13$

Note from Definition 9.7 that the product $A_{1 \times p} B_{p \times 1}$ is always a real number. However, the product of matrices with more than one row or column is a matrix.

Definition 9.8   The product of matrices $A_{m \times p}$ and $B_{p \times n}$ is the $m \times n$ matrix whose i, jth entry is the product of the ith row of A and the jth column of B. This product is denoted AB or $A \cdot B$.

Example

Multiply: $\begin{bmatrix} 3 & 0 & 1 \\ 0 & 1 & 2 \end{bmatrix} \begin{bmatrix} 1 & -2 \\ -1 & 2 \\ 1 & 1 \end{bmatrix}$.

Solution

Since the matrix on the left is 2×3 and the one on the right is 3×2, the product will be 2×2.

The 1, 1 entry is $\begin{bmatrix} 3 & 0 & 1 \end{bmatrix} \begin{bmatrix} 1 \\ -1 \\ 1 \end{bmatrix} = 3 \cdot 1 + 0(-1) + 1 \cdot 1 = 4;$

the 1, 2 entry is $\begin{bmatrix} 3 & 0 & 1 \end{bmatrix} \begin{bmatrix} -2 \\ 2 \\ 1 \end{bmatrix} = 3(-2) + 0(2) + 1 \cdot 1 = -5;$

the 2, 1 entry is $\begin{bmatrix} 0 & 1 & 2 \end{bmatrix} \begin{bmatrix} 1 \\ -1 \\ 1 \end{bmatrix} = 0 \cdot 1 + 1(-1) + 2 \cdot 1 = 1;$

the 2, 2 entry is $\begin{bmatrix} 0 & 1 & 2 \end{bmatrix} \begin{bmatrix} -2 \\ 2 \\ 1 \end{bmatrix} = 0(-2) + 1 \cdot 2 + 2 \cdot 1 = 4.$

Therefore, $\begin{bmatrix} 3 & 0 & 1 \\ 0 & 1 & 2 \end{bmatrix} \begin{bmatrix} 1 & -2 \\ -1 & 2 \\ 1 & 1 \end{bmatrix} = \begin{bmatrix} 4 & -5 \\ 1 & 4 \end{bmatrix}.$

Example

If $A = \begin{bmatrix} 1 & 2 \\ -1 & 3 \end{bmatrix}$ and $B = \begin{bmatrix} 2 & 1 \\ 1 & 1 \end{bmatrix}$, find AB and BA.

Solution

$AB = \begin{bmatrix} 1 & 2 \\ -1 & 3 \end{bmatrix} \begin{bmatrix} 2 & 1 \\ 1 & 1 \end{bmatrix} = \begin{bmatrix} 2+2 & 1+2 \\ -2+3 & -1+3 \end{bmatrix} = \begin{bmatrix} 4 & 3 \\ 1 & 2 \end{bmatrix};$

$BA = \begin{bmatrix} 2 & 1 \\ 1 & 1 \end{bmatrix} \begin{bmatrix} 1 & 2 \\ -1 & 3 \end{bmatrix} = \begin{bmatrix} 2-1 & 4+3 \\ 1-1 & 2+3 \end{bmatrix} = \begin{bmatrix} 1 & 7 \\ 0 & 5 \end{bmatrix}$

9.2 Matrix Multiplication

Properties of products

The preceding example shows very clearly that the multiplication of matrices, in general, is *not commutative*. Thus, when discussing products of matrices, we must specify the *order* in which the matrices are to be considered as factors. For the product AB, we say that A is *right-multiplied* by B, and that B is *left-multiplied* by A.

Note that the definition of the product of two matrices, A and B, requires that the matrix A have the same number of *columns* as B has *rows*; the result, AB, then has the same number of rows as A and the same number of columns as B. Such matrices A and B are said to be **conformable** for multiplication. The fact that two matrices are conformable in the order AB, however, does not mean that they necessarily are conformable in the order BA.

Example

If $A = \begin{bmatrix} 3 & 1 \\ 1 & 0 \\ 2 & 1 \end{bmatrix}$ and $B = \begin{bmatrix} 1 & -1 \\ 2 & 1 \end{bmatrix}$, find AB.

Solution

Since A is a 3×2 matrix, and B is a 2×2 matrix, they are conformable for multiplication in the order AB. We have

$$AB = \begin{bmatrix} 3 & 1 \\ 1 & 0 \\ 2 & 1 \end{bmatrix} \begin{bmatrix} 1 & -1 \\ 2 & 1 \end{bmatrix} = \begin{bmatrix} 3+2 & -3+1 \\ 1+0 & -1+0 \\ 2+2 & -2+1 \end{bmatrix} = \begin{bmatrix} 5 & -2 \\ 1 & -1 \\ 4 & -1 \end{bmatrix}.$$

Note that the matrices A and B in the example above are not conformable in the order BA.

In much of the matrix work in this book, we shall focus our attention on matrices having the same number of rows as columns. For brevity, a matrix of order $n \times n$ is often called a **square matrix** of order n. Although many of the ideas we shall discuss are applicable to conformable matrices of any order, we shall apply the notions only to square matrices.

Theorem 9.3 If A, B, and C are $n \times n$ square matrices, then

$$(AB)C = A(BC).$$

If A is a square matrix, then A^2, A^3, etc. denote AA, $(AA)A$, etc.

Theorem 9.4 If A, B, and C are $n \times n$ square matrices, then

$$A(B + C) = AB + AC$$

and

$$(B + C)A = BA + CA.$$

The proofs of these theorems involve some complicated symbolism and are omitted here, but you will be asked to show their validity for the case of 2 × 2 matrixes in the exercises. Observe that, because matrix multiplication is not, in general, commutative, we must establish both the left-hand and the right-hand distributive property.

Definition 9.9   The **principal diagonal** of a square matrix is the ordered set of entries a_{jj}, extending from the upper left-hand corner to the lower right hand corner of the matrix. Thus, the principal diagonal contains a_{11}, a_{22}, a_{33}, etc.

For example, the principal diagonal of

$$\begin{bmatrix} 1 & 3 & -1 \\ 5 & 2 & 3 \\ 6 & 4 & 0 \end{bmatrix}$$

consists of 1, 2, and 0, in that order.

Definition 9.10   A **diagonal matrix** is a square matrix in which all entries not in the principal diagonal are 0.

Thus,

$$\begin{bmatrix} 4 & 0 \\ 0 & 2 \end{bmatrix} \quad \text{and} \quad \begin{bmatrix} 1 & 0 & 0 \\ 0 & 1 & 0 \\ 0 & 0 & 0 \end{bmatrix}$$

are diagonal matrices.

Definition 9.11 $I_{n \times n}$ denotes the diagonal matrix having 1's for entries on the principal diagonal.

For example,

$$I_{2 \times 2} = \begin{bmatrix} 1 & 0 \\ 0 & 1 \end{bmatrix} \quad \text{and} \quad I_{4 \times 4} = \begin{bmatrix} 1 & 0 & 0 & 0 \\ 0 & 1 & 0 & 0 \\ 0 & 0 & 1 & 0 \\ 0 & 0 & 0 & 1 \end{bmatrix}.$$

The following properties are consequences of the definitions we have adopted.

Theorem 9.5   For each matrix $A_{n \times n}$,

$$A_{n \times n} I_{n \times n} = I_{n \times n} A_{n \times n} = A_{n \times n}.$$

Furthermore, $I_{n \times n}$ is the unique matrix having this property for all matrices $A_{n \times n}$.

9.2 Matrix Multiplication

Accordingly, $I_{n \times n}$ is the **identity element for multiplication** in the set of $n \times n$ square matrices. The proof of this theorem, for the illustrative case $n = 2$, is left as an exercise.

Order of multiplication

The following result relates the order in which matrices can be multiplied by real numbers and by other matrices.

Theorem 9.6 *If A and B are $n \times n$ square matrices, and a is a real number, then*
$$a(AB) = (aA)B = A(aB).$$

The proof for the case $n = 2$ is left as an exercise.

Example Let
$$A = \begin{bmatrix} 1 & 3 \\ 2 & -4 \end{bmatrix} \quad \text{and} \quad B = \begin{bmatrix} 4 & 1 \\ -2 & -1 \end{bmatrix}.$$

Then
$$4(AB) = 4\left(\begin{bmatrix} 1 & 3 \\ 2 & -4 \end{bmatrix}\begin{bmatrix} 4 & 1 \\ -2 & -1 \end{bmatrix}\right) = 4\begin{bmatrix} 4-6 & 1-3 \\ 8+8 & 2+4 \end{bmatrix}$$
$$= 4\begin{bmatrix} -2 & -2 \\ 16 & 6 \end{bmatrix} = \begin{bmatrix} -8 & -8 \\ 64 & 24 \end{bmatrix},$$

$$(4A)B = \begin{bmatrix} 4 & 12 \\ 8 & -16 \end{bmatrix}\begin{bmatrix} 4 & 1 \\ -2 & -1 \end{bmatrix} = \begin{bmatrix} -8 & -8 \\ 64 & 24 \end{bmatrix},$$

and
$$A(4B) = \begin{bmatrix} 1 & 3 \\ 2 & -4 \end{bmatrix}\begin{bmatrix} 16 & 4 \\ -8 & -4 \end{bmatrix} = \begin{bmatrix} -8 & -8 \\ 64 & 24 \end{bmatrix}.$$

Thus, $4(AB) = (4A)B = A(4B)$.

Exercise 9.2

A Write each product as a single matrix.

Examples

a. $3\begin{bmatrix} 2 & 1 \\ -1 & 3 \\ 2 & 0 \end{bmatrix}$

b. $\begin{bmatrix} 3 & 1 & -1 \\ 0 & -1 & 2 \end{bmatrix} \cdot \begin{bmatrix} 1 & -1 \\ 0 & 2 \\ 1 & 0 \end{bmatrix}$

Solutions

a. $\begin{bmatrix} 6 & 3 \\ -3 & 9 \\ 6 & 0 \end{bmatrix}$

b. $\begin{bmatrix} 3+0-1 & -3+2+0 \\ 0+0+2 & 0-2+0 \end{bmatrix} = \begin{bmatrix} 2 & -1 \\ 2 & -2 \end{bmatrix}$

1. $-5\begin{bmatrix} 0 & 1 & -1 \\ 3 & -1 & 2 \end{bmatrix}$

2. $2\begin{bmatrix} 2 & 1 & 3 & -2 \\ 4 & 2 & 0 & -1 \\ 0 & 0 & -1 & 2 \end{bmatrix}$

3. $\begin{bmatrix} 1 & -2 \end{bmatrix} \cdot \begin{bmatrix} 3 \\ 2 \end{bmatrix}$

4. $\begin{bmatrix} 3 & -2 & 2 \end{bmatrix} \cdot \begin{bmatrix} 1 \\ 0 \\ -2 \end{bmatrix}$

5. $\begin{bmatrix} 3 & -1 \\ 2 & 1 \end{bmatrix} \cdot \begin{bmatrix} 1 & -4 \\ 2 & 1 \end{bmatrix}$

6. $\begin{bmatrix} 1 & -5 \\ 0 & 2 \end{bmatrix} \cdot \begin{bmatrix} 3 & 1 \\ -1 & 2 \end{bmatrix}$

7. $\begin{bmatrix} 4 & -5 \\ 7 & 3 \end{bmatrix} \cdot \begin{bmatrix} 5 & -1 \\ -2 & 7 \end{bmatrix}$

8. $\begin{bmatrix} 1 & -2 \\ -3 & 1 \end{bmatrix} \cdot \begin{bmatrix} 5 & 1 \\ 0 & 2 \end{bmatrix}$

9. $\begin{bmatrix} -3 & 1 & 0 \\ 2 & 1 & 1 \end{bmatrix} \cdot \begin{bmatrix} 2 & 0 \\ 1 & -1 \\ 3 & 0 \end{bmatrix}$

10. $\begin{bmatrix} 1 & -1 & 0 \\ 2 & 1 & 3 \end{bmatrix} \cdot \begin{bmatrix} 4 & -1 \\ 2 & 0 \\ 1 & 1 \end{bmatrix}$

11. $\begin{bmatrix} -1 & 0 & 1 \\ 2 & 1 & 0 \\ 1 & 0 & 0 \end{bmatrix} \cdot \begin{bmatrix} 0 & 1 & 3 \\ 1 & 0 & 2 \\ -1 & 1 & 1 \end{bmatrix}$

12. $\begin{bmatrix} 2 & -3 & 1 \\ 0 & 1 & -1 \\ 2 & 0 & 0 \end{bmatrix} \cdot \begin{bmatrix} 1 & 0 & 0 \\ 0 & 1 & 0 \\ 0 & 0 & 1 \end{bmatrix}$

13. $\begin{bmatrix} 2 & -2 & -1 \\ 1 & 1 & -2 \\ 1 & 0 & -1 \end{bmatrix} \cdot \begin{bmatrix} -1 & -2 & 5 \\ -1 & -1 & 3 \\ -1 & -2 & 4 \end{bmatrix}$

14. $\begin{bmatrix} -1 & -2 & 5 \\ -1 & -1 & 3 \\ -1 & -2 & 4 \end{bmatrix} \cdot \begin{bmatrix} 2 & -2 & -1 \\ 1 & 1 & -2 \\ 1 & 0 & -1 \end{bmatrix}$

Let $A = \begin{bmatrix} 1 & -2 \\ 1 & 0 \end{bmatrix}$ and $B = \begin{bmatrix} -1 & 2 \\ -1 & 1 \end{bmatrix}$. Compute each of the following products.

15. AB

16. BA

17. $(AB)A$

18. $(BA)B$

19. $A^t B$

20. AB^t

Find a matrix X satisfying each matrix equation.

21. $3X + \begin{bmatrix} 1 & 0 \\ 2 & 1 \end{bmatrix} = \begin{bmatrix} -2 & 3 \\ -1 & -2 \end{bmatrix}$

22. $2X + 3\begin{bmatrix} 1 & 1 \\ 0 & 1 \end{bmatrix} = \begin{bmatrix} 7 & -1 \\ 3 & -5 \end{bmatrix}$

23. $X + 2I = \begin{bmatrix} 3 & -1 \\ 1 & 2 \end{bmatrix}$

24. $3X - 2I = \begin{bmatrix} 7 & 3 \\ 6 & 4 \end{bmatrix}$

B 25. Show that if $A = \begin{bmatrix} -1 & 2 \\ 0 & 1 \end{bmatrix}$ and $B = \begin{bmatrix} 1 & 0 \\ -1 & 2 \end{bmatrix}$, then

   a. $(A + B)(A + B) \neq A^2 + 2AB + B^2$,
   b. $(A + B)(A - B) \neq A^2 - B^2$.

26. a. Show that for each matrix $A_{2 \times 2}$,

$$A_{2 \times 2} \cdot I_{2 \times 2} = I_{2 \times 2} \cdot A_{2 \times 2} = A_{2 \times 2}.$$

b. Show that if, for a given matrix $B_{2 \times 2}$ and for *all* $A_{2 \times 2}$,

$$A_{2 \times 2} \cdot B_{2 \times 2} = B_{2 \times 2} \cdot A_{2 \times 2} = A_{2 \times 2},$$

then $B_{2 \times 2} = I_{2 \times 2}$.

27. Show that

$$\left(\begin{bmatrix} a_{11} & a_{12} \\ a_{21} & a_{22} \end{bmatrix} \cdot \begin{bmatrix} b_{11} & b_{12} \\ b_{21} & b_{22} \end{bmatrix} \right) \cdot \begin{bmatrix} c_{11} & c_{12} \\ c_{21} & c_{22} \end{bmatrix} = \begin{bmatrix} a_{11} & a_{12} \\ a_{21} & a_{22} \end{bmatrix} \cdot \left(\begin{bmatrix} b_{11} & b_{12} \\ b_{21} & b_{22} \end{bmatrix} \cdot \begin{bmatrix} c_{11} & c_{12} \\ c_{21} & c_{22} \end{bmatrix} \right).$$

28. Show that

$$\begin{bmatrix} a_{11} & a_{12} \\ a_{21} & a_{22} \end{bmatrix} \cdot \left(\begin{bmatrix} b_{11} & b_{12} \\ b_{21} & b_{22} \end{bmatrix} + \begin{bmatrix} c_{11} & c_{12} \\ c_{21} & c_{22} \end{bmatrix} \right)$$

$$= \begin{bmatrix} a_{11} & a_{12} \\ a_{21} & a_{22} \end{bmatrix} \cdot \begin{bmatrix} b_{11} & b_{12} \\ b_{21} & b_{22} \end{bmatrix} + \begin{bmatrix} a_{11} & a_{12} \\ a_{21} & a_{22} \end{bmatrix} \cdot \begin{bmatrix} c_{11} & c_{12} \\ c_{21} & c_{22} \end{bmatrix}.$$

29. Show that

$$\left(\begin{bmatrix} b_{11} & b_{12} \\ b_{21} & b_{22} \end{bmatrix} + \begin{bmatrix} c_{11} & c_{12} \\ c_{21} & c_{22} \end{bmatrix} \right) \cdot \begin{bmatrix} a_{11} & a_{12} \\ a_{21} & a_{22} \end{bmatrix}$$

$$= \begin{bmatrix} b_{11} & b_{12} \\ b_{21} & b_{22} \end{bmatrix} \cdot \begin{bmatrix} a_{11} & a_{12} \\ a_{21} & a_{22} \end{bmatrix} + \begin{bmatrix} c_{11} & c_{12} \\ c_{21} & c_{22} \end{bmatrix} \cdot \begin{bmatrix} a_{11} & a_{12} \\ a_{21} & a_{22} \end{bmatrix}.$$

30. Show that $\begin{bmatrix} 0 & a \\ a & 0 \end{bmatrix}^2 = a^2 I.$

31. Show that $(A_{2 \times 2} \cdot B_{2 \times 2})^t = B_{2 \times 2}^t \cdot A_{2 \times 2}^t.$

32. Show that $A_{2 \times 2}^2 = (-A_{2 \times 2})^2.$

In Exercises 33–39, prove the specified part of Theorem 9.2 for 2×2 matrices.

33. Part I **34.** Part II **35.** Part III **36.** Part V
37. Part VI **38.** Part VII **39.** Part VIII

40. Prove that if A and B are 2×2 matrices and a is a real number, then

$$a(AB) = (aA)B = A(aB).$$

9.3 Solution of Linear Systems by Using Row-Equivalent Matrices

Elementary transformations

An **elementary transformation** of a matrix $A_{n \times m}$ is one of the following three operations upon the rows of the matrix:

1. Multiply the entries of any row of $A_{n \times m}$ by k, where $k \in R$, $k \neq 0$.
2. Interchange any two rows of $A_{n \times m}$.
3. Multiply the entries of any row of $A_{n \times m}$ by k, where $k \in R$, and add to the corresponding entries of any other row.

Examples   Let $A = \begin{bmatrix} 1 & 3 \\ -2 & 4 \end{bmatrix}$.

a. The matrix $B = \begin{bmatrix} 1 & 3 \\ -1 & 2 \end{bmatrix}$ is obtained from A by multiplying each entry of Row 2 by 1/2.

b. The matrix $C = \begin{bmatrix} -2 & 4 \\ 1 & 3 \end{bmatrix}$ is obtained from A by interchanging Rows 1 and 2.

c. The matrix $D = \begin{bmatrix} 0 & 5 \\ -2 & 4 \end{bmatrix}$ is obtained from A by multiplying each entry of Row 2 by 1/2 and adding the result to the corresponding entry of Row 1.

Notice that the inverse of an elementary transformation is an elementary transformation. That is, you can undo an elementary transformation by means of an elementary transformation. For example, if A is transformed into B by interchanging two rows, then you can regain A from B by again interchanging the same two rows. It follows that if B results from performing a *succession* of elementary transformations on A, then A can similarly be obtained from B by performing the inverse operations in reverse order.

Row-equivalent matrices

Definition 9.12   If B is a matrix resulting from a succession of a finite number of elementary transformations on a matrix A, then A and B are **row-equivalent** matrices. This is expressed by writing $A \sim B$ or $B \sim A$.

Example   Show that $\begin{bmatrix} 1 & -2 & -1 \\ 1 & 0 & 2 \\ -4 & 3 & 1 \end{bmatrix} \sim \begin{bmatrix} 3 & -6 & -3 \\ 1 & 0 & 2 \\ -4 & 3 & 1 \end{bmatrix}$.

Solution   Multiplying each entry of the first row of the left-hand matrix by 3, we obtain the right-hand matrix.

Example   Show that $\begin{bmatrix} 1 & -2 & -1 \\ 1 & 0 & 2 \\ -4 & 3 & 1 \end{bmatrix} \sim \begin{bmatrix} -4 & 3 & 1 \\ 1 & 0 & 2 \\ 1 & -2 & -1 \end{bmatrix}$.

Solution   Interchanging the first and third row of the left-hand matrix, we obtain the right-hand matrix.

Sometimes it is convenient to make several elementary transformations on the same matrix.

9.3 Solution of Linear Systems by Using Row-Equivalent Matrices

Example Show that $\begin{bmatrix} 1 & -2 & -1 \\ 1 & 0 & 2 \\ -4 & 3 & 1 \end{bmatrix} \sim \begin{bmatrix} 1 & -2 & -1 \\ 0 & 2 & 3 \\ 0 & -5 & -3 \end{bmatrix}$.

Solution Multiplying the entries of the first row of the left-hand matrix by -1 and adding these products to the corresponding entries of the second row, and then multiplying the entries of the first row by 4 and adding these products to the corresponding entries of the third row, we obtain the right-hand matrix.

The $n \times n$ matrices that are row-equivalent to $I_{n \times n}$ are called **nonsingular** matrices. If an $n \times n$ matrix is not row-equivalent to $I_{n \times n}$, it is called **singular**.

Example Show that $A = \begin{bmatrix} 1 & 2 & 1 \\ -1 & 1 & 0 \\ 1 & 0 & 1 \end{bmatrix}$ is a nonsingular matrix.

Solution We first make appropriate transformations to obtain "0" elements in each entry (except for the principal diagonal) in columns 1, 2, and 3. Each reference to a row indicates the row of the preceding matrix.

$$\begin{bmatrix} 1 & 2 & 1 \\ -1 & 1 & 0 \\ 1 & 0 & 1 \end{bmatrix} \sim \begin{bmatrix} 1 & 2 & 1 \\ 0 & 3 & 1 \\ 0 & -2 & 0 \end{bmatrix} \begin{array}{l} \\ \text{Row 2 + Row 1} \\ \text{Row 3 + }[(-1) \times \text{Row 1}] \end{array}$$

$$\sim \begin{bmatrix} 1 & 0 & 1 \\ 0 & 3 & 1 \\ 0 & 0 & \frac{2}{3} \end{bmatrix} \begin{array}{l} \text{Row 1 + Row 3} \\ \\ \text{Row 3 + }\left[\left(\frac{2}{3}\right) \times \text{Row 2}\right] \end{array}$$

$$\sim \begin{bmatrix} 1 & 0 & 0 \\ 0 & 3 & 0 \\ 0 & 0 & \frac{2}{3} \end{bmatrix} \begin{array}{l} \text{Row 1 + }\left[\left(-\frac{3}{2}\right) \times \text{Row 3}\right] \\ \text{Row 2 + }\left[\left(-\frac{3}{2}\right) \times \text{Row 3}\right] \\ \\ \end{array}$$

$$\sim \begin{bmatrix} 1 & 0 & 0 \\ 0 & 1 & 0 \\ 0 & 0 & 1 \end{bmatrix} \begin{array}{l} \\ \left(\frac{1}{3}\right) \times \text{Row 2} \\ \left(\frac{3}{2}\right) \times \text{Row 3} \end{array}$$

Since the last operation yields the matrix $I_{3 \times 3}$, A is nonsingular.

Matrix solution of linear systems

In a linear system of the form

$$a_{11}x + a_{12}y + a_{13}z = c_1$$
$$a_{21}x + a_{22}y + a_{23}z = c_2$$
$$a_{31}x + a_{32}y + a_{33}z = c_3,$$

the matrices

$$\begin{bmatrix} a_{11} & a_{12} & a_{13} \\ a_{21} & a_{22} & a_{23} \\ a_{31} & a_{32} & a_{33} \end{bmatrix} \quad \text{and} \quad \begin{bmatrix} a_{11} & a_{12} & a_{13} & c_1 \\ a_{21} & a_{22} & a_{23} & c_2 \\ a_{31} & a_{32} & a_{33} & c_3 \end{bmatrix}$$

are called the **coefficient matrix** and the **augmented matrix**, respectively. Similar definitions hold for a system of n linear equations.

Starting with the augmented matrix of a linear system, and generating a sequence of row-equivalent matrices, we can obtain a matrix from which the solution set of the system is evident simply by inspection. The validity of the method, which is illustrated by example below, follows from the fact that performing elementary transformations on the augmented matrix of a system corresponds to performing the same sorts of operations on the equations of the system itself. Neither multiplying an equation by a nonzero constant nor interchanging two equations has any effect on the solution set of the system. Multiplying one equation by a constant and adding it to another equation in effect replaces the other by a linear combination of equations; a generalization of Theorems 8.1 and 8.2 ensures that the solution set remains the same.

Example

Solve the system

$$x + 2y - 3z = -4$$
$$2x - y + z = 3$$
$$3x + 2y + z = 10.$$

Solution

The augmented matrix of the system is

$$\begin{bmatrix} 1 & 2 & -3 & -4 \\ 2 & -1 & 1 & 3 \\ 3 & 2 & 1 & 10 \end{bmatrix}.$$

We perform elementary transformations on this matrix as follows.

$$\begin{bmatrix} 1 & 2 & -3 & -4 \\ 0 & -5 & 7 & 11 \\ 3 & 2 & 1 & 10 \end{bmatrix} \quad \text{Row 2} + [-2 \times \text{Row 1}]$$

9.3 Solution of Linear Systems by Using Row-Equivalent Matrices

$$\begin{bmatrix} 1 & 2 & 3 & \vdots & -4 \\ 0 & -5 & 7 & \vdots & 11 \\ 0 & -4 & 10 & \vdots & 22 \end{bmatrix} \quad \text{Row 3} + [-3 \times \text{Row 1}]$$

$$\begin{bmatrix} 1 & 2 & -3 & \vdots & -4 \\ 0 & -5 & 7 & \vdots & 11 \\ 0 & 0 & -22 & \vdots & -66 \end{bmatrix} \quad -5 \times \text{Row 3, then Row 3} + [4 \times \text{Row 2}]$$

The resulting system is

$$x + 2y - 3z = -4$$
$$-5y + 7z = 11$$
$$-22z = -66$$

which may be solved by reverse substitution. This example was solved in Section 8.2 and the augmented matrices here correspond exactly to the equivalent systems obtained there. Matrix notation, however, so facilitates the operations on the equations (rows) that it is easy to obtain even simpler equivalent systems. We can rework the above example as follows; the system of equations corresponding to each successive matrix is shown on the right of the matrix in the following solution.

$$\begin{array}{r} \\ \text{Row 2} + [-2 \times \text{Row 1}] \\ \text{Row 3} + [-3 \times \text{Row 1}] \end{array} \begin{bmatrix} 1 & 2 & -3 & \vdots & -4 \\ 0 & -5 & 7 & \vdots & 11 \\ 0 & -4 & 10 & \vdots & 22 \end{bmatrix} \begin{array}{l} x + 2y - 3z = -4 \\ 0x - 5y + 7z = 11 \\ 0x - 4y + 10z = 22 \end{array}$$

$$\text{Row 1} + \left[\frac{2}{5} \times \text{Row 2}\right] \begin{bmatrix} 1 & 0 & -\frac{1}{5} & \vdots & \frac{2}{5} \\ 0 & -5 & 7 & \vdots & 11 \\ & & & & \end{bmatrix} \begin{array}{l} x + 0y - \frac{1}{5}z = \frac{2}{5} \\ 0x - 5y + 7z = 11 \end{array}$$

$$\text{Row 3} + \left[-\frac{4}{5} \times \text{Row 2}\right] \begin{bmatrix} 0 & 0 & \frac{22}{5} & \vdots & \frac{66}{5} \end{bmatrix} \quad 0x + 0y + \frac{22}{5}z = \frac{66}{5}$$

$$5 \times \text{Row 1} \begin{bmatrix} 5 & 0 & -1 & \vdots & 2 \\ 0 & -5 & 7 & \vdots & 11 \\ & & & & \end{bmatrix} \begin{array}{l} 5x + 0y - z = 2 \\ 0x - 5y + 7z = 11 \end{array}$$

$$\frac{5}{22} \times \text{Row 3} \begin{bmatrix} 0 & 0 & 1 & \vdots & 3 \end{bmatrix} \quad 0x + 0y + z = 3$$

$$\begin{array}{r} \text{Row 1} + \text{Row 3} \\ \text{Row 2} + [-7 \times \text{Row 3}] \end{array} \begin{bmatrix} 5 & 0 & 0 & \vdots & 5 \\ 0 & -5 & 0 & \vdots & -10 \\ 0 & 0 & 1 & \vdots & 3 \end{bmatrix} \begin{array}{l} 5x + 0y + 0z = 5 \\ 0x - 5y + 0z = -10 \\ 0x + 0y + z = 3 \end{array}$$

$$\frac{1}{5} \times \text{Row 1} \begin{bmatrix} 1 & 0 & 0 & \vdots & 1 \\ & & & & \\ & & & & \end{bmatrix} \quad x + 0y + 0z = 1$$

$$-\frac{1}{5} \times \text{Row 2} \begin{bmatrix} 0 & 1 & 0 & \vdots & 2 \\ & & & & \\ 0 & 0 & 1 & \vdots & 3 \end{bmatrix} \begin{array}{l} 0x + y + 0z = 2 \\ \\ 0x + 0y + z = 3 \end{array}$$

Solution continued on overleaf

The last system is equivalent to

$$x = 1$$
$$y = 2$$
$$z = 3.$$

From this, the solution set, $\{(1, 2, 3)\}$, for the given system is evident by inspection.

For any given $n \times n$ linear system with *nonsingular* coefficient matrix, there are many sequences of row operations which will transform the augmented matrix of a system equivalently to one of the form

$$\begin{bmatrix} 1 & 0 & 0 & \cdots & 0 & \vdots & x_1 \\ 0 & 1 & 0 & & 0 & \vdots & \cdot \\ 0 & 0 & 1 & & 0 & \vdots & \cdot \\ \vdots & & & & \vdots & \vdots & \vdots \\ 0 & 0 & 0 & \cdots & 1 & \vdots & x_n \end{bmatrix}, \quad (1)$$

from which the solution set, $\{(x_1, \ldots, x_n)\}$, of the original system is evident by inspection. Finding the most efficient sequence depends on experience and insight, but several good systematic procedures exist. The procedure used above was, briefly, the following one: obtain a 1 on the diagonal using type 1 or 2 transformations if necessary; then "clear out" the remainder of the column using type 3 transformations to obtain 0's; then proceed to the next column.

Systems with singular coefficient matrices

If the row-reduction process yields a row of the form

$$0 \quad 0 \quad \cdots \quad 0 \quad a,$$

then that row in the system corresponds to an equation of the form

$$0 \cdot x_1 + 0 \cdot x_2 + \cdots + 0 \cdot x_n = a.$$

This equation has a solution only if $a = 0$, and in that case has an infinite number of solutions. Thus, the system will have no solution when the row-reduction process yields a row of the form

$$0 \quad 0 \quad \cdots \quad 0 \quad a,$$

where $a \neq 0$. The system will have an infinite number of solutions when the process does not yield such a row but does yield a row of zeros.

Example

Solve the system

$$2x + y = 1$$
$$4x + 2y = 0.$$

Solutions The augmented matrix of the system is
$$\begin{bmatrix} 2 & 1 & \vdots & 1 \\ 4 & 2 & \vdots & 0 \end{bmatrix}.$$

Thus, we have
$$\begin{bmatrix} 2 & 1 & \vdots & 1 \\ 4 & 2 & \vdots & 0 \end{bmatrix} \sim \begin{bmatrix} 2 & 1 & \vdots & 1 \\ 0 & 0 & \vdots & -2 \end{bmatrix} \quad \text{Row } 2 + [-2 \times \text{Row } 1]$$

Since the last row in the matrix on the right is of the form
$$0 \quad 0 \quad a$$
with $a = -2$, the system has no solution.

Example Solve the system
$$2x + y = 1$$
$$4x + 2y = 2.$$

Solution The augmented matrix of this system is
$$\begin{bmatrix} 2 & 1 & \vdots & 1 \\ 4 & 2 & \vdots & 2 \end{bmatrix}.$$

Thus, we have
$$\begin{bmatrix} 2 & 1 & \vdots & 1 \\ 4 & 2 & \vdots & 2 \end{bmatrix} \sim \begin{bmatrix} 2 & 1 & \vdots & 1 \\ 0 & 0 & \vdots & 0 \end{bmatrix} \quad \text{Row } 2 + [-2 \times \text{Row } 1].$$

Since the last row in the matrix on the right is of the form
$$0 \quad 0 \quad a$$
with $a = 0$ and there are no rows of this form with $a \neq 0$, the system has an infinite number of solutions.

The system corresponding to the matrix
$$\begin{bmatrix} 2 & 1 & \vdots & 1 \\ 0 & 0 & \vdots & 0 \end{bmatrix}$$
is
$$2x + y = 1$$
$$0x + 0y = 0.$$

Since the second equation is satisfied by any ordered pair, the solution set of the system is the same as that of the first equation. When this occurs we usually write the general form of the solution by expressing one variable in terms of the other. In this case, we have $y = 1 - 2x$ and we write the solution to this system as $(x, 1 - 2x)$, $x \in R$.

Exercise 9.3

A Use row transformations of the augmented matrix to solve each system of equations. Note the elementary transformations that are used. (Each coefficient matrix is nonsingular.)

Example

$x + 2y + z = 4$
$2x + y - z = -1$
$-x + y + z = 2$

Solution

$$\begin{bmatrix} 1 & 2 & 1 & \vdots & 4 \\ 2 & 1 & -1 & \vdots & -1 \\ -1 & 1 & 1 & \vdots & 2 \end{bmatrix} \text{ (augmented matrix)}$$

$$\begin{bmatrix} 1 & 2 & 1 & \vdots & 4 \\ 0 & -3 & -3 & \vdots & -9 \\ 0 & 3 & 2 & \vdots & 6 \end{bmatrix} \begin{matrix} \text{Row 2} + [-2 \times \text{Row 1}] \\ \text{Row 1} + \text{Row 3} \end{matrix}$$

$$\begin{bmatrix} 1 & 2 & 1 & \vdots & 4 \\ 0 & 1 & 1 & \vdots & 3 \\ 0 & 3 & 2 & \vdots & 6 \end{bmatrix} -\frac{1}{3} \times \text{Row 2}$$

$$\begin{bmatrix} 1 & 0 & -1 & \vdots & -2 \\ 0 & 1 & 1 & \vdots & 3 \\ 0 & 0 & -1 & \vdots & -3 \end{bmatrix} \begin{matrix} \text{Row 1} + [-2 \times \text{Row 2}] \\ \text{Row 3} + [-3 \times \text{Row 2}] \end{matrix}$$

$$\begin{bmatrix} 1 & 0 & -1 & \vdots & -2 \\ 0 & 1 & 1 & \vdots & 3 \\ 0 & 0 & 1 & \vdots & 3 \end{bmatrix} -1 \times \text{Row 3}$$

$$\begin{bmatrix} 1 & 0 & 0 & \vdots & 1 \\ 0 & 1 & 0 & \vdots & 0 \\ 0 & 0 & 1 & \vdots & 3 \end{bmatrix} \begin{matrix} \text{Row 1} + \text{Row 3} \\ \text{Row 2} + [-1 \times \text{Row 3}] \end{matrix}$$

The solution is (1, 0, 3). This same example was also solved in Exercise 8.2.

1. $x - 2y = 4$
   $x + 3y = -1$

2. $x + y = -1$
   $x - 4y = -14$

3. $3x - 2y = 13$
   $4x - y = 19$

4. $4x - 3y = 16$
   $2x + y = 8$

5. $x - 2y = 6$
   $3x + y = 25$

6. $x - y = -8$
   $x + 2y = 9$

7. $x + y - z = 0$
   $2x - y + z = -6$
   $x + 2y - 3z = 2$

8. $2x - y + 3z = 1$
   $x + 2y - z = -1$
   $3x + y + z = 2$

9. $2x - y = 0$
   $3y + z = 7$
   $2x + 3z = 1$

10. $3x - z = 7$
    $2x + y = 6$
    $3y - z = 7$

9.3 Solution of Linear Systems by Using Row-Equivalent Matrices

11. $2x - 5y + 3z = -1$
        $-3x - y + 2z = 11$
        $-2x + 7y + 5z = 9$

12. $2x + y + z = 4$
        $3x - z = 3$
        $2x + 3z = 13$

Use row transformations of the augmented matrix to solve each system of equations. Note the elementary transformations that are used. (The coefficient matrix may be singular.)

Example

$x - y + 2z = 3$
$2x + 3y - 6z = 1$
$4x + y - 2z = 7$

Solution

$$\begin{bmatrix} 1 & -1 & 2 & \vdots & 3 \\ 2 & 3 & -6 & \vdots & 1 \\ 4 & 1 & -2 & \vdots & 7 \end{bmatrix} \quad \text{(augmented matrix)}$$

$$\begin{bmatrix} 1 & -1 & 2 & \vdots & 3 \\ 0 & 5 & -10 & \vdots & -5 \\ 0 & 5 & -10 & \vdots & -5 \end{bmatrix} \quad \begin{array}{l} \text{Row } 2 + [-2 \times \text{Row } 1] \\ \text{Row } 3 + [-4 \times \text{Row } 1] \end{array}$$

$$\begin{bmatrix} 1 & -1 & 2 & \vdots & 3 \\ 0 & 1 & -2 & \vdots & -1 \\ 0 & 5 & -10 & \vdots & -5 \end{bmatrix} \quad \frac{1}{5} \times \text{Row } 2$$

$$\begin{bmatrix} 1 & 0 & 0 & \vdots & 2 \\ 0 & 1 & -2 & \vdots & -1 \\ 0 & 0 & 0 & \vdots & 0 \end{bmatrix} \quad \begin{array}{l} \text{Row } 1 + \text{Row } 2 \\ \\ \text{Row } 3 + [-5 \times \text{Row } 2] \end{array}$$

The zero row indicates there are an infinite number of solutions. The final equivalent system is

$$x = 2 \qquad \qquad x = 2$$
$$y - 2z = -1 \quad \text{or} \quad y = 2z - 1.$$
$$0x + 0y + 0z = 0$$

For any real-number z the triple $(2, 2z - 1, z)$ is a solution of the system, and conversely. Thus the solution set may be described as the set of all triples of the form $(2, 2z - 1, z), z \in R$.

13. $x + 2y + 2z = 3$
        $2x + 5y + 5z = 7$
        $y + z = 1$

14. $x - y + z = -1$
        $3x - 2y + 2z = 1$
        $2x - 4y + 4z = -10$

15. $x + 3y + 7z = 5$
        $-x + y + z = -1$
        $x + 11y + 22z = 14$

16. $x + 2y = -1$
        $3x + y - 5z = 5$
        $2x + 9y + 5z = -10$

17. $x + 2y + 2z = 5$
        $x + y + z = 4$
        $2x + 3y + 3z = 8$

18. $x - y + z = 4$
        $2x + 3y - z = 5$
        $x - 6y + 4z = 6$

19. $\begin{aligned} x + 2y - z &= 1 \\ 3x + y + z &= 3 \\ 2x - y + 2z &= -1 \end{aligned}$

20. $\begin{aligned} x + 3y + z &= 1 \\ 2x + 4y - z &= -1 \\ 2y + 3z &= 3 \end{aligned}$

B Show that each product is a matrix that is row-equivalent to $\begin{bmatrix} a & b \\ c & d \end{bmatrix}$ if $k \neq 0$.

21. $\begin{bmatrix} k & 0 \\ 0 & 1 \end{bmatrix} \begin{bmatrix} a & b \\ c & d \end{bmatrix}$

22. $\begin{bmatrix} 1 & 0 \\ 0 & k \end{bmatrix} \begin{bmatrix} a & b \\ c & d \end{bmatrix}$

23. $\begin{bmatrix} 0 & 1 \\ 1 & 0 \end{bmatrix} \begin{bmatrix} a & b \\ c & d \end{bmatrix}$

24. $\begin{bmatrix} 1 & k \\ 0 & 1 \end{bmatrix} \begin{bmatrix} a & b \\ c & d \end{bmatrix}$

25. $\begin{bmatrix} 1 & 0 \\ k & 1 \end{bmatrix} \begin{bmatrix} a & b \\ c & d \end{bmatrix}$

9.4 The Determinant Function

Associated with each square matrix A having real-number entries is a real number called the **determinant** of A and denoted by δA or $\delta(A)$ (read "the determinant of A"). Thus we have a function, δ (delta), with domain the set of all square matrices having real-number entries, and with range the set of all real numbers; $\delta(A_{n \times n})$ is called a determinant of **order** n.

Let us begin by examining δ over the set $S_{2 \times 2}$ of 2×2 matrices.

Definition 9.13 The **determinant** of the matrix

$$\begin{bmatrix} a_{11} & a_{12} \\ a_{21} & a_{22} \end{bmatrix}$$

is the number $a_{11}a_{22} - a_{12}a_{21}$.

The determinant of a square matrix is customarily displayed in the same form as the matrix, but with vertical bars in lieu of brackets. Thus,

$$\delta \begin{bmatrix} a_{11} & a_{12} \\ a_{21} & a_{22} \end{bmatrix} = \begin{vmatrix} a_{11} & a_{12} \\ a_{21} & a_{22} \end{vmatrix} = a_{11}a_{22} - a_{12}a_{21}.$$

Examples Compute the determinant of each matrix.

a. $\begin{bmatrix} 2 & 4 \\ 1 & 0 \end{bmatrix}$

b. $\begin{bmatrix} 4 & -3 \\ 1 & 2 \end{bmatrix}$

9.4 The Determinant Function

Solutions

a. $\begin{vmatrix} 2 & 4 \\ 1 & 0 \end{vmatrix} = (2)(0) - (4)(1)$
$= -4$

b. $\begin{vmatrix} 4 & -3 \\ 1 & 2 \end{vmatrix} = (4)(2) - (-3)(1)$
$= 11$

Minors and cofactors

Before we define the determinant of a higher-order matrix, we need to define two terms. Throughout the following discussion, A is the $n \times n$ matrix

$$A = \begin{bmatrix} a_{11} & \cdots & a_{1n} \\ \vdots & & \vdots \\ a_{n1} & \cdots & a_{nn} \end{bmatrix}.$$

Definition 9.14 The **minor** M_{ij} of a_{ij} is the determinant of the $(n-1) \times (n-1)$ matrix obtained by deleting the ith row and the jth column of the matrix A.

Examples Let $A = \begin{bmatrix} 1 & 4 & -2 \\ 2 & 3 & -4 \\ 0 & 2 & 3 \end{bmatrix}$. Compute **a.** M_{23} and **b.** M_{31}.

Solutions

a. Write the matrix A and cross out the second row and third column.

$$M_{23} = \delta \begin{bmatrix} 1 & 4 & -2 \\ \cancel{2} & \cancel{3} & \cancel{-4} \\ 0 & 2 & \cancel{3} \end{bmatrix} = \begin{vmatrix} 1 & 4 \\ 0 & 2 \end{vmatrix} = (1)(2) - (4)(0) = 2$$

b. Write the matrix A and cross out the third row and first column.

$$M_{31} = \delta \begin{bmatrix} \cancel{1} & 4 & -2 \\ \cancel{2} & 3 & 4 \\ \cancel{0} & \cancel{2} & \cancel{3} \end{bmatrix} = \begin{vmatrix} 4 & -2 \\ 3 & 4 \end{vmatrix} = (4)(4) - (-2)(3) = 22$$

Definition 9.15 The **cofactor** A_{ij} of the entry a_{ij} is

$$A_{ij} = (-1)^{i+j} M_{ij}.$$

Examples Let $A = \begin{bmatrix} 1 & 4 & -2 \\ 2 & 3 & -4 \\ 0 & 2 & 3 \end{bmatrix}$. Compute **a.** A_{23} and **b.** A_{31}.

Solutions

The matrix A is the same one that was used in the foregoing example. Thus, from that example, we have $M_{23} = 2$ and $M_{31} = 22$. Hence by Definition 9.15

a. $A_{23} = (-1)^{2+3} M_{23}$
$= (-1)(2) = -2$

b. $A_{31} = (-1)^{3+1} M_{31}$
$= (1)(22) = 22$

Determinants of n × n matrices

We can use the cofactors of the entries of a square matrix A with order greater than 2×2 to define the determinant of A.

Definition 9.16   The **determinant** of the square matrix

$$\begin{bmatrix} a_{11} & a_{12} & \cdots & a_{1n} \\ a_{21} & a_{22} & \cdots & a_{2n} \\ \vdots & & & \vdots \\ a_{n1} & a_{n2} & \cdots & a_{nn} \end{bmatrix}$$

is the sum of the n products formed by multiplying each entry in any single row (or any single column) by its cofactor.

In applying Definition 9.16 the determinant is said to be **expanded** about whatever row (or column) is chosen.

It can be shown, although we shall not do so here, that the value of the determinant is independent of the row or column about which the determinant is expanded.

Example   Let $A = \begin{bmatrix} 1 & 0 & 1 \\ 2 & 3 & 0 \\ 3 & 0 & 4 \end{bmatrix}$. Compute $\delta(A)$.

Solution   We shall expand about column 2.

$$\delta(A) = 0 \cdot A_{12} + 3 \cdot A_{22} + 0 \cdot A_{32}$$

$$= 3(-1)^{2+2} \begin{vmatrix} 1 & 1 \\ 3 & 4 \end{vmatrix} = 3(1)(1)$$

$$= 3$$

Choosing a row or column

Note that each term in the sum used to compute the determinant of a matrix is of the form $a_{ij} A_{ij}$. Thus, if $a_{ij} = 0$ then that term does not contribute to the sum and therefore the cofactor of that entry need not be computed. Therefore to minimize the number of operations necessary to compute the determinant of a matrix we expand about the row or column with the most zero entries. Thus, in the preceding example we expanded about the second column which has the largest number of zero entries.

Example   Let $A = \begin{bmatrix} 1 & 0 & 1 & 1 \\ 4 & -2 & 0 & 0 \\ -3 & 1 & 1 & 0 \\ 0 & 2 & -4 & 1 \end{bmatrix}$. Compute $\delta(A)$.

9.4 The Determinant Function

Solution Observe that no row or column has more than two zero entries and both Row 2 and Column 4 have 2 zero entries. We therefore expand the determinant about either Row 2 or Column 4. We shall use Row 2.

$$\delta(A) = 4A_{21} + (-2)A_{22} + 0 \cdot A_{23} + 0 \cdot A_{24}$$

$$= 4(-1)^{2+1} \begin{vmatrix} 0 & 1 & 1 \\ 1 & 1 & 0 \\ 2 & -4 & 1 \end{vmatrix} + (-2)(-1)^{2+2} \begin{vmatrix} 1 & 1 & 1 \\ -3 & 1 & 0 \\ 0 & -4 & 1 \end{vmatrix}$$

Now, using cofactors we compute the two 3×3 determinants in the right-hand member above. We shall expand each of the determinants about the first column.

$$\begin{vmatrix} 0 & 1 & 1 \\ 1 & 1 & 0 \\ 2 & -4 & 1 \end{vmatrix} = 0 \cdot \begin{vmatrix} 1 & 0 \\ -4 & 1 \end{vmatrix} + 1(-1) \begin{vmatrix} 1 & 1 \\ -4 & 1 \end{vmatrix} + 2 \cdot \begin{vmatrix} 1 & 1 \\ 1 & 0 \end{vmatrix}$$

$$= 0(1) + (-1)(5) + 2(-1)$$

$$= -7$$

and

$$\begin{vmatrix} 1 & 1 & 1 \\ -3 & 1 & 0 \\ 0 & -4 & 1 \end{vmatrix} = 1 \cdot \begin{vmatrix} 1 & 0 \\ -4 & 1 \end{vmatrix} + (-3)(-1) \begin{vmatrix} 1 & 1 \\ -4 & 1 \end{vmatrix} + 0 \cdot \begin{vmatrix} 1 & 1 \\ 1 & 0 \end{vmatrix}$$

$$= 1(1) + 3(5) + 0(-1)$$

$$= 16$$

Thus,

$$\delta(A) = 4(-1)(-7) + (-2)(1)(16) = -4.$$

Exercise 9.4

A *Compute the determinant of each matrix.*

Examples

a. $\begin{bmatrix} -2 & \frac{1}{2} \\ 3 & 1 \end{bmatrix}$

b. $\begin{bmatrix} 3 & 2 \\ -1 & 0 \end{bmatrix}$

Solutions

a. $\begin{vmatrix} -2 & \frac{1}{2} \\ 3 & 1 \end{vmatrix} = (-2)(1) - \left(\frac{1}{2}\right)(3) = -\frac{7}{2}$

b. $\begin{vmatrix} 3 & 2 \\ -1 & 0 \end{vmatrix} = (3)(0) - (2)(-1) = 2$

1. $\begin{bmatrix} 3 & 0 \\ 0 & 1 \end{bmatrix}$

2. $\begin{bmatrix} 4 & 0 \\ 0 & -\frac{1}{2} \end{bmatrix}$

3. $\begin{bmatrix} 2 & -1 \\ 0 & 2 \end{bmatrix}$

4. $\begin{bmatrix} 3 & -5 \\ 0 & 10 \end{bmatrix}$
5. $\begin{bmatrix} -2 & 4 \\ -1 & 2 \end{bmatrix}$
6. $\begin{bmatrix} 1 & 1 \\ 2 & 1 \\ 4 & 8 \end{bmatrix}$

7. $\begin{bmatrix} 3 & 1 \\ 2 & -3 \end{bmatrix}$
8. $\begin{bmatrix} 4 & -2 \\ 5 & -4 \end{bmatrix}$

Specify the minor and the cofactor (in determinant form) of the indicated entry of

$$A = \begin{bmatrix} 2 & 1 & -2 & 0 \\ 1 & 0 & 3 & -1 \\ -2 & 1 & 2 & 2 \\ 1 & -1 & 3 & 1 \end{bmatrix}.$$

Example a_{21}

Solution The minor is $\begin{vmatrix} 1 & -2 & 0 \\ 1 & 2 & 2 \\ -1 & 3 & 1 \end{vmatrix}$. Since $2 + 1$ is odd, the cofactor is $-\begin{vmatrix} 1 & -2 & 0 \\ 1 & 2 & 2 \\ -1 & 3 & 1 \end{vmatrix}$.

9. a_{11} 10. a_{13} 11. a_{23} 12. a_{41}
13. a_{31} 14. a_{33} 15. a_{44} 16. a_{14}

Compute the determinant of the given matrix.

Example $\begin{bmatrix} 1 & 2 & 1 \\ 0 & -1 & 2 \\ 3 & 4 & 1 \end{bmatrix}$

Solution Expand about Column 1.

$$\delta(A) = (1)A_{11} + (0)A_{21} + (3)A_{31}$$

Compute the necessary cofactors.

$$A_{11} = (-1)^{1+1} \begin{vmatrix} -1 & 2 \\ 4 & 1 \end{vmatrix} = (1)[(-1)(1) - (2)(4)] = -9$$

$$A_{31} = (-1)^{3+1} \begin{vmatrix} 2 & 1 \\ -1 & 2 \end{vmatrix} = (1)[(2)(2) - (1)(-1)] = 5$$

Therefore,

$$\delta(A) = (1)(-9) + (3)(5) = 6.$$

17. $\begin{bmatrix} 2 & 1 & 3 \\ 0 & 4 & 1 \\ 0 & 0 & 2 \end{bmatrix}$

18. $\begin{bmatrix} -3 & 0 & 0 \\ -1 & 4 & 0 \\ 5 & 6 & -\frac{1}{4} \end{bmatrix}$

19. $\begin{bmatrix} 0 & 1 & 1 \\ 1 & 2 & 1 \\ 3 & -1 & 0 \end{bmatrix}$

20. $\begin{bmatrix} 2 & -4 & 1 \\ 1 & 1 & 0 \\ 0 & 2 & 0 \end{bmatrix}$

21. $\begin{bmatrix} 2 & 1 & 1 \\ 3 & -4 & 2 \\ 4 & 2 & 2 \end{bmatrix}$

22. $\begin{bmatrix} 1 & -4 & 2 \\ 2 & -1 & 4 \\ -3 & 2 & -6 \end{bmatrix}$

Example

$\begin{bmatrix} 0 & 1 & 0 & 2 \\ 0 & 1 & 1 & 4 \\ 2 & 0 & 0 & 3 \\ 1 & 1 & 1 & 2 \end{bmatrix}$

Solution

Expand about Column 3.

$$\delta(A) = (0)A_{13} + (1)A_{23} + (0)A_{33} + (1)A_{43}$$

Compute the necessary cofactors.

$$A_{23} = (-1)^{2+3} \begin{vmatrix} 0 & 1 & 2 \\ 2 & 0 & 3 \\ 1 & 1 & 2 \end{vmatrix} = -3$$

$$A_{43} = (-1)^{4+3} \begin{vmatrix} 0 & 1 & 2 \\ 0 & 1 & 4 \\ 2 & 0 & 3 \end{vmatrix} = -4$$

Thus,

$$\delta(A) = (1)(-3) + (1)(-4) = -7.$$

23. $\begin{bmatrix} 1 & 0 & 0 & 0 \\ 0 & 3 & 0 & 0 \\ 0 & 0 & -2 & 0 \\ 0 & 0 & 0 & 4 \end{bmatrix}$

24. $\begin{bmatrix} 1 & 0 & 2 & 0 \\ 0 & 1 & 3 & 1 \\ 0 & 0 & 2 & -1 \\ 0 & 0 & 0 & 1 \end{bmatrix}$

25. $\begin{bmatrix} 0 & 1 & 0 & 2 \\ 1 & 2 & 1 & 0 \\ 3 & -4 & 0 & 0 \\ 2 & 0 & 1 & 0 \end{bmatrix}$

26. $\begin{bmatrix} 1 & 4 & 1 & 1 \\ 0 & 2 & -3 & -4 \\ 2 & 0 & 0 & 4 \\ 1 & 1 & 1 & 2 \end{bmatrix}$

B 27. Show that $\begin{vmatrix} x & y & 1 \\ x_1 & y_1 & 1 \\ x_2 & y_2 & 1 \end{vmatrix} = 0$ represents an equation of the line through the points (x_1, y_1) and (x_2, y_2).

28. Use the results in Exercise 27 to find an equation of the line through the points $(3, -1)$ and $(-2, 5)$.

29. Show that if A is an $n \times n$ matrix with one row (or column) identically 0 then $\delta(A) = 0$.

30. In accordance with Definitions 9.14, 9.15, and 9.16, the determinant of an $n \times n$ matrix is the sum of a certain number of products of the entries. What is this number for $n = 2$? For $n = 3$? For $n = 4$?

31. Show that for any 2×2 matrix A, $\delta(aA) = a^2 \delta(A)$.

32. Show that for any 2×2 matrix A, $\delta(A^t) = \delta(A)$.

9.5 Properties of Determinants

Determinants have some properties that are useful by virtue of the fact that they permit us to generate equivalent determinants (name the same number) with different and simpler configurations of entries. This, in turn, helps us find values for determinants. We shall state these properties without proof. Some of the proofs are left as exercises.

Theorem 9.7 If each entry in any row, or each entry in any column, of a determinant is 0, then the determinant is equal to 0.

Examples   **a.** $\begin{vmatrix} 0 & 0 \\ 1 & 2 \end{vmatrix} = 0$ **b.** $\begin{vmatrix} 1 & 1 & 0 \\ 3 & 5 & 0 \\ 2 & 7 & 0 \end{vmatrix} = 0$ **c.** $\begin{vmatrix} 0 & 1 & 0 & 0 \\ 1 & 0 & 0 & 0 \\ 0 & 0 & 0 & 1 \\ 0 & 0 & 0 & 1 \end{vmatrix} = 0$

Theorem 9.8 If any two rows (or any two columns) of a determinant are interchanged, then the resulting determinant is the negative of the original determinant.

Examples   **a.** $\begin{vmatrix} 1 & 2 \\ 3 & 4 \end{vmatrix} = -\begin{vmatrix} 3 & 4 \\ 1 & 2 \end{vmatrix}$ **b.** $\begin{vmatrix} 1 & 2 & 3 \\ 4 & 5 & 6 \\ 7 & 8 & 9 \end{vmatrix} = -\begin{vmatrix} 3 & 2 & 1 \\ 6 & 5 & 4 \\ 9 & 8 & 7 \end{vmatrix}$

In (a), Rows 1 and 2 were interchanged. In (b), Columns 1 and 3 were interchanged.

9.5 Properties of Determinants

Theorem 9.9 *If two rows (or two columns) in a determinant have corresponding entries that are equal, the determinant is equal to 0.*

Examples

a. $\begin{vmatrix} 1 & 1 \\ 3 & 3 \end{vmatrix} = 0$
b. $\begin{vmatrix} 1 & 2 & 1 \\ 3 & 1 & 0 \\ 1 & 2 & 1 \end{vmatrix} = 0$
c. $\begin{vmatrix} 1 & 2 & 3 & 4 \\ 5 & 6 & 7 & 8 \\ 0 & 0 & 1 & 0 \\ 1 & 2 & 3 & 4 \end{vmatrix} = 0$

In (**a**) Columns 1 and 2 are identical. In (**b**) Rows 1 and 3 are identical. In (**c**) Rows 1 and 4 are identical.

Theorem 9.10 *If each of the entries of one row (or column) of a determinant is multiplied by k, the determinant is multiplied by k.*

Examples

a. $\begin{vmatrix} 1 & 0 & 0 \\ 2 & 1 & 3 \\ 1 \times 2 & 3 \times 2 & 4 \times 2 \end{vmatrix} = 2 \begin{vmatrix} 1 & 0 & 0 \\ 2 & 1 & 3 \\ 1 & 3 & 4 \end{vmatrix}$
b. $\begin{vmatrix} 4 & 5 & 8 \\ 1 & 1 & 2 \\ 3 & 1 & 6 \end{vmatrix} = 2 \begin{vmatrix} 4 & 5 & 4 \\ 1 & 1 & 1 \\ 3 & 1 & 3 \end{vmatrix}$

Note that this process is different from that of the multiplication of a matrix by a real number. In the latter, each entry in the matrix is multiplied by the real number, rather than, as here, only the entries in a single row or column being so multiplied.

Theorem 9.11 *If each entry of one row (or column) of a determinant is multiplied by a real number k and the resulting product is added to the corresponding entry in another row (or column, respectively) in the determinant, the resulting determinant is equal to the original determinant.*

Examples

a. $\begin{vmatrix} 1 & 1 \\ 2 & 1 \end{vmatrix} = \begin{vmatrix} 1 & 1 \\ 2 + 3(1) & 1 + 3(1) \end{vmatrix} = \begin{vmatrix} 1 & 1 \\ 5 & 4 \end{vmatrix}$

b. $\begin{vmatrix} 1 & 2 & 3 \\ 4 & 5 & 6 \\ 7 & 8 & 9 \end{vmatrix} = \begin{vmatrix} 1 + 2(3) & 2 & 3 \\ 4 + 2(6) & 5 & 6 \\ 7 + 2(9) & 8 & 9 \end{vmatrix} = \begin{vmatrix} 7 & 2 & 3 \\ 16 & 5 & 6 \\ 25 & 8 & 9 \end{vmatrix}$

Evaluation of determinants

The preceding theorems can be used to write sequences of equal determinants, leading from one form of a determinant to another and more useful form.

Example

Evaluate

$$D = \begin{vmatrix} 2 & -1 & 1 & -3 \\ 1 & 3 & -4 & 2 \\ 1 & 0 & -2 & 1 \\ 3 & -1 & 5 & 2 \end{vmatrix}.$$

Solution As a step toward evaluating the determinant, we shall use Theorem 9.11 to produce an equal determinant with a row or a column containing zero entries in all but one place. Let us arbitrarily select the second column for this role, because one entry is already zero. Multiplying a_{1j} by 3 and adding the result to a_{2j}, we obtain

$$D = \begin{vmatrix} 2 & -1 & 1 & -3 \\ 1+3(2) & 3+3(-1) & -4+3(1) & 2+3(-3) \\ 1 & 0 & -2 & 1 \\ 3 & -1 & 5 & 2 \end{vmatrix}$$

$$= \begin{vmatrix} 2 & -1 & 1 & -3 \\ 7 & 0 & -1 & -7 \\ 1 & 0 & -2 & 1 \\ 3 & -1 & 5 & 2 \end{vmatrix}.$$

Next, multiplying a_{1j} by -1 and adding the result to a_{4j}, we find that

$$D = \begin{vmatrix} 2 & -1 & 1 & -3 \\ 7 & 0 & -1 & -7 \\ 1 & 0 & -2 & 1 \\ 3-1(2) & -1-1(-1) & 5-1(1) & 2-1(-3) \end{vmatrix}$$

$$= \begin{vmatrix} 2 & -1 & 1 & -3 \\ 7 & 0 & -1 & -7 \\ 1 & 0 & -2 & 1 \\ 1 & 0 & 4 & 5 \end{vmatrix}.$$

If we now expand the determinant about the second column, we have

$$D = \begin{vmatrix} 2 & -1 & 1 & -3 \\ 7 & 0 & -1 & -7 \\ 1 & 0 & -2 & 1 \\ 1 & 0 & 4 & 5 \end{vmatrix} = -(-1)\begin{vmatrix} 7 & -1 & -7 \\ 1 & -2 & 1 \\ 1 & 4 & 5 \end{vmatrix} + 0A_{22} + 0A_{32} + 0A_{42}.$$

From this point, we can reduce the third-order determinant to a second-order determinant by a similar procedure or, alternatively, expand directly about the elements in any row or column. Expanding about the elements of the first row, we obtain

$$D = \begin{vmatrix} 7 & -1 & -7 \\ 1 & -2 & 1 \\ 1 & 4 & 5 \end{vmatrix} = 7\begin{vmatrix} -2 & 1 \\ 4 & 5 \end{vmatrix} - (-1)\begin{vmatrix} 1 & 1 \\ 1 & 5 \end{vmatrix} + (-7)\begin{vmatrix} 1 & -2 \\ 1 & 4 \end{vmatrix},$$

from which

$$D = 7(-14) + (4) - 7(6)$$
$$= -98 + 4 - 42 = -136.$$

9.5 Properties of Determinants

Exercise 9.5

A *Without evaluating, state why each statement is true. Verify selected examples by expansion.*

Examples

a. $\begin{vmatrix} 1 & 0 & 3 \\ 2 & 1 & 7 \\ 1 & 0 & 2 \end{vmatrix} = \begin{vmatrix} 1 & 0 & 3 \\ 2 & 1 & 7 \\ 3 & 1 & 9 \end{vmatrix}$

b. $\begin{vmatrix} 1 & 2 & 1 \\ 0 & 0 & 2 \\ 1 & 2 & 1 \end{vmatrix} = 0$

Solutions

a. The right-hand determinant is obtained from the left by adding Row 2 to Row 3.

b. Rows 1 and 3 have corresponding entries that are equal.

1. $\begin{vmatrix} 2 & 3 & 1 \\ 0 & 0 & 0 \\ -1 & 2 & 0 \end{vmatrix} = 0$

2. $\begin{vmatrix} 3 & 1 & 3 \\ 0 & 1 & 0 \\ 1 & 2 & 1 \end{vmatrix} = 0$

3. $\begin{vmatrix} 2 & 3 & 1 & 1 \\ 2 & 0 & 1 & 2 \\ 2 & 3 & 1 & 1 \\ 0 & 1 & 2 & 0 \end{vmatrix} = 0$

4. $\begin{vmatrix} 7 & 3 & 2 & 0 \\ 2 & 1 & 2 & 0 \\ 4 & 1 & 1 & 0 \\ 0 & 2 & 1 & 0 \end{vmatrix} = 0$

5. $\begin{vmatrix} 4 & 2 & 1 \\ 0 & -1 & -2 \\ 1 & 0 & 2 \end{vmatrix} = -\begin{vmatrix} 4 & 2 & 1 \\ 0 & 1 & 2 \\ 1 & 0 & 2 \end{vmatrix}$

6. $\begin{vmatrix} -2 & 3 & 1 \\ -1 & 0 & 1 \\ -2 & 1 & 0 \end{vmatrix} = -\begin{vmatrix} 2 & 3 & 1 \\ 1 & 0 & 1 \\ 2 & 1 & 0 \end{vmatrix}$

7. $2\begin{vmatrix} 1 & 0 & 2 \\ -1 & 2 & 0 \\ 1 & 1 & 1 \end{vmatrix} = \begin{vmatrix} 1 & 0 & 2 \\ -1 & 2 & 0 \\ 2 & 2 & 2 \end{vmatrix}$

8. $\begin{vmatrix} 3 & -4 & 2 \\ 1 & -2 & 0 \\ 0 & 8 & 1 \end{vmatrix} = -2\begin{vmatrix} 3 & 2 & 2 \\ 1 & 1 & 0 \\ 0 & -4 & 1 \end{vmatrix}$

9. $\begin{vmatrix} 1 & 2 \\ 3 & 4 \end{vmatrix} = \begin{vmatrix} 1+2 & 2 \\ 3+4 & 4 \end{vmatrix}$

10. $\begin{vmatrix} 1 & 2 \\ 3 & 4 \end{vmatrix} = \begin{vmatrix} 1+4 & 2 \\ 3+8 & 4 \end{vmatrix}$

11. $\begin{vmatrix} 1 & 2 & 1 \\ 0 & 2 & 3 \\ 2 & -1 & 2 \end{vmatrix} = \begin{vmatrix} 1 & 2 & 1 \\ 0 & 2 & 3 \\ 0 & -5 & 0 \end{vmatrix}$

12. $\begin{vmatrix} -1 & 1 & 0 \\ 2 & 3 & -1 \\ 2 & 1 & 2 \end{vmatrix} = \begin{vmatrix} 0 & 1 & 0 \\ 5 & 3 & -1 \\ 3 & 1 & 2 \end{vmatrix}$

Theorem 9.11 was used on the left-hand member of each of the following equalities to produce the elements in the right-hand member. Complete the entries.

Example

$\begin{vmatrix} 1 & 5 \\ 4 & 3 \end{vmatrix} = \begin{vmatrix} 1 & 5 \\ 0 & \end{vmatrix}$

Solution

To obtain the 0 in the 2, 1 position using Theorem 9.11, it must be that -4 times row 1 was added to row 2. The missing entry is therefore -17.

13. $\begin{vmatrix} 1 & 3 \\ 2 & 2 \end{vmatrix} = \begin{vmatrix} 1 & 3 \\ 0 & 0 \end{vmatrix}$ 14. $\begin{vmatrix} 2 & -1 \\ 3 & 1 \end{vmatrix} = \begin{vmatrix} 0 & 0 \\ 3 & 1 \end{vmatrix}$

15. $\begin{vmatrix} 1 & -2 & 1 \\ 3 & 1 & 4 \\ 0 & 2 & 1 \end{vmatrix} = \begin{vmatrix} 1 & -2 & 1 \\ 0 & 7 & 0 \\ 0 & 2 & 1 \end{vmatrix}$ 16. $\begin{vmatrix} 3 & -1 & 0 \\ 1 & 2 & 1 \\ 2 & 3 & 1 \end{vmatrix} = \begin{vmatrix} 3 & -1 & 0 \\ 1 & 2 & 1 \\ 1 & 1 & 0 \end{vmatrix}$

17. $\begin{vmatrix} 2 & 3 & 1 & 4 \\ 0 & 2 & 1 & 2 \\ 1 & 1 & 2 & 3 \\ 0 & 1 & 1 & 1 \end{vmatrix} = \begin{vmatrix} 0 & 1 & & -2 \\ 0 & 2 & 1 & 2 \\ 1 & 1 & 2 & 3 \\ 0 & 1 & 1 & 1 \end{vmatrix}$

18. $\begin{vmatrix} 2 & 1 & 1 & 0 \\ 1 & 2 & 0 & 2 \\ 3 & 1 & 0 & 3 \\ 2 & 1 & 4 & 2 \end{vmatrix} = \begin{vmatrix} 2 & 1 & 1 & 0 \\ 1 & 2 & 0 & 2 \\ 3 & 1 & 0 & 3 \\ -3 & 0 & 0 & 2 \end{vmatrix}$

First reduce each determinant to an equal 2 × 2 determinant and then evaluate.

Examples

a. $\begin{vmatrix} 0 & 3 & 2 \\ 1 & 7 & 8 \\ 0 & 5 & 4 \end{vmatrix}$ b. $\begin{vmatrix} 1 & 0 & 0 \\ 1 & 1 & 2 \\ 1 & -2 & 3 \end{vmatrix}$

Solutions

a. Expanding by Column 1,

$-1 \cdot \begin{vmatrix} 3 & 2 \\ 5 & 4 \end{vmatrix} = -1(12 - 10) = -2.$

b. Expanding by Row 1,

$1 \cdot \begin{vmatrix} 1 & 2 \\ -2 & 3 \end{vmatrix} = 1(3 - (-4)) = 7.$

19. $\begin{vmatrix} 2 & 1 & 0 \\ 3 & 2 & 1 \\ -1 & 2 & 0 \end{vmatrix}$ 20. $\begin{vmatrix} 1 & 2 & 1 \\ 2 & -1 & 2 \\ 0 & 1 & 0 \end{vmatrix}$ 21. $\begin{vmatrix} 1 & 0 & 3 \\ 2 & -1 & 1 \\ 1 & 2 & 1 \end{vmatrix}$

22. $\begin{vmatrix} 1 & 2 & -1 \\ 2 & 1 & 3 \\ 0 & 1 & 2 \end{vmatrix}$ 23. $\begin{vmatrix} 1 & 2 & 1 \\ -1 & 2 & 3 \\ 2 & -1 & 1 \end{vmatrix}$ 24. $\begin{vmatrix} 3 & -1 & 2 \\ 1 & 2 & 1 \\ -2 & 1 & 3 \end{vmatrix}$

25. $\begin{vmatrix} 0 & 0 & 1 & 2 \\ 6 & 0 & 0 & 1 \\ 6 & 1 & 0 & -1 \\ 6 & 1 & 0 & 2 \end{vmatrix}$ 26. $\begin{vmatrix} 4 & 2 & 0 & 2 \\ -1 & 0 & 2 & 1 \\ 3 & 0 & -1 & 1 \\ 0 & 0 & 2 & 1 \end{vmatrix}$

27. $\begin{vmatrix} 0 & 1 & 0 & 2 \\ 0 & 2 & 0 & 3 \\ 2 & -1 & 1 & 0 \\ 0 & 0 & 8 & 8 \end{vmatrix}$ 28. $\begin{vmatrix} 0 & 2 & -1 & 3 \\ 0 & 0 & 2 & 1 \\ 3 & 0 & 1 & 0 \\ -6 & 6 & 0 & 0 \end{vmatrix}$

29. $\begin{vmatrix} 1 & 2 & 3 & -1 \\ 0 & 4 & 8 & 4 \\ -2 & 0 & 1 & 1 \\ 2 & 1 & 0 & 1 \end{vmatrix}$

30. $\begin{vmatrix} 1 & 2 & 1 & 1 \\ 2 & -1 & 0 & 1 \\ 0 & 6 & 3 & 9 \\ 2 & 0 & -1 & 1 \end{vmatrix}$

B 31. Show that $\begin{vmatrix} 1 & a & a^2 \\ 1 & b & b^2 \\ 1 & c & c^2 \end{vmatrix} = (b-c)(c-a)(a-b)$.

32. Show that $\begin{vmatrix} a_{11} & a_{12} & a_{13} & a_{14} \\ a_{21} & a_{22} & a_{23} & a_{24} \\ 0 & 0 & a_{33} & a_{34} \\ 0 & 0 & a_{43} & a_{44} \end{vmatrix} = \begin{vmatrix} a_{11} & a_{12} \\ a_{21} & a_{22} \end{vmatrix} \cdot \begin{vmatrix} a_{33} & a_{34} \\ a_{43} & a_{44} \end{vmatrix}$.

33. Prove Theorem 9.9.

34. Prove Theorem 9.10.

9.6 The Inverse of a Square Matrix

In the field of real numbers, every element a except 0 has a multiplicative inverse $1/a$ with the property that $a \cdot 1/a = 1$. The question should (and does) arise, "Does every square matrix A have a multiplicative inverse A^{-1}?"

Definition 9.17 *For a given square matrix A of order n, if there is a square matrix A^{-1} of order n such that*

$$AA^{-1} = I \quad \text{and} \quad A^{-1}A = I,$$

*where I is the multiplicative identity matrix of order n, then A^{-1} is the **multiplicative inverse** of A.*

Inverse of a 2 × 2 matrix

To answer the question about the existence of a multiplicative inverse for a matrix, we shall begin by considering the simple case of 2 × 2 matrices. If we let

$$A = \begin{bmatrix} a_{11} & a_{12} \\ a_{21} & a_{22} \end{bmatrix},$$

we must see whether or not there exists a 2 × 2 matrix A^{-1} such that $AA^{-1} = I$. If so, let $A^{-1} = \begin{bmatrix} b & c \\ d & e \end{bmatrix}$. We wish to have

$$\begin{bmatrix} a_{11} & a_{12} \\ a_{21} & a_{22} \end{bmatrix} \begin{bmatrix} b & c \\ d & e \end{bmatrix} = \begin{bmatrix} 1 & 0 \\ 0 & 1 \end{bmatrix}.$$

This leads to
$$\begin{bmatrix} a_{11}b + a_{12}d & a_{11}c + a_{12}e \\ a_{21}b + a_{22}d & a_{21}c + a_{22}e \end{bmatrix} = \begin{bmatrix} 1 & 0 \\ 0 & 1 \end{bmatrix},$$
which is true if and only if
$$a_{11}b + a_{12}d = 1, \quad a_{11}c + a_{12}e = 0,$$
$$a_{21}b + a_{22}d = 0, \quad a_{21}c + a_{22}e = 1. \tag{1}$$
Solving these equations for b, c, d, and e, we have
$$(a_{11}a_{22} - a_{12}a_{21})b = a_{22}, \quad (a_{11}a_{22} - a_{12}a_{21})c = -a_{12},$$
$$(a_{11}a_{22} - a_{12}a_{21})d = -a_{21}, \quad (a_{11}a_{22} - a_{12}a_{21})e = a_{11},$$
from which
$$b = \frac{a_{22}}{a_{11}a_{22} - a_{12}a_{21}}, \quad c = \frac{-a_{12}}{a_{11}a_{22} - a_{12}a_{21}},$$
$$d = \frac{-a_{21}}{a_{11}a_{22} - a_{12}a_{21}}, \quad e = \frac{a_{11}}{a_{11}a_{22} - a_{12}a_{21}},$$
provided $a_{11}a_{22} - a_{12}a_{21} \neq 0$. Now the denominator of each of these fractions is just $\delta(A)$, so that
$$A^{-1} = \begin{bmatrix} b & c \\ d & e \end{bmatrix} = \begin{bmatrix} \dfrac{a_{22}}{\delta(A)} & \dfrac{-a_{12}}{\delta(A)} \\ \dfrac{-a_{21}}{\delta(A)} & \dfrac{a_{11}}{\delta(A)} \end{bmatrix} = \frac{1}{\delta(A)} \begin{bmatrix} a_{22} & -a_{12} \\ -a_{21} & a_{11} \end{bmatrix}.$$
By direct multiplication, it can be verified not only that
$$AA^{-1} = I,$$
but also (surprisingly, since matrix multiplication is not always commutative) that
$$A^{-1}A = I.$$
Thus, to write the inverse of a 2×2 square matrix A for which $\delta(A) \neq 0$, we interchange the entries on the principal diagonal, replace each of the other two entries with its negative, and multiply the result by $1/\delta(A)$.

Example If $A = \begin{bmatrix} 1 & 3 \\ 2 & -1 \end{bmatrix}$, find A^{-1}.

Solution We first observe that $\delta(A) = (1)(-1) - (3)(2) = -7$. Hence,
$$A^{-1} = -\frac{1}{7} \begin{bmatrix} -1 & -3 \\ -2 & 1 \end{bmatrix} = \begin{bmatrix} \dfrac{1}{7} & \dfrac{3}{7} \\ \dfrac{2}{7} & -\dfrac{1}{7} \end{bmatrix}.$$

9.6 The Inverse of a Square Matrix

It is a good idea always to check the result when finding A^{-1}, because there is much room for blundering in the process of determining the inverse. In the present example, we have

$$A^{-1}A = -\frac{1}{7}\begin{bmatrix} -1 & -3 \\ -2 & 1 \end{bmatrix}\begin{bmatrix} 1 & 3 \\ 2 & -1 \end{bmatrix} = -\frac{1}{7}\begin{bmatrix} -7 & 0 \\ 0 & -7 \end{bmatrix} = \begin{bmatrix} 1 & 0 \\ 0 & 1 \end{bmatrix}.$$

Matrices with no inverse

We have now arrived at a position where we can answer the question, "Does every 2×2 square matrix A have an inverse?" The answer is "No," for if $\delta(A)$ is 0, then the foregoing Equations (1) for b, c, d, e would have no solution.

Example

The matrix $\begin{bmatrix} 3 & 5 \\ 6 & 10 \end{bmatrix}$ has no inverse because

$$\delta(A) = 3(10) - 6(5) = 0.$$

Inverse of an $n \times n$ matrix

More generally, and without proving it, we have the following result.

Theorem 9.12 If

$$A = \begin{bmatrix} a_{11} & a_{12} & \cdots & a_{1n} \\ a_{21} & a_{22} & \cdots & a_{2n} \\ \vdots & \vdots & & \vdots \\ a_{n1} & a_{n2} & \cdots & a_{nn} \end{bmatrix},$$

and if $\delta(A) \neq 0$, then A has an inverse A^{-1} given by

$$A^{-1} = \frac{1}{\delta(A)}\begin{bmatrix} A_{11} & A_{21} & \cdots & A_{n1} \\ A_{12} & A_{22} & \cdots & A_{n2} \\ \vdots & \vdots & & \vdots \\ A_{1n} & A_{2n} & \cdots & A_{nn} \end{bmatrix},$$

where A_{ij} is the cofactor of a_{ij} in A. If $\delta(A) = 0$, then A has no inverse.

Square matrices A for which $\delta(A) \neq 0$ are *nonsingular* (see page 297) for it can be shown that A is row-equivalent to the identity if and only if $\delta(A) \neq 0$. Thus by Theorem 9.12, A has an inverse if and only if A is nonsingular.

Observe that A^{-1} is $1/\delta(A)$ times the transpose of the matrix obtained by replacing each entry of A with its cofactor.

Example

If $A = \begin{bmatrix} 1 & 0 & 1 \\ 2 & 1 & 0 \\ 1 & -1 & 1 \end{bmatrix}$, find A^{-1}.

Solution on overleaf

Solution   We first observe that $\delta(A) = -2$, and since $\delta(A)$ is not zero, A has an inverse. Next, replacing each entry in A with its cofactor, we obtain the matrix

$$\begin{bmatrix} 1 & -2 & -3 \\ -1 & 0 & 1 \\ -1 & 2 & 1 \end{bmatrix}, \text{ whose transpose is } \begin{bmatrix} 1 & -1 & -1 \\ -2 & 0 & 2 \\ -3 & 1 & 1 \end{bmatrix},$$

so that

$$A^{-1} = -\frac{1}{2}\begin{bmatrix} 1 & -1 & -1 \\ -2 & 0 & 2 \\ -3 & 1 & 1 \end{bmatrix}.$$

As a check, we have

$$A^{-1}A = -\frac{1}{2}\begin{bmatrix} 1 & -1 & -1 \\ -2 & 0 & 2 \\ -3 & 1 & 1 \end{bmatrix}\begin{bmatrix} 1 & 0 & 1 \\ 2 & 1 & 0 \\ 1 & -1 & 1 \end{bmatrix}$$

$$= -\frac{1}{2}\begin{bmatrix} -2 & 0 & 0 \\ 0 & -2 & 0 \\ 0 & 0 & -2 \end{bmatrix} = \begin{bmatrix} 1 & 0 & 0 \\ 0 & 1 & 0 \\ 0 & 0 & 1 \end{bmatrix}.$$

Theorem 9.12 is applicable to $n \times n$ square matrices, although, clearly, the process of actually determining A^{-1} by the formula given in that theorem becomes very laborious for matrices much larger than 3×3.

Elementary transformations   If the inverse of an $n \times n$ matrix A exists, it can be obtained by using elementary transformations. This is the method applied when using a computer to find the inverse of a matrix. To see this, we need the following result which we state without proof.

Theorem 9.13   If A is an $n \times n$ nonsingular matrix and if $[A \mid I]$ is the $n \times 2n$ matrix obtained by adjoining the $n \times n$ identity matrix to A, then

$$[A \mid I] \sim [I \mid A^{-1}].$$

The above theorem is used to compute A^{-1} by using elementary transformations to obtain $[I \mid A^{-1}]$ from $[A \mid I]$.

Example   If $A = \begin{bmatrix} 1 & 2 \\ 2 & 0 \end{bmatrix}$, find A^{-1}.

Solution We first observe that $\delta(A) = -4$, and since $\delta(A)$ is not zero, A has an inverse.

$$[A \mid I] = \begin{bmatrix} 1 & 2 & \vdots & 1 & 0 \\ 2 & 0 & \vdots & 0 & 1 \end{bmatrix}$$

$$\sim \begin{bmatrix} 0 & 2 & \vdots & 1 & -\frac{1}{2} \\ 2 & 0 & \vdots & 0 & 1 \end{bmatrix} \quad \text{Row } 1 + \left[-\frac{1}{2} \times \text{Row } 2 \right]$$

$$\sim \begin{bmatrix} 0 & 1 & \vdots & \frac{1}{2} & -\frac{1}{4} \\ 1 & 0 & \vdots & 0 & \frac{1}{2} \end{bmatrix} \quad \begin{array}{l} \frac{1}{2} \times \text{Row } 1 \\ \frac{1}{2} \times \text{Row } 2 \end{array}$$

$$\sim \begin{bmatrix} 1 & 0 & \vdots & 0 & \frac{1}{2} \\ 0 & 1 & \vdots & \frac{1}{2} & -\frac{1}{4} \end{bmatrix} \quad \begin{array}{l} \text{Interchange} \\ \text{Rows 1 and 2} \end{array}$$

$$= [I \mid A^{-1}]$$

Thus, $A^{-1} = \begin{bmatrix} 0 & \frac{1}{2} \\ \frac{1}{2} & -\frac{1}{4} \end{bmatrix}$.

As a check, we have

$$A^{-1}A = \begin{bmatrix} 0 & \frac{1}{2} \\ \frac{1}{2} & -\frac{1}{4} \end{bmatrix} \begin{bmatrix} 1 & 2 \\ 2 & 0 \end{bmatrix} = \begin{bmatrix} 1 & 0 \\ 0 & 1 \end{bmatrix}.$$

Properties of matrices and their inverses

There are a number of useful properties associated with matrices and their inverses. For example, let A and B be $n \times n$ nonsingular matrices. If we right-multiply AB by $B^{-1}A^{-1}$, and apply Theorem 9.3, we have

$$AB \cdot B^{-1}A^{-1} = A \cdot I \cdot A^{-1} = A \cdot A^{-1} = I.$$

Moreover, if we left-multiply AB by $B^{-1}A^{-1}$, we have

$$B^{-1}A^{-1} \cdot AB = B^{-1} \cdot I \cdot B = B^{-1} \cdot B = I.$$

Thus, since $(AB)(B^{-1}A^{-1}) = (B^{-1}A^{-1})(AB) = I$, by the definition of the inverse of a matrix we have

$$(AB)^{-1} = B^{-1}A^{-1}.$$

This proves the following theorem.

Theorem 9.14 *If A and B are $n \times n$ nonsingular square matrices, then AB has an inverse, namely*

$$(AB)^{-1} = B^{-1}A^{-1}.$$

This theorem can be used to find the inverse of products of any number of nonsingular matrices. For example, if there are three factors A, B, and C in a product,

$$(ABC)^{-1} = [(AB)C]^{-1} = C^{-1}(AB)^{-1} = C^{-1}B^{-1}A^{-1}.$$

Exercise 9.6

A Find the inverse of each matrix if one exists.

Example

$$B = \begin{bmatrix} 1 & 0 & -1 \\ 1 & 3 & 1 \\ 0 & 1 & 2 \end{bmatrix}$$

Solution 1 The determinant $\delta(B)$ is given by

$$\delta \begin{bmatrix} 1 & 0 & -1 \\ 1 & 3 & 1 \\ 0 & 1 & 2 \end{bmatrix} = 1(5) - 0 - 1(1) = 4.$$

Replacing each entry of B with its cofactor gives

$$\begin{bmatrix} 5 & -2 & 1 \\ -1 & 2 & -1 \\ 3 & -2 & 3 \end{bmatrix};$$

$$B^{-1} = \frac{1}{\delta(B)} \begin{bmatrix} 5 & -2 & 1 \\ -1 & 2 & -1 \\ 3 & -2 & 3 \end{bmatrix}^t = \frac{1}{4} \begin{bmatrix} 5 & -1 & 3 \\ -2 & 2 & -2 \\ 1 & -1 & 3 \end{bmatrix} = \begin{bmatrix} \frac{5}{4} & -\frac{1}{4} & \frac{3}{4} \\ -\frac{2}{4} & \frac{2}{4} & -\frac{2}{4} \\ \frac{1}{4} & -\frac{1}{4} & \frac{3}{4} \end{bmatrix}.$$

9.6 The Inverse of a Square Matrix

Solution 2

$$\begin{bmatrix} 1 & 0 & -1 & | & 1 & 0 & 0 \\ 1 & 3 & 1 & | & 0 & 1 & 0 \\ 0 & 1 & 2 & | & 0 & 0 & 1 \end{bmatrix}$$

$$\sim \begin{bmatrix} 1 & 0 & -1 & | & 1 & 0 & 0 \\ 0 & 3 & 2 & | & -1 & 1 & 0 \\ 0 & 1 & 2 & | & 0 & 0 & 1 \end{bmatrix} \text{Row 2} + [-1 \times \text{Row 1}]$$

$$\sim \begin{bmatrix} 1 & 0 & -1 & | & 1 & 0 & 0 \\ 0 & 3 & 2 & | & -1 & 1 & 0 \\ 0 & 0 & \frac{4}{3} & | & \frac{1}{3} & -\frac{1}{3} & 1 \end{bmatrix} \text{Row 3} + \left[-\frac{1}{3} \times \text{Row 2} \right]$$

$$\sim \begin{bmatrix} 1 & 0 & 0 & | & \frac{5}{4} & -\frac{1}{4} & \frac{3}{4} \\ 0 & 3 & 0 & | & -\frac{3}{2} & \frac{3}{2} & -\frac{3}{2} \\ 0 & 0 & \frac{4}{3} & | & \frac{1}{3} & -\frac{1}{3} & 1 \end{bmatrix} \begin{matrix} \text{Row 1} + \left[\frac{3}{4} \times \text{Row 3} \right] \\ \text{Row 2} + \left[-\frac{3}{2} \times \text{Row 3} \right] \\ \end{matrix}$$

$$\sim \begin{bmatrix} 1 & 0 & 0 & | & \frac{5}{4} & -\frac{1}{4} & \frac{3}{4} \\ 0 & 1 & 0 & | & -\frac{1}{2} & \frac{1}{2} & -\frac{1}{2} \\ 0 & 0 & 1 & | & \frac{1}{4} & -\frac{1}{4} & \frac{3}{4} \end{bmatrix} \begin{matrix} \frac{1}{3} \times \text{Row 2} \\ \frac{3}{4} \times \text{Row 3} \end{matrix}$$

Thus, $A^{-1} = \begin{bmatrix} \frac{5}{4} & -\frac{1}{4} & \frac{3}{4} \\ -\frac{1}{2} & \frac{1}{2} & -\frac{1}{2} \\ \frac{1}{4} & -\frac{1}{4} & \frac{3}{4} \end{bmatrix}$.

1. $\begin{bmatrix} 1 & 2 \\ 1 & 3 \end{bmatrix}$ 2. $\begin{bmatrix} 3 & 1 \\ 2 & -1 \end{bmatrix}$ 3. $\begin{bmatrix} 2 & -3 \\ 1 & 1 \end{bmatrix}$

4. $\begin{bmatrix} 3 & -2 \\ 2 & 1 \end{bmatrix}$ 5. $\begin{bmatrix} -2 & -1 \\ 4 & 2 \end{bmatrix}$ 6. $\begin{bmatrix} 3 & 1 \\ 9 & 3 \end{bmatrix}$

7. $\begin{bmatrix} 5 & 7 \\ 3 & 4 \end{bmatrix}$ 8. $\begin{bmatrix} 5 & -4 \\ 4 & -3 \end{bmatrix}$ 9. $\begin{bmatrix} 7 & 4 \\ -4 & -2 \end{bmatrix}$

10. $\begin{bmatrix} -9 & 5 \\ -4 & 2 \end{bmatrix}$ 11. $\begin{bmatrix} -2 & -6 \\ -3 & -9 \end{bmatrix}$ 12. $\begin{bmatrix} 21 & 7 \\ 9 & 3 \end{bmatrix}$

13. $\begin{bmatrix} 1 & -1 & 2 \\ 2 & 1 & 3 \\ 0 & 0 & 2 \end{bmatrix}$
14. $\begin{bmatrix} 0 & 4 & 2 \\ 1 & 0 & 2 \\ 0 & -1 & 1 \end{bmatrix}$
15. $\begin{bmatrix} 2 & -1 & 1 \\ 3 & 0 & 1 \\ 2 & 2 & 1 \end{bmatrix}$

16. $\begin{bmatrix} 1 & 2 & 1 \\ 0 & 2 & 1 \\ -2 & 2 & 3 \end{bmatrix}$
17. $\begin{bmatrix} 2 & 1 & 1 \\ 1 & 0 & 2 \\ 4 & 2 & 2 \end{bmatrix}$
18. $\begin{bmatrix} -3 & 1 & -6 \\ 2 & 1 & 4 \\ 2 & 0 & 4 \end{bmatrix}$

19. $\begin{bmatrix} 1 & 2 & -3 \\ 3 & -1 & 0 \\ 5 & 3 & -6 \end{bmatrix}$
20. $\begin{bmatrix} 2 & 4 & -1 \\ 1 & 6 & 2 \\ 5 & 14 & 0 \end{bmatrix}$
21. $\begin{bmatrix} 2 & -1 & -5 \\ 1 & 3 & 4 \\ 0 & 1 & 2 \end{bmatrix}$

22. $\begin{bmatrix} 2 & 1 & -8 \\ 1 & 1 & -2 \\ 1 & 2 & 3 \end{bmatrix}$
23. $\begin{bmatrix} 0 & 0 & 1 \\ 0 & 1 & 0 \\ 1 & 0 & 0 \end{bmatrix}$
24. $\begin{bmatrix} 1 & 0 & 1 \\ 0 & 1 & 0 \\ 1 & 0 & 0 \end{bmatrix}$

25. Verify that

$$\left(\begin{bmatrix} 2 & 3 \\ 1 & -1 \end{bmatrix} \cdot \begin{bmatrix} 0 & 1 \\ 3 & 1 \end{bmatrix}\right)^{-1} = \begin{bmatrix} 0 & 1 \\ 3 & 1 \end{bmatrix}^{-1} \cdot \begin{bmatrix} 2 & 3 \\ 1 & -1 \end{bmatrix}^{-1}.$$

26. Verify that

$$\left(\begin{bmatrix} 1 & 2 \\ -1 & 0 \end{bmatrix} \cdot \begin{bmatrix} 1 & 1 \\ 2 & 0 \end{bmatrix} \cdot \begin{bmatrix} 2 & -1 \\ 0 & 1 \end{bmatrix}\right)^{-1} = \begin{bmatrix} 2 & -1 \\ 0 & 1 \end{bmatrix}^{-1} \cdot \begin{bmatrix} 1 & 1 \\ 2 & 0 \end{bmatrix}^{-1} \cdot \begin{bmatrix} 1 & 2 \\ -1 & 0 \end{bmatrix}^{-1}.$$

27. Verify that

$$\left(\begin{bmatrix} 3 & 0 & 1 \\ 2 & 1 & 0 \\ 0 & 1 & 2 \end{bmatrix} \cdot \begin{bmatrix} 2 & 1 & 0 \\ 1 & 1 & 2 \\ 0 & 1 & 0 \end{bmatrix}\right)^{-1} = \begin{bmatrix} 2 & 1 & 0 \\ 1 & 1 & 2 \\ 0 & 1 & 0 \end{bmatrix}^{-1} \cdot \begin{bmatrix} 3 & 0 & 1 \\ 2 & 1 & 0 \\ 0 & 1 & 2 \end{bmatrix}^{-1}.$$

B

28. Show that $[A^t]^{-1} = [A^{-1}]^t$ for each nonsingular 2×2 matrix.

29. Show that $\delta(A^{-1}) = 1/\delta(A)$ for each nonsingular 2×2 matrix.

30. Prove that if a and b are real numbers, then $\delta(aA^2 + bA) = \delta(aA + bI) \times \delta(A)$ for all 2×2 matrices A.

31. Prove that $\delta(B^{-1}AB) = \delta(A)$ for all nonsingular 2×2 matrices A and B.

32. Prove that if A is a 2×2 matrix and a, b, and c are real numbers, with $c \neq 0$, and if $aA^2 + bA + cI = 0$, then A has an inverse.

9.7 Solution of Linear Systems Using Matrix Inverses

In Section 9.3 we solved linear systems using row-equivalent matrices. The solution for a linear system can also be found by using the inverse of a matrix.

9.7 Solution of Linear Systems Using Matrix Inverses

We first verify the matrix product equation

$$\begin{bmatrix} a_{11} & a_{12} & \cdots & a_{1n} \\ \vdots & \vdots & & \vdots \\ a_{n1} & a_{n2} & \cdots & a_{nn} \end{bmatrix} \begin{bmatrix} x_1 \\ \vdots \\ x_n \end{bmatrix} = \begin{bmatrix} a_{11}x_1 + a_{12}x_2 + \cdots + a_{1n}x_n \\ \vdots \\ a_{n1}x_1 + a_{n2}x_2 + \cdots + a_{nn}x_n \end{bmatrix},$$

and hence note that the linear system

$$\begin{aligned} a_{11}x_1 + a_{12}x_2 + \cdots + a_{1n}x_n &= c_1 \\ a_{21}x_1 + a_{22}x_2 + \cdots + a_{2n}x_n &= c_2 \\ \vdots \quad \vdots \quad \quad \vdots \quad \quad \vdots& \\ a_{n1}x_1 + a_{n2}x_2 + \cdots + a_{nn}x_n &= c_n \end{aligned} \qquad (1)$$

can be written as the matrix equation

$$\begin{bmatrix} a_{11} & a_{12} & \cdots & a_{1n} \\ \vdots & \vdots & & \vdots \\ a_{n1} & a_{n2} & \cdots & a_{nn} \end{bmatrix} \begin{bmatrix} x_1 \\ \vdots \\ x_n \end{bmatrix} = \begin{bmatrix} c_1 \\ \vdots \\ c_n \end{bmatrix},$$

where the first factor in the left-hand member is the coefficient matrix for the system. In more concise notation, this latter equation can be written

$$AX = B,$$

where A is an $n \times n$ square matrix, and X and B are $n \times 1$ column matrices.

Solution of systems

If A in the foregoing equation is nonsingular, we can left-multiply both members of this equation by A^{-1} to obtain the equivalent matrices

$$A^{-1}AX = A^{-1}B,$$
$$IX = A^{-1}B,$$
$$X = A^{-1}B,$$

where $A^{-1}B$ is an $n \times 1$ column matrix. Since X and $A^{-1}B$ are equal, each entry in X is equal to the corresponding entry in $A^{-1}B$, and hence these latter entries constitute the components of the solution of the given linear system. If A is a singular matrix, then of course it has no inverse, and either the system has no solution or the solution is not unique.

Example

Use matrices to find the solution set of

$$\begin{aligned} 2x + y + z &= 1 \\ x - 2y - 3z &= 1 \\ 3x + 2y + 4z &= 5. \end{aligned}$$

Solution on overleaf

Solution We first write this as a matrix equation of the form $AX = B$, thus:

$$\begin{bmatrix} 2 & 1 & 1 \\ 1 & -2 & -3 \\ 3 & 2 & 4 \end{bmatrix} \begin{bmatrix} x \\ y \\ z \end{bmatrix} = \begin{bmatrix} 1 \\ 1 \\ 5 \end{bmatrix}.$$

We next determine $\delta(A)$, obtaining

$$\delta \begin{bmatrix} 2 & 1 & 1 \\ 1 & -2 & -3 \\ 3 & 2 & 4 \end{bmatrix} = 2(-2) - 1(13) + 1(8) = -9.$$

Observing that A is nonsingular, we then find A^{-1} by either of the methods of Section 9.6. We shall use the first one discussed.

$$A^{-1} = \begin{bmatrix} 2 & 1 & 1 \\ 1 & -2 & -3 \\ 3 & 2 & 4 \end{bmatrix}^{-1} = \frac{1}{9}\begin{bmatrix} -2 & -2 & -1 \\ -13 & 5 & 7 \\ 8 & -1 & -5 \end{bmatrix}$$

As a matter of routine, we check the latter by verifying that $A^{-1}A = I$.

$$A^{-1}A = -\frac{1}{9}\begin{bmatrix} -2 & -2 & -1 \\ -13 & 5 & 7 \\ 8 & -1 & -5 \end{bmatrix}\begin{bmatrix} 2 & 1 & 1 \\ 1 & -2 & -3 \\ 3 & 2 & 4 \end{bmatrix}$$

$$= -\frac{1}{9}\begin{bmatrix} -9 & 0 & 0 \\ 0 & -9 & 0 \\ 0 & 0 & -9 \end{bmatrix} = \begin{bmatrix} 1 & 0 & 0 \\ 0 & 1 & 0 \\ 0 & 0 & 1 \end{bmatrix}$$

Now, since $X = A^{-1}B$, we have

$$\begin{bmatrix} x \\ y \\ z \end{bmatrix} = -\frac{1}{9}\begin{bmatrix} -2 & -2 & -1 \\ -13 & 5 & 7 \\ 8 & -1 & -5 \end{bmatrix}\begin{bmatrix} 1 \\ 1 \\ 5 \end{bmatrix} = -\frac{1}{9}\begin{bmatrix} -9 \\ 27 \\ -18 \end{bmatrix} = \begin{bmatrix} 1 \\ -3 \\ 2 \end{bmatrix}.$$

Hence, $x = 1$, $y = -3$, and $z = 2$, and the solution set is $\{(1, -3, 2)\}$.

The computation of A^{-1} is laborious when A is a square matrix containing many rows and columns. The foregoing method is not always the easiest to use in solving systems. It is, however, very useful when solving several systems of the form $AX = B$ having the same coefficient matrix A.

Example Solve.

a. $\begin{bmatrix} 2 & 1 & 1 \\ 1 & -2 & -3 \\ 3 & 2 & 4 \end{bmatrix}\begin{bmatrix} x \\ y \\ z \end{bmatrix} = \begin{bmatrix} 8 \\ 5 \\ 10 \end{bmatrix}$ b. $\begin{bmatrix} 2 & 1 & 1 \\ 1 & -2 & -3 \\ 3 & 2 & 4 \end{bmatrix}\begin{bmatrix} x \\ y \\ z \end{bmatrix} = \begin{bmatrix} 0 \\ 10 \\ -11 \end{bmatrix}$

9.7 Solution of Linear Systems Using Matrix Inverses

Solution In both examples we have $A = \begin{bmatrix} 2 & 1 & 1 \\ 1 & -2 & -3 \\ 3 & 2 & 4 \end{bmatrix}$, and from the previous example,

$$A^{-1} = -\frac{1}{9}\begin{bmatrix} -2 & -2 & -1 \\ -13 & 5 & 7 \\ 8 & -1 & -5 \end{bmatrix}.$$

a. The solution is

$$\begin{bmatrix} x \\ y \\ z \end{bmatrix} = A^{-1}\begin{bmatrix} 8 \\ 5 \\ 10 \end{bmatrix} = \begin{bmatrix} 4 \\ 1 \\ -1 \end{bmatrix}.$$

b. The solution is

$$\begin{bmatrix} x \\ y \\ z \end{bmatrix} = A^{-1}\begin{bmatrix} 0 \\ 10 \\ -11 \end{bmatrix} = \begin{bmatrix} 1 \\ 3 \\ -5 \end{bmatrix}.$$

Exercise 9.7

Find the solution set of the given system by using matrices. If the system has no unique solution, so state.

1. $2x - 3y = -1$
   $x + 4y = 5$

2. $3x - 4y = -2$
   $x - 2y = 0$

3. $3x + 6y = -2$
   $6x + 12y = 36$

4. $2x - 4y = 7$
   $x - 2y = 1$

5. $2x - 3y = 0$
   $2x + y = 16$

6. $2x + 3y = 3$
   $3x - 4y = 0$

7. $3x + y = -5$
   $2x - 4y = -16$

8. $2x - 3y = -8$
   $x - 4y = -9$

9. $3x + 9y = 2$
   $6x + 18y = 4$

10. $2x + y = 1$
    $4x + 2y = 2$

11. $x - 4y = -6$
    $4x - y = 6$

12. $2x + 3y = 3$
    $3x - 2y = -2$

13. $x + y = 2$
    $2x - z = 1$
    $2y - 3z = -1$

14. $2x - 6y + 3z = -12$
    $3x - 2y + 5z = -4$
    $4x + 5y - 2z = 10$

15. $x - 2y + z = -1$
    $3x + y - 2z = 4$
    $y - z = 1$

16. $2x + 5z = 9$
    $4x + 3y = -1$
    $3y - 4z = -13$

17. $2x + 2y + z = 1$
    $x - y + 6z = 21$
    $3x + 2y - z = -4$

18. $4x + 8y + z = -6$
    $2x - 3y + 2z = 0$
    $x + 7y - 3z = -8$

19. $x + y + z = 0$
    $2x - y - 4z = 15$
    $x - 2y - z = 7$

20. $x + y - 2z = 3$
    $3x - y + z = 5$
    $3x + 3y - 6z = 9$

21. Find the solution set of each system.

   a. $x + y + z = 3$
      $2x - y - 4z = 4$
      $x - 2y - z = -1$

   b. $x + y + z = -2$
      $2x - y - 4z = 1$
      $x - 2y - z = 0$

   c. $x + y + z = 1$
      $2x - y - 4z = 0$
      $x - 2y - z = 1$

22. Find the solution set of each system.

   a. $2x + 5z = 1$
      $4x + 3y = 1$
      $3y - 4z = 1$

   b. $2x + 5z = 2$
      $4x + 3y = -1$
      $3y - 4z = 1$

   c. $2x + 5z = 0$
      $4x + 3y = 2$
      $3y - 4z = 1$

9.8 Cramer's Rule

In Section 9.7, we obtained the solution set of the linear system (1) on page 323 with nonsingular coefficient matrix by first expressing the system in the matrix form $AX = B$ and then left-multiplying both members of the equation by A^{-1} to obtain

$$A^{-1}AX = X = A^{-1}B.$$

If, now, this technique is viewed in terms of determinants, we arrive at a general solution for such systems.

If the coefficient matrix A is nonsingular, then its inverse, A^{-1}, is

$$A^{-1} = \frac{1}{\delta(A)} \begin{bmatrix} A_{11} & A_{21} & \cdots & A_{n1} \\ \vdots & \vdots & & \vdots \\ A_{1n} & A_{2n} & \cdots & A_{nn} \end{bmatrix}.$$

Now, since $B = \begin{bmatrix} c_1 \\ c_2 \\ \vdots \\ c_n \end{bmatrix}$, we have

$$X = A^{-1}B = \frac{1}{\delta(A)} \begin{bmatrix} c_1 A_{11} + c_2 A_{21} + \cdots + c_n A_{n1} \\ c_1 A_{12} + c_2 A_{22} + \cdots + c_n A_{n2} \\ \vdots \\ c_1 A_{1n} + c_2 A_{2n} + \cdots + c_n A_{nn} \end{bmatrix}.$$

Each entry in $X = A^{-1}B$ can be seen to be of the form

$$\frac{c_1 A_{1j} + c_2 A_{2j} + \cdots + c_n A_{nj}}{\delta(A)}.$$

9.8 Cramer's Rule

But $c_1 A_{1j} + c_2 A_{2j} + \cdots + c_n A_{nj}$ is just the expansion of the determinant

$$\begin{array}{c} \text{jth} \\ \text{column} \\ \downarrow \end{array}$$

$$\begin{vmatrix} a_{11} & a_{12} & \cdots & c_1 & \cdots & a_{1n} \\ a_{21} & a_{22} & \cdots & c_2 & \cdots & a_{2n} \\ \vdots & \vdots & & \vdots & & \vdots \\ a_{n1} & a_{n2} & \cdots & c_n & \cdots & a_{nn} \end{vmatrix}$$

about the jth column, which has entries $c_1, c_2, \ldots, c_n$ in place of $a_{1j}, a_{2j}, \ldots, a_{nj}$.

Thus, each entry x_j in the matrix $X = \begin{bmatrix} x_1 \\ x_2 \\ \vdots \\ x_n \end{bmatrix} = A^{-1}B$ is

$$x_j = \frac{\delta(A_j)}{\delta(A)} = \frac{\begin{vmatrix} a_{11} & a_{12} & \cdots & c_1 & \cdots & a_{1n} \\ a_{21} & a_{22} & \cdots & c_2 & \cdots & a_{2n} \\ \vdots & \vdots & & \vdots & & \vdots \\ a_{n1} & a_{n2} & \cdots & c_n & \cdots & a_{nn} \end{vmatrix}}{\begin{vmatrix} a_{11} & a_{12} & \cdots & & \cdots & a_{1n} \\ a_{21} & a_{22} & \cdots & & \cdots & a_{2n} \\ \vdots & \vdots & & & & \vdots \\ a_{n1} & a_{n2} & \cdots & & \cdots & a_{nn} \end{vmatrix}}.$$

Application of Cramer's rule

This relationship expresses **Cramer's rule**. Cramer's rule is the assertion that if the determinant of the coefficient matrix of an $n \times n$ linear system *is not* 0, then the equations are consistent (the system has a solution) and have a unique solution which can be found as follows.

To find x_j in solving the matrix equation $AX = B$:

1. Write the determinant of the coefficient matrix for the system.
2. Replace each entry in the jth column of the coefficient matrix A with the corresponding entry from the column matrix B, and find the determinant of the resulting matrix.
3. Divide the result in Step 2 by the result in Step 1.

Hence, in a nonsingular 3×3 system:

$$x = \frac{\delta(A_x)}{\delta(A)}, \quad y = \frac{\delta(A_y)}{\delta(A)}, \quad \text{and} \quad z = \frac{\delta(A_z)}{\delta(A)}.$$

Example Use Cramer's rule to solve the system
$$-4x + 2y - 9z = 2$$
$$3x + 4y + z = 5$$
$$x - 3y + 2z = 8.$$

Solution By inspection,
$$\delta(A) = \begin{vmatrix} -4 & 2 & -9 \\ 3 & 4 & 1 \\ 1 & -3 & 2 \end{vmatrix}$$
$$= -4(11) - 2(5) - 9(-13)$$
$$= -44 - 10 + 117 = 63.$$

Replacing the entries in the first column of A with corresponding constants 2, 5, and 8, we have
$$\delta(A_x) = \begin{vmatrix} 2 & 2 & -9 \\ 5 & 4 & 1 \\ 8 & -3 & 2 \end{vmatrix}$$
$$= 2(11) - 2(2) - 9(-47)$$
$$= 22 - 4 + 423 = 441.$$

Hence, by Cramer's rule,
$$x = \frac{\delta(A_x)}{\delta(A)} = \frac{441}{63} = 7.$$

Similarly, by replacing, in turn, the entries of the second and third columns of A with the corresponding constants, 2, 5, and 8, we have
$$\delta(A_y) = \begin{vmatrix} -4 & 2 & -9 \\ 3 & 5 & 1 \\ 1 & 8 & 2 \end{vmatrix} \text{ and } \delta(A_z) = \begin{vmatrix} -4 & 2 & 2 \\ 3 & 4 & 5 \\ 1 & -3 & 8 \end{vmatrix}.$$

Now,
$$\delta(A_y) = -4(2) - 2(5) - 9(19)$$
$$= -8 - 10 - 171 = -189$$

and
$$\delta(A_z) = -4(47) - 2(19) + 2(-13)$$
$$= -188 - 38 - 26 = -252,$$

so that
$$y = \frac{\delta(A_y)}{\delta(A)} = \frac{-189}{63} = -3$$

9.8 Cramer's Rule

and

$$z = \frac{\delta(A_z)}{\delta(A)} = \frac{-252}{63} = -4,$$

and the solution set of the system is $\{(7, -3, -4)\}$.

If $\delta(A) = 0$ for a linear system, then the system either has infinitely many solutions (the equations are consistent and one of them can be obtained from the others by linear combinations) or has no solutions (the equations are inconsistent). The distinction can be determined using methods from Section 9.3.

Exercise 9.8

A *Find the solution set of each of the following systems by Cramer's rule. If $\delta(A) = 0$ in any of the systems, use the methods in Section 9.3 to determine whether or not the equations in the system are consistent.*

1. $x - y = 2$
   $x + 4y = 5$

2. $x + y = 4$
   $x - 2y = 0$

3. $3x - 4y = -2$
   $x + y = 6$

4. $2x - 4y = 7$
   $x - 2y = 1$

5. $\frac{1}{3}x - \frac{1}{2}y = 0$
   $\frac{1}{2}x + \frac{1}{4}y = 4$

6. $\frac{2}{3}x + y = 1$
   $x - \frac{4}{3}y = 0$

7. $x - 2y = 6$
   $\frac{2}{3}x - \frac{4}{3}y = 6$

8. $\frac{1}{2}x + y = 3$
   $-\frac{1}{4}x - y = -3$

9. $x - 3y = 1$
   $y = 1$

10. $2x - 3y = 12$
    $x = 4$

11. $ax + by = 1$
    $bx + ay = 1$

12. $x + y = a$
    $x - y = b$

13. $x - 2y + z = -1$
    $3x + y - 2z = 4$
    $y - z = 1$

14. $2x + 5z = 9$
    $4x + 3y = -1$
    $3y - 4z = -13$

15. $2x + 2y + z = 1$
    $x - y + 6z = 21$
    $3x + 2y - z = -4$

16. $4x + 8y + z = -6$
    $2x - 3y + 2z = 0$
    $x + 7y - 3z = -8$

17. $x + y + z = 0$
    $2x - y - 4z = 15$
    $x - 2y - z = 7$

18. $x + y - 2z = 2$
    $3x - y + z = 5$
    $3x + 3y - 6z = 6$

19. $x - 2y - 2z = 3$
    $2x - 4y + 4z = 1$
    $3x - 3y - 3z = 4$

20. $3x - 2y + 5z = 6$
    $4x - 4y + 3z = 0$
    $5x - 4y + z = -5$

21. $x - 4z = -1$
$3x + 3y = 2$
$3x + 4z = 5$

22. $2x - \frac{2}{3}y + z = 2$
$6x - 4y - 3z = 0$
$4x + 5y - 3z = -1$

23. $x + y + z = 0$
$w + 2y - z = 4$
$2w - y + 2z = 3$
$-2w + 2y - z = -2$

24. $x + y + z = 0$
$x + z + w = 0$
$x + y + w = 0$
$y + z + w = 0$

B 25. For the system
$$a_1 x + b_1 y + c_1 = 0$$
$$a_2 x + b_2 y + c_2 = 0,$$
show that if both $\delta(A_y) = 0$ and $\delta(A_x) = 0$, and if c_1 and c_2 are not both 0, then $\delta(A) = 0$, and the equations are consistent. *Hint*: Show that the first two determinant equations imply that $a_1 c_2 = a_2 c_1$ and $b_1 c_2 = b_2 c_1$ and that the rest follows from the formation of a proportion with these equations.

26. Show that if $\delta(A) = 0$ and $\delta(A_x) = 0$, and if a_1 and a_2 are not both 0, then $\delta(A_y) = 0$, where $\delta(A)$ is the determinant of the coefficient matrix of the system in Exercise 25.

Chapter Review

[9.1] *Write each sum or difference as a single matrix.*

1. $\begin{bmatrix} 4 & -7 \\ 2 & 1 \end{bmatrix} + \begin{bmatrix} -3 & 6 \\ -1 & 0 \end{bmatrix}$

2. $\begin{bmatrix} 3 & -1 & 7 \\ 6 & 2 & 5 \end{bmatrix} + \begin{bmatrix} -1 & 6 & -9 \\ 8 & -3 & 7 \end{bmatrix}$

3. $\begin{bmatrix} 2 & -4 & 3 \\ 6 & 1 & 7 \\ 2 & 8 & 0 \end{bmatrix} - \begin{bmatrix} 4 & -1 & 2 \\ 3 & 8 & 1 \\ 7 & 6 & -5 \end{bmatrix}$

4. $\begin{bmatrix} -11 & 2 & -6 \\ 7 & 1 & 2 \\ -3 & 4 & 8 \end{bmatrix} - \begin{bmatrix} -3 & 5 & 0 \\ 1 & 4 & 2 \\ 6 & -1 & 3 \end{bmatrix}$

[9.2] *Write each product as a single matrix.*

5. $-7 \begin{bmatrix} 3 & -1 \\ 2 & 0 \\ 1 & 1 \end{bmatrix}$

6. $\begin{bmatrix} 3 & -1 & 2 \end{bmatrix} \begin{bmatrix} 4 \\ -1 \\ 0 \end{bmatrix}$

7. $\begin{bmatrix} 3 & -1 \\ 6 & 5 \end{bmatrix} \cdot \begin{bmatrix} -4 & 2 \\ 1 & 3 \end{bmatrix}$

8. $\begin{bmatrix} -1 & 7 & 6 \\ 3 & 1 & 2 \\ 1 & 0 & 1 \end{bmatrix} \cdot \begin{bmatrix} 1 & -1 & 2 \\ 1 & 0 & 3 \\ 2 & 1 & 1 \end{bmatrix}$

Review Exercises

[9.3] Use row transformations on the augmented matrix to solve each system of equations.

9. $2x - y = 5$
   $x + 3y = -1$

10. $2x - y + z = 4$
    $x + 3y - z = 4$
    $x + 2y + z = 5$

[9.4] Evaluate each determinant.

11. $\begin{vmatrix} -3 & 0 \\ 2 & 1 \end{vmatrix}$

12. $\begin{vmatrix} 1 & 5 \\ -1 & 2 \end{vmatrix}$

13. $\begin{vmatrix} 3 & 1 & 0 \\ 2 & 0 & 1 \\ 1 & 2 & -1 \end{vmatrix}$

14. $\begin{vmatrix} 3 & 1 & -2 \\ -1 & 2 & 1 \\ 1 & -2 & 1 \end{vmatrix}$

15. $\begin{vmatrix} 3 & 0 & 1 & 1 \\ 0 & 2 & -1 & 0 \\ 0 & 1 & 0 & 2 \\ 1 & 0 & 0 & 1 \end{vmatrix}$

16. $\begin{vmatrix} -2 & 1 & 4 & 0 \\ 2 & 0 & 4 & 1 \\ 1 & 1 & 0 & 0 \\ 2 & -1 & 3 & 1 \end{vmatrix}$

[9.5] Reduce each determinant to an equal 2×2 determinant and evaluate.

17. $\begin{vmatrix} 3 & -1 & 2 \\ 1 & -2 & 0 \\ 2 & 1 & -1 \end{vmatrix}$

18. $\begin{vmatrix} 1 & 7 & -11 \\ 12 & 10 & 15 \\ 5 & 11 & 14 \end{vmatrix}$

[9.6] Find the inverse of each nonsingular matrix.

19. $\begin{bmatrix} -4 & 2 \\ 11 & 3 \end{bmatrix}$

20. $\begin{bmatrix} 1 & -1 & 2 \\ 3 & 1 & 0 \\ 2 & 1 & 1 \end{bmatrix}$

[9.7] Use matrices to solve each system.

21. $x - y = -3$
    $2x + 3y = -1$

22. $2x + z = 7$
    $y + 2z = 1$
    $3x + y + z = 9$

23. a. $x - y = 1$
       $3x + y = 1$
    b. $x - y = 2$
       $3x + y = -1$
    c. $x - y = 3$
       $3x + y = 1$

[9.8] Use Cramer's rule to solve each system.

24. $3x - y = -5$
    $x + 2y = -6$

25. $x - y + 2z = 3$
    $2x + y - z = 3$
    $x - 2y + 2z = 4$

Supplemental Exercises for Chapters 8–9

Each of the exercises in this set is more difficult than those at the end of each section and may require the use of several of the techniques learned in the preceding chapters.

1. Find constants a, b, and c so that
$$\frac{2x-3}{x^3-2x^2+4x} = \frac{a}{x} + \frac{bx+c}{x^2-2x+4}.$$

2. Find constants a, b, and c so that
$$\frac{3x-2}{x^3-5x^2+9x-6} = \frac{a}{x-2} + \frac{bx+c}{x^2-3x+3}.$$

3. Find constants a, b, c, and d so that
$$\frac{3x^3-x-2}{x^4+x^3+x^2} = \frac{a}{x} + \frac{b}{x^2} + \frac{cx+d}{x^2+x+1}.$$

4. Find constants a, b, c, d and e so that
$$\frac{5x^4-19x^3+36x^2-33x+13}{(x-1)(x^2-2x+2)^2} = \frac{a}{x-1} + \frac{bx+c}{x^2-2x+2} + \frac{dx+e}{(x^2-2x+2)^2}.$$

Solve each system of equations.

5. $2x + 3y + z + w = 2$
   $x + 2z - w = -4$
   $x + y + 2z + w = 0$
   $2x - y + 2z + 2w = -24$

6. $x + 3y + z - 2w = 4$
   $2x - y + 2z + w = 10$
   $ - y + 2z - 3w = 10$
   $x + 2z + w = 7$

7. $\dfrac{1}{x^2} + \dfrac{3}{y} + z = 2$

   $\dfrac{2}{x^2} - \dfrac{1}{y} + z = 7$

   $\dfrac{2}{x^2} + \dfrac{1}{y} + 2z = 9$

8. $\dfrac{2}{x^2} - y^2 + 2z = 0$

   $\dfrac{4}{x^2} + 4y^2 - 8z = 3$

   $\dfrac{1}{x^2} + 2y^2 - 3z = \dfrac{5}{2}$

9. $2x + 3y - z = 1$

   $x + 2y - 3z = 1$

10. $-x + 2y = 4$

    $2y - z = 1$

11. Find all complex numbers $z = a + bi$ with the property that $z^2 = i$.

12. Show that the determinant of a 4×4 diagonal matrix is the product of the entries on the principal diagonal. Does this result generalize to $n \times n$ diagonal matrices?

13. We say that a square matrix A is **upper triangular** if $a_{ij} = 0$ for $i > j$. Show that the determinant of a 4×4 upper-triangular matrix is the product of the entries on the principal diagonal. Does this result generalize to $n \times n$ upper-triangular matrices?

14. Show that if A is a 3×3 nonsingular upper-triangular matrix, then A^{-1} is an upper-triangular matrix. Does this result generalize to $n \times n$ matrices?

15. State conditions under which

    $$\begin{bmatrix} a_1 & 0 & 0 & 0 \\ 0 & a_2 & 0 & 0 \\ 0 & 0 & a_3 & 0 \\ 0 & 0 & 0 & a_4 \end{bmatrix}$$

    is nonsingular, and find the inverse assuming those conditions are satisfied. Does this result generalize to $n \times n$ diagonal matrices?

16. Is it possible to find a column matrix A and a row matrix B so that $AB = I_{2 \times 2}$? Why or why not?

17. Show that if A and B are $n \times n$ matrices with $\delta(AB) = \delta(A)\delta(B)$ and $AB = I$, then $BA = I$.

18. Assume $\delta(AB) = \delta(A) \cdot \delta(B)$ and show that if $A_{4 \times 4}$ is nonsingular, then

    $$\delta(A_{4 \times 4}^{-1}) = \dfrac{1}{\delta(A_{4 \times 4})}.$$

10 Sequences and Series

A function with domain $\{1, 2, 3, \ldots\}$ will relate some object with 1, some object with 2, and so forth. Thus, such a function puts objects in sequence: something goes first, second, and so on. The notion of a sequence is formalized in this chapter. The chapter concludes with a discussion of mathematical induction, a most useful technique for proving results about sequences.

10.1 Sequences

Let us consider a class of functions in which each function has as its domain either the set N of positive integers or a subset of successive members of N.

Definition 10.1 A **sequence function** is a function having as its domain the set N of positive integers $1, 2, 3, \ldots$. A **finite-sequence function** has as its domain the set of positive integers $1, 2, 3, \ldots, n$, for some fixed n.

For example, the function defined by

$$s(n) = n + 3, \quad n \in \{1, 2, 3, \ldots\}, \tag{1}$$

is a sequence function. The elements in the range of such a function, considered in the order

$$s(1), s(2), s(3), s(4), \ldots,$$

are said to form a **sequence**. Similarly, the elements of a finite-sequence function, considered in order, constitute a **finite sequence**.

10.1 Sequences

For example, the sequence associated with (1) is found by successively substituting the numbers 1, 2, 3, ..., for n:

$$s(1) = 1 + 3 = 4,$$
$$s(2) = 2 + 3 = 5,$$
$$s(3) = 3 + 3 = 6,$$
$$s(4) = 4 + 3 = 7,$$

and so on. Thus, the first four terms of (1) are 4, 5, 6, and 7. The nth term, commonly called the **general term**, is $n + 3$.

Examples Find the first three terms and the twenty-fifth term of the given sequence.

a. $s(n) = n^2$ **b.** $s(n) = \dfrac{1}{n}$

Solutions

a. $s(1) = 1^2 = 1$
$s(2) = 2^2 = 4$
$s(3) = 3^2 = 9$
$s(25) = 25^2 = 625$

b. $s(1) = \dfrac{1}{1} = 1$
$s(2) = \dfrac{1}{2}$
$s(3) = \dfrac{1}{3}$
$s(25) = \dfrac{1}{25}$

Given several terms in a sequence, we are often able to construct an expression for a general term of a sequence to which they belong. Such a sequence is not unique. Thus, if the first three terms in a sequence are 2, 4, 6, ..., we may *surmise* that the general term is $s(n) = 2n$. Note, however, that the sequences for both

$$s(n) = 2n$$

and

$$s(n) = 2n + (n-1)(n-2)(n-3)$$

start with 2, 4, 6, but that the two sequences differ for terms following the third.

Sequence notation

The notation ordinarily used for the terms in a sequence is not function notation as such. It is customary to denote a term in a sequence by means of a subscript. Thus, the sequence $s(1), s(2), s(3), s(4), \ldots$ would appear as $s_1, s_2, s_3, s_4, \ldots$. For example, instead of writing $s(n) = 1/n$, we write $s_n = 1/n$; the first three terms of this sequence are $s_1 = 1$, $s_2 = 1/2$, and $s_3 = 1/3$.

Arithmetic progressions

Let us next consider two special kinds of sequences that have many applications. The first kind can be defined as follows:

Definition 10.2   An **arithmetic progression** is a sequence defined by equations of the form
$$s_1 = a, \quad s_{n+1} = s_n + d,$$
where $a, d \in R$, and $n \in N$.

Since each term in such a sequence is obtained from the preceding term by adding d, d is called the **common difference.**

Examples

Find the first four terms of the arithmetic progression.

  a. $s_1 = 1, \; d = 3$ **b.** $s_1 = -3, \; d = 2$

Solutions

a. $s_1 = 1$
$s_2 = s_1 + 3 = 4$
$s_3 = s_2 + 3 = 7$
$s_4 = s_3 + 3 = 10;$
1, 4, 7, 10

b. $s_1 = -3$
$s_2 = s_1 + 2 = -1$
$s_3 = s_2 + 2 = 1$
$s_4 = s_3 + 2 = 3;$
$-3, -1, 1, 3$

In Definition 10.2, each term of the sequence, after the first term, is defined by its relation to previous terms. Such a formulation is called a **recursive** definition. To obtain the general term of an arithmetic progression with $s_1 = a$, observe that since the sequence progresses from term to term by adding the common difference, any term is obtained by adding an appropriate number of differences to a. The second term requires one difference, the third requires two differences and, in general, s_n requires $n - 1$ differences added to a. This indicates that the following theorem is true. A complete proof requires the technique of mathematical induction which is considered in Section 10.5.

Theorem 10.1   The nth term in the sequence defined by
$$s_1 = a, \quad s_{n+1} = s_n + d,$$
where $a, d \in R$, and $n \in N$, is
$$s_n = a + (n-1)d. \tag{2}$$

Examples

Find the general term and the one-hundredth term of the arithmetic progression.

  a. $s_1 = 4, \; d = 5$ **b.** $s_1 = -3, \; d = 4$

Solutions

a. $s_n = 4 + 5(n-1) = 5n - 1;$
$s_{100} = 5(100) - 1 = 499$

b. $s_n = -3 + 4(n-1) = 4n - 7;$
$s_{100} = 4(100) - 7 = 393$

Geometric progressions

The second kind of sequence we shall consider can be defined as follows:

Definition 10.3 *A **geometric progression** is a sequence defined by equations of the form*

$$s_1 = a, \quad s_{n+1} = rs_n,$$

where $a, r \in R$, $a, r \neq 0$, and $n \in N$.

Thus, 3, 9, 27, 81, ... is a geometric progression in which each term except the first is obtained by multiplying the preceding term by 3. Since the effect of multiplying the terms in this way is to produce a fixed ratio between any two successive terms, the multiplier, r, is called the **common ratio**.

Examples   Find the first four terms in the geometric progression.

   a. $s_1 = -3$, $r = 2$ **b.** $s_1 = 2$, $r = -1$

Solutions

   a. $s_1 = -3$
   $s_2 = rs_1 = -6$
   $s_3 = rs_2 = -12$
   $s_4 = rs_3 = -24$;
   $-3, -6, -12, -24$

   b. $s_1 = 2$
   $s_2 = rs_1 = -2$
   $s_3 = rs_2 = 2$
   $s_4 = rs_3 = -2$;
   $2, -2, 2, -2$

Definition 10.3 is a recursive definition of a geometric progression. To obtain the general term of a geometric progression with $s_1 = a$, observe that the sequence progresses from term to term by multiplying by the common ratio. Thus any term is obtained by multiplying by the common ratio an appropriate number of times. The second term requires one factor of the common ratio, the third requires two factors and in general the nth term, s_n, requires $n - 1$ factors of the common ratio multiplied by a. This indicates that the following theorem is true. The proof by mathematical induction is left as an exercise in Section 10.5.

Theorem 10.2   *The nth term in the sequence defined by*

$$s_1 = a, \quad s_{n+1} = rs_n,$$

where $a, r \in R$, $a \neq 0$, $r \neq 0$, and $n \in N$, is

$$s_n = ar^{n-1}.$$

Examples Find the general term and the ninth term of the geometric progression.

    a. $s_1 = 3, r = 2$ **b.** $s_1 = 1, r = -3$

Solutions

    a. $s_n = 3(2)^{n-1}$;
        $s_9 = 3(2^8) = 768$

    b. $s_n = 1(-3)^{n-1}$;
        $s_9 = (-3)^8 = 6561$

Exercise 10.1

A Find the first four terms in the sequence with the general term as given.

Examples

    a. $s_n = \dfrac{n(n+1)}{2}$ **b.** $s_n = (-1)^n 2^n$

Solutions

    a. $s_1 = \dfrac{1(1+1)}{2} = 1$

        $s_2 = \dfrac{2(2+1)}{2} = 3$

        $s_3 = \dfrac{3(3+1)}{2} = 6$

        $s_4 = \dfrac{4(4+1)}{2} = 10$;

        1, 3, 6, 10

    b. $s_1 = (-1)^1 2^1 = -2$

        $s_2 = (-1)^2 2^2 = 4$

        $s_3 = (-1)^3 2^3 = -8$

        $s_4 = (-1)^4 2^4 = 16$;

        $-2, 4, -8, 16$

1. $s_n = n - 5$ 2. $s_n = 2n - 3$ 3. $s_n = \dfrac{n^2 - 2}{2}$

4. $s_n = \dfrac{3}{n^2 + 1}$ 5. $s_n = 1 + \dfrac{1}{n}$ 6. $s_n = \dfrac{n}{2n - 1}$

7. $s_n = \dfrac{n(n-1)}{2}$ 8. $s_n = \dfrac{5}{n(n+1)}$ 9. $s_n = (-1)^n$

10. $s_n = (-1)^{n+1}$ 11. $s_n = \dfrac{(-1)^n (n-2)}{n}$ 12. $s_n = (-1)^{n-1} 3^{n+1}$

Write the next three terms in each of the following arithmetic progressions.

Examples

    a. 5, 9, ...   **b.** $x, x - a, \ldots$

Solutions Find the common difference and then continue the sequence.

    a. $d = 9 - 5 = 4$;
        13, 17, 21

    b. $d = (x - a) - x = -a$;
        $x - 2a, x - 3a, x - 4a$

10.1 Sequences

13. $3, 7, \ldots$
14. $-6, -1, \ldots$
15. $-4, -1, \ldots$
16. $-8, 5, \ldots$
17. $x, x + 1, \ldots$
18. $a, a + 5, \ldots$
19. $2x + 1, 2x + 4, \ldots$
20. $3a, 5a, \ldots$

Write the next four terms in each of the following geometric progressions.

Examples

a. $3, 6, \ldots$

b. $x, 2, \ldots$

Solutions

Find the common ratio, and then continue the sequence.

a. $r = \dfrac{6}{3} = 2$;

  $12, 24, 48, 96$

b. $r = \dfrac{2}{x}$ $(x \neq 0)$;

  $\dfrac{4}{x}, \dfrac{8}{x^2}, \dfrac{16}{x^3}, \dfrac{32}{x^4}$

21. $2, 8, \ldots$
22. $4, 8, \ldots$
23. $\dfrac{2}{3}, \dfrac{4}{3}, \ldots$
24. $\dfrac{1}{2}, -\dfrac{3}{2}, \ldots$
25. $\dfrac{a}{x}, -1, \ldots$
26. $\dfrac{a}{b}, \dfrac{a}{bc}, \ldots$
27. $-4, x, \ldots$
28. $-x, x^2, \ldots$

Example

Find the general term and the fourteenth term of the arithmetic progression $-6, -1, \ldots$.

Solution

Find the common difference.
$$d = -1 - (-6) = 5$$

Use $s_n = a + (n - 1)d$.
$$s_n = -6 + (n - 1)5 = 5n - 11;$$
$$s_{14} = 5(14) - 11 = 59$$

29. Find the general term and the seventh term in the arithmetic progression $7, 11, \ldots$.
30. Find the general term and the twelfth term in the arithmetic progression $2, \dfrac{5}{2}, \ldots$.
31. Find the general term and the twentieth term in the arithmetic progression $3, -2, \ldots$.
32. Find the general term and the ninth term in the arithmetic progression $\dfrac{3}{4}, 2, \ldots$.
33. Find the general term and the eighth term in the arithmetic progression $x, 5x, \ldots$.
34. Find the general term and the twelfth term in the arithmetic progression $2a, -2a, \ldots$

Example

Find the general term and the ninth term of the geometric progression $-24, 12, \ldots$.

Solution

Find the common ratio.

$$r = \frac{12}{-24} = -\frac{1}{2}$$

Use $s_n = ar^{n-1}$.

$$s_n = -24\left(-\frac{1}{2}\right)^{n-1}; \quad s_9 = -24\left(-\frac{1}{2}\right)^8 = -\frac{3}{32}$$

35. Find the general term and the sixth term in the geometric progression $48, 96, \ldots$.

36. Find the general term and the eighth term in the geometric progression $-3, \frac{3}{2}, \ldots$.

37. Find the general term and the seventh term in the geometric progression $-\frac{1}{3}, 1, \ldots$.

38. Find the general term and the ninth term in the geometric progression $-81, 27, \ldots$.

B 39. Find the general term and the fifth term in the geometric progression $x, x^3, \ldots$.

40. Find the general term and the tenth term in the geometric progression $x, x(1 + y), \ldots$.

41. Find the general term and the sixth term in the arithmetic progression $x, y, \ldots$.

42. Find the general term and the tenth term of the arithmetic progression $x, x^2, \ldots$.

43. Find the general term and the eighth term of the geometric progression $x, y, \ldots$.

44. Find the general term and the tenth term of the geometric progression $x, x + 2, \ldots$.

Example

Find the first term in an arithmetic progression in which the third term is 7 and the eleventh term is 55.

Solution

Find the common difference by considering an arithmetic progression with first term 7 and with ninth term 55. Use $s_n = a + (n - 1)d$.

$$s_9 = 7 + (9 - 1)d$$
$$55 = 7 + 8d$$
$$d = 6$$

Use the difference to find the first term in an arithmetic progression in which the third term is 7. Again use $s_n = a + (n - 1)d$.

$$s_3 = a + (3 - 1)6$$
$$7 = a + 12$$
$$a = -5$$

10.2 Series

45. If the third term in an arithmetic progression is 7 and the eighth term is 17, find the common difference. What are the first and the twentieth terms?

46. If the fifth term of an arithmetic progression is -16 and the twentieth term is -46, what is the twelfth term?

47. Which term in the arithmetic progression $4, 1, \ldots$ is -77?

48. What is the twelfth term in an arithmetic progression in which the second term is x and the third term is y?

49. Find the first term of a geometric progression with fifth term 48 and ratio 2.

50. Find two different values for x so that $-\frac{3}{2}, x, -\frac{8}{27}$ will be in geometric progression.

10.2 Series

Associated with any sequence is a *series*.

Definition 10.4 A **series** is the indicated sum of the terms in a sequence.

We shall ordinarily denote a series of n terms by S_n. For example, with the finite sequence

$$4, 7, 10, \ldots, 3n + 1,$$

for a given counting number n, there is associated the finite series

$$S_n = 4 + 7 + 10 + \cdots + (3n + 1);$$

similarly, with the finite sequence

$$x, x^2, x^3, x^4, \ldots, x^n,$$

there is associated the finite series

$$S_n = x + x^2 + x^3 + x^4 + \cdots + x^n.$$

Since the terms in the series are the same as those in the sequence, we can refer to the first term or the second term or the general term of a series in the same manner as we do for a sequence.

Sum of the first n terms of an arithmetic progression

Consider the series S_n of the first n terms of the general arithmetic progression,

$$S_n = a + (a + d) + (a + 2d) + \cdots + [a + (n - 1)d], \quad (1)$$

and then consider the same series written as

$$S_n = s_n + (s_n - d) + (s_n - 2d) + \cdots + [s_n - (n - 1)d], \quad (2)$$

where the terms are displayed in reverse order. Adding (1) and (2) term by term, we have

$$S_n + S_n = (a + s_n) + (a + s_n) + (a + s_n) + \cdots + (a + s_n),$$

where the term $(a + s_n)$ occurs n times. Then

$$2S_n = n(a + s_n),$$

$$S_n = \frac{n}{2}(a + s_n). \tag{3}$$

If (3) is rewritten as

$$S_n = n\left(\frac{a + s_n}{2}\right),$$

we observe that the sum is given by the product of the number of terms in the series and the average of the first and last terms. The validity of (3) can be established by mathematical induction and is deferred until Section 10.5.

An alternative form for (3) is obtained by substituting $a + (n - 1)d$ for s_n in (3) to obtain

$$S_n = \frac{n}{2}(a + [a + (n - 1)d]),$$

$$S_n = \frac{n}{2}[2a + (n - 1)d], \tag{3'}$$

where the sum is now expressed in terms of a, n, and d.

Example Compute $1 + 3 + \cdots + 99$.

Solution The sequence $1, 3, \ldots$ is an arithmetic progression with $s_1 = 1$, $d = 2$. We have $s_n = 99$, $a = 1$, and $d = 2$. Thus, from Equation (2), page 336, we have

$$99 = 1 + (n - 1)2.$$

Solving for n, we obtain $n = 50$. Now we use Equation (3') to compute

$$S_{50} = \frac{50}{2}[2a + (50 - 1)d]$$

$$= \frac{50}{2}[2(1) + (49)(2)] = 2500.$$

Sum of the first n terms of a geometric progression

To find an explicit representation for the sum of a given number of terms in a geometric progression in terms of a, r, and n, we employ a device somewhat similar to the one used in finding the sum in an arithmetic progression. Consider the geometric series (4) containing n terms, and the series (5) obtained by multiplying both members of (4) by r:

$$S_n = a + ar + ar^2 + ar^3 + \cdots + ar^{n-2} + ar^{n-1}, \tag{4}$$

$$rS_n = ar + ar^2 + ar^3 + ar^4 + \cdots + ar^{n-1} + ar^n. \tag{5}$$

10.2 Series

When we subtract (5) from (4), all terms in the right-hand members except the first term in (4) and the last term in (5) vanish, yielding

$$S_n - rS_n = a - ar^n.$$

Factoring S_n from the left-hand member gives

$$(1 - r)S_n = a - ar^n,$$

$$S_n = \frac{a - ar_n}{1 - r}, \tag{6}$$

if $r \neq 1$, and we have a formula for the sum of the first n terms of a geometric progression. Establishing the validity of Equation (6), which can be accomplished by mathematical induction, is left to the exercises in Section 10.5.

Examples Find the sum of the first ten terms in the geometric progression.

a. $a = 2, r = \dfrac{1}{2}$ **b.** $a = \dfrac{1}{2}, r = -2$

Solutions

a. $S_{10} = \dfrac{a - ar^{10}}{1 - r}$ **b.** $S_{10} = \dfrac{a - ar^{10}}{1 - r}$

$= \dfrac{2 - 2(1/2)^{10}}{1 - (1/2)} = \dfrac{1023}{256}$ $= \dfrac{(1/2) - (1/2)(-2)^{10}}{1 - (-2)} = \dfrac{1023}{6}$

An alternative expression for (6) can be obtained by first writing

$$S_n = \frac{a - r(ar^{n-1})}{1 - r},$$

and then, since $s_n = ar^{n-1}$, expressing this as

$$S_n = \frac{a - rs_n}{1 - r}, \tag{7}$$

where the sum is now given in terms of a, s_n, and r.

Examples Find the sum of the terms in the geometric progression up to the *n*th term.

a. $a = 2, r = \dfrac{1}{2}, s_n = \dfrac{1}{8}$ **b.** $a = \dfrac{1}{2}, r = -3, s_n = \dfrac{81}{2}$

Solutions

a. $S_n = \dfrac{a - rs_n}{1 - r}$ **b.** $S_n = \dfrac{a - rs_n}{1 - r}$

$= \dfrac{2 - (1/2)(1/8)}{1 - (1/2)} = \dfrac{31}{8}$ $= \dfrac{(1/2) - (-3)(81/2)}{1 - (-3)} = \dfrac{61}{2}$

Sigma notation

A series for which the general term is known can be represented in a very convenient, compact way by means of what is called **sigma**, or **summation**, **notation**. The Greek letter $\sum$ (sigma) is used to denote a sum. For example,

$$S_n = 4 + 7 + 10 + \cdots + (3n + 1)$$

can be written

$$S_n = \sum_{j=1}^{n} (3j + 1),$$

where we understand that S_n is the series having terms obtained by replacing j in the expression $3j + 1$ with the numbers $1, 2, 3, \ldots, n$, successively. Similarly,

$$S = \sum_{j=3}^{6} j^2$$

appears in expanded form as

$$S = 3^2 + 4^2 + 5^2 + 6^2,$$

where the first value for j is 3 and the last is 6.

The variable used in conjunction with summation notation is called the **index of summation**, and the set of integers over which we sum (in this case, $\{3, 4, 5, 6\}$) is called the **range of summation**.

Notation for an infinite sum

To indicate that a series has an infinite number of terms, we cannot use the notation S_n for the sum, because there is no value to substitute for n. We therefore adopt notation such as

$$S_\infty = \sum_{j=1}^{\infty} \frac{1}{2^j} \tag{8}$$

to indicate that there is no last term in a series. In expanded form, the infinite series (8) is given by

$$S_\infty = \frac{1}{2} + \frac{1}{4} + \frac{1}{8} + \cdots.$$

The meaning of such an infinite sum will be discussed in Section 10.3.

Exercise 10.2

A *Write each series in expanded form.*

Examples

a. $\sum_{j=2}^{5} (j^2 + 1)$

b. $\sum_{k=1}^{\infty} (-1)^k 2^{k+1}$

10.2 Series

Solutions

a. $j = 2,\ 2^2 + 1 = 5;$
$j = 3,\ 3^2 + 1 = 10;$
$j = 4,\ 4^2 + 1 = 17;$
$j = 5,\ 5^2 + 1 = 26.$

$$\sum_{j=2}^{5} (j^2 + 1) = 5 + 10 + 17 + 26$$

b. $k = 1,\ (-1)^1 2^{1+1} = (-1)(4) = -4;$
$k = 2,\ (-1)^2 2^{2+1} = (1)(8) = 8;$
$k = 3,\ (-1)^3 2^{3+1} = (-1)(16) = -16.$

$$\sum_{k=1}^{\infty} (-1)^k 2^{k+1} = -4 + 8 - 16 + \cdots$$

1. $\sum_{j=1}^{4} j^2$

2. $\sum_{j=1}^{4} (3j - 2)$

3. $\sum_{j=1}^{3} \frac{(-1)^j}{2^j}$

4. $\sum_{j=3}^{5} \frac{(-1)^{j+1}}{j-2}$

5. $\sum_{j=-3}^{4} (2j + 1)$

6. $\sum_{j=-2}^{2} j^3$

7. $\sum_{j=0}^{\infty} \frac{1}{2^j}$

8. $\sum_{j=0}^{\infty} \frac{j}{1+j}$

Write each series in sigma notation with the range of summation starting with 1.

Examples

a. $5 + 8 + 11 + 14$

b. $x^2 + x^4 + x^6$

c. $\frac{3}{5} + \frac{5}{7} + \frac{7}{9} + \cdots$

Solutions

Find an expression for the general term and write in sigma notation.

a. $3j + 2$

$\sum_{j=1}^{4} (3j + 2)$

b. x^{2j}

$\sum_{j=1}^{3} x^{2j}$

c. $\frac{2j+1}{2j+3}$

$\sum_{j=1}^{\infty} \frac{2j+1}{2j+3}$

9. $x + x^3 + x^5 + x^7$

10. $x^3 + x^5 + x^7 + x^9 + x^{11}$

11. $1 + 4 + 9 + 16 + 25$

12. $\frac{1}{3} + \frac{1}{9} + \frac{1}{27} + \frac{1}{81}$

13. $1 \cdot 2 + 2 \cdot 3 + 3 \cdot 4 + 4 \cdot 5 + \cdots$

14. $\frac{1}{2} + \frac{2}{3} + \frac{3}{4} + \frac{4}{5} + \cdots$

15. $\frac{2}{1} + \frac{3}{2} + \frac{4}{3} + \frac{5}{4} + \cdots$

16. $\frac{1}{1} + \frac{2}{3} + \frac{3}{5} + \frac{4}{7} + \cdots$

Find each of the following sums.

Example

$$\sum_{j=1}^{12} (4j + 1)$$

Solution on overleaf

Solution Write the first two or three terms in expanded form:

$$5 + 9 + 13 + \cdots.$$

This is an arithmetic series. The first term is 5 and the common difference is 4. Therefore we can use

$$S_n = \frac{n}{2}[2a + (n-1)d]$$

to obtain

$$S_{12} = \frac{12}{2}[2(5) + (12-1)4] = 324.$$

17. $\sum_{j=1}^{7}(2j+1)$ 18. $\sum_{j=1}^{21}(3j-2)$ 19. $\sum_{j=3}^{15}(7j-1)$

20. $\sum_{j=10}^{20}(2j-3)$ 21. $\sum_{k=1}^{8}\left(\frac{1}{2}k - 3\right)$ 22. $\sum_{k=1}^{100}k$

23. $\sum_{k=1}^{100}(2k-1)$ 24. $\sum_{k=1}^{100}(10k-2)$

Example $\sum_{j=2}^{5}\left(\frac{1}{3}\right)^j$

Solution Write the first two or three terms in expanded form:

$$\left(\frac{1}{3}\right)^2 + \left(\frac{1}{3}\right)^3 + \cdots.$$

This is a geometric series in which the first term is $\frac{1}{9}$, the ratio is $\frac{1}{3}$, and $n = 4$. Therefore, we can use $S_n = \dfrac{a - ar^n}{1-r}$ to obtain

$$S_4 = \frac{\frac{1}{9} - \frac{1}{9}\left(\frac{1}{3}\right)^4}{1 - \frac{1}{3}} = \frac{40}{243}.$$

25. $\sum_{j=1}^{6} 3^j$ 26. $\sum_{j=1}^{4} 2^j$ 27. $\sum_{k=3}^{7}\left(\frac{1}{2}\right)^{k-2}$

28. $\sum_{j=3}^{12} 2^{j-5}$ 29. $\sum_{j=1}^{6}\left(\frac{1}{3}\right)^{j+1}$ 30. $\sum_{j=1}^{4} 2^{j+3}$

31. $\sum_{j=4}^{8}(-2)^{j+4}$ 32. $\sum_{j=5}^{10}\left(-\frac{1}{2}\right)^{j+2}$

B 33. Find the sum of all even integers n, for $13 < n < 29$.

34. Find the sum of all integral multiples of 7 between 8 and 110.

35. Find the sum of the powers of 2 between 5 and 2000.

36. Find the sum of the powers of 1/2 between 1/300 and 1.

37. How many bricks will there be in a wall one brick in thickness if there are 27 bricks in the bottom row, 25 in the second row, and so forth, to the top row, which has one brick?

38. If there are a total of 256 bricks in a wall arranged in the manner of those in Exercise 37, how many bricks are there in the tenth row from the bottom?

39. If ten dollars is deposited each month in an account which draws 12% annual interest and which is compounded monthly, how much will be in the account at the end of ten years?

40. How much has to be deposited monthly in an account earning 9% annual interest compounded monthly to guarantee that the account will be worth one million dollars at the end of thirty years?

41. Find $\sum_{j=1}^{n} \left(\frac{1}{2}\right)^j$ for $n = 2, 3, 4,$ and 5. What value do you think $\sum_{j=1}^{n} \left(\frac{1}{2}\right)^j$ approximates as n becomes greater and greater?

42. Find p if $\sum_{j=1}^{5} pj = 14$.

43. Find p and q if $\sum_{j=1}^{4} (pj + q) = 28$ and $\sum_{j=2}^{5} (pj + q) = 44$.

44. Consider $S_n = \sum_{j=1}^{n} f(j)$. Explain why this equation defines a sequence function. What is the variable denoting an element in the domain? The range?

10.3 Limits of Sequences and Series

A sequence that is strictly increasing but bounded

Consider the sequence function defined by

$$S_n = \frac{n}{n+1}, \quad n \in N. \tag{1}$$

If we write the range of (1) in the form

$$\frac{1}{2}, \frac{2}{3}, \frac{3}{4}, \frac{4}{5}, \ldots, \frac{n}{n+1}, \ldots,$$

then it is clear that each of the terms is greater than the preceding term; indeed, the difference of consecutive terms is

$$\frac{n+1}{n+2} - \frac{n}{n+1} = \frac{(n^2 + 2n + 1) - (n^2 + 2n)}{(n+1)(n+2)} = \frac{1}{(n+1)(n+2)} > 0.$$

Such a sequence is said to be **strictly increasing**. On the other hand, it is also clear that, no matter how large a value is assigned to n, we have

$$\frac{n}{n+1} < 1,$$

because the denominator is one larger than the numerator; in fact, we have

$$1 - \frac{n}{n+1} = \frac{(n+1) - n}{n+1} = \frac{1}{n+1} > 0.$$

Thus we have a sequence in which each term is greater than the preceding term and yet no term is equal to or greater than 1. We note, however—and this is a very basic consideration—that the value of $n/(n+1)$ is as close to 1 as we please if n is large enough.

Example

The value of $\frac{n}{n+1}$ is within $1/1000$ of 1 if $n > 999$, because

$$1 - \frac{n}{n+1} = \frac{1}{n+1}, \quad \text{and we have} \quad \frac{1}{n+1} < \frac{1}{1000}$$

provided $n + 1 > 1000$—that is, $n > 999$.

Limit of a sequence

If it is true that the nth term in a sequence differs from the number L by as little as we please for all sufficiently large n, we say that **the sequence approaches the number L as a limit**. The symbolism

$$\lim_{n \to \infty} s_n = L$$

(read "the limit, as n increases without bound, of s_n is L") is used to denote this situation. A thorough discussion of the notion of a limit is included in courses in calculus, and will not be attempted here. A few elementary ideas, however, are in order.

A sequence in which the nth term approaches a number L as $n \to \infty$ is said to be a **convergent sequence**, and the sequence is said to **converge** to L. It is not necessary for convergence that a sequence be strictly increasing. For example,

$$1, \frac{1}{2}, \frac{1}{3}, \frac{1}{4}, \ldots, \frac{1}{n}, \ldots$$

converges to 0, but each term in the sequence is less than, instead of greater than, the term that precedes it. Again, the sequence

$$-1, \frac{1}{2}, -\frac{1}{3}, \frac{1}{4}, \ldots, \frac{(-1)^n}{n}, \ldots$$

converges to 0 but is neither increasing nor decreasing. We can rephrase the definition of convergence of a sequence as follows:

10.3 Limits of Sequences and Series

Definition 10.5 A sequence $s_1, s_2, \ldots, s_n, \ldots$ **converges** to the number L,

$$\lim_{n \to \infty} s_n = L,$$

if and only if the absolute value of the difference between the nth term in the sequence and the number L is as small as we please for all sufficiently large n. Thus the sequence converges to the number L if and only if

$$\lim_{n \to \infty} |L - s_n| = 0.$$

For example, the **alternating** (because the signs alternate) **sequence**

$$-\frac{1}{3}, \frac{1}{9}, -\frac{1}{27}, \ldots, \frac{(-1)^n}{3^n}, \ldots$$

converges to 0 since the absolute value of the difference between $\dfrac{(-1)^n}{3^n}$ and 0 is as small as we please for n large enough. We express this by writing

$$\lim_{n \to \infty} \left(\frac{-1}{3}\right)^n = 0.$$

On the other hand, the alternating sequence

$$\frac{1}{2}, -\frac{2}{3}, \frac{3}{4}, -\frac{4}{5}, \ldots, (-1)^{n+1} \frac{n}{n+1}, \ldots$$

does not converge. As n increases, the nth term oscillates back and forth from the neighborhood of $+1$ to the neighborhood of -1, and we cannot find a number L such that $\lim_{n \to \infty} |L - s_n| = 0$. Such a sequence is said to **diverge**. A sequence such as

$$1, 2, 3, \ldots, n, \ldots$$

also is said to diverge. An answer to the logical question of what we mean by "enough" when we say "n large enough" requires a more precise definition of limit than we have given here. As remarked earlier, a course in the calculus will treat this in detail.

Sequence of partial sums of a series

For an infinite series,

$$S_\infty = \sum_{j=1}^{\infty} s_j,$$

we can consider the infinite sequence of **partial sums**:

$$S_1 = s_1$$
$$S_2 = s_1 + s_2$$
$$\cdot\ \cdot\ \cdot$$
$$S_n = s_1 + s_2 + \cdots + s_n$$
$$\cdot\ \cdot\ \cdot$$

Definition 10.6 An infinite series

$$S_\infty = \sum_{j=1}^{\infty} s_j$$

converges if and only if $S_1, S_2, \ldots, S_n, \ldots$, the corresponding sequence of partial sums, converges.

If the sequence of partial sums converges to the number L,

$$\lim_{n \to \infty} S_n = L,$$

then L is said to be the **sum** of the infinite series, and we write

$$S_\infty = \sum_{j=1}^{\infty} s_j = L.$$

If the sequence of partial sums diverges, then the series is said to **diverge**.

Sum of an infinite geometric progression

We recall from Section 10.2 that the sum of n terms (the nth partial sum) of a geometric progression is given, for $r \neq 1$, by

$$S_n = \frac{a - ar^n}{1 - r}. \tag{2}$$

If $|r| < 1$, that is, if $-1 < r < 1$, then $|r|^n$ becomes smaller and smaller for increasingly large n. For example, if $r = 1/2$, then

$$r^2 = \frac{1}{4}, \quad r^3 = \frac{1}{8}, \quad r^4 = \frac{1}{16},$$

and so forth; and $(1/2)^n$ is as small as we please if n is sufficiently large. Writing (2) as

$$S_n = \frac{a}{1 - r}(1 - r^n), \tag{3}$$

we see that the value of the factor $(1 - r^n)$ is as close as we please to 1 provided $|r| < 1$ and n is taken large enough. Since this argument shows that the sequence of partial sums (3) converges to

$$\frac{a}{1 - r},$$

we have the following result.

Theorem 10.3 The sum of an infinite geometric progression,

$$a + ar + ar^2 + \cdots + ar^n + \cdots,$$

Theorem continued

10.3 Limits of Sequences and Series

with $|r| < 1$, is

$$S_\infty = \lim_{n \to \infty} S_n = \frac{a}{1-r}.$$

An interesting application of this sum arises in connection with repeating decimals—that is, decimal numerals that, after a finite number of decimal places, have endlessly repeating groups of digits. For example,

$$0.2121\overline{21},$$

$$0.138512512\overline{512}$$

are repeating decimals. The bar denotes that the numerals appearing under it are repeated endlessly. Consider the problem of expressing such a decimal fraction as an arithmetic fraction. We illustrate the process involved with the first example above. The decimal $0.2121\overline{21}$ can be written as

$$0.21 + 0.0021 + 0.000021 + \cdots, \qquad (4)$$

which is a geometric progression with ratio $r = 0.01$. Since the ratio is less than 1 in absolute value, we can use Theorem 10.3 to find the sum of the infinite series (4). Thus,

$$S_\infty = \frac{a}{1-r} = \frac{0.21}{1-0.01} = \frac{21}{99} = \frac{7}{33},$$

and the given decimal fraction is equivalent to 7/33.

Exercise 10.3

A *Discuss the limiting behavior of each expression as $n \to \infty$.*

Example

$$\frac{n^2 + 3}{n^2}$$

Solution By writing $\dfrac{n^2 + 3}{n^2}$ as $\dfrac{n^2}{n^2} + \dfrac{3}{n^2}$ and then as $1 + \dfrac{3}{n^2}$, we observe that

$$\lim_{n \to \infty} \frac{n^2 + 3}{n^2} = \lim_{n \to \infty} \left(1 + \frac{3}{n^2}\right) = 1 + 0 = 1.$$

1. $\dfrac{1}{n}$
2. $1 + \dfrac{1}{n^2}$
3. $\dfrac{n+1}{n}$
4. $\dfrac{n+3}{n^2}$

5. $2n$
6. $(-1)^n$
7. $\dfrac{1}{2^n}$
8. $(-1)^n \dfrac{1}{n}$

State which of the following sequences are convergent.

Example

$$1, \frac{3}{2}, \frac{7}{4}, \frac{15}{8}, \ldots, \frac{2^n - 1}{2^{n-1}}, \ldots$$

Solution

Writing the general term as $\frac{2^n}{2^{n-1}} - \frac{1}{2^{n-1}}$, or $2 - \frac{1}{2^{n-1}}$, we observe that

$$\lim_{n \to \infty} \frac{2^n - 1}{2^{n-1}} = \lim_{n \to \infty} \left(2 - \frac{1}{2^{n-1}}\right) = 2 - 0 = 2.$$

The sequence is convergent.

9. $\frac{1}{2}, \frac{1}{4}, \frac{1}{8}, \frac{1}{16}, \ldots, \frac{1}{2^n}, \ldots$ 10. $2, \frac{3}{2}, \frac{4}{3}, \frac{5}{4}, \ldots, \frac{n+1}{n}, \ldots$

11. $1, 2, 3, 4, 5, \ldots, n, \ldots$ 12. $2, 4, 6, 8, \ldots, 2n, \ldots$

13. $1, -\frac{1}{2}, \frac{1}{4}, -\frac{1}{8}, \ldots, (-1)^{n-1}\frac{1}{2^{n-1}}, \ldots$ 14. $1, -1, 1, -1, \ldots, (-1)^{n+1}, \ldots$

Find the sum of each of the following infinite geometric series. If the series has no sum, so state.

Examples

a. $3 + 2 + \cdots$ b. $\frac{1}{81} - \frac{1}{54} + \cdots$

Solutions

a. $r = \frac{2}{3}$; series has a sum since $|r| < 1$. Using Theorem 10.3, we have

$$S_\infty = \frac{a}{1 - r} = \frac{3}{1 - \frac{2}{3}} = 9.$$

b. $r = -\frac{1}{54} \div \frac{1}{81} = -\frac{3}{2}$; series does not have a sum since $|r| > 1$.

15. $12 + 6 + \cdots$ 16. $2 + 1 + \cdots$ 17. $\frac{1}{36} + \frac{1}{30} + \cdots$

18. $\frac{1}{16} - \frac{1}{8} + \cdots$ 19. $\sum_{j=1}^{\infty} \left(\frac{2}{3}\right)^j$ 20. $\sum_{j=1}^{\infty} \left(-\frac{1}{4}\right)^j$

21. $\sum_{j=2}^{\infty} \left(\frac{2}{3}\right)^{j+1}$ 22. $\sum_{j=-1}^{\infty} \left(\frac{-1}{3}\right)^{j+2}$

10.4 The Binomial Theorem

Find an arithmetic fraction equal to each of the given decimal numerals.

Example $0.81\overline{8181}$

Solution Rewrite as a series: $0.81 + 0.0081 + 0.000081 + \cdots$.

Find the common ratio: $r = 0.01$. Use $S_\infty = \dfrac{a}{1-r}$.

$$S_\infty = \frac{0.81}{1 - 0.01} = \frac{81}{99} = \frac{9}{11}.$$

23. $0.31\overline{3131}$ 24. $0.45\overline{4545}$ 25. $2.41\overline{0410}$
26. $3.02\overline{7027}$ 27. $0.128\overline{888}$ 28. $0.8\overline{3333}$

B 29. A force is applied to a particle moving in a straight line in such a fashion that each second it moves only one half of the distance it moved the preceding second. If the particle moves ten centimeters the first second, approximately how far will it move before coming to rest?

30. The arc length through which the bob on a pendulum moves is nine-tenths of its preceding arc length. Approximately how far will the bob move before coming to rest if the first arc length is 12 inches?

10.4 The Binomial Theorem

Factorial notation

There are situations in which it is necessary to write the product of several consecutive positive integers. To facilitate writing products of this type, we use a special symbol $n!$ (read "n factorial" or "factorial n"), which is defined by

$$n! = n(n-1)(n-2) \cdots (3)(2)(1).$$

Thus,

$$5! = 5 \cdot 4 \cdot 3 \cdot 2 \cdot 1 \quad \text{(read "five factorial")},$$

and

$$8! = 8 \cdot 7 \cdot 6 \cdot 5 \cdot 4 \cdot 3 \cdot 2 \cdot 1 \quad \text{(read "eight factorial")}.$$

Factorial notation can also be used to represent products of consecutive positive integers, beginning with integers different from 1. For example,

$$8 \cdot 7 \cdot 6 \cdot 5 = \frac{8!}{4!},$$

because

$$\frac{8!}{4!} = \frac{8 \cdot 7 \cdot 6 \cdot 5 \cdot 4 \cdot 3 \cdot 2 \cdot 1}{4 \cdot 3 \cdot 2 \cdot 1} = 8 \cdot 7 \cdot 6 \cdot 5.$$

Since

$$n! = n(n-1)(n-2)(n-3) \cdots 5 \cdot 4 \cdot 3 \cdot 2 \cdot 1$$

and

$$(n-1)! = (n-1)(n-2)(n-3) \cdots 5 \cdot 4 \cdot 3 \cdot 2 \cdot 1,$$

for $n > 1$ we can write the recursive relationship

$$n! = n(n-1)!. \tag{1}$$

For example,

$$7! = 7 \cdot 6!,$$
$$27! = 27 \cdot 26!,$$
$$(n+2)! = (n+2)(n+1)!.$$

If (1) is to hold also for $n = 1$, then we must have

$$1! = 1 \cdot (1-1)!$$

or

$$1! = 1 \cdot 0!.$$

Therefore, for consistency, we define $0!$ by

$$0! = 1.$$

A special case of the use of factorial notation occurs in the binomial expansion that follows. We make the following definition:

$$\binom{n}{r} = \frac{n!}{r!(n-r)!}. \tag{2}$$

The symbol $\binom{n}{r}$ is read "the binomial coefficient n, r," or simply, "n, r." Some examples are

$$\binom{5}{3} = \frac{5!}{3!(5-3)!} = \frac{5!}{3!2!} = \frac{5 \cdot 4 \cdot 3!}{3!2 \cdot 1} = 10,$$

$$\binom{5}{1} = \frac{5!}{1!(5-1)!} = \frac{5!}{4!} = \frac{5 \cdot 4!}{4!} = 5,$$

and

$$\binom{5}{0} = \frac{5!}{0!(5-0)!} = \frac{5!}{5!} = 1.$$

10.4 The Binomial Theorem

Binomial expansions

The series obtained by expanding a binomial of the form

$$(a + b)^n$$

is particularly useful in certain branches of mathematics. Starting with familiar examples, where n takes the values 1, 2, 3, 4, and 5 in turn, we can show by direct multiplication that

$$(a + b)^1 = a + b$$
$$(a + b)^2 = a^2 + 2ab + b^2$$
$$(a + b)^3 = a^3 + 3a^2b + 3ab^2 + b^3$$
$$(a + b)^4 = a^4 + 4a^3b + 6a^2b^2 + 4ab^3 + b^4$$
$$(a + b)^5 = a^5 + 5a^4b + 10a^3b^2 + 10a^2b^3 + 5ab^4 + b^5.$$

We observe that in each case:

1. The first term is a^n.

2. The variable factors of the second term are $a^{n-1}b^1$, and the coefficient is n, which can be written in the form

$$\frac{n}{1!}.$$

3. The variable factors of the third term are $a^{n-2}b^2$, and the coefficient can be written in the form

$$\frac{n(n-1)}{2!}.$$

4. The variable factors of the fourth term are $a^{n-3}b^3$, and the coefficient can be written in the form

$$\frac{n(n-1)(n-2)}{3!}.$$

The foregoing expansions suggest the following result, known as the **binomial theorem**. Its proof, which requires the use of mathematical induction, is omitted.

Theorem 10.4   *For each natural number n,*

$$(a+b)^n = a^n + \frac{n}{1!}a^{n-1}b + \frac{n(n-1)}{2!}a^{n-2}b^2 + \frac{n(n-1)(n-2)}{3!}a^{n-3}b^3$$

$$+ \cdots + \frac{n(n-1)(n-2)\cdots(n-r+2)}{(r-1)!}a^{n-r+1}b^{r-1} + \cdots + b^n, \quad (3)$$

where r is the number of the term.

For example,

$$(x - 2)^4 = x^4 + \frac{4}{1!}x^3(-2)^1 + \frac{4\cdot 3}{2!}x^2(-2)^2 + \frac{4\cdot 3\cdot 2}{3!}x(-2)^3$$
$$+ \frac{4\cdot 3\cdot 2\cdot 1}{4!}(-2)^4$$
$$= x^4 - 8x^3 + 24x^2 - 32x + 16.$$

In this case, $a = x$ and $b = -2$ in the binomial expansion.

Observe that the coefficients of the terms in the binomial expansion (3) can be represented as follows.

1st term: $1 = \dfrac{n!}{0!n!} = \dbinom{n}{0}$,

2nd term: $\dfrac{n}{1!} = \dfrac{n\cdot(n-1)!}{1!(n-1)!} = \dfrac{n!}{1!(n-1)!} = \dbinom{n}{1}$,

3rd term: $\dfrac{n(n-1)}{2!} = \dfrac{n(n-1)(n-2)!}{2!(n-2)!} = \dfrac{n!}{2!(n-2)!} = \dbinom{n}{2}$,

rth term: $\dfrac{n(n-1)(n-2)\cdots(n-r+2)}{(r-1)!}$

$$= \dfrac{n(n-1)(n-2)\cdots(n-r+2)(n-r+1)!}{(r-1)!(n-r+1)!}$$

$$= \dfrac{n!}{(r-1)!(n-r+1)!} = \dbinom{n}{r-1}.$$

Hence, the binomial expansion (3) can be represented by the expression

$$(a + b)^n = \binom{n}{0}a^n + \binom{n}{1}a^{n-1}b + \binom{n}{2}a^{n-2}b^2 + \binom{n}{3}a^{n-3}b^3 + \cdots$$
$$+ \binom{n}{r-1}a^{n-r+1}b^{r-1} + \cdots + \binom{n}{n}b^n. \quad (4)$$

For example,

$$(x - 2)^4 = \binom{4}{0}x^4 + \binom{4}{1}x^3(-2)^1 + \binom{4}{2}x^2(-2)^2 + \binom{4}{3}x(-2)^3 + \binom{4}{4}(-2)^4.$$

Using Formula (2) for $\dbinom{n}{r}$ to find the coefficients, we obtain the same result as above,

$$(x - 2)^4 = x^4 - 8x^3 + 24x^2 - 32x + 16.$$

10.4 The Binomial Theorem

rth term in a binomial expansion

Note that the rth term in a binomial expansion is given by

$$\binom{n}{r-1} a^{n-r+1} b^{r-1} = \frac{n!}{(r-1)!(n-r+1)!} a^{n-r+1} b^{r-1}$$

$$= \frac{n(n-1)(n-2)\cdots(n-r+2)}{(r-1)!} a^{n-r+1} b^{r-1}. \quad (5)$$

For example, by the left-hand member of (5), the seventh term of $(x-2)^{10}$ is

$$\binom{10}{6} x^4 (-2)^6 = \frac{10!}{6!4!} x^4 (-2)^6 = \frac{10 \cdot 9 \cdot 8 \cdot 7 \cdot 6!}{6! \cdot 4 \cdot 3 \cdot 2 \cdot 1} x^4 (64)$$

$$= 13440 x^4,$$

while, by the right-hand member of (5), we have

$$\frac{10 \cdot 9 \cdot 8 \cdot 7 \cdot 6 \cdot 5}{6 \cdot 5 \cdot 4 \cdot 3 \cdot 2 \cdot 1} x^4 (64) = 13440 x^4.$$

Exercise 10.4

A
1. Write $(2n)!$ in expanded form for $n = 4$.
2. Write $2n!$ in expanded form for $n = 4$.
3. Write $n(n-1)!$ in expanded form for $n = 6$.
4. Write $2n(2n-1)!$ in expanded form for $n = 2$.

Write in expanded form and simplify.

Examples a. $\dfrac{7!}{4!}$ b. $\dfrac{4!6!}{8!}$

Solutions a. $\dfrac{7 \cdot 6 \cdot 5 \cdot 4!}{4!}$ b. $\dfrac{4 \cdot 3 \cdot 2 \cdot 1 \cdot 6!}{8 \cdot 7 \cdot 6!}$

210

$\dfrac{3}{7}$

5. $5!$
6. $7!$
7. $\dfrac{9!}{7!}$
8. $\dfrac{12!}{11!}$
9. $\dfrac{5!7!}{8!}$
10. $\dfrac{12!8!}{16!}$
11. $\dfrac{8!}{2!(8-2)!}$
12. $\dfrac{10!}{4!(10-4)!}$

Write each product in factorial notation.

Examples   **a.** $1 \cdot 2 \cdot 3 \cdot 4 \cdot 5 \cdot 6$ **b.** $11 \cdot 12 \cdot 13 \cdot 14$ **c.** 150

Solutions   **a.** $6!$ **b.** $\dfrac{14!}{10!}$ **c.** $\dfrac{150!}{149!}$

13. $1 \cdot 2 \cdot 3$ **14.** $1 \cdot 2 \cdot 3 \cdot 4 \cdot 5$ **15.** $3 \cdot 4 \cdot 5 \cdot 6$

16. 7 **17.** $8 \cdot 7 \cdot 6$ **18.** $28 \cdot 27 \cdot 26 \cdot 25 \cdot 24$

Write each expression in factorial notation and simplify.

Examples   **a.** $\binom{6}{2}$ **b.** $\binom{4}{4}$

Solutions   **a.** $\binom{6}{2} = \dfrac{6!}{2!(6-2)!} = \dfrac{6!}{2!4!} = \dfrac{6 \cdot 5 \cdot 4!}{2 \cdot 1 \cdot 4!} = 15$ **b.** $\binom{4}{4} = \dfrac{4!}{4!(4-4)!} = \dfrac{4!}{4!0!} = 1$

19. $\binom{6}{5}$ **20.** $\binom{4}{2}$ **21.** $\binom{3}{3}$ **22.** $\binom{5}{5}$

23. $\binom{7}{0}$ **24.** $\binom{2}{0}$ **25.** $\binom{5}{2}$ **26.** $\binom{5}{3}$

Write each expression in factored form and show the first three factors and the last three factors.

Example $(2n + 1)!$

Solution $(2n + 1)(2n)(2n - 1) \cdot \cdots \cdot 3 \cdot 2 \cdot 1$

27. $n!$ **28.** $(n + 4)!$ **29.** $(3n)!$

30. $3n!$ **31.** $(n - 2)!$ **32.** $(3n - 2)!$

Simplify each expression.

Examples   **a.** $\dfrac{(n-1)!}{(n-3)!}$ **b.** $\dfrac{(n-1)!(2n)!}{2n!(2n-2)!}$

10.4 The Binomial Theorem

Solutions

a. $\dfrac{(n-1)(n-2)(n-3)!}{(n-3)!}$

$(n-1)(n-2)$

b. $\dfrac{(n-1)!(2n)(2n-1)(2n-2)!}{2(n)(n-1)!(2n-2)!}$

$2n - 1$

33. $\dfrac{(n+2)!}{n!}$

34. $\dfrac{(n+2)!}{(n-1)!}$

35. $\dfrac{(n+1)(n+2)!}{(n+3)!}$

36. $\dfrac{(2n+4)!}{(2n+2)!}$

37. $\dfrac{(2n)!(n-2)!}{4(2n-2)!(n)!}$

38. $\dfrac{(2n+1)!(2n-1)!}{[(2n)!]^2}$

Expand.

Example $(a - 3b)^4$

Solution   From the binomial expansion (3) on page 355,

$$(a-3b)^4 = a^4 + \frac{4}{1!}a^3(-3b) + \frac{4\cdot 3}{2!}a^2(-3b)^2 + \frac{4\cdot 3\cdot 2}{3!}a(-3b)^3$$

$$+ \frac{4\cdot 3\cdot 2\cdot 1}{4!}(-3b)^4$$

$$= a^4 - 12a^3b + 54a^2b^2 - 108ab^3 + 81b^4.$$

Alternatively, from the abbreviation of the binomial expansion (4) on page 356, we have

$$(a-3b)^4 = \binom{4}{0}a^4 + \binom{4}{1}a^3(-3b) + \binom{4}{2}a^2(-3b)^2 + \binom{4}{3}a(-3b)^3 + \binom{4}{4}(-3b)^4,$$

which also simplifies to the expression obtained above.

39. $(x + 3)^5$ 40. $(2x + y)^4$ 41. $(x - 3)^4$ 42. $(2x - 1)^5$

43. $\left(2x - \dfrac{y}{2}\right)^3$ 44. $\left(\dfrac{x}{3} + 3\right)^5$ 45. $\left(\dfrac{x}{2} + 2\right)^6$ 46. $\left(\dfrac{2}{3} - a^2\right)^4$

Write the first four terms in each expansion. Do not simplify the terms.

Example $(x + 2y)^{15}$

Solution   From the binomial expansion (3), we have

$$x^{15} + \frac{15}{1!}x^{14}(2y) + \frac{15\cdot 14}{2!}x^{13}(2y)^2 + \frac{15\cdot 14\cdot 13}{3!}x^{12}(2y)^3,$$

or alternatively from the form (4),

$$\binom{15}{0}x^{15} + \binom{15}{1}x^{14}(2y) + \binom{15}{2}x^{13}(2y)^2 + \binom{15}{3}x^{12}(2y)^3.$$

47. $(x + y)^{20}$ 48. $(x - y)^{15}$ 49. $(a - 2b)^{12}$

50. $(2a - b)^{12}$ 51. $(x - \sqrt{2})^{10}$ 52. $\left(\dfrac{x}{2} + 2\right)^8$

B *Find each power to the nearest hundredth.*

Example $(0.97)^7$

Solution Either form of the binomial expansion (3) or (4) can be used. We shall use (3).

$$(0.97)^7 = (1 - 0.03)^7$$

$$= 1^7 + \frac{7}{1!}(1)^6(-0.03)^1 + \frac{7 \cdot 6}{2!}(1)^5(-0.03)^2 + \frac{7 \cdot 6 \cdot 5}{3!}(1)^4(-0.03)^3 + \cdots$$

$$= 1 - 0.21 + 0.0189 - 0.000945 + \cdots$$

$$= 0.807955^+$$

Hence, to the nearest hundredth, $(0.97)^7 = 0.81$.

53. $(1.02)^{10}$ *Hint*: $1.02 = (1 + 0.02)$ 54. $(1.01)^{15}$

55. If an amount of money (P) is invested at 4% compounded annually, the amount (A) present at the end of (n) years is given by $A = P(1 + 0.04)^n$. Find the amount A (to the nearest dollar) if \$1000 was invested for 10 years.

56. In Exercise 55, find the amount present at the end of 20 years.

Find each specified term.

Example $(x - 2y)^{12}$, the seventh term.

Solution In Equation (5), page 357, use $n = 12$ and $r = 7$.

$$\frac{12!}{6!(12 - 6)!} x^6(-2y)^6 = \frac{12 \cdot 11 \cdot 10 \cdot 9 \cdot 8 \cdot 7}{6!} x^6(-2y)^6 = 59{,}136 x^6 y^6$$

57. $(a - b)^{15}$, the sixth term   58. $(x + 2)^{12}$, the fifth term
59. $(x - 2y)^{10}$, the fifth term   60. $(a^3 - b)^9$, the seventh term

61. Given that the binomial formula holds for $(1 + x)^n$ where n is a negative integer:
    a. Write the first four terms of $(1 + x)^{-1}$.
    b. Find the first four terms of the quotient $1/(1 + x)$, by dividing 1 by $(1 + x)$. Compare the results of (a) and (b).

62. Given that the binomial formula holds as an infinite "sum" for $(1 + x)^n$, where n is a rational number and $|x| < 1$, find to two decimal places.

  a. $\sqrt{1.02}$ **b.** $\sqrt{0.99}$

63. Show that

$$\binom{k}{r} + \binom{k}{r-1} = \binom{k+1}{r}.$$

64. Show that

$$n\binom{n-1}{k} = (n-k)\binom{n}{k}.$$

10.5 Mathematical Induction

The material in the present section depends on a special property of the set of natural numbers, or positive integers, $N = \{1, 2, 3, \ldots\}$:

  a. $1 \in N$
  b. If $k \in N$, then $k + 1 \in N$.

In addition, *N contains no elements not implied by properties a and b.* These properties underlie the following theorem, called the **principle of mathematical induction**, which we state without proof.

Theorem 10.5 *If a given sentence involving natural numbers n is true for $n = 1$, and if its truth for $n = k$ implies its truth for $n = k + 1$, then it is true for every natural number n.*

Requirements of a proof by mathematical induction

We can exploit Theorem 10.5 to prove a number of assertions. Although the technique we shall use is called **proof by mathematical induction**, the argument we shall employ is deductive, as have been all of the other arguments in this book. Proofs by mathematical induction require two things:

  a. A demonstration that the assertion to be proved is true for the natural number 1.
  b. A demonstration that the truth of the assertion for a natural number k implies its truth for $k + 1$.

When these two demonstrations have been made, the principle of mathematical induction assures us that the assertion is true for every natural number.

Example   Prove that the sum of the first n natural numbers is $\dfrac{n(n+1)}{2}$.

Solution   We wish to show that

$$1 + 2 + 3 + \cdots + n = \frac{n(n+1)}{2}.$$

We must do two things:

1. We must first show that the assertion is true for $n = 1$, that is, that

$$1 = \frac{1(1+1)}{2},$$

   which is true.

2. We must next show that the truth of the assertion for $n = k$ implies its truth for $n = k + 1$. That is, we must show that the truth of

$$1 + 2 + 3 + \cdots + k = \frac{k(k+1)}{2} \qquad (1)$$

   implies the truth of

$$1 + 2 + 3 + \cdots + k + (k+1) = \frac{(k+1)[(k+1)+1]}{2}. \qquad (2)$$

   Assuming the truth of (1), we add $k + 1$ to each member of this equation, obtaining

$$1 + 2 + 3 + \cdots + k + (k+1) = \frac{k(k+1)}{2} + (k+1)$$

$$= (k+1)\left(\frac{k}{2} + 1\right)$$

$$= (k+1)\left(\frac{k+2}{2}\right),$$

   from which we obtain

$$1 + 2 + 3 + \cdots + k + (k+1) = \frac{(k+1)[(k+1)+1]}{2},$$

   which is (2).

Thus the second fact necessary for our proof is established. By the principle of mathematical induction, the assertion is true for every natural number n.

Example   Prove that for any arithmetic progression,

$$S_n = \frac{n}{2}(2a + (n-1)d).$$

Solution

For $n = 1$, the assertion is that

$$S_1 = \frac{1}{2}(2a + 0 \cdot d) = a,$$

which is true. We next must show that

$$S_k = \frac{k}{2}(2a + (k-1)d)$$

implies that

$$S_{k+1} = \frac{k+1}{2}[2a + (k+1-1)d] = \frac{k+1}{2}(2a + kd).$$

Now since S_{k+1} is the sum of the first $k+1$ terms, S_{k+1} is the sum of the first k terms and the $k+$ 1st term. That is,

$$S_{k+1} = S_k + s_{k+1}.$$

Therefore

$$S_{k+1} = \frac{k}{2}(2a + (k-1)d) + a + kd$$

$$= ka + a + \frac{k(k-1)}{2}d + kd$$

$$= (k+1)a + \frac{k(k+1)}{2} \cdot d = \frac{k+1}{2}(2a + kd).$$

Thus, by the principle of mathematical induction, the assertion is true for every natural number n.

This method of proof is often compared to lining up a row of dominoes, with the assumption that whenever one domino is toppled, the one following will topple. One then needs only to topple the first domino ($n = 1$) and the whole row behind it will topple, as indicated in Figure 10.1.

Figure 10.1

Exercise 10.5

A By mathematical induction, prove the validity of the formulas in Exercises 1–14 for all positive integral values of n.

1. $\dfrac{1}{2} + \dfrac{2}{2} + \dfrac{3}{2} + \cdots + \dfrac{n}{2} = \dfrac{n(n+1)}{4}$

2. $1 + 3 + 5 + \cdots + (2n-1) = n^2$

3. $2 + 4 + 6 + \cdots + 2n = n(n+1)$

4. $2 + 6 + 10 + \cdots + (4n-2) = 2n^2$

5. $2 + 8 + 14 + \cdots + (6n-4) = n(3n-1)$

6. $1 + 5 + 9 + \cdots + (4n-3) = n(2n-1)$

7. $7 + 10 + 13 + \cdots + (3n+4) = \dfrac{n(3n+11)}{2}$

8. $6 + 11 + 16 + \cdots + (5n+1) = \dfrac{n(5n+7)}{2}$

9. $1^2 + 2^2 + 3^2 + \cdots + n^2 = \dfrac{n(n+1)(2n+1)}{6}$

10. $2 + 2^2 + 2^3 + \cdots + 2^n = 2^{n+1} - 2$

11. $1^3 + 3^3 + 5^3 + \cdots + (2n-1)^3 = n^2(2n^2-1)$

12. $\dfrac{1}{1 \cdot 2} + \dfrac{1}{2 \cdot 3} + \dfrac{1}{3 \cdot 4} + \cdots + \dfrac{1}{n(n+1)} = \dfrac{n}{n+1}$

13. $1 \cdot 2 + 2 \cdot 3 + 3 \cdot 4 + \cdots + n(n+1) = \dfrac{n(n+1)(n+2)}{3}$

14. $1 \cdot 4 + 2 \cdot 9 + 3 \cdot 16 + \cdots + n(n+1)^2 = \dfrac{1}{12} n(n+1)(n+2)(3n+5)$

B
15. Show that $n^3 + 2n$ is divisible by 3 for every $n \in N$.

16. Show that $3^{2n} - 1$ is divisible by 8 for every $n \in N$.

17. Show that if $2 + 4 + 6 + \cdots + 2n = n(n+1) + 2$ is true for $n = k$, then it is true for $n = k + 1$. Is it true for every $n \in N$?

18. Show that $n^3 + 11n = 6(n^2 + 1)$ is true for $n = 1, 2,$ and 3. Is it true for every $n \in N$?

19. Use mathematical induction to prove Theorem 10.1.

20. Use mathematical induction to prove Theorem 10.2.

21. Use mathematical induction to prove that for a geometric progression, $S_n = \dfrac{a - ar^n}{1 - r}$ for all $n \in N$, provided $r \neq 1$.

Chapter Review

[10.1] Write the first three terms of the sequence with the general term as given.

1. $s_n = n^2 + 1$
2. $s_n = \dfrac{1}{n+1}$

Write the next three terms of each of the following arithmetic progressions.

3. $7, 10, \ldots$
4. $a, a - 2, \ldots$

Write the next three terms in each of the following geometric progressions.

5. $-2, 6, \ldots$
6. $\dfrac{2}{3}, 1, \ldots$

7. Find the general term and the seventh term of the arithmetic progression $-3, 2, \ldots$.

8. Find the general term and the fifth term of the geometric progression $-2, \dfrac{2}{3}, \ldots$.

9. If the fourth term of an arithmetic progression is 13 and the ninth term is 33, find the seventh term.

10. Which term in a geometric progression $-\dfrac{2}{9}, \dfrac{2}{3}, \ldots$ is 54?

[10.2] 11. Write $\sum_{k=2}^{5} k(k-1)$ in expanded form.

12. Write $x^2 + x^3 + x^4 + \cdots$ in sigma notation.

13. Find the value for $\sum_{j=3}^{9} (3j - 1)$.

14. Find the value for $\sum_{j=1}^{5} \left(\dfrac{1}{3}\right)^j$.

[10.3] 15. Specify the limit of $\dfrac{3n^2 - 1}{n^2}$ as $n \to \infty$.

16. Find the value of the geometric series $4 - 2 + 1 - \dfrac{1}{2} + \cdots$.

17. Find the value of $\sum_{i=1}^{\infty} \left(\dfrac{1}{3}\right)^i$.

18. Find a fraction equivalent to $0.44\overline{4}$.

[10.4] 19. Write $n(n-3)!$ in expanded form for $n=5$.

Write each of the following expressions in expanded form and simplify.

20. $\dfrac{8!3!}{7!}$ 21. $\dbinom{7}{2}$ 22. $\dfrac{(n-1)!}{n!(n+1)!}$

23. Write the first four terms of the binomial expansion of $(x-2y)^{10}$.

24. Find the eighth term in the expansion of $(x-2y)^{10}$.

[10.5] *By mathematical induction, prove each formula for all positive integral values of n.*

25. $3+6+9+\cdots+3n = \dfrac{3n(n+1)}{2}$ 26. $\dfrac{1}{2}+\dfrac{1}{4}+\dfrac{1}{8}+\cdots+\dfrac{1}{2^n} = 1-\dfrac{1}{2^n}$

11 Counting and Probability

In this final chapter we consider several important counting techniques, permutations and combinations, and functions $P_{n,r}$ and $C_{n,r}$ related to them. These notions are then applied to computing probabilities.

11.1 Basic Counting Principles; Permutations

Associated with each finite set A is a nonnegative integer n, namely the number of elements in A. Hence, we have a function from the set of all finite sets to the set of nonnegative integers. The symbolism $n(A)$ is used to denote elements in the range of this set function n. For example, if

$$A = \{5, 7, 9\}, \quad B = \{1/2, 0, 3, -5, 7\}, \quad C = \emptyset,$$

then

$$n(A) = 3, \quad n(B) = 5, \quad \text{and} \quad n(C) = 0.$$

Counting properties

All the sets with which we shall hereafter be concerned are assumed to be finite sets. We then have the following properties, called **counting properties**, for the function n.

  I $n(A \cup B) = n(A) + n(B)$, if $A \cap B = \emptyset$.

Thus, if A and B are disjoint sets, then the number of elements in their union is the sum of the number of elements in A and the number of elements in B.

For example, suppose there are five roads from town R to town S, and two railroads from town R to town S. If A is the set of roads and B the set of railroads from R to S, then $n(A) = 5$, $n(B) = 2$, and $n(A \cup B) = 5 + 2 = 7$; thus there are seven ways one can go from town R to town S by driving or riding on a train.

  II $n(A \cup B) = n(A) + n(B) - n(A \cap B)$, if $A \cap B \neq \emptyset$.

That is, if A and B have a nonnull intersection, then to count the number of elements in $A \cup B$, we might add the number of elements in A to the number of elements in B. But since any elements in the intersection of A and B are counted twice in this process (once in A and once in B), we must subtract the number of such elements from the sum $n(A) + n(B)$ to obtain the number of elements in $A \cup B$, as suggested in Figure 11.1.

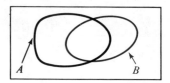

Figure 11.1

For example, suppose there are fifteen unrelated girls and seventeen unrelated boys in a mathematics class, and suppose that there are precisely two brother-sister pairs in the class. If A denotes the set of different families represented by the girls and B denotes the set of different families represented by the boys, then the number of different families represented by all of the members of the class is

$$n(A) + n(B) - n(A \cap B) = 15 + 17 - 2 = 30.$$

Actually, Property I is a consequence of Property II.

Definition 11.1 The **Cartesian product** of two sets A and B, denoted by $A \times B$, is the set of all ordered pairs (a, b) such that $a \in A$ and $b \in B$.

The following property relates $n(A \times B)$ to $n(A)$ and $n(B)$.

III $n(A \times B) = n(A) \cdot n(B)$

This asserts that the number of elements in the Cartesian product of sets A and B is the product of the number of elements in A and the number of elements in B.

For example, suppose that there are five roads from town R to town S (set A), and further suppose that there are three roads from town S to town T (set B). Then for each element of A there are three elements of B, and the total possible ways one can drive from R to T via S is

$$n(A \times B) = n(A) \cdot n(B) = 5 \cdot 3 = 15.$$

Property III can be stated in an equivalent way as follows. Suppose the first of two operations (for example choosing a road from R to S) can be done in a ways. Suppose the second operation (choosing a road from S to T) can be done in b ways and each possible second operation can succeed every first operation. Then the two operations in sequence can be done in $a \cdot b$ ways. This formulation has the following very useful generalization.

IV Suppose the first of several operations can be done in a ways, the second in b ways, no matter what came first, the third in c ways, no matter what came prior, and so on. Then the number of ways the operation can be done in sequence is $a \cdot b \cdot c \cdot \cdots$.

11.1 Basic Counting Principles; Permutations

Example Given the set of digits $A = \{1, 2, 3\}$, how many different three-digit numerals can be constructed from the members of A if no member is used more than once?

Solution We apply the fourth counting property. The first digit can be any of the three numerals 1, 2, or 3. No matter what comes first, 2 choices remain for the second digit and then 1 choice remains for the third digit. By the fourth counting property, the number of ways of choosing the three digits in order is $3 \cdot 2 \cdot 1 = 6$.

Permutations Notice that in the preceding example the order in which the digits were listed was important when forming the three-digit numerals.

Definition 11.2 *A **permutation** of a set A is an ordering (first, second, etc.) of the members of A.*

To count the number of permutations of a set with n members, note that we have n choices for the first member in the permutation, $n - 1$ for the second, and so on until only 1 remains for the last. By Property IV, the total number of possible permutations is

$$n(n - 1) \cdot (n - 2) \cdot \cdots \cdot 1 = n!$$

This proves the following theorem.

Theorem 11.1 *Let $P_{n,n}$ denote the number of distinct permutations of the members of a set A containing n members. Then*

$$P_{n,n} = n! \tag{1}$$

The symbol $P_{n,n}$ (or sometimes $_nP_n$ or P_n^n) is read "the number of permutations of n things taken n at a time."

Example In how many ways can nine men be assigned positions to form distinct baseball teams?

Solution Let A denote the set of men, so that $n(A) = 9$. The total number of ways in which 9 men can be assigned 9 positions on a team, or, in other words, the number of possible permutations of the members of a 9-element set, is, by Equation (1),

$$P_{9,9} = 9! = 9 \cdot 8 \cdot 7 \cdots 1 = 362{,}880.$$

In many applications there are more members from which to choose than there are places to fill, as the following theorem and example illustrate.

Theorem 11.2 Let $P_{n,r}$ denote the number of permutations of the members, taken r at a time, of a set A containing n members; that is, let $P_{n,r}$ be the number of distinct orderings of r elements when there is a set A of n elements from which to choose. Then

$$P_{n,r} = n(n-1)(n-2)\cdots(n-r+1). \qquad (2)$$

The proof follows the proof of Theorem 11.1 except that only r choices are made so the last choice involves $n - (r-1) = n - r + 1$ possibilities.

Example

In how many ways can a basketball team be formed by choosing players for the five positions from a set of ten players?

Solution

Let A denote the set of players, so that $n(A) = 10$. Then from (2) and the fact that a basketball team consists of 5 players, we have

$$P_{10,5} = 10 \cdot 9 \cdot 8 \cdots (10 - 5 + 1)$$
$$= 10 \cdot 9 \cdot 8 \cdot 7 \cdot 6 = 30{,}240.$$

An alternative expression for $P_{n,r}$ can be obtained by observing that

$$P_{n,r} = n(n-1)(n-2)\cdots(n-r+1)$$
$$= \frac{n(n-1)(n-2)\cdots(n-r+1)(n-r)!}{(n-r)!},$$

so that

$$P_{n,r} = \frac{n!}{(n-r)!}. \qquad (3)$$

Distinguishable permutations

The problem of finding the number of distinguishable permutations of n objects taken n at a time, if some of the objects are identical, requires a little more careful analysis. As an example, consider the number of permutations of the letters of the word *DIVISIBLE*. We can make a distinction between the three *I*'s by assigning subscripts to each so that we have nine distinct letters,

$$D, I_1, V, I_2, S, I_3, B, L, E.$$

The number of permutations of these nine letters is of course 9!. If the letters other than I_1, I_2, and I_3 are retained in the positions they occupy in a permutation of the above nine letters, I_1, I_2, and I_3 can be permuted among themselves 3! ways. Thus, if P is the number of *distinguishable* permutations of the letters

$$D, I, V, I, S, I, B, L, E,$$

then, since for each of these there are 3! ways in which the *I*'s can be permuted without otherwise changing the order of the other letters, it follows that

$$3! \cdot P = 9!,$$

11.1 Basic Counting Principles; Permutations

from which

$$P = \frac{9!}{3!}.$$

As another example, consider the letters of the word *MISSISSIPPI*. There would exist 11! distinguishable permutations of the letters in this word if each letter were distinct. Note, however, that the letters *S* and *I* each appear four times and the letter *P* appears twice. Reasoning as we did in the previous example, we see that the number *P* of distinguishable permutations of the letters in *MISSISSIPPI* is given by

$$4!4!2!P = 11!,$$

from which

$$P = \frac{11!}{4!4!2!}.$$

These two examples suggest the following theorem, which we state without proof.

Theorem 11.3 *If there are n_1 identical objects of a first kind, n_2 of a second, ..., and n_k of a kth, with $n_1 + n_2 + \cdots + n_k = n$, then the number of distinguishable permutations of the n objects is given by*

$$P = \frac{n!}{n_1! n_2! \cdots n_k!}.$$

Exercise 11.1

A *Given the following sets, find $n(A \cap B)$, $n(A \cup B)$, and $n(A \times B)$.*

Example $\quad A = \{a, b, c\}, \; B = \{c, d\}$

Solution $\quad A \cap B = \{c\}$. Therefore $n(A \cap B) = 1$.
$n(A \cup B) = n(A) + n(B) - n(A \cap B) = 3 + 2 - 1 = 4$
$n(A \times B) = n(A) \cdot n(B) = 3 \cdot 2 = 6$

1. $A = \{d, e\}, \; B = \{e, f, g, h\}$
2. $A = \{e\}, \; B = \{a, b, c, d\}$
3. $A = \{1, 2, 3, 4\}, \; B = \{3, 4, 5, 6\}$
4. $A = \{1, 2\}, \; B = \{3, 4, 5\}$
5. $A = \{1, 2\}, \; B = \{1, 2\}$
6. $A = \emptyset, \; B = \{2, 3, 4\}$

Example In how many ways can three members of a class be assigned a grade of *A, B, C,* or *D*?

Solution Sometimes a simple diagram, such as ——, ——, ——, designating a sequence, is a helpful preliminary device. Since each of the students may receive any one of four different grades, the sequence would appear as

$$\underline{4}, \underline{4}, \underline{4}.$$

From counting Property IV, there are $4 \cdot 4 \cdot 4$, or 64, possible ways the grades may be assigned.

Example In how many different ways can three members of a class be assigned a grade of *A, B, C,* or *D* so that no two members receive the same grade?

Solution Since the first student may receive any one of four different grades, the second student may then receive any one of three different grades, and the third student may then receive any one of two different grades, the sequence would appear as

$$\underline{4}, \underline{3}, \underline{2}.$$

From counting Property IV, there are $4 \cdot 3 \cdot 2$, or 24, possible ways the grades may be assigned. In this case, we could have obtained the same result from Theorem 11.2, since $P_{4,3} = 4 \cdot 3 \cdot 2 = 24$.

In each problem, a digit or letter may be used more than once unless stated otherwise.

7. How many different two-digit numerals can be formed from the digits 5 and 6?
8. How many different two-digit numerals can be formed from the digits 7, 8, and 9?
9. In how many different ways can four students be seated in a row?
10. In how many different ways can five students be seated in a row?
11. In how many different ways can four questions on a true-false test be answered?
12. In how many different ways can five questions on a true-false test be answered?
13. In how many ways can you write different three-digit numerals from $\{2, 3, 4, 5\}$?
14. How many different seven-digit telephone numbers can be formed from the set of digits $\{1, 2, 3, 4, 5, 6, 7, 8, 9, 0\}$?
15. In how many ways can you write different three-digit numerals, using $\{2, 3, 4, 5\}$, if no digit is to be used more than once in each numeral?
16. How many different seven-digit telephone numbers can be formed from the set of digits $\{1, 2, 3, 4, 5, 6, 7, 8, 9, 0\}$ if no digit is to be used more than once in any number?
17. How many three-letter permutations can be formed from $\{A, N, S, W, E, R\}$?
18. If no letter is to be used more than once in any permutation, how many different three-letter permutations can be formed from $\{A, N, S, W, E, R\}$?
19. How many four-digit numerals for positive odd integers can be formed from $\{1, 2, 3, 4, 5\}$?
20. How many four-digit numerals for positive even integers can be formed from $\{1, 2, 3, 4, 5\}$?

11.1 Basic Counting Principles; Permutations

21. How many numerals for positive integers less than 500 can be formed from $\{3, 4, 5\}$?
22. How many numerals for positive odd integers less than 500 can be formed from $\{3, 4, 5\}$?
23. How many numerals for positive even integers less than 500 can be formed from $\{3, 4, 5\}$?
24. How many numerals for positive even integers between 400 and 500, inclusive, can be formed from $\{3, 4, 5\}$?
25. How many permutations of the elements of $\{P, R, I, M, E\}$ end in a vowel?
26. How many permutations of the elements of $\{P, R, O, D, U, C, T\}$ end in a vowel?
27. Find the number of distinguishable permutations of the letters in the word

    LIMIT.

28. Find the number of distinguishable permutations of the letters in the word

    COMBINATION.

29. Find the number of distinguishable permutations of the letters in the word

    COLORADO.

30. Find the number of distinguishable permutations of the letters in the word

    TALLAHASSEE.

B Show that each of the following is true.

31. $P_{5,3} = 5(P_{4,2})$
32. $P_{5,r} = 5(P_{4,r-1})$
33. $P_{n,3} = n(P_{n-1,2})$
34. $P_{n,3} - P_{n,2} = (n-3)(P_{n,2})$
35. Solve for n: $P_{n,5} = 5(P_{n,4})$.
36. Solve for n: $P_{n,5} = 9(P_{n-1,4})$.

Example In how many ways can four students be seated around a circular table?

Solution In any such arrangement (which is called a **circular permutation**), there is no first position. Each person can take four different initial positions without affecting the arrangement. Thus, there are

$$\frac{4!}{4} = 6 \quad \text{arrangements.}$$

(In general, there are $n!/n$, or $(n-1)!$, circular permutations of n things taken n at a time.)

37. In how many ways can five students be seated around a circular table?
38. In how many ways can six students be seated around a circular table?
39. In how many ways can six students be seated around a circular table if a certain two must be seated together?

40. In how many ways can three different keys be arranged on a key ring? *Hint*: Arrangements should be considered identical if one can be obtained from the other by turning the ring over. In general, there are only $(1/2)(n - 1)!$ distinct arrangements of n keys on a ring $(n \geq 3)$.

11.2 Combinations

An additional counting concept is needed before we turn our attention to probability—namely, finding the number of distinct r-element subsets of an n-element set with no reference to relative order of the elements in the subset. For example, five different cards can be arranged in 5! permutations, but to a poker player they represent the same hand. The set of five cards (with no reference to the arrangement of the cards) is called a *combination*.

Definition 11.3   *A subset of an n-element set A is called a **combination**.*

The counting of combinations is related to the counting of permutations. From Theorem 11.2, we know that the number of distinct permutations of n elements of a set A taken r at a time is given by

$$P_{n,r} = \frac{n!}{(n-r)!}.$$

With this in mind, consider the following result concerning the number $C_{n,r}$ of combinations of n things taken r at a time. The symbol $C_{n,r}$ is read "the number of combinations of n things taken r at a time."

Theorem 11.4   *The number $C_{n,r}$ of distinct combinations of the members, taken r at a time, of a set containing n members is given by*

$$C_{n,r} = \frac{P_{n,r}}{r!}. \tag{1}$$

To see that this is true, we observe that there are, by definition, $C_{n,r}$ r-element subsets of the set A, where $n(A) = n$. Also, from Theorem 11.1, each of these subsets has $r!$ permutations of its members. There are therefore $C_{n,r} r!$ permutations of n elements of A taken r at a time. Thus

$$P_{n,r} = C_{n,r} r!,$$

from which we obtain

$$C_{n,r} = \frac{P_{n,r}}{r!}.$$

Thus, to find the number of r-element subsets of an n-element set A, we count the number of permutations of the elements of A taken r at a time, and then divide

11.2 Combinations

by the number of possible permutations of an r-element set. This seems very much like counting a set of people by counting the number of arms and legs and dividing the result by 4, but this approach gives us a very useful expression for the number we seek, $C_{n,r}$. Since

$$P_{n,r} = n(n-1)(n-2)\cdots(n-r+1),$$

it follows that

$$C_{n,r} = \frac{P_{n,r}}{r!} = \frac{n(n-1)(n-2)\cdots(n-r+1)}{r!}. \quad (2)$$

Example In how many ways can a committee of five be selected from a set of twelve persons?

Solution What we wish here is the number of 5-element subsets of a 12-element set. From (2), we have

$$C_{12,5} = \frac{12 \cdot 11 \cdot 10 \cdot 9 \cdot 8}{5 \cdot 4 \cdot 3 \cdot 2 \cdot 1} = 792.$$

By Equation (3) on page 370, we have the alternative expression

$$C_{n,r} = \frac{P_{n,r}}{r!} = \frac{n!}{r!(n-r)!}. \quad (3)$$

Notice that the value of $C_{n,r}$ given in the right-hand member of (3) is the same as the value of the binomial coefficient $\binom{n}{r}$ given in the right-hand member of (2) on page 354. Thus, $C_{n,r}$ and $\binom{n}{r}$ are equal and therefore are interchangeable in mathematical formulas:

$$C_{n,r} = \binom{n}{r}.$$

Since the numbers $\binom{n}{r}$, or $C_{n,r}$, are the coefficients in the binomial expansion, and since these coefficients are symmetric, we have the following plausible assertion.

Theorem 11.5 $C_{n,r} = C_{n,n-r}$

To verify this, note that from (3) we have both

$$C_{n,r} = \frac{n!}{r!(n-r)!}$$

and

$$C_{n,n-r} = \frac{n!}{(n-r)![n-(n-r)]!} = \frac{n!}{(n-r)!r!} = C_{n,r}.$$

Theorem 11.5 is plausible also since each time a distinct set of r objects is chosen, a distinct set of $n - r$ objects remains unchosen.

Exercise 11.2

Example How many different amounts of money can be formed from a penny, a nickel, a dime and a quarter?

Solution We want to find the total number of combinations that can be formed by taking the coins 1, 2, 3, and 4 at a time. By Equation (3) we have

$$C_{4,1} = \frac{4!}{1!3!} = 4, \quad C_{4,2} = \frac{4!}{2!2!} = 6, \quad C_{4,3} = \frac{4!}{3!1!} = 4, \quad C_{4,4} = \frac{4!}{4!0!} = 1,$$

and the total number of combinations is 15. Clearly each combination gives a different amount.

A
1. How many different amounts of money can be formed from a penny, a nickel, and a dime?
2. How many different amounts of money can be formed from a penny, a nickel, a dime, a quarter, and a half-dollar?
3. How many different committees of four persons each can be chosen from a group of six persons?
4. How many different committees of four persons each can be chosen from a group of ten persons?
5. In how many different ways can a set of five cards be selected from a standard bridge deck containing 52 cards?
6. In how many different ways can a set of 13 cards be selected from a standard bridge deck of 52 cards?
7. In how many different ways can a hand consisting of five spades, five hearts, and three diamonds be selected from a standard bridge deck of 52 cards?
8. In how many different ways can a hand consisting of ten spades, one heart, one diamond, and one club be selected from a standard bridge deck of 52 cards?
9. In how many different ways can a hand consisting of either five spades, five hearts, five diamonds, or five clubs be selected from a standard bridge deck?
10. In how many different ways can a hand consisting of three aces and two cards that are not aces be selected from a standard bridge deck?
11. A combination of three balls is picked at random from a box containing five red, four white, and three blue balls. In how many ways can the set chosen contain at least one white ball?
12. In Exercise 11, in how many ways can the set chosen contain at least one white and one blue ball?
13. A set of five distinct points lies on a circle. How many inscribed triangles can be drawn having all their vertices in this set?
14. A set of ten distinct points lies on a circle. How many inscribed triangles can be drawn having all of their vertices in this set?

11.3 Probability Functions

B 15. A set of ten distinct points lies on a circle. How many inscribed hexagons can be drawn having all their vertices in this set?

16. A set of ten distinct points lies on a circle. How many inscribed quadrilaterals can be drawn having all their vertices in this set?

17. Given $C_{n,7} = C_{n,5}$, find n.

18. Given $C_{n,3} = C_{50,47}$, find n.

11.3 Probability Functions

Sample spaces, outcomes, and events

When an experiment of some kind is undertaken, associated with the experiment is a set of possible results. For example, when a die is rolled, it will come to a stop with the number of spots on its upper face corresponding to one of the numerals 1, 2, 3, 4, 5, 6. This exhausts all possibilities. Of course, in the absence of chicanery, exactly which one of these random results will occur cannot be precisely specified in advance of the experiment. A question of great practical importance regarding the result of casting a die, and the result of any comparable experiment in which the outcome is uncertain, is "Can we assign some kind of measure to the degree of uncertainty involved?" The answer is "Yes," but before we assign a measure, let us define some necessary terms.

Definition 11.4 *The set of all possible results of an experiment is called a **sample space** for the experiment, and is commonly denoted by the letter S.*

Definition 11.5 *Each element of a sample space is called an **outcome**, or **sample point**.*

Definition 11.6 *Any subset of a sample space is called an **event**, and is commonly denoted by the letter E.*

The reason for the terminology in Definition 11.6 is that, in conducting an experiment, one may be interested in sets of outcomes rather than in individual outcomes. In the tossing of a die, for example, if the sample space is taken as {1, 2, 3, 4, 5, 6}, then the event that an outcome (a numeral) denotes an even integer is the set {2, 4, 6}, which is a subset of the sample space. The event that an outcome denotes an odd integer is the set {1, 3, 5}. These two events are complements of each other, and are examples of **complementary events**, that is, two events whose intersection is ∅ and whose union is the entire sample space.

The likelihood of any outcome in a sample space is given by a **probability function**. Such a function assigns to each outcome a number called the probability of the outcome. Probabilities are numbers in [0, 1], and the sum of the probabilities of all outcomes must be 1. If the probability of an outcome is 0.25, this means that the outcome should occur 1/4 or 25% of the time in the long run. Probabilities can be assigned by either a priori or a posteriori considerations, as follows.

A priori considerations involve physical, geometrical, and other inherent properties of the experiment in question. They involve no sampling of outcomes. One way to assign a probability to an event is simply to use the ratio of the number of outcomes in the event to the number of possible outcomes. Thus, when a die is cast and we admit as outcomes the die's stopping with any of its six different faces uppermost, then without making any trial throws we would assign the value 1/6 as the probability of each of the six possible outcomes.

In general, we have the following definition.

Definition 11.7 *If E is any subset containing $n(E)$ members (outcomes) of a sample space containing $n(S)$ equally likely outcomes, then the (a priori)* **probability** *of the occurrence of E, $P(E)$, is given by*

$$P(E) = \frac{n(E)}{n(S)}. \tag{1}$$

Example

If two dice are cast, what is the (a priori) probability that the sum of the number of dots showing on the top faces of the dice is less than 6?

Solution

For our sample spaces, let us consider A the set of possible outcomes for one die and B the set for the other. Then

$$n(A) = 6 \quad \text{and} \quad n(B) = 6.$$

The possible outcomes for both would be $S = A \times B$, the Cartesian product of A and B, and $n(A \times B) = n(S) = 36$. Each outcome here is an ordered pair (a, b), where a is the numeral on the upper face of the first die, and b is that of the second die. The event we seek is $\{(a, b) \mid a + b < 6\}$. The lattice in the figure shows the situation schematically. Since

$$E = \{(1, 1), (1, 2), (1, 3), (1, 4), (2, 1),$$
$$(2, 2), (2, 3), (3, 1), (3, 2), (4, 1)\},$$

we have $n(E) = 10$, and

$$P(E) = \frac{n(E)}{n(S)} = \frac{10}{36} = \frac{5}{18}.$$

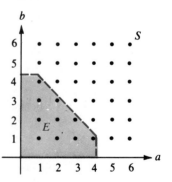

Care must be taken in interpreting the meaning of "the probability of an event." In the foregoing example, 5/18 does not assure us, for instance, that E will occur 5 times out of 18 casts, or, indeed, that even one cast out of 18 will produce the event described. What it does imply, however, is that if you cast the dice a very great number of times, then you can *expect* the sum on the exposed faces to be less than 6 about 5/18 of the time.

11.3 Probability Functions

A posteriori considerations involve testing the experiment a certain number of times. Mortality tables give probability functions of this sort. Actually, (1) can still be used in defining such probability functions, provided we interpret the function $n(E)$ as being the number of times the event E occurred in the test and $n(S)$ as being the total number of times the experiment was performed in the test.

Although diagrams are often useful in illustrating an event and a sample space, the number of outcomes $n(E)$ in the event and the number $n(S)$ of possible outcomes in the sample space ordinarily are computed directly.

Example If three marbles are drawn at random from an urn containing six white marbles and four blue marbles, what is the probability that the three marbles are all white?

Solution Since the number of ways of drawing three white marbles from six white marbles is $C_{6,3}$, and the number of ways of drawing three marbles from the total number of marbles is $C_{10,3}$, we have

$$P(E) = \frac{n(E)}{n(S)} = \frac{C_{6,3}}{C_{10,3}} = \frac{\frac{6!}{3!3!}}{\frac{10!}{3!7!}} = \frac{1}{6}.$$

Exercise 11.3

1. A die is cast. List the outcomes in the sample space. List the outcomes in the event E that the number on the upper face of the die is greater than 2. Determine $P(E)$.

2. A coin is tossed. List the outcomes in the sample space. List the outcomes in the event E that a head appears. Determine $P(E)$.

3. Two coins are tossed. List the outcomes in the sample space. List the outcomes in the event E that both coins show the same face. Determine $P(E)$.

4. A die is cast and a coin is tossed. List the outcomes in the sample space. List the outcomes in the event E that the coin shows a head and the die shows a numeral greater than 4. Determine $P(E)$.

In Exercises 5–10, consider an experiment in which two dice are cast. Determine the probability that the specified event occurs.

5. The sum of the numbers of dots shown is 7.
6. The sum of the numbers of dots shown is 8.
7. The sum of the numbers of dots shown is 3.
8. The sum of the numbers of dots shown is 12.
9. At least one of the numbers of dots shown is less than 3.
10. Both dice show the same number of dots.

In Exercises 11–14, consider an experiment in which a single card is drawn at random from a standard deck of 52 cards. Determine the probability that the specified event occurs.

11. The card is the King of hearts.
12. The card is an ace.
13. The card is a black face card.
14. The card is a 5, 6, 7, or 8.

In Exercises 15–18, consider an experiment in which two cards are drawn at random from a standard deck of 52 cards. Determine the probability that the specified event occurs.

15. Both cards are spades.
16. Both cards are red.
17. Both cards are face cards.
18. Both cards are aces.

In Exercises 19–22, consider an experiment in which 2 marbles are drawn at random from an urn containing 8 red, 6 blue, 4 green, and 2 white marbles. Determine the probability that the specified event occurs.

19. Both marbles are white.
20. Both marbles are green.
21. Both marbles are blue.
22. Both marbles are red.

11.4 Probability of the Union of Events

Probability of the complement of an event

Recall that we defined the **complement** of a subset E of a set S to be the set of all members of S that do not belong to E. If we denote the complement of an event E in a sample space S by E', then $P(E')$ denotes the probability of the occurrence of E'. Since an outcome in S must lie in either E or E', but not both, and $P(S) = 1$, it follows that

$$P(E) + P(E') = 1,$$

or

$$P(E') = 1 - P(E).$$

We can often use this fact to simplify the computing of a probability.

Example

If two marbles are drawn at random from an urn containing 6 white, 2 red, and 5 green marbles, what is the probability that not both are white, that is, that at least one is not white?

Solution

Rather than consider all of the possible pairs of marbles in which not both are white, we simply compute the probability of the event E that they *are* both white and subtract the result from 1. We have

$$P(E) = \frac{C_{6,2}}{C_{13,2}} = \frac{15}{78} = \frac{5}{26}.$$

11.4 Probability of the Union of Events

Then

$$P(E') = 1 - \frac{5}{26} = \frac{21}{26},$$

and this is just the probability that not both marbles are white.

Probability of the union of events

By the second counting property on page 367, if E_1 and E_2 are events in a sample space S, then

$$n(E_1 \cup E_2) = n(E_1) + n(E_2) - n(E_1 \cap E_2).$$

This leads directly to the following theorem.

Theorem 11.6 *If S is a sample space, and E_1 and E_2 are any events in S, then*

$$P(E_1 \text{ or } E_2) = P(E_1 \cup E_2) = P(E_1) + P(E_2) - P(E_1 \cap E_2).$$

Example

If two cards are drawn from a standard deck of playing cards, what is the probability that either both are red or both are Jacks?

Solution

A deck of cards contains 52 cards, and an outcome here consists of two cards. Hence, the number of elements of the sample space is the number of ways (combinations) one can draw two cards from 52, which is $C_{52,2}$. Let E_1 be the event that both are red. Since there are 26 red cards in a deck, the number of outcomes (combinations) in the event E_1 is

$$n(E_1) = C_{26,2}.$$

Let E_2 be the event that both cards are Jacks. Then, because there are four Jacks,

$$n(E_2) = C_{4,2}.$$

Since there is only one red pair of Jacks, $n(E_1 \cap E_2) = C_{2,2} = 1$. We then have

$$P(E_1 \cup E_2) = P(E_1) + P(E_2) - P(E_1 \cap E_2)$$

$$= \frac{C_{26,2}}{C_{52,2}} + \frac{C_{4,2}}{C_{52,2}} - \frac{1}{C_{52,2}}$$

$$= \frac{C_{26,2} + C_{4,2} - 1}{C_{52,2}}$$

$$= \frac{\frac{26 \cdot 25}{1 \cdot 2} + \frac{4 \cdot 3}{1 \cdot 2} - \frac{2 \cdot 1}{1 \cdot 2}}{\frac{52 \cdot 51}{1 \cdot 2}}$$

$$= \frac{325 + 6 - 1}{1326} = \frac{330}{1326} = \frac{165}{663} = \frac{55}{221}.$$

Probability of the union of disjoint events

Of course, if E_1 and E_2 are *disjoint*, then $E_1 \cap E_2 = \emptyset$, so that $P(E_1 \cap E_2) = 0$, and then the equation in Theorem 11.6 reduces to

$$P(E_1 \cup E_2) = P(E_1) + P(E_2).$$

Disjoint events are said to be **mutually exclusive**.

Example

A card is drawn at random from a standard deck of 52 cards. What is the probability that the card is either a face card (Jack, Queen, or King) or a four?

Solution

Let E_1 be the event that the card is a four. There are four such cards in a deck, so that $n(E_1) = C_{4,1} = 4$. Let E_2 be the event that the card is a Jack, Queen, or King. There are twelve such cards in a deck. Hence, $n(E_2) = C_{12,1} = 12$. Since the sample space is just the entire deck, $n(S) = C_{52,1} = 52$, and since E_1 and E_2 are mutually exclusive,

$$P(E_1 \cup E_2) = P(E_1) + P(E_2)$$
$$= \frac{4}{52} + \frac{12}{52} = \frac{16}{52} = \frac{4}{13}.$$

Therefore, the probability is 4/13.

Exercise 11.4

A Two dice are cast. Let E be the event that both dice show the same numeral. Let F be the event that the sum of the numbers thrown is greater than eight. Find each of the following probabilities.

1. $P(E)$
2. $P(F)$
3. $P(E \cup F)$
4. $P(E')$
5. $P(F')$
6. $P(E' \cup F')$

A box contains five red, four white, and three blue marbles. Two marbles are drawn from the box. Let RR be the event that both marbles are red, WW that both marbles are white, BB that both marbles are blue, and RW, RB, BW that a red and a white, a red and a blue, and a blue and a white are drawn, respectively. Find each probability.

7. $P(RR)$
8. $P(BB)$
9. $P(WW)$
10. $P(RW)$
11. $P(RB)$
12. $P(BW)$
13. What is the probability that neither is white?
14. What is the probability that neither is blue?
15. What is the probability that at least one is red?
16. What is the probability that either one is red or else both are white?

17. What is the probability that by drawing a single card from a standard deck of 52 cards, one will get a 2, 3, or 4?

18. What is the probability that if two cards are drawn from a standard deck of 52 cards they will be of the same suit? Different suits?

B *If the probability of the event E that a person will receive k dollars is P(E), then the person's* **mathematical expectation relative to this event** *is kP(E).*

19. A lottery offers a prize of $50, and 70 tickets are sold. What is the mathematical expectation of a person who buys three tickets? If each ticket costs $1, is the person's expectation greater or less than his outlay?

20. The probability that a certain horse will win the Irish Sweepstakes is 2/9. If you hold a ticket on this horse to pay $100,000 if he wins, what is your mathematical expectation?

If E_1, E_2, E_3, etc., are mutually exclusive events, and the return to a person is k_1 if E_1 occurs, k_2 if E_2 occurs, etc., then the person's mathematical expectation is $\sum_{i=1}^{n} k_i P(E_i)$.

21. One coin is selected at random from a collection containing a penny, a nickel, and a dime. What is the expectation?

22. One coin is selected at random from a collection containing a dime, a quarter, and a half-dollar. What is the expectation?

23. Three $1 bills and four $5 bills are hidden from view. What is the expectation on a single selection?

24. Three $1 bills, four $5 bills, and one $10 bill are hidden from view. What is the expectation on a single draw?

11.5 Probability of the Intersection of Events

In some experiments, we may be interested in events that are not dependent on each other, in the sense that the occurrence of one may have no effect on the probability of the occurrence or nonoccurrence of the other.

Independent events

Consider an experiment in which two cards are drawn at random, one after the other, from a deck of ten cards, six of which are red and four blue. We can inquire into the probability that the first card drawn is red and the second blue. The simplest such situation would be one in which the first card is drawn, observed, and returned to the deck, which is then shuffled thoroughly before the second card is drawn. In this case, the sample space would consist of a set of ordered pairs (x, y), where x is the result of the first draw and y the result of the second draw.

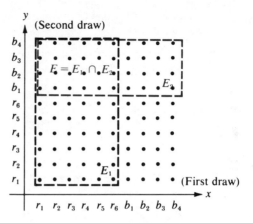

Figure 11.2

Since there are ten possibilities in each case, the sample space consists of $10 \times 10 = 100$ ordered pairs. In the Cartesian graph of Figure 11.2, r_i and b_i are used to designate the drawing of red and blue cards, respectively.

The events E_1 that the first card drawn is a red card and E_2 that the second card drawn is a blue card are outlined in the figure. The event E that both E_1 and E_2 occur is the intersection of E_1 and E_2; that is, $E = E_1 \cap E_2$. By inspection,

$$P(E_1) = \frac{60}{100} = \frac{3}{5}, \qquad P(E_2) = \frac{40}{100} = \frac{2}{5},$$

and

$$P(E) = P(E_1 \cap E_2) = \frac{24}{100} = \frac{6}{25}.$$

Moreover, in this example, it is evident that

$$P(E) = P(E_1 \cap E_2) = P(E_1) \cdot P(E_2).$$

Definition 11.8 *If E_1 and E_2 are events in a sample space, and if*

$$P(E_1 \cap E_2) = P(E_1) \cdot P(E_2),$$

*then E_1 and E_2 are **independent events**. If two events are not independent, then they are said to be **dependent**.*

Dependent events

Now consider the same experiment, except that this time the first card is not returned to the deck before the second is taken. Then there will be ten possible first draws, but only nine possible second draws. The sample space will therefore contain 10×9 ordered pairs, such that no ordered pair with first and second components the same remains in the set.

Figure 11.3 shows a graph of the sample space, which is the same as that in the preceding figure except that one diagonal is missing. Again, the figure shows

11.5 Probability of the Intersection of Events

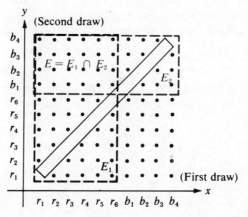

Figure 11.3

E_1 and E_2, the events that a red and a blue are obtained on the first and second draw, respectively. The event that both occur is $E = E_1 \cap E_2$, which is also shown in the figure. By inspection,

$$P(E_1) = \frac{54}{90} = \frac{3}{5}, \quad P(E_2) = \frac{36}{90} = \frac{2}{5}, \quad \text{and} \quad P(E) = P(E_1 \cap E_2) = \frac{24}{90} = \frac{4}{15}.$$

This time,

$$P(E_1 \cap E_2) \neq P(E_1) \cdot P(E_2),$$

so that the events are dependent.

Conditional probability

Suppose that in the experiment described above we wanted to know the probability of the second draw producing a blue card, given that the first draw produced a red card. We would then be interested in the probability of the occurrence of the event E_2 given the occurrence of the event E_1. This is called the **conditional probability of E_2 given E_1** and is denoted $P(E_2|E_1)$.

To compute $P(E_2|E_1)$, observe that since we are assuming E_1 has occurred we do not consider outcomes outside of E_1 to be possible. Thus,

$$P(E_2|E_1) = \frac{n(E_1 \cap E_2)}{n(E_1)}, \quad n(E_1) \neq 0. \tag{2}$$

To state this result in terms of probabilities of E_1, E_2 and $E_1 \cap E_2$, we divide numerator and denominator of the right-hand member of Equation (2) by $n(S)$ to obtain

$$P(E_2|E_1) = \frac{n(E_1 \cap E_2)/n(S)}{n(E_1)/n(S)}$$

$$= \frac{P(E_1 \cap E_2)}{P(E_1)}, \quad P(E_1) \neq 0. \tag{3}$$

Example   Find the conditional probability of drawing an ace on the second draw from a standard deck of 52 cards, given that the first draw produced a King which was not replaced.

Solution   Let E_1 be the event that the first draw produces a King and E_2 be the event that the second draw produces an ace. Then

$$P(E_1 \cap E_2) = \frac{16}{(52)(51)}, \quad P(E_1) = \frac{4}{52},$$

and

$$P(E_2 | E_1) = \frac{\frac{16}{(52)(51)}}{\frac{4}{52}} = \frac{4}{51}.$$

Conditional probability and independence   If E_1 and E_2 are independent and $P(E_1) \neq 0$, then by Definition 11.8 and Equation (3)

$$P(E_2 | E_1) = \frac{P(E_1 \cap E_2)}{P(E_1)} = \frac{P(E_1) \cdot P(E_2)}{P(E_1)} = P(E_2).$$

For example, in the experiment described on page 383, the probability that the second card is blue, given that the first is red, is

$$P(E_2 | E_1) = \frac{P(E_1 \cap E_2)}{P(E_1)} = \frac{\frac{6}{25}}{\frac{3}{5}} = \frac{2}{5} = P(E_2)$$

provided the first card is returned to the deck before the second is drawn; but it is

$$P(E_2 | E_1) = \frac{P(E_1 \cap E_2)}{P(E_1)} = \frac{\frac{4}{15}}{\frac{3}{5}} = \frac{4}{9} \neq P(E_2)$$

provided the first card is not returned to the deck before the second is drawn.

Exercise 11.5

A   1. A bag contains four red marbles and ten blue marbles. If two marbles are drawn in succession, and if the first is not replaced, what is the probability that the first is red and the second is blue? Are the two draws independent events?

2. A red die and a green die are cast. What is the probability of obtaining a sum greater than 9, given that the green die shows 4?

3. A bag contains four white and six red marbles. Two marbles are drawn from the bag and replaced, and two more marbles are then drawn from the bag. What is the probability of drawing:
   a. Two red marbles on the first draw and two white ones on the second draw?
   b. A total of two white marbles?
   c. Four white marbles?
   d. Four red marbles?

4. In Exercise 3, what is the probability of drawing:
   a. Exactly three white marbles in the two draws?
   b. At least three white marbles in the two draws?
   c. Exactly two red marbles in the two draws?
   d. At least two red marbles in the two draws?

5. A red and a green die are cast. Let E_1 be the event that at least one die shows 3 and E_2 be the event that the sum of the two numbers thrown is 8.
   a. Find $P(E_1)$.
   b. Find $P(E_2)$.
   c. Find $P(E_2|E_1)$.
   d. Are E_1 and E_2 independent?

6. In Exercise 5, let E_1 be the event that neither die shows a result larger than 4 and let E_2 be the event that the dice do not show the same number.
   a. Find $P(E_1)$.
   b. Find $P(E_2)$.
   c. Find $P(E_2|E_1)$.
   d. Are E_1 and E_2 independent?

7. A coin is tossed three consecutive times. What is the probability that:
   a. The second toss is a head?
   b. The third toss is a head?
   c. Both the second and third tosses are heads?
   d. The first and third tosses are heads?
   e. The first and third tosses are heads but the second is not?

8. In Exercise 7, state whether each of the following pairs of events is independent.
   a. a and b
   b. a and c
   c. a and d
   d. a and e
   e. d and e

9. The probability that A will pass a course in college algebra is 5/6, that B will pass 3/4, and that C will pass, 2/3. What is the probability of each of the following events?
   a. At least one of the three will pass.
   b. At least A and C will pass.
   c. A and C will pass but B will not.
   d. At least two of the three will pass.

10. In Exercise 9, what is the probability of **c**, given the occurrence of **a**? Are the events **a** and **c** independent?

11. A day is selected at random in some fashion such that any day of the week is an equally likely choice. Let the probability be 1/30 that a day selected at random will be a rainy day.
    a. What is the probability that a rainy Wednesday will be selected?
    b. What is the probability that a dry Thursday will be selected?
    c. What is the probability that either Monday, Tuesday, or Wednesday will be selected, and that it will not rain that day?
    d. Let E_1 be the selection of Sunday, and let E_2 be the event that it does not rain on the day selected. What is the conditional probability of E_2, given the occurrence of E_1? Are E_1 and E_2 independent?

12. Two identical urns contain marbles. One urn contains all red marbles, while half of the contents of the other urn are white and half are red. If a marble is drawn at random from one of the urns and found to be red, what is the probability that it was drawn from the urn containing only red marbles?

13. Of two dice, one is normal, but the other has three faces showing 4 spots and three faces showing 3 spots. If one of the dice is chosen at random and a 4 is thrown with it, what is the probability that the normal die was chosen?

14. Argue that if E_1 and E_2 are mutually exclusive events with nonzero probabilities, then E_1 and E_2 are not independent.

Chapter Review

[11.1]
1. How many different two-digit numerals can be formed from $\{6, 7, 8, 9\}$?
2. How many different ways can six questions on a true-false test be answered?
3. How many four-digit numerals for positive odd integers can be formed from $\{3, 4, 5, 6\}$?
4. How many distinguishable permutations can be formed using the letters in the word TENNIS?

[11.2]
5. How many different committees of 5 persons can be formed from a group of 12 persons?
6. In how many different ways can a hand consisting of 3 spades, 5 hearts, 4 diamonds, and 1 club be selected from a standard bridge deck of 52 cards?
7. A box contains 4 red, 6 white, and 2 blue marbles. In how many ways can you select 3 marbles from the box if at least one of the marbles chosen is white?
8. How many hexagons can be drawn whose vertices are members of a set of 9 fixed points on the circle?

[11.3] *In Exercises 9–15, consider an experiment in which two cards are drawn at random from a deck of 52 cards. What is the probability that:*

9. Both cards are clubs?
10. Both cards are tens?
11. Neither card is a face card?
12. Both cards are black?

[11.4]
13. One card is red and one is black?
14. One card is a face card and the other is not?
15. Both cards are face cards or one card is a ten and the other an ace?
16. What is the probability of randomly drawing one red and one blue marble from a box containing 4 red and 6 blue marbles?

Review Exercises

[11.5] *In Exercises 17–20, consider an experiment in which two cards are drawn at random from a deck of 52 cards. What is the probability that:*

17. One card is black and the other is the ace of spades?

18. Both cards are red and one (but not the other) is a face card?

19. Both cards are red and at least one is a face card?

20. Neither card is a face card and at least one is red?

Supplemental Exercises for Chapters 10–11

Each of the exercises in this set is more difficult than those at the end of each section and may require the use of several of the techniques learned in the preceding chapters.

Find the limit of each of the following sequences if it exists.

1. $s_n = \dfrac{2n^2 + 1}{3n^2 - 2n + 1}$

2. $s_n = \dfrac{n^{1/2} + 1}{2n^{3/2} + n^{1/2} - 1}$

3. $s_n = \dfrac{2n}{\sqrt{n^2 + 1}}$

4. $s_n = \dfrac{\sqrt{n^2 + 2n + 4}}{n}$

Compute the value of each series. Hint: Use partial fractions to rewrite each term in the sum and then try to find a general expression for each partial sum.

5. $\displaystyle\sum_{n=1}^{\infty} \dfrac{2}{n^2 + n}$

6. $\displaystyle\sum_{n=1}^{\infty} \dfrac{1/3}{n^2 + 7n + 12}$

An experiment (such as the toss of a coin) with only two possible outcomes is called a **Bernoulli trial**. The outcomes in a Bernoulli trial are usually designated as success (S) or failure (F). In Exercises 7–12, assume that n independent Bernoulli trials have been performed and for each trial $P(S) = p$ and $P(F) = 1 - p$.

7. Find a formula for P (exactly k successes).

8. Find a formula for P (exactly k failures).

Supplemental Exercises for Chapters 10–11

9. Find a formula for P (at least k successes).

10. Find a formula for P (at least k failures).

11. Show that $\sum_{k=0}^{n} P(\text{exactly } k \text{ successes}) = 1$.

12. Assume that for every success a person is paid one dollar. Show that the person's expectation is np dollars.

For Exercises 13 and 14, assume that a fair coin is tossed until a head occurs and the number of tosses is recorded.

13. What is the sample space for the above experiment? Compute the probability that exactly n tosses occur before a head is obtained.

14. Compute the probability that at least n tosses occur before a head is obtained.

Using the principle of mathematical induction, prove each of the following.

15. For any natural number n, 3 is a factor of $n(n + 1)(n + 2)$.

16. If I is the $n \times n$ identity matrix, then $\delta(I) = 1$.

17. If A is a square matrix and $X_0 \neq 0$ satisfies $AX = \lambda X$, then X_0 satisfies $A^n X = \lambda^n X$ for every $n \in N$.

18. The Binomial Theorem (Theorem 10.4).

Appendices

A Translation of Axes; Parametric Equations

A.1 Translation of Axes

The identification of the graph of $Ax^2 + By^2 = C \ (A^2 + B^2 \neq 0)$ as discussed in Section 6.2 depends on the fact that the graph is symmetric with respect to the origin. By using a process called **translation of axes**, it is possible to discuss conic sections that do not have this property of symmetry. Figure A.1.1 shows a point $P(x, y)$ in the plane together with two sets of coordinate axes, an xy-system and an $x'y'$-system, whose respective axes are parallel. Each set of axes can be viewed as the result of "sliding" the other set across the plane while the axes remain parallel to their original position—hence the terminology "translation of axes."

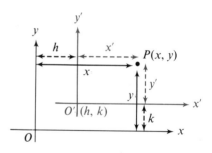

Figure A.1.1

Equations of translations

If the origin O' of the $x'y'$-system in Figure A.1.1 corresponds to the point (h, k) in the xy-system, then the (x', y') coordinates of any point P in the plane are related to the (x, y) coordinates of P by

$$x = x' + h$$
$$y = y' + k \tag{1}$$

We can use these equations of translation together with the familiar process of completing the square in a quadratic expression to identify graphs of equations of the form

$$Ax^2 + By^2 + Dx + Ey + F = 0 \quad (A^2 + B^2 \neq 0). \tag{2}$$

A.1 Translation of Axes

We do this by first completing the square in x and y in (2) to obtain an equation of the form

$$A(x - h)^2 + B(y - k)^2 = C,$$

and then use Equations (1) to express this in the form

$$Ax'^2 + By'^2 = C,$$

which is an equation of a conic in the $x'y'$-system. The center of the graph is at (h, k) in the xy-system.

Example Find an equation of the form $Ax'^2 + By'^2 = C$ for the graph of

$$8x^2 + 4y^2 + 24x - 4y + 1 = 0.$$

Solution We first rewrite the given equation equivalently as

$$8(x^2 + 3x \quad) + 4(y^2 - y \quad) = -1,$$

and then complete the square in x and y:

$$8\left(x^2 + 3x + \frac{9}{4}\right) + 4\left(y^2 - y + \frac{1}{4}\right) = -1 + 18 + 1,$$

$$8\left(x + \frac{3}{2}\right)^2 + 4\left(y - \frac{1}{2}\right)^2 = 18. \qquad (3)$$

If now we take $h = -3/2$ and $k = 1/2$ in Equations (1), we find, upon substituting $x' - 3/2$ for x and $y' + 1/2$ for y, that Equation (3) becomes

$$8x'^2 + 4y'^2 = 18,$$

from which we obtain

$$4x'^2 + 2y'^2 = 9.$$

We recognize this as an equation of an ellipse.

Sketching graphs

It is sometimes convenient to set up an auxiliary coordinate system to sketch a graph. For example, to sketch the graph of

$$8x^2 + 4y^2 + 24x - 4y + 1 = 0$$

above, we can use the $x'y'$-system with its origin at $(-3/2, 1/2)$ and graph

$$4x'^2 + 2y'^2 = 9.$$

First we sketch the x'- and y'-axes (colored lines) with origin at $(-3/2, 1/2)$, and then complete the graph. In the

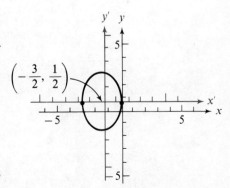

Figure A.1.2

$x'y'$-system, the intercepts on the x'-axis are $3/2$ and $-3/2$, and the intercepts on the y'-axis are $3/\sqrt{2}$ and $-3/\sqrt{2}$. The result is the graph of

$$8x^2 + 4y^2 + 24x - 4y + 1 = 0$$

in the xy-system, as shown in Figure A.1.2.

Exercise A.1

Find an equation of the form $x'^2 = Ay'$ or $y'^2 = Ax'$ for the graph of each given equation. Sketch the graph in the xy-system and show the x'- and y'-axes.

Example $\quad y^2 + 4x - 4y + 8 = 0$

Solution Rewrite the equation equivalently as

$$y^2 - 4y = -4x - 8$$

and complete the square in y as follows:

$$y^2 - 4y + 4 = -4x - 8 + 4,$$
$$(y - 2)^2 = -4(x + 1).$$

Hence, in the $x'y'$-system we have

$$y'^2 = -4x'.$$

The graph appears as shown.

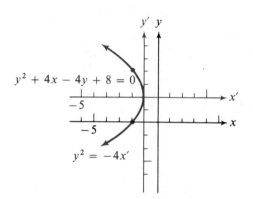

Use Equations (1) on page 394 with $h = -1$ and $k = 2$ to translate the axes so that the origin of the $x'y'$-sytem is at $(-1, 2)$ in the xy-system.

1. $y^2 - 8x - 2y + 17 = 0$
2. $y^2 - 10y + 16x + 25 = 0$
3. $x^2 - 8x + 4y + 12 = 0$
4. $x^2 - 12x + 8y + 36 = 0$
5. $4x^2 - 12x - 24y - 15 = 0$
6. $4y^2 - 4y + 32x + 33 = 0$

Find an equation of the form $Ax'^2 + By'^2 = C$ for the graph of the given equation. Sketch the graph in the xy-system, and show the x'- and y'-axes. Identify the curve.

7. $4x^2 + 9y^2 - 8x - 18y - 23 = 0$
8. $x^2 + 4y^2 + 4x - 24y + 24 = 0$
9. $9x^2 + y^2 - 36x - 2y + 28 = 0$
10. $25x^2 + 4y^2 - 50x + 16y - 59 = 0$
11. $4x^2 - 9y^2 - 8x + 36y - 68 = 0$
12. $16x^2 - y^2 + 64x + 6y + 39 = 0$
13. $3x^2 - 2y^2 - 12x - 4y + 16 = 0$
14. $9x^2 - y^2 + 18x - 12y - 18 = 0$

A.2 Parametric Representation of Relations

Find an equation in the $x'y'$-system with origin at the given point in the xy-system for the graph whose equation in the xy-system is given.

Example (4, 1); $3xy - 3x - 12y + 13 = 0$

Solution We take $h = 4$ and $k = 1$ in Equations (1) on page 394 to obtain
$$x = x' + 4 \quad \text{and} \quad y = y' + 1.$$
Replacing x and y in the given equation, we have
$$3(x' + 4)(y' + 1) - 3(x' + 4) - 12(y' + 1) + 13 = 0,$$
from which
$$3x'y' + 3x' + 12y' + 12 - 3x' - 12 - 12y' - 12 + 13 = 0,$$
or
$$3x'y' + 1 = 0.$$

15. $(3, -2);\ 2xy + 4x - 6y - 15 = 0$
16. $(5, -2);\ xy + 2x - 5y - 11 = 0$
17. $(-1, 2);\ x^2 + 3xy - 4x + 3y - 5 = 0$
18. $(2, -3);\ x^2 - 4x - y + 1 = 0$
19. $(-1, -3);\ x^2 - 3xy + 2y^2 - 7x + 9y + 7 = 0$
20. $(-3, 1);\ 2x^2 - 4xy + 3y^2 + 16x - 18y + 27 = 0$
21. $(h, k);\ x^2 + y^2 - 2hx - 2ky + h^2 + k^2 - r^2 = 0$
22. $(h, k);\ Ax^2 + By^2 - 2Ahx - 2Bky + Ah^2 + Bk^2 - C = 0$

A.2 Parametric Representations of Relations

Relations are sometimes defined by systems of equations of the form
$$x = g(t), \quad y = h(t),$$
where the variables representing elements in the domain and range, in this case x and y, respectively, are each related to a third variable, in this case t. Such equations are called **parametric equations** of the relation, and the variable t is called a **parameter**. For any arbitrary value of t in the common domain of g and h, the values for x and y are determined. For example, we can obtain some ordered pairs (x, y) and the graph of the relation defined by
$$x = t \quad \text{and} \quad y = 1 - t$$

(Figure A.2.1), by assigning some values to t, as shown in the table below, and obtaining the corresponding values for x and y.

t	$x = t$	$y = 1 - t$	(x, y)
-3	-3	4	$(-3, 4)$
-2	-2	3	$(-2, 3)$
-1	-1	2	$(-1, 2)$
0	0	1	$(0, 1)$
1	1	0	$(1, 0)$
2	2	-1	$(2, -1)$
3	3	-2	$(3, -2)$

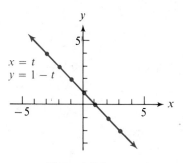

Figure A.2.1

Parametric to Cartesian form

Sometimes a Cartesian equation for a curve can be obtained if we are given a set of parametric equations. Thus, if $x = t$ and $y = 1 - t$, as in the foregoing example, we can eliminate the parameter by substituting x for t in $y = 1 - t$, to obtain

$$y = 1 - x \quad \text{or} \quad x + y = 1,$$

the Cartesian form of a linear equation.

Cartesian to parametric form

You can always obtain one set of parametric equations for a curve (or parts of a curve) having a given Cartesian equation by using $y = tx$, where t is a parameter.

Example

Find a set of parametric equations for the curve with equation $x^2 + y^2 - 4x = 0$.

Solution

Let $y = tx$, and replace y in the Cartesian equation with tx.

$$x^2 + (tx)^2 - 4x = 0$$
$$x^2 + t^2 x^2 - 4x = 0$$
$$(1 + t^2)x^2 - 4x = 0$$
$$x[(1 + t^2)x - 4] = 0$$

We thus have either $x = 0$ or $(1 + t^2)x - 4 = 0$. If $x = 0$, then $y = tx = 0$. If $(1 + t^2)x - 4 = 0$, then

$$x = \frac{4}{1 + t^2}.$$

Since

$$y = tx = t\left(\frac{4}{1 + t^2}\right) = \frac{4t}{1 + t^2},$$

A.2 Parametric Representation of Relations

we have

$$x = \frac{4}{1+t^2} \quad \text{and} \quad y = \frac{4t}{1+t^2} \quad \text{and} \quad x = 0, \quad y = 0$$

as parametric equations for the Cartesian equation $x^2 + y^2 - 4x = 0$.

Exercise A.2

A a. Sketch the graph of the given parametric equations.
   b. Eliminate the parameter and give a Cartesian equation for the graph.

Example $x = t^2 - 3; \; y = t + 1$

Solution a. Since we need ordered pairs (x, y) in order to determine the graph, we can begin by making a table of values using appropriate replacements for t.

t	-4	-3	-2	-1	0	1	2	3	4
x	13	6	1	-2	-3	-2	1	6	13
y	-3	-2	-1	0	1	2	3	4	5

We can thus locate the points associated with the ordered pairs (x, y), connect them in order of increasing t, and sketch the graph. (It appears to be a parabola.)

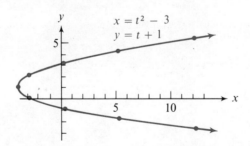

b. We can eliminate the parameter t from the original equations by first writing the equation $y = t + 1$ equivalently as

$$t = y - 1$$

and then substituting $y - 1$ for t in the equation $x = t^2 - 3$.

$$x = (y - 1)^2 - 3$$
$$x = y^2 - 2y - 2$$

From this Cartesian form and our earlier work, we can now assert that the graph is indeed a parabola.

1. $x = 2 - 3t$; $y = 3t - 1$
2. $x = 4t + 6$; $y = 2t - 3$
3. $x = 3t - 2$; $y = 3 - 2t$
4. $x = t + 2$; $y = 2t - 3$
5. $x = t^2$; $y = 1 - t$
6. $x = t + 1$; $y = t^2 - 1$
7. $x = 2 + \dfrac{1}{t}$; $y = \dfrac{1}{t} + 1$
8. $x = \dfrac{2}{t} + 1$; $y = \dfrac{3}{t} - 1$

Use the relationship $y = tx$ to obtain parametric equations for the graph of the given Cartesian equation.

9. $3x - y = 4$
10. $x^2 - y^2 = 9$
11. $2x + 5y = -1$
12. $x^2 + 4y^2 = 100$
13. $x^2 + y^2 = 16$
14. $y = 2x^2 - 3x + 1$

B 15. Show that the equations
$$x = x_1 + at, \quad y = y_1 + bt \quad (a \neq 0)$$
are parametric equations for the line passing through the point (x_1, y_1) with slope b/a.

16. Show that the equations
$$x = (1 - t)x_1 + tx_2, \quad y = (1 - t)y_1 + ty_2$$
are parametric equations for the line passing through the points (x_1, y_1) and (x_2, y_2).

B Mathematical Structure

Here we present the basic substance of the text in compact form. It is a handy guide for following the development of the course and for reviewing material in preparation for tests. Also, you may find it a convenient reference for mathematics courses that you take in the future.

Axioms for Real Numbers

Equality axioms   [1.2]

- E-1 $a = a$
- E-2   If $a = b$, then $b = a$.
- E-3   If $a = b$ and $b = c$, then $a = c$.
- E-4   If $a = b$, then a may be **replaced by** b and b by a in any mathematical statement without altering the truth or falsity of the statement.

Field axioms   [1.2]

- F-1 $a + b$ is a unique element of R.
- F-2 $(a + b) + c = a + (b + c)$
- F-3   There exists an element $0 \in R$ with the property
$$a + 0 = a \quad \text{and} \quad 0 + a = a$$
for all $a \in R$.
- F-4   For each $a \in R$, there exists an element $-a \in R$ with the property
$$a + (-a) = 0 \quad \text{and} \quad (-a) + a = 0.$$
- F-5 $a + b = b + a$

401

F-6 $a \cdot b$ is a unique element of R.

F-7 $(a \cdot b) \cdot c = a \cdot (b \cdot c)$

F-8   There exists an element $1 \in R$, $1 \neq 0$, with the property

$$a \cdot 1 = a \quad \text{and} \quad 1 \cdot a = a$$

for all $a \in R$.

F-9   For each element $a \in R$ ($a \neq 0$), there exists an element $a^{-1} \in R$ with the property

$$a \cdot (a^{-1}) = 1 \quad \text{and} \quad (a^{-1}) \cdot a = 1.$$

F-10 $a \cdot b = b \cdot a$

F-11 $a \cdot (b + c) = a \cdot b + a \cdot c \quad \text{and} \quad (b + c) \cdot a = b \cdot a + c \cdot a.$

[1.4]

Order axioms

O-1   If a is a real number, then exactly one of the following statements is true:

a is negative, a is zero,   or a is positive.

O-2   If a and b are positive real numbers, then $a + b$ is positive and $a \cdot b$ is positive.

O-3   There is a one-to-one correspondence between the set of real numbers and the set of points on a geometric line.

Properties of Numbers

[1.3]   Assume all variables represent arbitrary real numbers unless otherwise noted.

If $a = b$, then
$a + c = b + c \quad \text{and} \quad a \cdot c = b \cdot c.$

If $a + b = 0$, then $a = -b$.

If $a \cdot b = 1$ $(a,b \neq 0)$, then $a = \dfrac{1}{b}$.

If $a + c = b + c$, then $a = b$.

If $a \cdot c = b \cdot c$ $(c \neq 0)$, then $a = b$.

For every a, $a \cdot 0 = 0$.

If $a \cdot b = 0$, then either $a = 0$ or $b = 0$, or both.

$-(-a) = a$

$(-a) + (-b) = -(a + b)$

$(-a)(b) = -(ab)$

$(-a)(-b) = ab$

$\dfrac{-a}{b} = \dfrac{a}{-b} = -\dfrac{a}{b}$ $(b \neq 0)$

$\dfrac{-a}{-b} = \dfrac{a}{b}$ $(b \neq 0)$

$\dfrac{ac}{bc} = \dfrac{a}{b}$ $(b,c \neq 0)$

$\dfrac{1}{a} \cdot \dfrac{1}{b} = \dfrac{1}{ab}$ $(a,b \neq 0)$

Mathematical Structure

$$\frac{a}{b} \cdot \frac{c}{d} = \frac{ac}{bd} \quad (b,d \neq 0) \qquad\qquad 1 \div \frac{a}{b} = \frac{b}{a} \quad (a,b \neq 0)$$

$$\frac{a}{c} + \frac{b}{c} = \frac{a+b}{c} \quad (c \neq 0) \qquad\qquad \frac{a}{b} \div \frac{c}{d} = \frac{ad}{bc} \quad (b,c,d \neq 0)$$

$$\frac{a}{b} + \frac{c}{d} = \frac{ad+bc}{bd} \quad (b,d \neq 0) \qquad\qquad \frac{a}{b} = \frac{c}{d} \quad (b,d \neq 0) \text{ if and only if } ad = bc.$$

$$\frac{a}{b} - \frac{c}{d} = \frac{ad-bc}{bd} \quad (b,d \neq 0)$$

[1.4] If $a < b$ and $b < c$, then $a < c$.

If $a < b$, then $a + c < b + c$.

If $a < b$ and $c > 0$, then $ac < bc$.

If $a < b$ and $c < 0$, then $ac > bc$.

[2.2] $a^m \cdot a^n = a^{m+n}$. $\quad (a^m)^n = a^{mn}$. $\quad (ab)^n = a^n b^n$.

[2.3] $(x + a)(x + b) = x^2 + (a + b)x + ab$

$(x + a)^2 = x^2 + 2ax + a^2$

$(x + a)(x - a) = x^2 - a^2$

$(a + b)(x + y) = ax + ay + bx + by$

$(x + a)(x^2 - ax + a^2) = x^3 + a^3$

$(x - a)(x^2 + ax + a^2) = x^3 - a^3$

[2.4] $\dfrac{a^m}{a^n} = a^{m-n} \quad (a \neq 0). \qquad \left(\dfrac{a}{b}\right)^m = \dfrac{a^m}{b^m} \quad (b \neq 0).$

[2.5] If $P(x)$ is a real polynomial and c is any real number, then there exists a unique real polynomial $Q(x)$ and a real number r such that

$$P(x) = (x - c)Q(x) + r.$$

[3.1] $a^0 = 1 \quad (a \neq 0). \qquad a^{-n} = \dfrac{1}{a^n} \quad (a \neq 0).$

[3.2] *Assume all quantities are defined.*

$$(a^n)^{1/n} = a \quad (n \text{ an odd natural number}),$$
$$(a^n)^{1/n} = |a| \quad (n \text{ an even natural number}),$$
$$(a^{1/n})^m = (a^m)^{1/n} \quad (n \in N, m \in J),$$
$$(a^{1/np})^{mp} = (a^{1/n})^m \quad (m \in J, n, p \in N),$$
$$a^{m/n} = (a^{1/n})^m \quad (m \in J, n \in N).$$

[3.3] $\sqrt[n]{a} = a^{1/n} \quad (a \geq 0),$

$\sqrt[n]{a^n} = a \quad (n \text{ an odd natural number}),$

$\sqrt[n]{a^n} = |a| \quad (n \text{ an even natural number}),$

$\sqrt[n]{a^m} = (\sqrt[n]{a})^m \quad (n \in N, m \in J),$

$\sqrt[n]{a}\sqrt[n]{b} = \sqrt[n]{ab} \quad (n \in N),$

$\dfrac{\sqrt[n]{a}}{\sqrt[n]{b}} = \sqrt[n]{\dfrac{a}{b}} \quad (b \neq 0, n \in N),$

$\sqrt[cn]{a^{cm}} = \sqrt[n]{a^m} \quad (m \in J; n, c \in N).$

[3.5] The complex numbers satisfy the field axioms F-1 through F-11 and the properties listed above for Sections 2.2, 2.4, and 3.1.

[4.1] If $P(x)$, $Q(x)$, and $R(x)$ are expressions, then for all values of x for which $P(x)$, $Q(x)$, and $R(x)$ are real numbers, the sentence
$$P(x) = Q(x)$$
is equivalent to each of the following:

    I $P(x) + R(x) = Q(x) + R(x),$

    II $P(x) \cdot R(x) = Q(x) \cdot R(x) \text{ for } x \in \{x \mid R(x) \neq 0\}.$

[4.2] $ab = 0$ if and only if $a = 0$ or $b = 0$ or both.

[4.3] If $a, b, c \in R, a \neq 0$, then $ax^2 + bx + c = 0$ is equivalent to
$$x = \frac{-b \pm \sqrt{b^2 - 4ac}}{2a}.$$

[4.4] The solution set of $U(x) = V(x)$ is a subset of but not necessarily equal to the solution set of $[U(x)]^n = [V(x)]^n$, for each natural number n.

[4.6] If $P(x)$, $Q(x)$, and $R(x)$ are expressions, then for all values of x for which $P(x)$, $Q(x)$, and $R(x)$ are real numbers, the sentence
$$P(x) < Q(x)$$
is equivalent to each of the following.

I $P(x) + R(x) < Q(x) + R(x)$,

II $P(x) \cdot R(x) < Q(x) \cdot R(x)$ for $x \in \{x | R(x) > 0\}$,

III $P(x) \cdot R(x) > Q(x) \cdot R(x)$ for $x \in \{x | R(x) < 0\}$.

Similarly, the sentence
$$P(x) \leq Q(x)$$
is equivalent to sentences of the form I-III, with $<$ (or $>$) replaced by $\leq$ (or $\geq$) under the same conditions, $R(x) > 0$ and $R(x) < 0$, as above.

[4.8] $|x| = a$ is equivalent to $x = a$ or $x = -a$.

$|x| < a$ is equivalent to $-a < x$ and $x < a$, i.e., $-a < x < a$.

$|x| > a$ is equivalent to $x < -a$ or $x > a$.

$|x| \leq a$ is equivalent to $-a \leq x$ and $x \leq a$, i.e., $-a \leq x \leq a$.

$|x| \geq a$ is equivalent to $x \leq -a$ or $x \geq a$.

[5.2] The length of the line segment between two points (x_1, y_1) and (x_2, y_2) is given by
$$d = \sqrt{(x_2 - x_1)^2 + (y_2 - y_1)^2},$$
and the slope is given by
$$m = \frac{y_2 - y_1}{x_2 - x_1} \quad (x_2 \neq x_1).$$

[5.3] Forms for linear equations:

$y - y_1 = m(x - x_1)$ point-slope form,

$y = mx + b$ slope-intercept form,

$\dfrac{x}{a} + \dfrac{y}{b} = 1$ intercept form.

[6.3] $y = kx$ direct variation,

$y = \dfrac{k}{x}$ inverse variation,

$y = kuvw$ joint variation.

[6.4] If $P(x) = a_n x^n + a_{n-1} x^{n-1} + \cdots + a_0$ $(a_n \neq 0)$, and if k is between $P(x_1)$ and $P(x_2)$, then there exists at least one c between x_1 and x_2 such that $P(c) = k$.

Let $P(x)$ be a polynomial over the field R of real numbers. If $x_1, x_2 \in R$, with $x_1 < x_2$, and $P(x_1)$ and $P(x_2)$ are opposite in sign, then there exists at least one $c \in R$, $x_1 < c < x_2$, such that $P(c) = 0$.

If $P(x)$ is a real polynomial, then for every real number c there exists a unique real polynomial $Q(x)$ such that

$$P(x) = (x - c)Q(x) + P(c).$$

If $P(x)$ is a polynomial with real-number coefficients and $P(r) = 0$, then $(x - r)$ is a factor of $P(x)$.

Let $P(x)$ be a polynomial with real-number coefficients.

I If $r_1 \geq 0$ and the coefficients of the terms in $Q(x)$ and the term $P(r_1)$ are all of the same sign in the right-hand member of

$$P(x) = (x - r_1)Q(x) + P(r_1),$$

then $P(x) = 0$ can have no solution greater than r_1.

II If $r_2 \leq 0$ and the coefficients of the terms in $Q(x)$ and the term $P(r_2)$ alternate in sign (zero suitably denoted by $+0$ and -0) in the right-hand member of

$$P(x) = (x - r_2)Q(x) + P(r_2),$$

then $P(x) = 0$ can have no solution less than r_2.

If $P(z)$ is a polynomial over the real numbers, and $P(z) = 0$ for some $z \in C$, then $P(\bar{z}) = 0$.

Every polynomial function of degree $n \geq 1$ over the complex numbers has at least one complex zero.

If $P(z)$ is a polynomial of degree $n \geq 1$ over the complex numbers, then $P(z)$ can be expressed as a product of a constant and n linear factors of the form $(z - z_k)$.

[6.5] If a is a composite number, then a is the product of only one set of prime factors; that is, the prime factorization of a is unique except for the ordering of the factors.

If the rational number p/q, in lowest terms, is a solution of

$$P(x) = a_n x^n + a_{n-1} x^{n-1} + \cdots + a_0 = 0,$$

where $a_j \in J$, then p is an integral factor of a_0 and q is an integral factor of a_n.

Mathematical Structure

[6.6] If $P(x) = a_n x^n + a_{n-1} x^{n-1} + \cdots + a_0$ is a real polynomial equation of degree n, then the graph of
$$P(x) = a_n x^n + a_{n-1} x^{n-1} + \cdots + a_0, \quad a_n \neq 0,$$
is a smooth curve that has at most $n - 1$ turning points.

The first components of the turning points of the graph of
$$P(x) = a_n x^n + a_{n-1} x^{n-1} + \cdots + a_1 x + a_0$$
are solutions of the equation
$$P'(x) = n a_n x^{n-1} + (n - 1) a_{n-1} x^{n-2} + \cdots + 2 a_2 x + a_1 = 0.$$

[6.7] The graph of the rational function over R defined by $y = P(x)/Q(x)$ has a vertical asymptote at $x = a$ for each value a at which $Q(x)$ vanishes and $P(x)$ does not vanish.

The graph of the rational function over R defined by
$$y = \frac{a_n x^n + a_{n-1} x^{n-1} + \cdots + a_0}{b_m x^m + b_{m-1} x^{m-1} + \cdots + b_0},$$
where $a_n, b_m \neq 0$ and n, m are nonnegative integers, has

    I   a horizontal asymptote at $y = 0$ if $n < m$,
    II a horizontal asymptote at $y = a_n/b_m$ if $n = m$,
    III no horizontal asymptotes if $n > m$.

[7.1] Let $x, y \in Q$, and let $x > y > 0$. Then
$$b^x > b^y \text{ if } b > 1, \quad b^x = b^y \text{ if } b = 1, \quad \text{and} \quad b^x < b^y \text{ if } 0 < b < 1.$$

[7.2] Assume x_1 and x_2 are values for which all expressions are defined.

$\log_b(x_1 x_2) = \log_b x_1 + \log_b x_2$

$\log_b \dfrac{x_2}{x_1} = \log_b x_2 - \log_b x_1$

$\log_b(x_1)^m = m \log_b x_1$

[8.1] Any ordered pair that satisfies both the equations
$$f(x, y) = 0 \quad \text{and} \quad g(x, y) = 0$$
will also satisfy the equation
$$a \cdot f(x, y) + b \cdot g(x, y) = 0,$$
for all real numbers a and b.

[8.2] If any equation in the system
$$f(x, y, z) = 0, \quad g(x, y, z) = 0, \quad \text{and} \quad h(x, y, z) = 0$$
is replaced by a linear combination, with nonzero coefficients, of itself and any one of the other equations in the system, then the result is an equivalent system.

[8.3] If $r_1 \neq r_2$, then there exist constants $c_1, c_2 \in R$ such that
$$\frac{ax + b}{(x - r_1)(x - r_2)} = \frac{c_1}{x - r_1} + \frac{c_2}{x - r_2}.$$

If $P(x)$ is of degree less than $Q(x)$, and if $Q(x) = (x - r_1)(x - r_2) \cdots (x - r_n)$, where no two factors are identical, then there exist constants $c_1, c_2, \ldots, c_n \in R$ such that
$$\frac{P(x)}{Q(x)} = \frac{c_1}{x - r_1} + \frac{c_2}{x - r_2} + \cdots + \frac{c_n}{x - r_n}.$$

If $P(x)$ is of degree less than $Q(x)$, and if $Q(x) = (x - r_1)^n$, then there exist constants $c_1, c_2, \ldots, c_n \in R$ such that
$$\frac{P(x)}{Q(x)} = \frac{c_1}{x - r_1} + \frac{c_2}{(x - r_1)^2} + \cdots + \frac{c_n}{(x - r_1)^n}.$$

[8.5] Any half-plane is a convex set.

The intersection of two convex sets is a convex set.

[9.1] If A, B, and C are $m \times n$ matrices with real-number entries, then:

    I $(A + B)_{m \times n}$ is a matrix with real-number entries,
    II $(A + B) + C = A + (B + C)$,
    III the matrix $\mathbf{0}_{m \times n}$ has the property that for every matrix $A_{m \times n}$,
$$A + \mathbf{0} = A \quad \text{and} \quad \mathbf{0} + A = A,$$
    IV for every matrix $A_{m \times n}$, the matrix $-A_{m \times n}$ has the property that
$$A + (-A) = \mathbf{0} \quad \text{and} \quad (-A) + A = \mathbf{0},$$
    V $A + B = B + A$.

[9.2] If A and B are $m \times n$ matrices and $c, d \in R$, then:

    I cA is an $m \times n$ matrix,   V $1A = A$,
    II $c(dA) = (cd)A$,   VI $(-1)A = -A$,
    III $(c + d)A = cA + dA$,   VII $0A = \mathbf{0}$,
    IV $c(A + B) = cA + cB$,   VIII $c\mathbf{0} = \mathbf{0}$.

If A, B, and C are $n \times n$ square matrices, then
$$(AB)C = A(BC).$$

If A, B, and C are $n \times n$ square matrices, then
$$A(B + C) = AB + AC \quad \text{and} \quad (B + C)A = BA + CA.$$

Mathematical Structure

For each matrix $A_{n \times n}$, we have

$$A_{n \times n} I_{n \times n} = I_{n \times n} A_{n \times n} = A_{n \times n}.$$

Furthermore, $I_{n \times n}$ is the unique matrix having this property for all matrices $A_{n \times n}$.

If A and B are $n \times n$ square matrices, and a is a real number, then

$$a(AB) = (aA)B = A(aB).$$

[9.5] If each entry in any row, or each entry in any column, of a determinant is 0, then the determinant is equal to 0.

If any two rows (or any two columns) of a determinant are interchanged, the resulting determinant is the negative of the original determinant.

If two rows (or two columns) in a determinant have corresponding entries that are equal, the determinant is equal to 0.

If each of the entries of one row (or column) of a determinant is multiplied by k, the determinant is multiplied by k.

If each entry of one row (or column) of a determinant is multiplied by a real number k and the resulting product is added to the corresponding entry in another row (or column, respectively) in the determinant, the resulting determinant is equal to the original determinant.

[9.6] If

$$A = \begin{bmatrix} a_{11} & a_{12} & \cdots & a_{1n} \\ a_{21} & a_{22} & \cdots & a_{2n} \\ \vdots & \vdots & & \vdots \\ a_{n1} & a_{n2} & \cdots & a_{nn} \end{bmatrix},$$

and if $\delta(A) \neq 0$, then

$$A^{-1} = \frac{1}{\delta(A)} \begin{bmatrix} A_{11} & A_{21} & \cdots & A_{n1} \\ A_{12} & A_{22} & \cdots & A_{n2} \\ \vdots & \vdots & & \vdots \\ A_{1n} & A_{2n} & \cdots & A_{nn} \end{bmatrix},$$

where A_{ij} is the cofactor of a_{ij} in A. If $\delta(A) = 0$, then A has no inverse.

If A and B are $n \times n$ nonsingular square matrices, then AB has an inverse, namely,

$$(AB)^{-1} = B^{-1} A^{-1}.$$

[9.8] Cramer's Rule:

$$x = \frac{\delta(A_x)}{\delta(A)}, \quad y = \frac{\delta(A_y)}{\delta(A)}, \quad \text{and} \quad z = \frac{\delta(A_z)}{\delta(A)}, \quad (\delta(A) \neq 0).$$

[10.1] The nth term in the arithmetic sequence defined by
$$s_1 = a, \quad s_{n+1} = s_n + d,$$
where $a, d \in R$, and $n \in N$, is
$$s_n = a + (n-1)d.$$
The nth term in the geometric sequence defined by
$$s_1 = a, \quad s_{n+1} = rs_n,$$
where $a, r \in R$, $a \neq 0$, $r \neq 0$, and $n \in N$, is
$$s_n = ar^{n-1}.$$

[10.2] The sum of the first n terms of an arithmetic progression is
$$S_n = \frac{n}{2}(a + s_n), \quad \text{or} \quad S_n = \frac{n}{2}[2a + (n-1)d].$$
The sum of the first n terms of a geometric progression is
$$S_n = \frac{a - ar^n}{1 - r}, \quad \text{or} \quad S_n = \frac{a - rs_n}{1 - r} \quad (r \neq 1).$$

[10.3] The sum of an infinite geometric progression, $a + ar + ar^2 + \cdots + ar^n + \cdots$, with $|r| < 1$, is
$$S_\infty = \lim_{n \to \infty} S_n = \frac{a}{1-r}.$$

[10.4] For each natural number n,
$$(a+b)^n = a^n + \frac{n}{1!}a^{n-1}b + \frac{n(n-1)}{2!}a^{n-2}b^2 + \frac{n(n-1)(n-2)}{3!}a^{n-3}b^3 + \cdots$$
$$+ \frac{n(n-1)(n-2) \cdots (n-r+2)}{(r-1)!} a^{n-r+1}b^{r-1} + \cdots + b^n,$$
where r is the number of the term. Alternatively,
$$(a+b)^n = \binom{n}{0}a^n + \binom{n}{1}a^{n-1}b + \binom{n}{2}a^{n-2}b^2 + \binom{n}{3}a^{n-3}b^3 + \cdots$$
$$+ \binom{n}{r-1}a^{n-r+1}b^{r-1} + \cdots + \binom{n}{n}b^n.$$
The rth term in a binomial expansion is given by
$$\binom{n}{r-1}a^{n-r+1}b^{r-1} = \frac{n!}{(r-1)!(n-r+1)!} a^{n-r+1}b^{r-1}$$
$$= \frac{n(n-1)(n-2) \cdots (n-r+2)}{(r-1)!} a^{n-r+1}b^{r-1}.$$

[10.5] If a given statement involving n is true for $n = 1$, and if its truth for $n = k$ implies its truth for $n = k+1$, then it is true for every natural number n.

Mathematical Structure

[11.1] Counting properties:

    I $n(A \cup B) = n(A) + n(B)$, if $A \cap B = \emptyset$,

    II $n(A \cup B) = n(A) + n(B) - n(A \cap B)$, if $A \cap B \neq \emptyset$,

    III $n(A \times B) = n(A) \cdot n(B)$,

    IV Suppose the first of several operations can be done in a ways, the second in b ways, no matter what came first, the third in c ways, no matter what came prior, and so on. Then the number of ways the operation can be done in sequence is $a \cdot b \cdot c \cdot \cdots$.

Let $P_{n,n}$ denote the number of distinct permutations of a set A, where $n(A) = n$. Then

$$P_{n,n} = n!.$$

[11.2] Let $P_{n,r}$ denote the number of permutations of the members, taken r at a time, of a set containing n members; that is, let $P_{n,r}$ be the number of distinct orderings of r elements when there is a set of n elements from which to choose. Then

$$P_{n,r} = n(n-1)(n-2) \cdots [n - (r-1)]$$
$$= n(n-1)(n-2) \cdots (n-r+1)$$
$$= \frac{n!}{(n-r)!}.$$

Let $C_{n,r}$ denote the number of distinct combinations of the members, taken r at a time, of a set containing n members. Then

$$C_{n,r} = \binom{n}{r} = \frac{P_{n,r}}{r!} = \frac{n!}{r!(n-r)!};$$

$$\binom{n}{r} = \binom{n}{n-r}.$$

[11.4] $P(E') = 1 - P(E)$

If S is a sample space, and E_1 and E_2 are any events in S, then

$$P(E_1 \text{ or } E_2) = P(E_1 \cup E_2) = P(E_1) + P(E_2) - P(E_1 \cap E_2),$$
$$P(E_1 \text{ or } E_2) = P(E_1 \cup E_2) = P(E_1) + P(E_2) \text{ if } E_1 \cap E_2 = \emptyset.$$

Mathematical Systems

Three operations appear over and over in the algebraic systems of this text, operations called "addition," "multiplication," and "multiplication by a scalar." Operations having the same properties as these occur also in many other systems in pure and applied mathematics.

As a consequence, theorems and the relevant skills developed for one of the systems carry over to any other system whose operations have properties that include the basic properties of the analogous operations in the first system.

Axioms F-1 to F-11 listed on pages 401–402 are the basic properties for addition and multiplication. Some important systems for which various ones of these axioms hold appear in the following table.

System	Binary Operations	Axioms
Group	one	1, 2, 3, 4
Abelian group	one	1, 2, 3, 4, 5
Ring	two	1, 2, 3, 4, 5, 6, 7, 11
Ring with identity	two	1, 2, 3, 4, 5, 6, 7, 11, 8
Commutative ring with identity	two	1, 2, 3, 4, 5, 6, 7, 11, 8, 10
Integral domain	two	1, 2, 3, 4, 5, 6, 7, 11, 8, 10, 9'
Field	two	1, 2, 3, 4, 5, 6, 7, 11, 8, 10, 9

Axiom 9', which applies to an integral domain, is a corollary of Axiom 9. It is stated as follows:

If $a \cdot c = b \cdot c$ (or $c \cdot a = c \cdot b$) and $c \neq 0$, then $a = b$.

The above listing of systems and the axioms which hold in the respective system shows that every field is an integral domain, every integral domain is a ring, every ring is an Abelian group, and every Abelian group is a group.

Some of the above systems also satisfy one or more of the order axioms listed on page 402. A system satisfying Axioms O-1 and O-2 is said to be **ordered**. If it also satisfies O-3 it is said to be **completely ordered**. The *set of complex numbers* constitutes a field but does not satisfy the order axioms; it is not an ordered field. The *set of rational numbers* is a field satisfying Axioms O-1 and O-2 but not O-3; it is an ordered field. The *set of real numbers* is also an ordered field, and it satisfies Axiom O-3; it is a complete ordered field, and is the only complete ordered field in the sense that any other complete ordered field would have to be "just like" (isomorphic with) the complete field of real numbers.

There are still other algebraic systems, with different sets of axioms, which we have not considered in this text. For example, in Boolean algebra, which is important in logic and in the theory of switching circuits, not only does multiplication distribute over addition,

$$A \times (B + C) = (A \times B) + (A \times C),$$

but also addition distributes over multiplication,

$$A + (B \times C) = (A + B) \times (A + C).$$

C Tables

Table I
Exponential Functions, Base e

x	e^x	e^{-x}	x	e^x	e^{-x}
0.00	1.0000	1.0000	1.5	4.4817	0.2231
0.01	1.0101	0.9901	1.6	4.9530	0.2019
0.02	1.0202	0.9802	1.7	5.4739	0.1827
0.03	1.0305	0.9705	1.8	6.0496	0.1653
0.04	1.0408	0.9608	1.9	6.6859	0.1496
0.05	1.0513	0.9512	2.0	7.3891	0.1353
0.06	1.0618	0.9418	2.1	8.1662	0.1225
0.07	1.0725	0.9324	2.2	9.0250	0.1108
0.08	1.0833	0.9331	2.3	9.9742	0.1003
0.09	1.0942	0.9139	2.4	11.023	0.0907
0.10	1.1052	0.9048	2.5	12.182	0.0821
0.11	1.1163	0.8958	2.6	13.464	0.0743
0.12	1.1275	0.8869	2.7	14.880	0.0672
0.13	1.1388	0.8781	2.8	16.445	0.0608
0.14	1.1503	0.8694	2.9	18.174	0.0550
0.15	1.1618	0.8607	3.0	20.086	0.0498
0.16	1.1735	0.8521	3.1	22.198	0.0450
0.17	1.1853	0.8437	3.2	24.533	0.0408
0.18	1.1972	0.8353	3.3	27.113	0.0369
0.19	1.2092	0.8270	3.4	29.964	0.0334
0.20	1.2214	0.8187	3.5	33.115	0.0302
0.21	1.2337	0.8106	3.6	36.598	0.0273
0.22	1.2461	0.8025	3.7	40.447	0.0247
0.23	1.2586	0.7945	3.8	44.701	0.0224
0.24	1.2712	0.7866	3.9	49.402	0.0202
0.25	1.2840	0.7788	4.0	54.598	0.0183
0.30	1.3499	0.7408	4.1	60.340	0.0166
0.35	1.4191	0.7047	4.2	66.686	0.0150
0.40	1.4918	0.6703	4.3	73.700	0.0136
0.45	1.5683	0.6376	4.4	81.451	0.0123
0.50	1.6487	0.6065	4.5	90.017	0.0111
0.55	1.7333	0.5769	4.6	99.484	0.0101
0.60	1.8221	0.5488	4.7	109.95	0.0091
0.65	1.9155	0.5220	4.8	121.51	0.0082
0.70	2.0138	0.4966	4.9	134.29	0.0074
0.75	2.1170	0.4724	5.0	148.41	0.0067
0.80	2.2255	0.4493	5.5	244.69	0.0041
0.85	2.3396	0.4274	6.0	403.43	0.0025
0.90	2.4596	0.4066	6.5	665.14	0.0015
0.95	2.5857	0.3867	7.0	1096.6	0.0009
1.0	2.7183	0.3679	7.5	1808.0	0.0006
1.1	3.0042	0.3329	8.0	2981.0	0.0003
1.2	3.3201	0.3012	8.5	4914.8	0.0002
1.3	3.6693	0.2725	9.0	8103.1	0.0001
1.4	4.0552	0.2466	10.0	22026	0.00005

Table II
Common Logarithms

x	0	1	2	3	4	5	6	7	8	9
1.0	0.0000	0.0043	0.0086	0.0128	0.0170	0.0212	0.0253	0.0294	0.0334	0.0374
1.1	0.0414	0.0453	0.0492	0.0531	0.0569	0.0607	0.0645	0.0682	0.0719	0.0755
1.2	0.0792	0.0828	0.0864	0.0899	0.0934	0.0969	0.1004	0.1038	0.1072	0.1106
1.3	0.1139	0.1173	0.1206	0.1239	0.1271	0.1303	0.1335	0.1367	0.1399	0.1430
1.4	0.1461	0.1492	0.1523	0.1553	0.1584	0.1614	0.1644	0.1673	0.1703	0.1732
1.5	0.1761	0.1790	0.1818	0.1847	0.1875	0.1903	0.1931	0.1959	0.1987	0.2014
1.6	0.2041	0.2068	0.2095	0.2122	0.2148	0.2175	0.2201	0.2227	0.2253	0.2279
1.7	0.2304	0.2330	0.2355	0.2380	0.2405	0.2430	0.2455	0.2480	0.2504	0.2529
1.8	0.2553	0.2577	0.2601	0.2625	0.2648	0.2672	0.2695	0.2718	0.2742	0.2765
1.9	0.2788	0.2810	0.2833	0.2856	0.2878	0.2900	0.2923	0.2945	0.2967	0.2989
2.0	0.3010	0.3032	0.3054	0.3075	0.3096	0.3118	0.3139	0.3160	0.3181	0.3201
2.1	0.3222	0.3243	0.3263	0.3284	0.3304	0.3324	0.3345	0.3365	0.3385	0.3404
2.2	0.3424	0.3444	0.3464	0.3483	0.3502	0.3522	0.3541	0.3560	0.3579	0.3598
2.3	0.3617	0.3636	0.3655	0.3674	0.3692	0.3711	0.3729	0.3747	0.3766	0.3784
2.4	0.3802	0.3820	0.3838	0.3856	0.3874	0.3892	0.3909	0.3927	0.3945	0.3962
2.5	0.3979	0.3997	0.4014	0.4031	0.4048	0.4065	0.4082	0.4099	0.4116	0.4133
2.6	0.4150	0.4166	0.4183	0.4200	0.4216	0.4232	0.4249	0.4265	0.4281	0.4298
2.7	0.4314	0.4330	0.4346	0.4362	0.4378	0.4393	0.4409	0.4425	0.4440	0.4456
2.8	0.4472	0.4487	0.4502	0.4518	0.4533	0.4548	0.4564	0.4579	0.4594	0.4609
2.9	0.4624	0.4639	0.4654	0.4669	0.4683	0.4698	0.4713	0.4728	0.4742	0.4757
3.0	0.4771	0.4786	0.4800	0.4814	0.4829	0.4843	0.4857	0.4871	0.4886	0.4900
3.1	0.4914	0.4928	0.4942	0.4955	0.4969	0.4983	0.4997	0.5011	0.5024	0.5038
3.2	0.5051	0.5065	0.5079	0.5092	0.5105	0.5119	0.5132	0.5145	0.5159	0.5172
3.3	0.5185	0.5198	0.5211	0.5224	0.5237	0.5250	0.5263	0.5276	0.5289	0.5302
3.4	0.5315	0.5328	0.5340	0.5353	0.5366	0.5378	0.5391	0.5403	0.5416	0.5428
3.5	0.5441	0.5453	0.5465	0.5478	0.5490	0.5502	0.5514	0.5527	0.5539	0.5551
3.6	0.5563	0.5575	0.5587	0.5599	0.5611	0.5623	0.5635	0.5647	0.5658	0.5670
3.7	0.5682	0.5694	0.5705	0.5717	0.5729	0.5740	0.5752	0.5763	0.5775	0.5786
3.8	0.5798	0.5809	0.5821	0.5832	0.5843	0.5855	0.5866	0.5877	0.5888	0.5899
3.9	0.5911	0.5922	0.5933	0.5944	0.5955	0.5966	0.5977	0.5988	0.5999	0.6010
4.0	0.6021	0.6031	0.6042	0.6053	0.6064	0.6075	0.6085	0.6096	0.6107	0.6117
4.1	0.6128	0.6138	0.6149	0.6160	0.6170	0.6180	0.6191	0.6201	0.6212	0.6222
4.2	0.6232	0.6243	0.6253	0.6263	0.6274	0.6284	0.6294	0.6304	0.6314	0.6325
4.3	0.6335	0.6345	0.6355	0.6365	0.6375	0.6385	0.6395	0.6405	0.6415	0.6425
4.4	0.6435	0.6444	0.6454	0.6464	0.6474	0.6484	0.6493	0.6503	0.6513	0.6522
4.5	0.6532	0.6542	0.6551	0.6561	0.6571	0.6580	0.6590	0.6599	0.6609	0.6618
4.6	0.6628	0.6637	0.6646	0.6656	0.6665	0.6675	0.6684	0.6693	0.6702	0.6712
4.7	0.6721	0.6730	0.6739	0.6749	0.6758	0.6767	0.6776	0.6785	0.6794	0.6803
4.8	0.6812	0.6821	0.6830	0.6839	0.6848	0.6857	0.6866	0.6875	0.6884	0.6893
4.9	0.6902	0.6911	0.6920	0.6928	0.6937	0.6946	0.6955	0.6964	0.6972	0.6981
5.0	0.6990	0.6998	0.7007	0.7016	0.7024	0.7033	0.7042	0.7050	0.7059	0.7067
5.1	0.7076	0.7084	0.7093	0.7101	0.7110	0.7118	0.7126	0.7135	0.7143	0.7152
5.2	0.7160	0.7168	0.7177	0.7185	0.7193	0.7202	0.7210	0.7218	0.7226	0.7235
5.3	0.7243	0.7251	0.7259	0.7267	0.7275	0.7284	0.7292	0.7300	0.7308	0.7316
5.4	0.7324	0.7332	0.7340	0.7348	0.7356	0.7364	0.7372	0.7380	0.7388	0.7396
x	0	1	2	3	4	5	6	7	8	9

Table II (*Continued*)

x	0	1	2	3	4	5	6	7	8	9
5.5	0.7404	0.7412	0.7419	0.7427	0.7435	0.7443	0.7451	0.7459	0.7466	0.7474
5.6	0.7482	0.7490	0.7497	0.7505	0.7513	0.7520	0.7528	0.7536	0.7543	0.7551
5.7	0.7559	0.7566	0.7574	0.7582	0.7589	0.7597	0.7604	0.7612	0.7619	0.7627
5.8	0.7634	0.7642	0.7649	0.7657	0.7664	0.7672	0.7679	0.7686	0.7694	0.7701
5.9	0.7709	0.7716	0.7723	0.7731	0.7738	0.7745	0.7752	0.7760	0.7767	0.7774
6.0	0.7782	0.7789	0.7796	0.7803	0.7810	0.7818	0.7825	0.7832	0.7839	0.7846
6.1	0.7853	0.7860	0.7868	0.7875	0.7882	0.7889	0.7896	0.7903	0.7910	0.7917
6.2	0.7924	0.7931	0.7938	0.7945	0.7952	0.7959	0.7966	0.7973	0.7980	0.7987
6.3	0.7993	0.8000	0.8007	0.8014	0.8021	0.8028	0.8035	0.8041	0.8048	0.8055
6.4	0.8062	0.8069	0.8075	0.8082	0.8089	0.8096	0.8102	0.8109	0.8116	0.8122
6.5	0.8129	0.8136	0.8142	0.8149	0.8156	0.8162	0.8169	0.8176	0.8182	0.8189
6.6	0.8195	0.8202	0.8209	0.8215	0.8222	0.8228	0.8235	0.8241	0.8248	0.8254
6.7	0.8261	0.8267	0.8274	0.8280	0.8287	0.8293	0.8299	0.8306	0.8312	0.8319
6.8	0.8325	0.8331	0.8338	0.8344	0.8351	0.8357	0.8363	0.8370	0.8376	0.8382
6.9	0.8388	0.8395	0.8401	0.8407	0.8414	0.8420	0.8426	0.8432	0.8439	0.8445
7.0	0.8451	0.8457	0.8463	0.8470	0.8476	0.8482	0.8488	0.8494	0.8500	0.8506
7.1	0.8513	0.8519	0.8525	0.8531	0.8537	0.8543	0.8549	0.8555	0.8561	0.8567
7.2	0.8573	0.8579	0.8585	0.8591	0.8597	0.8603	0.8609	0.8615	0.8621	0.8627
7.3	0.8633	0.8639	0.8645	0.8651	0.8657	0.8663	0.8669	0.8675	0.8681	0.8686
7.4	0.8692	0.8698	0.8704	0.8710	0.8716	0.8722	0.8727	0.8733	0.8739	0.8745
7.5	0.8751	0.8756	0.8762	0.8768	0.8774	0.8779	0.8785	0.8791	0.8797	0.8802
7.6	0.8808	0.8814	0.8820	0.8825	0.8831	0.8837	0.8842	0.8848	0.8854	0.8859
7.7	0.8865	0.8871	0.8876	0.8882	0.8887	0.8893	0.8899	0.8904	0.8910	0.8915
7.8	0.8921	0.8927	0.8932	0.8938	0.8943	0.8949	0.8954	0.8960	0.8965	0.8971
7.9	0.8976	0.8982	0.8987	0.8993	0.8998	0.9004	0.9009	0.9015	0.9020	0.9025
8.0	0.9031	0.9036	0.9042	0.9047	0.9053	0.9058	0.9063	0.9069	0.9074	0.9079
8.1	0.9085	0.9090	0.9096	0.9101	0.9106	0.9112	0.9117	0.9122	0.9128	0.9133
8.2	0.9138	0.9143	0.9149	0.9154	0.9159	0.9165	0.9170	0.9175	0.9180	0.9186
8.3	0.9191	0.9196	0.9201	0.9206	0.9212	0.9217	0.9222	0.9227	0.9232	0.9238
8.4	0.9243	0.9248	0.9253	0.9258	0.9263	0.9269	0.9274	0.9279	0.9284	0.9289
8.5	0.9294	0.9299	0.9304	0.9309	0.9315	0.9320	0.9325	0.9330	0.9335	0.9340
8.6	0.9345	0.9350	0.9355	0.9360	0.9365	0.9370	0.9375	0.9380	0.9385	0.9390
8.7	0.9395	0.9400	0.9405	0.9410	0.9415	0.9420	0.9425	0.9430	0.9435	0.9440
8.8	0.9445	0.9450	0.9455	0.9460	0.9465	0.9469	0.9474	0.9479	0.9484	0.9489
8.9	0.9494	0.9499	0.9504	0.9509	0.9513	0.9518	0.9523	0.9528	0.9533	0.9538
9.0	0.9542	0.9547	0.9552	0.9557	0.9562	0.9566	0.9571	0.9576	0.9581	0.9586
9.1	0.9590	0.9595	0.9600	0.9605	0.9609	0.9614	0.9619	0.9624	0.9628	0.9633
9.2	0.9638	0.9643	0.9647	0.9652	0.9657	0.9661	0.9666	0.9671	0.9675	0.9680
9.3	0.9685	0.9689	0.9694	0.9699	0.9703	0.9708	0.9713	0.9717	0.9722	0.9727
9.4	0.9731	0.9736	0.9741	0.9745	0.9750	0.9754	0.9759	0.9763	0.9768	0.9773
9.5	0.9777	0.9782	0.9786	0.9791	0.9795	0.9800	0.9805	0.9809	0.9814	0.9818
9.6	0.9823	0.9827	0.9832	0.9836	0.9841	0.9845	0.9850	0.9854	0.9859	0.9863
9.7	0.9868	0.9872	0.9877	0.9881	0.9886	0.9890	0.9894	0.9899	0.9903	0.9908
9.8	0.9912	0.9917	0.9921	0.9926	0.9930	0.9934	0.9939	0.9943	0.9948	0.9952
9.9	0.9956	0.9961	0.9965	0.9969	0.9974	0.9978	0.9983	0.9987	0.9991	0.9996
x	0	1	2	3	4	5	6	7	8	9

Table III
Natural Logarithms

n	$\log_e n$	n	$\log_e n$	n	$\log_e n$
	*	4.5	1.5041	9.0	2.1972
0.1	7.6974	4.6	1.5261	9.1	2.2083
0.2	8.3906	4.7	1.5476	9.2	2.2192
0.3	8.7960	4.8	1.5686	9.3	2.2300
0.4	9.0837	4.9	1.5892	9.4	2.2407
0.5	9.3069	5.0	1.6094	9.5	2.2513
0.6	9.4892	5.1	1.6292	9.6	2.2618
0.7	9.6433	5.2	1.6487	9.7	2.2721
0.8	9.7769	5.3	1.6677	9.8	2.2824
0.9	9.8946	5.4	1.6864	9.9	2.2925
1.0	0.0000	5.5	1.7047	10	2.3026
1.1	0.0953	5.6	1.7228	11	2.3979
1.2	0.1823	5.7	1.7405	12	2.4849
1.3	0.2624	5.8	1.7579	13	2.5649
1.4	0.3365	5.9	1.7750	14	2.6391
1.5	0.4055	6.0	1.7918	15	2.7081
1.6	0.4700	6.1	1.8083	16	2.7726
1.7	0.5306	6.2	1.8245	17	2.8332
1.8	0.5878	6.3	1.8405	18	2.8904
1.9	0.6419	6.4	1.8563	19	2.9444
2.0	0.6931	6.5	1.8718	20	2.9957
2.1	0.7419	6.6	1.8871	25	3.2189
2.2	0.7885	6.7	1.9021	30	3.4012
2.3	0.8329	6.8	1.9169	35	3.5553
2.4	0.8755	6.9	1.9315	40	3.6889
2.5	0.9163	7.0	1.9459	45	3.8067
2.6	0.9555	7.1	1.9601	50	3.9120
2.7	0.9933	7.2	1.9741	55	4.0073
2.8	1.0296	7.3	1.9879	60	4.0943
2.9	1.0647	7.4	2.0015	65	4.1744
3.0	1.0986	7.5	2.0149	70	4.2485
3.1	1.1314	7.6	2.0281	75	4.3175
3.2	1.1632	7.7	2.0412	80	4.3820
3.3	1.1939	7.8	2.0541	85	4.4427
3.4	1.2238	7.9	2.0669	90	4.4998
3.5	1.2528	8.0	2.0794	100	4.6052
3.6	1.2809	8.1	2.0919	110	4.7005
3.7	1.3083	8.2	2.1041	120	4.7875
3.8	1.3350	8.3	2.1163	130	4.8676
3.9	1.3610	8.4	2.1282	140	4.9416
4.0	1.3863	8.5	2.1401	150	5.0106
4.1	1.4110	8.6	2.1518	160	5.0752
4.2	1.4351	8.7	2.1633	170	5.1358
4.3	1.4586	8.8	2.1748	180	5.1930
4.4	1.4816	8.9	2.1861	190	5.2470

*Subtract 10 for $n < 1$. Thus, $\log_e 0.1 = 7.6974 - 10 = -2.3026$.

Table IV
Squares, Square Roots, and Prime Factors

No.	Sq.	Sq. Root	Prime Factors	No.	Sq.	Sq. Root	Prime Factors
1	1	1.000		51	2,601	7.141	$3 \cdot 17$
2	4	1.414	2	52	2,704	7.211	$2^2 \cdot 13$
3	9	1.732	3	53	2,809	7.280	53
4	16	2.000	2^2	54	2,916	7.348	$2 \cdot 3^3$
5	25	2.236	5	55	3,025	7.416	$5 \cdot 11$
6	36	2.449	$2 \cdot 3$	56	3,136	7.483	$2^3 \cdot 7$
7	49	2.646	7	57	3,249	7.550	$3 \cdot 19$
8	64	2.828	2^3	58	3,364	7.616	$2 \cdot 29$
9	81	3.000	3^2	59	3,481	7.681	59
10	100	3.162	$2 \cdot 5$	60	3,600	7.746	$2^2 \cdot 3 \cdot 5$
11	121	3.317	11	61	3,721	7.810	61
12	144	3.464	$2^2 \cdot 3$	62	3,844	7.874	$2 \cdot 31$
13	169	3.606	13	63	3,969	7.937	$3^2 \cdot 7$
14	196	3.742	$2 \cdot 7$	64	4,096	8.000	2^6
15	225	3.873	$3 \cdot 5$	65	4,225	8.062	$5 \cdot 13$
16	256	4.000	2^4	66	4,356	8.124	$2 \cdot 3 \cdot 11$
17	289	4.123	17	67	4,489	8.185	67
18	324	4.243	$2 \cdot 3^2$	68	4,624	8.246	$2^2 \cdot 17$
19	361	4.359	19	69	4,761	8.307	$3 \cdot 23$
20	400	4.472	$2^2 \cdot 5$	70	4,900	8.367	$2 \cdot 5 \cdot 7$
21	441	4.583	$3 \cdot 7$	71	5,041	8.426	71
22	484	4.690	$2 \cdot 11$	72	5,184	8.485	$2^3 \cdot 3^2$
23	529	4.796	23	73	5,329	8.544	73
24	576	4.899	$2^3 \cdot 3$	74	5,476	8.602	$2 \cdot 37$
25	625	5.000	5^2	75	5,625	8.660	$3 \cdot 5^2$
26	676	5.099	$2 \cdot 13$	76	5,776	8.718	$2^2 \cdot 19$
27	729	5.196	3^3	77	5,929	8.775	$7 \cdot 11$
28	784	5.292	$2^2 \cdot 7$	78	6,084	8.832	$2 \cdot 3 \cdot 13$
29	841	5.385	29	79	6,241	8.888	79
30	900	5.477	$2 \cdot 3 \cdot 5$	80	6,400	8.944	$2^4 \cdot 5$
31	961	5.568	31	81	6,561	9.000	3^4
32	1,024	5.657	2^5	82	6,724	9.055	$2 \cdot 41$
33	1,089	5.745	$3 \cdot 11$	83	6,889	9.110	83
34	1,156	5.831	$2 \cdot 17$	84	7,056	9.165	$2^2 \cdot 3 \cdot 7$
35	1,225	5.916	$5 \cdot 7$	85	7,225	9.220	$5 \cdot 17$
36	1,296	6.000	$2^2 \cdot 3^2$	86	7,396	9.274	$2 \cdot 43$
37	1,369	6.083	37	87	7,569	9.327	$3 \cdot 29$
38	1,444	6.164	$2 \cdot 19$	88	7,744	9.381	$2^3 \cdot 11$
39	1,521	6.245	$3 \cdot 13$	89	7,921	9.434	89
40	1,600	6.325	$2^3 \cdot 5$	90	8,100	9.487	$2 \cdot 3^2 \cdot 5$
41	1,681	6.403	41	91	8,281	9.539	$7 \cdot 13$
42	1,764	6.481	$2 \cdot 3 \cdot 7$	92	8,464	9.592	$2^2 \cdot 23$
43	1,849	6.557	43	93	8,649	9.644	$3 \cdot 31$
44	1,936	6.633	$2^2 \cdot 11$	94	8,836	9.695	$2 \cdot 47$
45	2,025	6.708	$3^2 \cdot 5$	95	9,025	9.747	$5 \cdot 19$
46	2,116	6.782	$2 \cdot 23$	96	9,216	9.798	$2^5 \cdot 3$
47	2,209	6.856	47	97	9,409	9.849	97
48	2,304	6.928	$2^4 \cdot 3$	98	9,604	9.899	$2 \cdot 7^2$
49	2,401	7.000	7^2	99	9,801	9.950	$3^2 \cdot 11$
50	2,500	7.071	$2 \cdot 5^2$	100	10,000	10.000	$2^2 \cdot 5^2$

Answers to Odd Numbered Exercises

> Have you obtained your *Student Guide to College Algebra, Fifth Edition*?
> It can help you with this course by acting as:
>
> 1. a tutor for those sections of the course with which you have difficulty,
> 2. an aid in catching up with work covered during an absence from class,
> 3. a vehicle for reviewing for examinations.
>
> You can get a copy of the *Student Guide to College Algebra, Fifth Edition*, at your college bookstore. If your bookstore doesn't have it in stock, please ask the bookstore manager to order you a copy.

Exercise 1.1 (page 4) **1.** $\{3, 4, 5, 6, 7, 8, 9, 10, 11\}$ **3.** $\{a, c, e, h, i, m, s, t\}$
5. {Monday, Tuesday, Wednesday, Thursday, Friday, Saturday, Sunday} **7.** $\in$ **9.** $\notin$ **11.** $\neq$
13. $=$ **15.** $\not\subset$ **17.** $\subset$ **19.** $\{1, 2, 3, 4, 6, 8\}$ **21.** $\{6, 8\}$ **23.** $\{5, 6, 7, 8\}$
25. $\{1, 3, 5, 6, 7, 8\}$ **27.** $\{5, 7\}$ **29.** $\{1, 3, 5, 6, 7, 8\}$ **31.** $\{5, 7\}$ **33.** $\{1, 3\}$ **35.** $\{1, 2, 3, 4\}$
37. $\{x \mid x = 2n - 1, n \in N\}$ **39.** $\{x \mid x = 3n, n \in N\}$ **41.** $\{x \mid x < 100, x \in N\}$ **43.** $B \cap C$
45. $A \cap (B \cup C)$ **47.** $A \cap (B \cup C)'$ **49.** $8, 16, 32, 2^n$
51. $x \in A \cap B$ implies $x \in A$ and $x \in B$; therefore $x \in A \cap B$ implies $x \in A$. Hence, $A \cap B \subset A$.

Exercise 1.2 (page 9) **1.** False **3.** True **5.** True **7.** False **9.** False **11.** False
13. E-2 **15.** E-3 **17.** E-4 **19.** E-1 **21.** F-2 **23.** F-10 **25.** F-3 **27.** F-7
29. F-11 **31.** F-8 **33.** F-4 **35.** F-11 **37.** $ab + a$ **39.** $pq + pr$ **41.** $d^2 + 3d$
43. $(de + g)f$ **45.** $13r + qr$ **47.** $s(t + u + v)$ **49.** $3x + 3y + 3z$

Exercise 1.3 (page 14) **1.** Theorem 1.11-II **3.** Theorem 1.1 or 1.4 **5.** Theorem 1.7
7. Theorem 1.11-VII **9.** Theorem 1.10 **11.** Theorem 1.9 **13.** Theorem 1.11-IV
15. Theorem 1.2 **17.** Theorem 1.9 **19.** Theorem 1.11-IV

Answers to Odd-Numbered Exercises

Exercise 1.4 (page 19)   **1.** Part II   **3.** Part I   **5.** Part III   **7.** Part IV   **9.** $x > 3$
11. $x \le 3$ **13.** $x \ge 3$ **15.** $1 \le x \le 3$ **17.** $1 \le x < 3$ **19.** $-3, -1, 1, 2$ **21.** $0, \frac{1}{2}, \frac{2}{3}, 1$
23. $-1, -\frac{2}{3}, -\frac{1}{4}, \frac{1}{8}$ **25.** 11   **27.** 11   **29.** -11 **31.** $\begin{cases} x & \text{if } x \ge 0 \\ -x & \text{if } x < 0 \end{cases}$
33. $\begin{cases} 1 - x & \text{if } x \ge 1 \\ x - 1 & \text{if } x < 1 \end{cases}$ **35.** $\begin{cases} 1 + x & \text{if } x \ge -1 \\ -(1 + x) & \text{if } x < -1 \end{cases}$ **37.** $|-2|$ **39.** 4   **41.** $|-1|$
43. $[-2, 0]$ **45.** $(-\infty, 4]$ **47.** $\{x \mid -3 < x \le 1\}$ **49.** $\{x \mid x \le 0\}$
51. $(0, 4]$ **53.** $(0, 1]$ **55.** The intersection is the empty set.
57.   **59.**   **61.**

Chapter 1 Review (page 22)   **1.** $\{1, 4, 6, 8, 9, 10, 12\}$ **2.** $\{1, 2, 4, 5, 6, 7, 8, 9, 10, 11, 12\}$
3. $\{2, 3, 5, 6, 7, 9, 11, 12\}$ **4.** $\{2, 5, 7, 11\}$ **5.** $\{3\}$ **6.** $\{1, 4, 6, 8, 9, 10, 12\}$ **7.** True
8. True   **9.** False   **10.** False   **11.** False   **12.** True   **13.** $\{2\}$ **14.** $\{-2, 0, 2\}$
15. $\left\{-2, 0, 2, -\frac{1}{2}, \frac{1}{2}\right\}$ **16.** $\left\{\sqrt{2}, -\sqrt{2}, \frac{1}{\sqrt{2}}\right\}$ **17.** $\left\{-2, -\frac{1}{2}, -\sqrt{2}\right\}$ **18.** $\left\{2, \frac{1}{2}, \sqrt{2}, \frac{1}{\sqrt{2}}\right\}$
19. F-11   **20.** F-5   **21.** F-10   **22.** F-10   **23.** Theorem 1.1
24. Theorem 1.8-IV   **25.** Theorem 1.11-VII   **26.** Theorem 1.8-II   **27.** Theorem 1.11-IV
28. Theorem 1.2   **29.** Part III   **30.** Part IV   **31.** Part II   **32.** Part III   **33.** $a < b$
34. $3 \le a \le 7$ **35.** $a \not< b$ or $a \ge b$ **36.** $a \ge 2$ **37.** $-1, -\frac{2}{3}, \frac{1}{2}, 1$ **38.** $-3, 0, \frac{1}{8}, \frac{2}{3}$
39. 5   **40.** $\begin{cases} x - 5 & \text{if } x \ge 5 \\ 5 - x & \text{if } x < 5 \end{cases}$ **41.** $\left[-\frac{1}{2}, 2\right)$ **42.** $\left[\frac{1}{2}, \infty\right)$

Exercise 2.1 (page 27)   **1.** $2x^2 + x$ **3.** $-2x^2 - 6x - 15$ **5.** $-x^3 - x^2 + x - 1$
7. $-2x^4 - 3x^3 + 2x^2 - x + 1$ **9.** $2x + 1$ **11.** $5x^2y$ **13.** $-2x^2yz + 2xy^2z - x^3y$
15. $-x^2 - x - 4$ **17.** $-4x - 8$ **19.** $-4x - 8$ **21.** $-2x^2 + 6x + 8$ **23.** $5x - 8$
25. $-2x + 1$ **27.** $3x^2 - 4x$ **29.** $x^3 - x^2 - 4x + 5$ **31.** $x^3 - x^2 + 2x - 1$ **33.** $2, 2, 8, 0$
35. $-1, 0, 1, 2$ **37.** $7, 4$ **39.** $25, 19$ **41.** 5 **43.** -8 **45.** $n, n, n - 2$, no degree

Exercise 2.2 (page 32)   **1.** $-2x^3y^4$ **3.** $-6x^5y^6$ **5.** $3x^{2n+2}$ **7.** $-2x^{n+2}y^{n+1}$ **9.** $5x - 10$
11. $2x^2 - 4x$ **13.** $x^4 + x^2 + x$ **15.** $x^2 + 5x + 6$ **17.** $3x^2 + 8x - 3$ **19.** $6x^2 + 3x - 3$
21. $-x^2 + 1$ **23.** $-2x^2 - 5x + 3$ **25.** $x^3 - 1$ **27.** $3x^3 + 5x^2 + 2x + 8$
29. $2x^3 + 5x^2 + 4x + 1$ **31.** $x^4 + 2x^3 + 2x^2 + x$ **33.** $x^4 + 2x^2 + x + 2$
35. $x^3 + x^2 + 2x + 2$ **37.** $-2x^5 + 5x^4 + 7x^3 - 23x^2 + 4x + 12$
39. $c^2 + c - 2, c^2 + 2ch + h^2 + c + h - 2, 2ch + h^2 + h$ **41.** $x^2 + 4x + 5, x^4 + 1, x^4 + 2x^2 + 1$
43. $m + n$ **45.** $2n, kn$

Exercise 2.3 (page 36) 1. $2x^2y(xy + 4 + 8y)$ 3. $(x - 4)(x + 4)$ 5. $3x^2(y^2 + 2)(y^2 - 2)$
7. $2(x + 3)^2$ 9. $3y^3(x + 1)^2$ 11. $5x(xyz + 2yz + 3)$ 13. $(x + 5)(x - 2)$
15. $(x - 5)(x - 3)$ 17. $(2x + 1)(x + 1)$ 19. $(3x - 1)(x - 2)$ 21. $(2x + 1)(x - 2)$
23. $(3x - 1)(2x - 3)$ 25. $(8x + 1)(2x + 1)$ 27. $(4x + 2)(3x - 1)$ 29. $3(x - 1)(x + 5)$
31. $x(y + 7)(y - 2)$ 33. $(y + 3)(x - 1)$ 35. $2(x - 2)(y + 3)$ 37. $(x + 3)(x^2 - 3x + 9)$
39. $(x + 2)(x^2 + x + 1)$ 41. $8(2x - y)(4x^2 + 2xy + y^2)$ 43. $(x^2 + y)(x + 3)$
45. $(x + 6)(x + 5)$ 47. $(x^2 + 1)(y^2 + 3)$ 49. $x(x^2 + 1)(x - 1)(x + 1)$
51. $(x^2 - xy + y^2)(x^2 + xy + y^2)$ 53. $(x^2 - x + 1)(x^2 + x + 1)$
55. $(x - y)(x^{n-1} + x^{n-2}y + \cdots + xy^{n-2} + y^{n-1})$ 57. $(x - 1)(x^{n-1} + x^{n-2} + \cdots + x + 1)(x^n + 1)$
59. $(x^2 - y)(x^2 + y)(x^{4n-4} + x^{4n-8}y^2 + \cdots + x^4y^{2n-4} + y^{2n-2})$ 61. $x^{2n}(x^{2n} + 1)^2$
63. $(x^{2n} + 1)(x^{4n} - x^{2n} + 1)$

Exercise 2.4 (page 40) 1. $5y^3z$ 3. $6x^3$ 5. $2x^3 + x$ 7. $3x - 2y + 1$ 9. $x + 2 + \dfrac{1}{x}$
11. $2x^2 - 1 + \dfrac{3}{x}$ 13. $x^3 + 2x + \dfrac{x - 4}{x^2}$ 15. $x + 1$ 17. $2x^2 + 3x + \dfrac{7}{2} + \dfrac{33/2}{2x - 3}$
19. $x^4 + 4x^3 + 18x^2 + 72x + 289 + \dfrac{1152}{x - 4}$ 21. $x^2 + 3 + \dfrac{4x - 4}{x^3 - x}$ 23. $2x^2 + \dfrac{3}{2}x - \dfrac{(1/2)x}{2x^2 + 1}$
25. $x^4 - x^3 + x^2 - x + 1$ 27. $x + \dfrac{x - 1}{x^3 + x - 1}$ 29. $1 + \dfrac{-(x + 1)}{(x + 1)^2 - 1}$ 31. $k - n$

Exercise 2.5 (page 44) 1. $x + 7$ 3. $x - 5 - \dfrac{5}{x + 5}$ 5. $2x^2 - 4x + 9$
7. $3x^3 + 7x^2 + 12x + 24 + \dfrac{54}{x - 2}$ 9. $x^5 + x^4 + x^3 + x^2 + x + 1$
11. $4x^4 + 13x^3 + 37x^2 + 112x + 336 + \dfrac{1008}{x - 3}$ 13. $x^2 - 4x + 7 - \dfrac{8}{x + 1}$
15. $x^3 - 3x^2 + 9x - 26 + \dfrac{64}{x + 3}$ 17. $x^2 - 5x + 25 - \dfrac{114}{x + 5}$ 19. $x^4 + 3x^2 + 3x - 6 + \dfrac{5}{x + 2}$
21. $\dfrac{1}{2}x + 1 + \dfrac{3}{2x - 2}$ 23. $x^2 - \dfrac{4}{3}x + 3 - \dfrac{19}{3x + 6}$ 25. $x^3 - \dfrac{1}{2}x^2 - \dfrac{1}{2}x - \dfrac{1}{2x - 2}$
27. $\dfrac{1}{2}x^3 - \dfrac{3}{4}x^2 - \dfrac{3}{8}x + \dfrac{5}{16} + \dfrac{37/16}{2x - 1}$ 29. $-\dfrac{1}{2}x^2 - \dfrac{1}{-2x + 2}$

Exercise 2.6 (page 48) 1. $4xy^3$; $x, y \neq 0$ 3. $2x$; $x \neq -\dfrac{1}{2}$ 5. $x + 1$; $x \neq 1$
7. $a - b$; $a \neq -b$ 9. $b^2(1 - ab)$; $a, b \neq 0$; $ab \neq -1$ 11. $x - 1$, $x \neq 1$
13. $\dfrac{x - 4}{x - 1}$; $x \neq 1, -1$ 15. $\dfrac{x^2 - x + 1}{(x + 1)^2}$, $x \neq -1$ 17. $\dfrac{x^2 - 2x + 4}{x + 3}$, $x \neq -3, -2$
19. $\dfrac{x^2 - 1}{2}$ 21. 1, $x \neq 1$ 23. $-(3x + 1)$, $x \neq \dfrac{1}{3}$ 25. $\dfrac{x^2 - 7}{x}$, $x \neq 0$

Answers to Odd-Numbered Exercises

27. $\dfrac{b - 2c}{a}$; $a \neq 0$, $b \neq -2c$ **29.** $x^4 + x^2 + 1$, $x \neq 1, -1$ **31.** $\dfrac{y^2 + 3}{x - y}$, $x \neq y$, $2x \neq y$

33. $\dfrac{x^2 - xy + y^2}{x - y}$, $x \neq y$ **35.** 21 **37.** $x^2 y^2 z^2$; $x, y, z \neq 0$ **39.** $x^2 + 2x + 1$; $x \neq 1, -1$

41. $\dfrac{x^2 - x + 1}{x^3 + 1}$, $x \neq -1$ **43.** $\dfrac{8(3x - 4)}{x^3}$ **45.** $\dfrac{(x + 1)(x - 1)}{4}$

Exercise 2.7 (page 52) **1.** 1890 **3.** 336 **5.** $x^3(x - 1)(x + 4)$ **7.** $\dfrac{x^2 + x - 1}{x^2(x + 1)}$

9. $\dfrac{y^2 + 3y - 2}{4y^2}$ **11.** $\dfrac{2}{x^2 - 1}$ **13.** $\dfrac{7x^2 + 2x + 3}{(x + 2)(x - 1)}$ **15.** $\dfrac{-4x + 5}{x - 3}$ **17.** $\dfrac{1 + x}{2 - x}$

19. $\dfrac{2w - 1}{(w - 1)(w - 2)(w + 1)}$ **21.** $\dfrac{x^2 - 3x - 1}{(x - 2)(x - 3)^2}$ **23.** $\dfrac{x^2 - 7x + 8}{(x + 1)(x - 3)^2}$ **25.** $\dfrac{x^2 + 2x - 2}{x^2 - 1}$

27. $\dfrac{z^3 - z - 1}{z^2}$ **29.** $\dfrac{x^2 - 2}{x^2 - 1}$ **31.** $\dfrac{4}{(x + 1)(x + 3)}$ **33.** $\dfrac{x^2 + a^2}{x^2 - a^2}$ **35.** $\dfrac{x^3 + 3x^2 + 2x + 2}{(x^3 - 1)(x^2 + 1)}$

37. $\dfrac{x^3 - 2x^2 + 2x}{(x - 1)(x^2 + x + 1)(x^2 - x + 1)}$

Exercise 2.8 (page 56) **1.** $\dfrac{acx}{y}$ **3.** $\dfrac{20bz}{9y}$ **5.** $\dfrac{x}{x - 1}$ **7.** $\dfrac{x^2 + x - 2}{x^2 - x - 2}$ **9.** $\dfrac{a^2 b^2}{a + b}$

11. $\dfrac{z^2 - 1}{z^2}$ **13.** $\dfrac{1}{x + 5}$ **15.** $\dfrac{x - 2}{x + 3}$ **17.** $\dfrac{x}{x + 1}$ **19.** $\dfrac{2x - 2}{2x - 1}$ **21.** $\dfrac{2x^2 + 2x}{2x^2 + 3x + 1}$

23. $\dfrac{x^2 + 2x}{x^2 - 1}$ **25.** $\dfrac{a + 1}{a + 3}$ **27.** $\dfrac{x - 3}{x + 3}$ **29.** $\dfrac{x + y}{x - y}$

Chapter 2 Review (page 57) **1.** $2x^2 - 2x - 1$ **2.** $6x^2$ **3.** 13 **4.** -3 **5.** $6x^2 + 10x + 4$
6. $3x^3 + 7x^2 + 3x + 2$ **7.** $(y - 5)(y - 3)$ **8.** $x(x + 2)^2$ **9.** $(x^2 + 4)(x - 2)(x + 2)$
10. $(z - 4)(z^2 + 4z + 16)$ **11.** $6xy^2 z$ **12.** $6y^2 + 7y + 1$ **13.** $2x - 5$ **14.** $3x + 2$
15. $x^3 + 2x^2 - x + 5$ **16.** $3x^2 - 4 + \dfrac{3}{x + 2}$ **17.** $2y + 11z$; $y, z \neq 0$ **18.** $x + 7$, $x \neq -3$

19. $\dfrac{x + 7}{x + 3}$; $x \neq 3, -3$ **20.** $\dfrac{x + 2}{x - 2}$; $x \neq 2, -2$ **21.** $\dfrac{7x + 3}{12}$ **22.** $\dfrac{6x}{x^2 + x - 2}$

23. $\dfrac{x^2 - 2x + 1}{x + 1}$ **24.** $\dfrac{y^2 - 4y + 8}{y^2 - y - 2}$ **25.** $\dfrac{x^4 z}{y^2}$ **26.** $\dfrac{x^2 + 8x + 7}{x + 5}$ **27.** $\dfrac{x^2 + x - 2}{x + 3}$

28. $\dfrac{x - 3}{x + 1}$

Exercise 3.1 (page 63) **1.** $\dfrac{1}{3}$ **3.** $-\dfrac{1}{8}$ **5.** 4 **7.** $\dfrac{1}{12}$ **9.** 3 **11.** $\dfrac{1}{2}$ **13.** x^3 **15.** $\dfrac{1}{x^2 y}$

17. $x^{10} y^5$ **19.** $\dfrac{x^6}{8y^3}$ **21.** $\dfrac{4x^2 z^2}{y^4}$ **23.** $\dfrac{z^8}{x^3 y^4}$ **25.** $\dfrac{y^2 + x}{xy^2}$ **27.** $\dfrac{x^2 - y^2}{xy}$ **29.** $\dfrac{x^3 y - 1}{xy^2}$

31. $y + x$ **33.** $x + y$ **35.** $\dfrac{x^4 y^2}{x^4 + 2x^2 y + y^2}$ **37.** $\dfrac{1 - x^2}{x^2}$ **39.** $\dfrac{x^2 + 2xy + y^2}{xy}$

41. $\dfrac{2x^5 - x^3 + 2x^2 - 1}{x^3}$ **43.** $y^3 + 4x^4$ **45.** $3x + 4y$ **47.** $y^3 z^6 - x$ **49.** 1.9711×10^4

51. 1.976×10^3 **53.** 5.86×10^{-2} **55.** 1.001×10^{-2} **57.** $248{,}000{,}000$ **59.** $-6{,}542{,}000$

61. 0.000365 **63.** -0.00000004625 **65.** 8.6×10^{-1} **67.** 6.7×10^{11} **69.** x^{2n-3} **71.** y^2

73. y^{5n+1} **75.** x^{2n+2} **77.** $x^{n+1} y^{n+3}$ **79.** $\dfrac{x^{n-1}}{y}$

Exercise 3.2 (page 69)

1. 3 **3.** $\dfrac{1}{8}$ **5.** 4 **7.** 4 **9.** $\dfrac{1}{4}$ **11.** $\dfrac{27}{8}$ **13.** $x^{5/6}$ **15.** $y^{4/3}$

17. $y^{1/3}$ **19.** $4x^{3/4}$ **21.** $x^{16} y^3$ **23.** $\dfrac{5}{3} x^3 y$ **25.** $x^{3/2} + x^2$ **27.** $x - x^{1/2}$ **29.** $y^{5/3} - y$

31. $x - y$ **33.** $x + y - (x + y)^{3/2}$ **35.** $x^{1/2} + 1$ **37.** $1 + x$ **39.** $(x + 1) - 1$

41. $x^{1/n^2} y z^n$ **43.** $x^{3n/2} y^{3n+3}$ **45.** a^{5n-3} **47.** 9 **49.** 8 **51.** $3|x|$ **53.** $\dfrac{1}{x^2(1 - x)^{1/2}}$

Exercise 3.3 (page 73)

1. $\sqrt{5}$ **3.** $2\sqrt[3]{x}$ **5.** $3\sqrt{y}$ **7.** $\sqrt[5]{x^3}$ **9.** $x\sqrt[3]{y}$ **11.** $\sqrt[3]{xy}$

13. $\sqrt{x^2 + y^2}$ **15.** $\dfrac{1}{\sqrt[4]{(x^2 - y^2)^3}}$ **17.** $2x^{4/3}$ **19.** $x^{1/2} y^{3/4}$ **21.** $(x + y)^{3/2}$ **23.** $(x^2 + 1)^{1/3}$

25. -3 **27.** -2 **29.** $x^3 y$ **31.** $3xy^2$ **33.** $x\sqrt[3]{x}$ **35.** $-3x\sqrt[3]{x}$ **37.** $x\sqrt{y}$

39. $4x^4$ **41.** x **43.** $2x\sqrt[3]{y}$ **45.** $\dfrac{\sqrt{2}}{2}$ **47.** $\dfrac{2\sqrt{2x}}{x}$ **49.** $\dfrac{\sqrt[3]{2}}{2}$ **51.** $\dfrac{\sqrt[3]{y}}{y}$ **53.** $\dfrac{1}{\sqrt{2}}$

55. $\dfrac{x}{y\sqrt[3]{x^2}}$ **57.** $\sqrt[3]{9}$ **59.** $2\sqrt{x}$ **61.** $\sqrt{2x}$ **63.** $(x - 1)^2$ **65.** $\dfrac{\sqrt[4]{(x - 1)^2 (x + 1)^3}}{x + 1}$

67. $\dfrac{\sqrt[3]{x^2 (y + 1)^2}}{x^2}$

69. No, because if $\sqrt{a^2 + b^2} = a + b$ then $a^2 + b^2 = (a + b)^2$, or $2ab = 0$. This is not true for every $a \geq 0$, $b \geq 0$.

Exercise 3.4 (page 77)

1. $7\sqrt{5}$ **3.** $5\sqrt{3}$ **5.** $-\sqrt{3}$ **7.** $8\sqrt{2}$ **9.** $7\sqrt[3]{2}$ **11.** $-\sqrt[4]{2}$

13. $6 - 3\sqrt{3}$ **15.** 2 **17.** $-3 - \sqrt{5}$ **19.** $3x\sqrt{2}$ **21.** $2x - 3 + \sqrt{x}$ **23.** $3 - x$

25. $\dfrac{1 + \sqrt{3}}{2}$ **27.** $\dfrac{x + x\sqrt{x}}{1 - x}$ **29.** $\dfrac{3x\sqrt{x} + 6x}{x - 4}$ **31.** $\dfrac{1}{\sqrt{3} - 1}$ **33.** $\dfrac{1 - x}{x^2 + x^2 \sqrt{x}}$

35. $\dfrac{-2x + 5}{2\sqrt{2x - 1} - 2x + 1}$ **37.** $\dfrac{x + y + 2\sqrt{xy}}{x - y}$ **39.** $2x + 1 + 2\sqrt{x^2 + x}$

41. $\dfrac{(x\sqrt{x} - x\sqrt{y} - x)(x - y - 1 + 2\sqrt{y})}{(x - y - 1)^2 - 4y}$ **43.** $\dfrac{x - 2}{x - 1 + \sqrt{x - 1}}$ **45.** $\dfrac{x - y}{x + y - 2\sqrt{xy}}$

47. $\dfrac{(x - y - 1)^2 - 4y}{(x\sqrt{x} - x\sqrt{y} + x)(x - y - 1 - 2\sqrt{y})}$

Answers to Odd-Numbered Exercises

Exercise 3.5 (page 84) 1. $1 - i$ 3. $9 - i$ 5. $4 + \frac{5}{2}i$ 7. $-2i$ 9. $-\frac{2}{3} + 3i$
11. $-2 + 6i$ 13. $7 + i$ 15. $\frac{1}{2} + \frac{1}{2}i$ 17. $6 + 7i$ 19. $\frac{2}{5} - \frac{9}{5}i$ 21. $-1 + 3i$
23. $\frac{1}{10} + \frac{3}{10}i$ 25. $4 + (3 - \sqrt{2})i$ 27. $(3 + 3\sqrt{2}) + (9 - \sqrt{2})i$
29. $\frac{3 - 3\sqrt{2}}{11} + \frac{9 + \sqrt{2}}{11}i$ 31. $(4 + 2\sqrt{3})i$ 33. 1 35. $-i$ 37. 1 39. $-i$
41. $-2 + 2i$ 43. $-1 + i$ 45. $-\frac{4}{25} - \frac{3}{25}i$
47. Let $z_1 = a_1 + b_1 i$ and $z_2 = a_2 + b_2 i$. Then
$$\overline{z_1 \pm z_2} = \overline{(a_1 \pm a_2) + (b_1 \pm b_2)i}$$
$$= (a_1 \pm a_2) - (b_1 \pm b_2)i$$
$$= (a_1 - b_1 i) \pm (a_2 - b_2 i)$$
$$= \bar{z}_1 \pm \bar{z}_2.$$
49. Repeated applications of Exercise 48 with $z = z_1 = z_2$ yields this result.

Chapter 3 Review (page 86) 1. $x^{12}y^8$ 2. $\frac{y^3}{x^3}$ 3. $\frac{y}{x^2}$ 4. x^3 5. $\frac{x^2 + 1}{x^3}$ 6. $xy^2 + x^2$
7. 3.51×10^4 8. 1.8×10^{-4} 9. $314{,}000{,}000$ 10. 0.00000675 11. $x^{7/6}$ 12. $\frac{y}{x^{1/2}}$
13. $x^4 y^8$ 14. $\frac{1}{xy}$ 15. $y^{4/3} + y$ 16. $4y^2 - y$ 17. $1 + x^{3/4}$ 18. $x^{4/3} - 1$
19. $2x^2 y\sqrt{xy}$ 20. $6x\sqrt{y}$ 21. $x\sqrt{3}$ 22. $3x$ 23. $\frac{\sqrt[3]{xy^2}}{y}$ 24. $\frac{1}{2\sqrt{3y}}$ 25. $\sqrt{5}$
26. $\sqrt{3xy}$ 27. $-3\sqrt{2}$ 28. $8\sqrt[3]{5}$ 29. 1 30. $2x - 5 - 3\sqrt{x}$ 31. $\frac{3 + \sqrt{2}}{7}$
32. $\frac{2x + 1 + 3\sqrt{x}}{4x - 1}$ 33. $\frac{1}{\sqrt{3} - 1}$ 34. $\frac{x - 1}{x + 2 + 3\sqrt{x}}$ 35. $4 - 2i$ 36. $2 + 4i$
37. $6 - 8i$ 38. i 39. $2i$ 40. $1 - 4i$ 41. i 42. $\frac{2}{3}$

Exercise 4.1 (page 92) 1. $\{11\}$ 3. $\{5\}$ 5. $\{2\}$ 7. $\left\{\frac{1}{4}\right\}$ 9. The empty set 11. $\left\{-\frac{1}{3}\right\}$
13. $\{9\}$ 15. $\{0\}$ 17. $\left\{\frac{1}{2}\right\}$ 19. The empty set 21. $\left\{\frac{2}{5}\right\}$ 23. Any x; $x \neq 1, -1$
25. $\{0\}$ 27. The empty set 29. Any x; $x \neq -2, 4$ 31. $r = \frac{S}{2\pi h}$, $h \neq 0$ 33. $k = v - gt$
35. $c = \frac{2A - bh}{h}$, $h \neq 0$ 37. $n = \frac{l - a + d}{d}$, $d \neq 0$ 39. $r^2 = \frac{V + \pi R^2 h}{-\pi h}$, $h \neq 0$
41. $y' = -\frac{x}{y}$, $y \neq 0$ 43. $y' = \frac{1}{x}$, $x \neq 0$, $y \neq 0$ 45. $y' = \frac{1 + 3x}{x^2 - 2y^3}$, $x^2 \neq 2y^3$

47. $x_1 = \dfrac{x_4}{x_2 - 2x_3}$, $x_2 \neq 2x_3$ **49.** $y = 6x - 6x_1 + y_1$ **51.** $k = \dfrac{-3}{5}$ **53.** $k = -13$
55. $k = 11$

Exercise 4.2 (page 98) **1.** $\{-2, 1\}$ **3.** $\{0, 5\}$ **5.** $\{1\}$ **7.** $\left\{-\dfrac{1}{2}, -\dfrac{1}{3}\right\}$ **9.** $\left\{\dfrac{1}{2}, 6\right\}$
11. $\left\{-\dfrac{1}{2}, -\dfrac{2}{3}\right\}$ **13.** $\left\{-\dfrac{3}{2}, 2\right\}$ **15.** $\left\{-\dfrac{3}{2}, -3\right\}$ **17.** $\left\{\dfrac{3}{5}, -2\right\}$ **19.** $\{2, -2\}$
21. $\{\sqrt{3}, -\sqrt{3}\}$ **23.** $\{2i, -2i\}$ **25.** $\{1, -3\}$ **27.** $\{3i, -3i\}$ **29.** $\{1+i, 1-i\}$
31. $\{1, -9\}$ **33.** $\{-6, -5\}$ **35.** $\left\{2, -\dfrac{1}{3}\right\}$ **37.** $\left\{\dfrac{1}{2} + \dfrac{\sqrt{17}}{2}, \dfrac{1}{2} - \dfrac{\sqrt{17}}{2}\right\}$
39. $\left\{\dfrac{3}{2} + \dfrac{i\sqrt{3}}{2}, \dfrac{3}{2} - \dfrac{i\sqrt{3}}{2}\right\}$ **41.** $\{-3 + 2i, -3 - 2i\}$

Exercise 4.3 (page 100) **1.** $\{3, 4\}$ **3.** $\left\{-2, \dfrac{3}{4}\right\}$ **5.** $\left\{-1, -\dfrac{1}{2}\right\}$ **7.** $\left\{\dfrac{-1+\sqrt{13}}{2}, \dfrac{-1-\sqrt{13}}{2}\right\}$
9. $\left\{-1, -\dfrac{1}{2}\right\}$ **11.** $\{-4+i, -4-i\}$ **13.** $\left\{\dfrac{-3+i}{2}, \dfrac{-3-i}{2}\right\}$
15. $\left\{\dfrac{-3+i\sqrt{11}}{2}, \dfrac{-3-i\sqrt{11}}{2}\right\}$ **17.** $\left\{\dfrac{1+i\sqrt{7}}{4}, \dfrac{1-i\sqrt{7}}{4}\right\}$ **19.** $\{-2, 1\}$ **21.** $\left\{\dfrac{1}{2} + \dfrac{1}{2}i, \dfrac{1}{2} - \dfrac{1}{2}i\right\}$
23. $\left\{-1 + \dfrac{i\sqrt{34}}{2}, -1 - \dfrac{i\sqrt{34}}{2}\right\}$ **25.** $\{2+\sqrt{5}, 2-\sqrt{5}\}$ **27.** $\{2i, -2i\}$ **29.** Two real roots
31. Two imaginary roots **33.** One real root of multiplicity 2 **35.** $k = \pm\dfrac{1}{\sqrt{x}}$
37. $k = 0$ or $k = \dfrac{1}{x}$ **39.** $k = -x$
41. If $ax^2 + bx + c = 0$ and $a \neq 0$ then completing the square we have

$$x^2 + \dfrac{b}{a}x + \dfrac{b^2}{4a^2} = \dfrac{-c}{a} + \dfrac{b^2}{4a^2},$$

$$\left(x + \dfrac{b}{2a}\right)^2 = \dfrac{b^2 - 4ac}{4a^2}.$$

Extracting roots, we obtain

$$x + \dfrac{b}{2a} = \dfrac{\pm\sqrt{b^2 - 4ac}}{2a}.$$

Thus,

$$x = \dfrac{-b \pm \sqrt{b^2 - 4ac}}{2a}.$$

Answers to Odd-Numbered Exercises

Exercise 4.4 (page 104)   **1.** $\{64\}$ **3.** $\{-7\}$ **5.** $\{4\}$ **7.** $\{-25\}$ **9.** $\{16\}$ **11.** $\{1\}$
13. $\{5\}$ **15.** $\left\{-\dfrac{7}{4}\right\}$ **17.** $\{4\}$ **19.** $\{1, 3\}$ **21.** $A = \pi r^2$ **23.** $y = \dfrac{1}{x^3}$
25. $y = \pm\sqrt{a^2 - x^2}$ **27.** $V = \pi h(R^2 - r^2)$ **29.** $\{16\}$ **31.** $\left\{\dfrac{1}{9}\right\}$

Exercise 4.5 (page 106)   **1.** $\{1, -1, 3, -3\}$ **3.** $\{2, -2, \sqrt{3}, -\sqrt{3}\}$ **5.** $\{2, -9\}$ **7.** $\{1\}$
9. $\{216, -27\}$ **11.** $\left\{\dfrac{1}{9}, -\dfrac{1}{2}\right\}$ **13.** $\left\{-\dfrac{6}{7}, -\dfrac{5}{4}\right\}$ **15.** $\{16, 256\}$ **17.** $\{87\}$
19. $\left\{-\dfrac{25}{8}, -\dfrac{11}{4}\right\}$ **21.** $\{0, -2, -1+\sqrt{2}, -1-\sqrt{2}\}$ **23.** $\{1, -1\}$ **25.** $\{64\}$
27. $\left\{-1, -\dfrac{35}{16}, \dfrac{19}{16}\right\}$

Exercise 4.6 (page 111)   **1.** $(-6, +\infty)$ **3.** $(-\infty, 3]$ **5.** $[-6, +\infty)$

7. $\left[\dfrac{5}{2}, +\infty\right)$ **9.** $(-\infty, -1)$ **11.** $[-2, +\infty)$ **13.** $[-3, +\infty)$

15. $\left(-\infty, \dfrac{8}{3}\right)$ **17.** $\left[-\dfrac{6}{5}, +\infty\right)$ **19.** $[-6, 1]$ **21.** $\left(-\dfrac{22}{3}, -4\right]$

23. $[-25, -16]$ **25.** $\left(\dfrac{1}{4}(5 - \varepsilon), +\infty\right)$ **27.** $\left(\dfrac{1}{3}(-2 - \varepsilon), \dfrac{1}{3}(\varepsilon - 2)\right)$

29. $\left(\dfrac{1}{3}(7 - \varepsilon), \dfrac{1}{3}(7 + \varepsilon)\right)$

Exercise 4.7 (page 116)   **1.** $(-\infty, -1) \cup (5, +\infty)$ **3.** $\left[-\dfrac{1}{2}, \dfrac{8}{3}\right]$ **5.** $(-\infty, -3) \cup (2, +\infty)$
7. $(-\infty, -2) \cup (4, +\infty)$ **9.** $(-\infty, -5) \cup (4, +\infty)$ **11.** $\left[-\dfrac{1}{3}, 2\right]$
13. $\left(-\infty, \dfrac{-3 - \sqrt{33}}{8}\right] \cup \left[\dfrac{-3 + \sqrt{33}}{8}, +\infty\right)$ **15.** No solutions
17. $(-\infty, 2 - \sqrt{3}] \cup [2 + \sqrt{3}, +\infty)$ **19.** No solution   **21.** $\left(-\infty, -\dfrac{1}{2}\right] \cup (0, +\infty)$
23. $(-\infty, 1)$ **25.** $(-2, 0)$ **27.** $(-1, 1) \cup (3, +\infty)$ **29.** $(-\infty, -1) \cup (0, 2)$
31. $[-3, -1] \cup [1, +\infty)$ **33.** $(-\infty, -4) \cup (-2, 0) \cup (2, +\infty)$ **35.** $(-\sqrt{\varepsilon}, \sqrt{\varepsilon})$
37. $(-\sqrt{1 + \varepsilon}, \sqrt{1 + \varepsilon})$

Exercise 4.8 (page 121) 1. $\{-5, 1\}$ 3. $\{-1, 7\}$ 5. $\{1, -2\}$ 7. $(-3, 7)$ 9. $\left(-\dfrac{1}{6}, \dfrac{7}{6}\right)$
11. $(-\infty, 3) \cup (5, +\infty)$ 13. $(-\infty, -10) \cup (4, +\infty)$ 15. $[3, 5]$ 17. $[0, 3]$
19. $\left(-\infty, -\dfrac{3}{2}\right] \cup \left[\dfrac{5}{2}, +\infty\right)$ 21. $\left(-\infty, -\dfrac{2}{3}\right] \cup \left[\dfrac{4}{3}, +\infty\right)$ 23. $[-3, 3]$ 25. $|x| \le 3$
27. $|x - 2| < 4$ 29. $|x + 5| \le 3$ 31. $\left|x + \dfrac{1}{3}\right| < 1$ 33. $(-1, 0) \cup (0, 1)$
35. $(-\infty, 2 - 2\sqrt{2}] \cup \{2\} \cup [2 + 2\sqrt{2}, +\infty)$ 37. $\left(-\infty, -\dfrac{3}{5}\right] \cup \left[-\dfrac{3}{7}, +\infty\right)$

Exercise 4.9 (page 124) 1. 15 dimes, 17 nickels   3. 15 pounds of 32% silver alloy   5. $158\frac{1}{3}$ gallons
7. 1 ounce   9. $2500 at 12%, $1500 at 14%   11. $11,000 in 8%, $33,000 in 12%
13. Rate of automobile is 60 miles per hour, rate of plane is 180 miles per hour   15. 400 miles
17. 5 ohms   19. 10 ohms   21. 11 and 14   23. 11 and 12; -11 and -12 25. $\sqrt{13{,}000}$ miles
27. $\dfrac{1}{2}$ second, $3\dfrac{1}{2}$ seconds   29. 6 days for the father alone, 12 days for the son alone   31. 5 ohms
33. $\dfrac{-5 + \sqrt{325}}{6}$ ohms   35. At least 94%   37. Between 14°F and 68°F
39. Between 1 second and 1.5 seconds and between 2 seconds and 2.5 seconds after the ball is thrown.

Chapter 4 Review (page 129)   1. $\left\{-11\dfrac{1}{2}\right\}$ 2. $\{3\}$ 3. $\left\{-\dfrac{5}{2}\right\}$ 4. $\left\{\dfrac{3}{5}\right\}$ 5. $y = \dfrac{x}{4}$
6. $x = 4y$ 7. $\{3, -2\}$ 8. $\{-7, 3\}$ 9. $\{\sqrt{5}, -\sqrt{5}\}$ 10. $\{5 + \sqrt{2}, 5 - \sqrt{2}\}$
11. $\left\{\dfrac{-5}{2} + \dfrac{\sqrt{33}}{2}, \dfrac{-5}{2} - \dfrac{\sqrt{33}}{2}\right\}$ 12. $\left\{\dfrac{1}{2}, -1\right\}$ 13. $\left\{\dfrac{1 \pm i\sqrt{15}}{4}\right\}$ 14. $\{-1 \pm \sqrt{5}\}$
15. $\left\{\dfrac{3 \pm \sqrt{9 - 4k}}{2k}\right\}$ 16. $\left\{\dfrac{-k \pm \sqrt{k^2 + 16}}{2}\right\}$ 17. $\{4, 9\}$ 18. $\{8\}$ 19. $\{1, -1, \sqrt{2}, -\sqrt{2}\}$
20. $\left\{\dfrac{1}{7}, -\dfrac{1}{6}\right\}$ 21. $[32, +\infty)$ 22. $\left(-\infty, -\dfrac{6}{5}\right)$ 23. $(-5, 2)$ 24. $\left(\dfrac{1}{3}, 1\right)$ 25. $\left\{-2, \dfrac{4}{3}\right\}$
26. $\left\{-1, \dfrac{7}{3}\right\}$ 27. $(-\infty, -1) \cup (7, +\infty)$ 28. $|x + 1| \le 4$ 29. 2113 and 2263 votes
30. 6 centimeters   31. 6 miles per hour going; 4 miles per hour returning   32. $950

Supplemental Exercises, Chapters 1-4 (page 131)   1. $2x^4 + 6x^3 - 7x^2 - 14x + 1$
2. $-6x^5 - 25x^4 - 14x^3 + 28x^2 + 7x - 2$ 3. $\dfrac{\sqrt{Ax - B^2}}{2x}$ 4. $\dfrac{3}{2}(24w + 1)\sqrt{w(9w + 1)^3}$
5. $\dfrac{-4(x^2 - 1 + hx)}{(1 + h^2 + 2xh + x^2)(1 + x^2)}$ 6. $\dfrac{-5}{(x + h - 1)(x - 1)}$ 7. $\dfrac{-6x(x^3 + 2)(-x^3 - x + 1)}{(x^2 + 1)^4}$

Answers to Odd-Numbered Exercises

8. $\dfrac{6x(x^2 + 1)^2(-x - 1)}{(x^3 - 1)^3}$ 9. $\dfrac{1}{(9 - x^2)^{3/2}}$ 10. $\dfrac{13x^4 + 9x^2 - 4x}{6(x^3 - 1)^{1/2}(x^2 + 1)^{2/3}}$

11. $\dfrac{-6x - 19}{6(2x + 3)^{2/3}(3x + 2)^{3/2}}$ 12. $-a^n(n - 1)x^{n-2}y^{-2n+1}$ 13. $\dfrac{10(x^2 - 3xy + y^2)}{(2y - 3x)^3}$

14. $5(y^2 - 4xy - x^2)(y - 2x)^{-3}$ 15. $\left\{-1, \dfrac{5}{7}, 2\right\}$ 16. $\left\{0, -1, -\dfrac{13}{12} + \dfrac{\sqrt{73}}{12}, -\dfrac{13}{12} - \dfrac{\sqrt{73}}{12}\right\}$

17. $\left\{\dfrac{7}{6}\right\}$ 18. $\left\{\dfrac{1}{\sqrt[3]{3}}\right\}$

19. If $R = 0$ then every $y \neq 0$ is a solution. If $R \neq 0$ then $y = -R$ and $y = \dfrac{R}{3}$ are solutions.

20. $x = \dfrac{L\sqrt[3]{\dfrac{A}{B}}}{1 + \sqrt[3]{\dfrac{A}{B}}}$ 21. $x = \dfrac{V_0}{C}$ or $x = \dfrac{V_0}{2 + C}$ 22. $x = \dfrac{B}{A}$ 23. $(-1, 0) \cup (0, +\infty)$

24. $(0, 12)$ 25. $(-\infty, +\infty)$ 26. $(-\sqrt{2}, \sqrt{2})$

27. $(-\infty, +\infty)$ 28. $(-\infty, +\infty)$

Exercise 5.1 (page 139) 1. $\left\{(0, 6), (1, 4), (2, 2), (-3, 12), \left(\dfrac{2}{3}, \dfrac{14}{3}\right)\right\}$

3. $\left\{(0, 0), (1, -3), (2, 3), \left(-3, -\dfrac{9}{7}\right), \left(\dfrac{2}{3}, -\dfrac{9}{7}\right)\right\}$

5. $\left\{(0, \sqrt{11}), (1, \sqrt{14}), (2, \sqrt{17}), (-3, \sqrt{2}), \left(\dfrac{2}{3}, \sqrt{13}\right)\right\}$

7. **a.** Domain: $\{2, 5, 7\}$; Range: $\{3, 7, 8\}$ **b.** A function
9. **a.** Domain: $\{2, 3\}$; Range: $\{-1, 4, 6\}$ **b.** Not a function
11. **a.** Domain: $\{5, 6, 7\}$; Range: $\{5, 6, 7\}$ **b.** A function
13. Domain: $\{x \mid x \in R\}$ 15. Domain: $\{x \mid x \in R\}$ 17. Domain: $\{x \mid x \in R, x \neq 2\}$
19. Domain: $\{x \mid x \in R, x \geq 0\}$ 21. Domain: $\{x \mid x \in R, -2 \leq x \leq 2\}$
23. Domain: $\{x \mid x \in R, x \neq 0, 1\}$ 25. Yes 27. Yes 29. No 31. No 33. 5
35. $-2a + 2$ 37. 0 39. $\dfrac{a^2 + 1}{a^2}$ 41. -1 43. $a^4 - 8a^3 + 22a^2 - 24a + 9$
45. $-\dfrac{1}{3}$ 47. $\sqrt{3}$ 49. $2h$ 51. $-3h$ 53. $-2ah - h^2$ 55. 3 57. 1
59. $2a + h - 1$ 61. **a.** 13 **b.** 49 63. **a.** $\dfrac{1}{a}$ **b.** $\dfrac{1}{a}$

Exercise 5.2 (page 146)   **1.** x-intercept is 2; y-intercept is 3   **3.** x-intercept is $-\frac{1}{2}$; y-intercept is 1
5. Both intercepts are 0   **7.** Both intercepts are 0   **9.** x-intercept is 3; no y-intercept
11. No x-intercept; y-intercept is -3 **13.**   **15.** 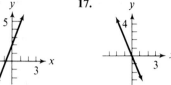   **17.**

19.   **21.** 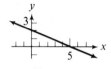   **23.**   **25.**

27. Distance, 5; slope, $\frac{4}{3}$ **29.** Distance, 13; slope, $\frac{12}{5}$ **31.** Distance, $\sqrt{2}$; slope, 1
33. Distance, $3\sqrt{5}$; slope, $\frac{1}{2}$ **35.** Distance, 5; slope, 0   **37.** Distance, 10; slope, undefined
39. 7, $\sqrt{68}$, $\sqrt{89}$ **41.** 10, 21, 17
43. From Exercise 40, the lengths of the sides of the triangle are 15, $3\sqrt{5}$, and $6\sqrt{5}$, respectively; since $15^2 = 225$, $(3\sqrt{5})^2 = 45$, and $(6\sqrt{5})^2 = 180$, it follows that $15^2 = (3\sqrt{5})^2 + (6\sqrt{5})^2$, and the converse of the Pythagorean theorem applies. Hence, the triangle is a right triangle.   **45.** $x - 2y = 8$
47. This follows from the fact that the midpoint of the segment PR is $\left(\frac{x_1 + x_2}{2}, y_1\right)$ and the midpoint of QR is $\left(x_2, \frac{y_1 + y_2}{2}\right)$. Then, since $\triangle POM$ is similar to $\triangle PQR$, $PO = \frac{1}{2}PQ$. Hence, $\left(\frac{x_1 + x_2}{2}, \frac{y_1 + y_2}{2}\right)$ is the midpoint of PQ.

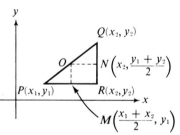

49. Range: $\{y \mid -1 \leq y \leq 3\}$

Exercise 5.3 (page 151)   **1.** $4x - y - 7 = 0$ **3.** $x + y - 10 = 0$ **5.** $3x - y = 0$
7. $x + 2y + 2 = 0$ **9.** $3x + 4y + 14 = 0$ **11.** $y - 2 = 0$
13. $y = -x + 3$; slope, -1; intercept, 3   **15.** $y = -\frac{3}{2}x + \frac{1}{2}$; slope, $-\frac{3}{2}$; intercept, $\frac{1}{2}$
17. $y = \frac{1}{3}x - \frac{2}{3}$; slope, $\frac{1}{3}$; intercept, $-\frac{2}{3}$ **19.** $y = \frac{4}{3}x - \frac{7}{3}$; slope, $\frac{4}{3}$; intercept, $-\frac{7}{3}$
21. $y = 6$; slope, 0; intercept, 6   **23.** $x = 5$; slope, undefined; no y-intercept
25. $3x + 2y - 6 = 0$ **27.** $5x + 2y + 10 = 0$ **29.** $6x - 2y + 3 = 0$ **31.** $3x - y - 5 = 0$
33. $2x - 3y + 19 = 0$ **35.** $x - 2 = 0$

37. Substituting $\dfrac{y_2 - y_1}{x_2 - x_1}$ for m in the formula $y - y_1 = m(x - x_1)$ gives the desired result.   **39.** $y = 0$

41. Given any four points $P_1(x_1, y_1)$, $P_2(x_2, y_2)$, $P_3(x_3, y_3)$ and $P_4(x_4, y_4)$ on the same line, the triangles $P_1Q_1P_2$ and $P_3Q_2P_4$ are similar. Thus,

$$\frac{y_4 - y_3}{x_4 - x_3} = \frac{y_2 - y_1}{x_2 - x_1},$$

and both pairs of points P_1, P_2 and P_3, P_4 yield the same value for the slope of the line.

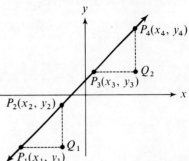

Exercise 5.4 (page 155)

1. 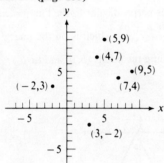   **3.**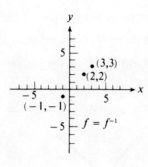

5. 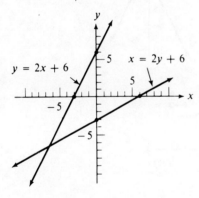   **7.**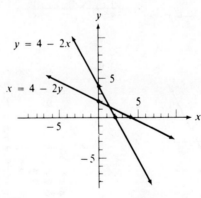

9. 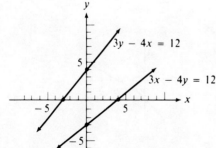   **11.**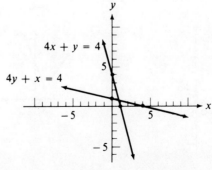

13. $f(x) = x$, $f^{-1}(x) = x$; $f[f^{-1}(x)] = f(x) = x$, $f^{-1}[f(x)] = f^{-1}(x) = x$

15. $f(x) = 4 - 2x$, $f^{-1}(x) = \dfrac{4-x}{2}$; $f[f^{-1}(x)] = f\left[\dfrac{4-x}{2}\right] = 4 - 2\left(\dfrac{4-x}{2}\right) = 4 - 4 + x = x$,

$f^{-1}[f(x)] = f^{-1}[4 - 2x] = \dfrac{4-(4-2x)}{2} = \dfrac{4-4+2x}{2} = \dfrac{2x}{2} = x$

17. $f(x) = \dfrac{3x-12}{4}$, $f^{-1}(x) = \dfrac{4x+12}{3}$; $f[f^{-1}(x)] = f\left[\dfrac{4x+12}{3}\right] = \dfrac{(3)\left(\dfrac{4x+12}{3}\right) - 12}{4} =$

$\dfrac{4x+12-12}{4} = \dfrac{4x}{4} = x$, $f^{-1}[f(x)] = f^{-1}\left[\dfrac{3x-12}{4}\right] = \dfrac{(4)\left(\dfrac{3x-12}{4}\right) + 12}{3} = \dfrac{3x-12+12}{3} = \dfrac{3x}{3} = x$

19. $f(x) = \dfrac{1}{x}$, $f^{-1}(x) = \dfrac{1}{x}$; $f[f^{-1}(x)] = \dfrac{1}{f^{-1}(x)} = \dfrac{1}{\frac{1}{x}} = x$, $f^{-1}[f(x)] = \dfrac{1}{f(x)} = \dfrac{1}{\frac{1}{x}} = x$ **21.** 4 **23.** 2

25. The line joining the points (b, a) and (a, b) has slope -1 and hence is perpendicular to the line $y = x$. The midpoint of the line segment joining the points (a, b) and (b, a) is $\left(\dfrac{a+b}{2}, \dfrac{a+b}{2}\right)$, which is on the line $y = x$. Thus, the line $y = x$ is the perpendicular bisector of the line segment joining the points (b, a) and (a, b).

Exercise 5.5 (page 159)

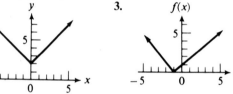

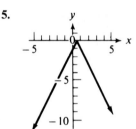

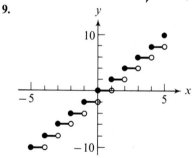

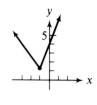

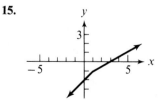

Answers to Odd-Numbered Exercises   431

17. 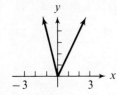   **19.** 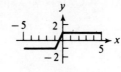   **21.**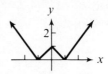

23. 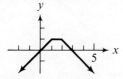   **25.** Cost in cents: $C = -c[-x]$, $x > 0$.

Exercise 5.6 (page 162)

1.   **3.** 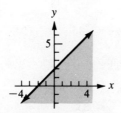 **5.** 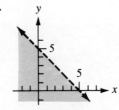 **7.**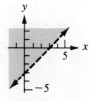

9. 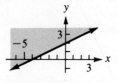 **11.** 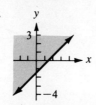 **13.**   **15.**

17. 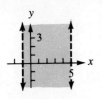 **19.** 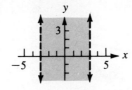 **21.**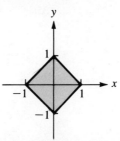

23. 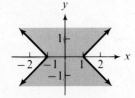 **25.** 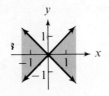 **27.**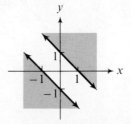

Chapter 5 Review (page 163) **1.** $\{4, 2, 3\}$ **2.** $\{1\}$ **3.** $\{x \mid x \neq -4\}$ **4.** $\{x \mid x \geq 6\}$ **5.** 13 **6.** -5 **7.** $x - h - 3$ **8.** $x^2 + 2xh + h^2 + 4$ **9.** 1 **10.** 0 **11.** 2 **12.** $\dfrac{1}{3}$ **13.** 12 **14.** $\dfrac{\sqrt{3}}{3} - 1$ **15.** 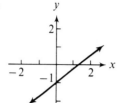 **16.** 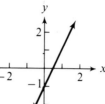 **17.**

18. **19.** **20.**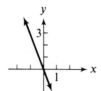

21. $\sqrt{41}$, $m = -\dfrac{4}{5}$ **22.** $\sqrt{5}$, $m = -\dfrac{1}{2}$ **23.** $4x - y - 15 = 0$ **24.** $x - 2y + 12 = 0$ **25.** $m = -4$, y-intercept $= 6$ **26.** $m = \dfrac{3}{2}$, y-intercept $= -8$ **27.** $2x - 3y + 6 = 0$ **28.** $12x - y - 4 = 0$ **29.** $4x - 3y + 11 = 0$ **30.** $x + 6y + 21 = 0$

31. 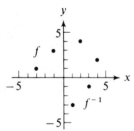 **32.** 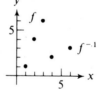 **33.**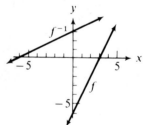

34. 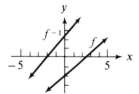 **35.** 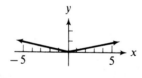 **36.**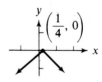

37. 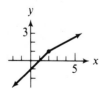 **38.** 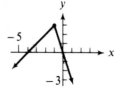 **39.**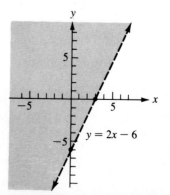

Answers to Odd-Numbered Exercises   433

40. 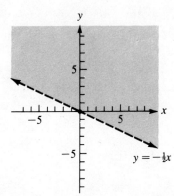   **41.** 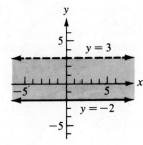   **42.**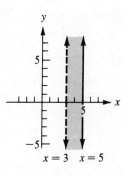

Exercise 6.1 (page 170) **1. a.** 1, 2 **b.** 2 **c.** $x = \dfrac{3}{2}$ **d.** $\left(\dfrac{3}{2}, -\dfrac{1}{4}\right)$

3. a. $-2, -\dfrac{1}{2}$ **b.** -2 **c.** $x = -\dfrac{5}{4}$ **d.** $\left(-\dfrac{5}{4}, \dfrac{9}{8}\right)$

5. a. 1 **b.** 1 **c.** $x = 1$ **d.** 0   **7. a.** $\dfrac{1}{2}$ **b.** -1 **c.** $x = \dfrac{1}{2}$ **d.** 0

9. a. None **b.** 1 **c.** $x = 0$ **d.** 1   **11. a.** None **b.** -9 **c.** $x = 0$ **d.** -9

13.   **15.**   **17.**   **19.**

21. 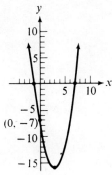   **23.** 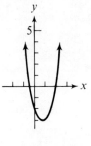   **25.** 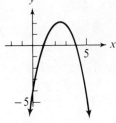   **27.**

29. $\dfrac{16}{3}$, slope of secant line 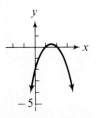   **31.**   **33.**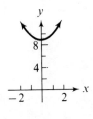

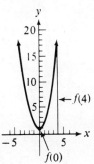
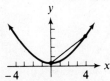
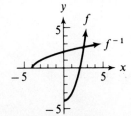
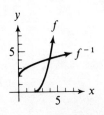

35. 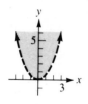 **37.** **39.** **41.**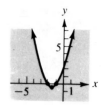

43. Varying k has the effect of translating the graph along the y-axis.

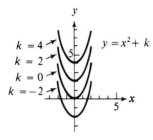

45. Varying k has the effect of translating the axis of symmetry along the x-axis and thus changes the vertex.

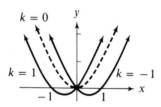

47. 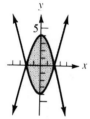   **49.** 4; 4   **51.** $2(x + a) - 1$, slope of the line joining $(x, f(x))$ and $(a, f(a))$

Exercise 6.2 (page 179)

1.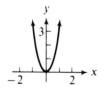
Parabola, intercepts $(0, 0)$

3.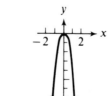
Parabola, intercepts $(0, 0)$

5.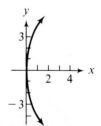
Parabola, intercepts $(0, 0)$

7.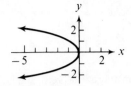
Parabola, intercepts $(0, 0)$

9.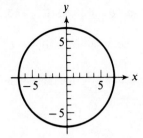
Circle,
x-intercepts $(\pm 7, 0)$,
y-intercepts $(0, \pm 7)$

11.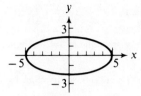
Ellipse,
x-intercepts $(\pm 5, 0)$,
y-intercepts $(0, \pm 2)$

13.
Two intersecting lines,
intercepts $(0, 0)$

15.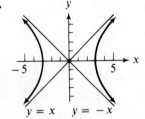
Hyperbola,
x-intercepts $(\pm 3, 0)$,
no y-intercepts,
asymptotes $y = \pm x$

17.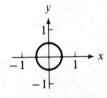
Circle,
x-intercepts $\left(\pm \dfrac{1}{2}, 0\right)$,
y-intercepts $\left(0, \pm \dfrac{1}{2}\right)$

19.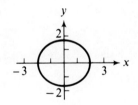
Ellipse,
x-intercepts $(\pm 2, 0)$,
y-intercepts $(0, \pm\sqrt{3})$

21.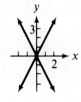
Two intersecting lines,
intercepts $(0, 0)$

23.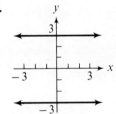
Two parallel lines,
intercepts $(0, \pm 3)$

25. Since A and B are positive, $Ax^2 - By^2 = 0$ is equivalent to $y = \pm\sqrt{\dfrac{A}{B}}\, x$. Thus, the graph of the relation is a pair of straight lines with slopes $\pm\sqrt{\dfrac{A}{B}}$ that intersect at $(0, 0)$.

27. Since $A \neq 0$, $Ax^2 = 0$ is equivalent to $x^2 = 0$, which is equivalent to $x = 0$. Thus, the graph of $Ax^2 = 0$ is a vertical straight line passing through the point $(0, 0)$.

29. Since $A, B > 0$ and $x^2, y^2 \geq 0$, the only solution of the equation $Ax^2 + By^2 = 0$ is $(0, 0)$ and the graph is a single point at the origin.

31. Let $P(x, y)$ be a point at distance 4 from $(2, -3)$; then by the distance formula, $\sqrt{(x-2)^2 + (y+3)^2} = 4$; squaring and simplifying yields $(x-2)^2 + (y+3)^2 = 16$.

33. Given $Ax^2 + By^2 = C$ with intercepts $(a, 0)$ and $(0, b)$. If $y = 0$, $x = a$ and $x^2 = \frac{C}{A}$; hence $a^2 = \frac{C}{A}$. If $x = 0$, $y = b$ and $y^2 = \frac{C}{B}$; hence $b^2 = \frac{C}{B}$. Because $Ax^2 + By^2 = C$, $\frac{x^2}{C/A} + \frac{y^2}{C/B} = 1$ and $\frac{x^2}{a^2} + \frac{y^2}{b^2} = 1$.

35. As $|x|$ increases, so do x^2 and Ax^2 of the fraction $\frac{C}{Ax^2}$; the value of this fraction approaches zero, and hence, $\sqrt{1 - \frac{C}{Ax^2}}$ approaches 1, and $y = \pm\sqrt{\frac{A}{B}}|x|\left(\sqrt{1 - \frac{C}{Ax^2}}\right)$ approaches $y = \pm\sqrt{\frac{A}{B}}|x|$.

37.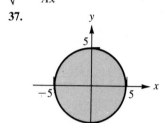

Exercise 6.3 (page 183) 1. 1, 16   3. 400 feet   5. 160 pounds per square foot   7. 1687.5 pounds
9. 16   11. 160 pounds per square foot   13. 1687.5 pounds
15. $k = 3$ 17. "Increases" the curvature

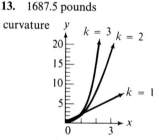

19. Let c_1 and d_1 be the circumference and diameter of one circle and c_2 and d_2 be corresponding parts in a second circle; then $c_2 = \pi d_2$ and $\pi = \frac{c_2}{d_2}$; similarly $c_1 = \pi d_1$, and substituting for π yields $c_1 = \frac{c_2}{d_2} d_1$; and since $c_2 \ne 0$, multiplying both sides by $\frac{1}{c_2}$ yields $\frac{c_1}{c_2} = \frac{d_1}{d_2}$.

21. Since y varies directly as x, there is $k_1 > 0$ so that $y = k_1 x$; and since z varies directly as x, there is $k_2 > 0$ so that $z = k_2 x$. Thus, $y + z = (k_1 + k_2)x$ and $y + z$ varies directly as x.

23. Yes. The constant of variation is $\frac{1}{k}$.

Exercise 6.4 (page 192) 1. $f(0) = 1$ and $f(1) = -1$ 3. $g(-3) = 10$ and $g(-2) = -33$
5. $P(-3) = 2$ and $P(-2) = -4$, $P(0) = -4$, and $P(1) = 2$ 7. 3; 17; 47   9. 56; 12; 326
11. -685; 95; 719   13. Upper bound 3; lower bound -4
15. Upper bound 4; lower bound -3 17. Upper bound 2; lower bound -3
19. Upper bound 1; lower bound -2 21. $-2i$ 23. $-i$ 25. $i, -2i$
27. $3i, \dfrac{-3 + \sqrt{29}}{2}, \dfrac{-3 - \sqrt{29}}{2}$ 29. $1 - i$; $x^3 - 2x + 4 = 0$

31. $P(x) = (2x - 3)(x - 2 + 2i)(x - 2 - 2i)$ **33.** $4 + i, 1 + i, 1 - i$

35. Let $P(x) = 0$ be a polynomial equation with real coefficients of degree n, where n is odd; then $P(x) = 0$ must have at least n zeros and, since $P(x)$ has only real coefficients, any complex zeros must occur as conjugate pairs; since there must always be at least one real zero, then $P(x) = 0$ must have at least one real root.

37. $-1, \dfrac{1 + \sqrt{3}i}{2}, \dfrac{1 - \sqrt{3}i}{2}$

39. Let $P(x) = a_n x^n + \cdots + a_0$ and suppose $P(z) = 0$; then, by Exercises 47–49, Section 3.5, we can write

$$\begin{aligned} \overline{P(z)} &= \overline{a_n z^n + \cdots + a_0} \\ &= \overline{a_n z^n} + \cdots + \overline{a_0} \quad \text{(since } a_i \in R\text{)} \\ &= a_n \overline{z}^n + \cdots + a_0 \\ &= P(\overline{z}) = \overline{0} = 0. \end{aligned}$$

41. Let $S(x) = P(x) - Q(x)$; then the degree of S is less than or equal to n. But there are more than n values of x for which $S(x) = 0$ (namely the values where $P(x) = Q(x)$). But the number of roots of $S(x) = 0$ must be the same as the degree of S if the degree of S is positive. Since the number of roots of $S(x) = 0$ is greater than the degree of S, it must be the case that the degree of S is 0 and $S(x)$ must be a constant. Since the equation $S(x) = 0$ has solutions, it must be the case that $P(x) - Q(x) = S(x) = 0$, or $P(x) = Q(x)$ for all x.

Exercise 6.5 (page 196) **1.** $\{4\}$ **3.** $\{1, -2\}$ **5.** None   **7.** $\left\{-\dfrac{3}{2}\right\}$ **9.** $\left\{2, -\dfrac{7}{4}\right\}$ **11.** $\left\{\dfrac{3}{2}\right\}$

13. $\left\{-\dfrac{1}{3}, 1 + \sqrt{5}, 1 - \sqrt{5}\right\}$ **15.** $\left\{-1, \dfrac{1}{2}, \dfrac{-1 + i\sqrt{3}}{2}, \dfrac{-1 - i\sqrt{3}}{2}\right\}$ **17.** $\left\{\dfrac{3}{4}, -\dfrac{4}{3}, i, -i\right\}$

19. $(2x + 3)(x - 1)(x + 1)$

21. Consider $x^2 - 3 = 0$. If the equation has real roots, then they are either rational or irrational. If rational, by Theorem 6.10, the only possibilities are ± 1 and ± 3; however, direct substitution shows none of these satisfies the equation. Now, $\sqrt{3}$ is a real number, and since $(\sqrt{3})^2 - 3 = 3 - 3 = 0$, $\sqrt{3}$ is a real zero of the equation, and since it is not among the rational zeros, it must be irrational.

23. By Theorem 6.10, the only possible rational roots of the equation $x^2 - p = 0$ are ± 1 and $\pm p$. However, since p is prime, $p \neq 1$ and thus $(\pm 1)^2 - p = 1 - p \neq 0$ and neither 1 nor -1 is a solution. If either p or $-p$ is a solution then $p^2 - p = 0$, and $p = 0$ or $p = 1$. Since p is prime, $p \neq 0$ and $p \neq 1$. Thus, neither p nor $-p$ is a solution. Finally, $\sqrt{p}$ is a real solution; since it is not rational, it must be irrational.

Exercise 6.6 (page 201) **1.** $(1, 15)$ is a local maximum; $(2, 14)$ is a local minimum.

3. $(-1, 0)$ and $(1, 0)$ are local minima; $(0, 1)$ is a local maximum.

5. $(0, 10)$ is neither a local maximum nor a local minimum; $(1, 8)$ is a local minimum.

7. 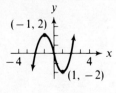   **9.** 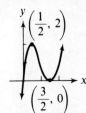   **11.**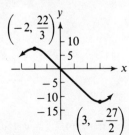

13. 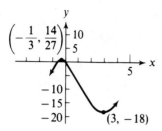 **15.** 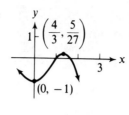 **17.**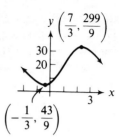

19. 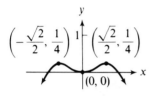 **21.** 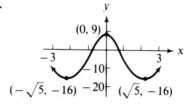 **23.**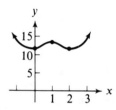

25.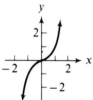

27. The vertex of the graph of $P(x) = ax^2 + bx + c$ is the only turning point. The only turning point occurs where $P'(x) = 2ax + b = 0$, or $x = \dfrac{-b}{2a}$. The y-coordinate is $P\left(-\dfrac{b}{2a}\right) = \dfrac{4ac - b^2}{4a}$.

Exercise 6.7 (page 206) **1.** $x = 3$ **3.** $x = 3$; $x = -2$ **5.** $x = -1$; $x = -4$ **7.** $x = -1$

9. 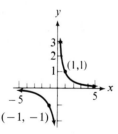 **11.** 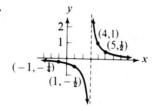 **13.**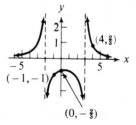

15.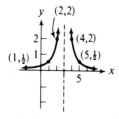

17. $x = 2$; $x = -2$; $y = 0$ **19.** $x = 4$; $y = x + 4$ **21.** $x = 4$; $x = -1$; $y = 1$

23.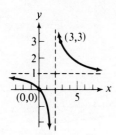
25.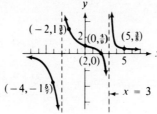
27.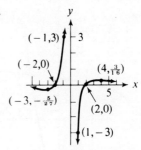

29.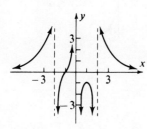
31.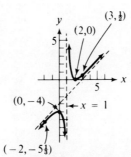
33.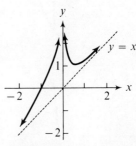

35.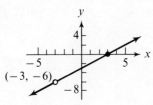
37.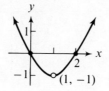
39.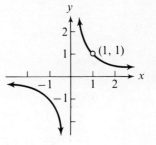

41. Behaves like $y = x^2$ as $|x|$ becomes large.
43. No, because if $x = a$ is a vertical asymptote, then the function is not defined at $x = a$; thus there is no y so that the point (a, y) is on the graph.
45. $(1, 3)$

Chapter 6 Review (page 208) 1. x-intercepts $3, -2$; axis of symmetry $x = \dfrac{1}{2}$; minimum pt. $\left(\dfrac{1}{2}, \dfrac{-25}{4}\right)$

2. x-intercepts $5, 2$; axis of symmetry $x = \dfrac{7}{2}$; maximum pt. $\left(\dfrac{7}{2}, \dfrac{9}{4}\right)$

3.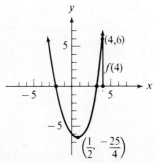
4.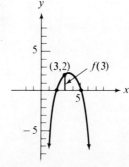

5.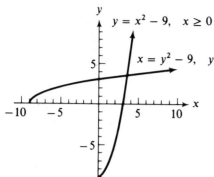

6.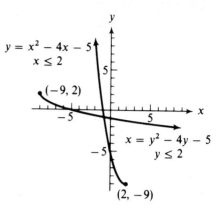

7.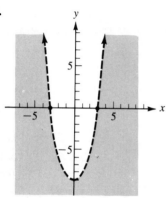

8.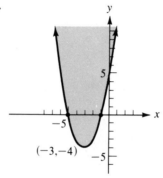

9.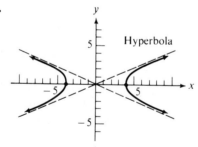
Hyperbola

10.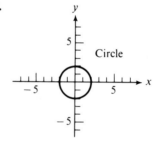
Circle

11.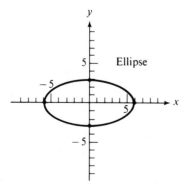
Ellipse

12.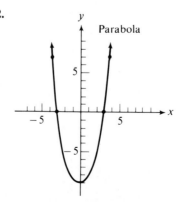
Parabola

Answers to Odd-Numbered Exercises 441

13. 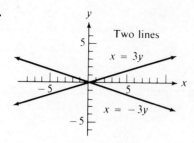 **14.**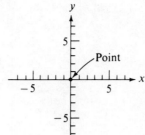

15. $y = 125$ **16.** $r = \dfrac{128}{243}$ **17.** 64 posts **18.** 432 rpm **19.** $P(1) = -3$ and $P(2) = 10$

20. $P\left(\dfrac{1}{2}\right) = \dfrac{-7}{16}$ and $P\left(\dfrac{3}{2}\right) = \dfrac{57}{16}$ **21.** 1 **22.** 53 **23.** Upper bound 2; lower bound -2

24. Upper bound 3; lower bound -4 **25.** $\{2, 2i, -2i\}$ **26.** $\left\{\dfrac{3}{2}, 2 + 2i, 2 - 2i\right\}$ **27.** $\{4\}$

28. $\left\{-\dfrac{1}{2}, 3\right\}$ **29.** $\{-2, i\sqrt{2}, -i\sqrt{2}\}$ **30.** $\left\{\dfrac{1}{2}, -\dfrac{1}{3}, \sqrt{\dfrac{2}{3}}, -\sqrt{\dfrac{2}{3}}\right\}$

31. (1, 5) is a local maximum; (2, 4) is a local minimum.

32. (0, 0) is a local maximum; $\left(\dfrac{2}{3}, \dfrac{-4}{27}\right)$ is a local minimum.

33. 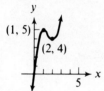 **34.** 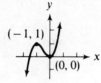 **35.**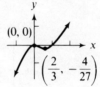

36. 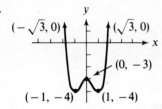 **37.**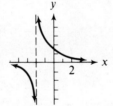

Vertical asymptote: $x = -2$
Horizontal asymptote: $y = 0$

38.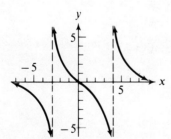

Vertical asymptotes: $x = -3$; $x = 4$
Horizontal asymptote: $y = 0$

39.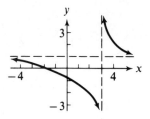
Vertical asymptote: $x = 3$
Horizontal asymptote: $y = 1$

40.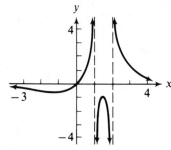
Vertical asymptotes: $x = 1$; $x = 2$
Horizontal asymptote: $y = 0$

Exercise 7.1 (page 213)   **1.** $(0, 1), (1, 3), (2, 9)$ **3.** $(0, -1), (1, -5), (2, -25)$ **5.** $(-3, 8), (0, 1), \left(3, \dfrac{1}{8}\right)$

7. $\left(-2, \dfrac{1}{100}\right), (1, 10), (0, 1)$ **9.** $(-1, 0.3679), (0, 1), (1, 2.7183)$

11. 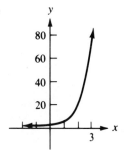   **13.** 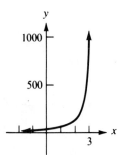   **15.**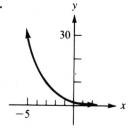

17. 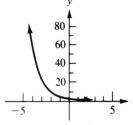   **19.** 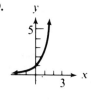   **21.**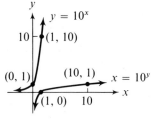

23. $\{-2\}$ **25.** $\left\{\dfrac{3}{4}\right\}$ **27.** 2 **29.** -2

31.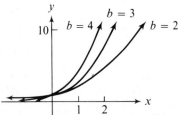

As $b > 1$ increases, the rate of increase of each curve becomes larger.

Answers to Odd-Numbered Exercises 443

Exercise 7.2 (page 217) 1. $\log_4 16 = 2$ 3. $\log_3 27 = 3$ 5. $\log_{1/2} \frac{1}{4} = 2$ 7. $\log_8 \frac{1}{2} = -\frac{1}{3}$
9. $\log_{10} 100 = 2$ 11. $\log_{10}(0.1) = -1$ 13. $2^6 = 64$ 15. $3^2 = 9$ 17. $\left(\frac{1}{3}\right)^{-2} = 9$
19. $10^3 = 1000$ 21. 2 23. 3 25. -1 27. 1 29. 2 31. -1 33. $\{2\}$
35. $\{2\}$ 37. $\{64\}$ 39. $\{-3\}$ 41. $\{100\}$ 43. $\{4\}$ 45. $\log_b x + \log_b y$
47. $\log_b x - \log_b y$ 49. $5\log_b x$ 51. $\frac{1}{3}\log_b x$ 53. $\frac{1}{2}(\log_b x - \log_b z)$
55. $\frac{1}{3}(\log_{10} x + 2\log_{10} y - \log_{10} z)$ 57. $\log_b 2xy^3$ 59. $\log_b x^{1/2} y^{2/3}$ 61. $\log_b \frac{x^3 y}{z^2}$
63. $\log_{10} \frac{x(x-2)}{z^2}$ 65. $\{500\}$ 67. $\{4\}$ 69. $\{3\}$
71. By definition, $\log_b 1$ is a number such that $b^{\log_b 1} = 1$; therefore $\log_b 1 = 0$.
73. Since $x_1 = b^{\log_b x_1}$ and $x_2 = b^{\log_b x_2}$, it follows that

$$\frac{x_2}{x_1} = \frac{b^{\log_b x_2}}{b^{\log_b x_1}} = b^{\log_b x_2 - \log_b x_1},$$

and by definition $\log_b\left(\frac{x_2}{x_1}\right) = \log_b x_2 - \log_b x_1$.

Exercise 7.3 (page 225) Note: The authors have used a calculator to obtain these answers. If the tables are used, slightly different values may be obtained.
1. 2 3. -3, or $7 - 10$ 5. -4, or $6 - 10$ 7. 4 9. 0.8280
11. $9.9101 - 10$, or -0.0899 13. $8.9031 - 10$, or -1.0969 15. 2.3945
17. $6.5328 - 10$, or -3.4672 19. 2.4330 21. 4.10 23. 3.67 25. 703 27. 2060
29. 0.0065 31. 0.0041 33. 1.0986 35. 2.8332 37. 6.1092 39. -2.7334 41. 1.649
43. 29.96 45. 1.259 47. 0.8607 49. 0.0821 51. 0.7788 53. 1.13 55. 0.923
57. 1.27
59. Since $e^N = x$, we have $\log_{10} e^N = \log_{10} x$; thus, by Theorem 7.2-III, $N \log_{10} e = \log_{10} x$. Hence,

$$N = \frac{\log_{10} x}{\log_{10} e}.$$

61. 0.2057, 0.0450, 0.0069, 0.0009 63. $-0.2303, -0.0461, -0.0069, -0.0009$

Exercise 7.4 (page 231) Note: The authors have used a calculator to obtain these answers. If tables are used, slightly different values may result.
1. 8.80 3. 220.26 5. 1.08 7. $\frac{\log_{10} 6}{3} = 0.2594$ 9. $-\frac{1}{3} + \frac{1}{3}\log_{10} 9 = -0.0152$
11. $\pm\sqrt{\log_{10} 150} = \pm 1.4752$ 13. $\frac{1}{3}\ln 5 = 0.5365$ 15. $-\frac{1}{2} + \frac{1}{2}\ln 25 = 1.1094$
17. $\pm\sqrt{2\ln 15} = \pm 2.3273$ 19. $\frac{\log_{10} 7}{\log_{10} 2} = 2.8074$ 21. $\frac{1}{2}\left(\frac{\log_{10} 3}{\log_{10} 7} + 1\right) = 0.7823$
23. $\pm\sqrt{\frac{\log_{10} 16}{\log_{10} 4}} = \pm 1.4142$ 25. $t = \frac{1}{k}\log_{10} A$ 27. $t = \frac{1}{k}\ln y$ 29. $n = \frac{\log_{10} y}{\log_{10} x}$ 31. 7%

33. $12,391.14; $12,648.84 **35.** $767.15 **37.** $27,782 **39.** 6.93% **41.** 7.7
43. 2.5×10^{-6} **45.** 30.0 inches; 16.1 inches **47.** 12.05 grams **49.** $k = 0.23026$

Chapter 7 Review (page 236) Note: Where possible, the authors have used a calculator to obtain these answers. If tables are used, slightly different values may result.

1. 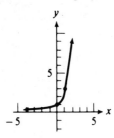   **2.**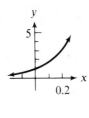

3. $\log_{16} \frac{1}{4} = -\frac{1}{2}$ **4.** $\log_7 343 = 3$ **5.** $2^3 = 8$ **6.** $10^{-4} = 0.0001$ **7.** $y = 4$
8. $x = 1000$ **9.** $\frac{1}{3} \log_{10} x + \frac{2}{3} \log_{10} y$ **10.** $\log_{10} 2 + 3 \log_{10} R - \frac{1}{2} \log_{10} P - \frac{1}{2} \log_{10} Q$
11. $\log_b \frac{x^2}{\sqrt[3]{y}}$ **12.** $\log_{10} \frac{\sqrt[3]{x^2 y}}{z^3}$ **13.** 1.6232   **14.** $7.4969 - 10$, or -2.5031 **15.** 2.8340
16. $8.6172 - 10$, or -1.3828 **17.** 67.4   **18.** 2.65   **19.** 0.0466   **20.** 0.6644   **21.** 1.9459
22. 3.1355   **23.** 5.4848   **24.** 6.2344   **25.** 403.4   **26.** 6.0496   **27.** 0.0302   **28.** 0.6570
29. 1.43   **30.** 0.696   **31.** $\frac{\log_{10} 8}{-2} = -0.4515$ **32.** $-2 - \log_{10} 25 = -3.3979$
33. $\frac{\ln 8}{4} = 0.5199$ **34.** $\frac{1}{2}(2 - \ln 16) = -0.3863$ **35.** $\frac{\log_{10} 2}{\log_{10} 3} = 0.6309$
36. $\frac{\log_{10} 80}{\log_{10} 3} - 1 = 2.9887$ **37.** $t = -\ln\left(\frac{y - A}{k}\right)$ **38.** $t = -c - \ln\left(\frac{y - A}{k}\right)$ **39.** 8%
40. 6.31×10^{-7} **41.** 7.13%   **42.** 0.65   **43.** $634.84   **44.** A little more than 223 months.

Supplemental Exercises. Chapters 5-7 (page 239)

1. 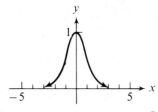   **2.** 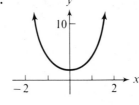   **3.**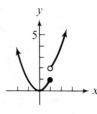

4. 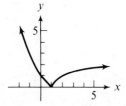   **5.** 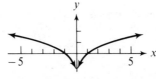   **6.**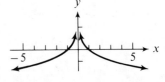

7. 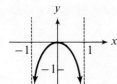 **8.**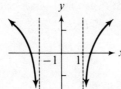

9. 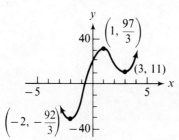 **10.** 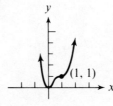 **11.**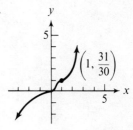

12. 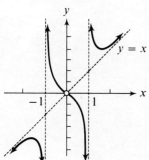 **13.**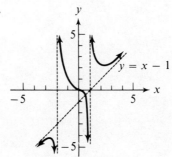

14. 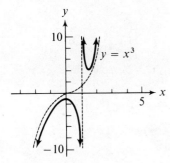 **15.**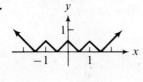

16.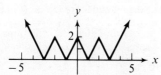

17. $\{y \mid y > -1\}$ **18.** $\{y \mid y > 1\}$ **19.** $\{y \mid 0 < y \le 1\}$ **20.** $\{y \mid y \le 0\}$ **21.** $[0, +\infty)$

22. $[0, +\infty)$ **23.** $y = \ln\left(\dfrac{x + \sqrt{x^2 + 4}}{2}\right)$ **24.** $y = \ln\left(\dfrac{x + \sqrt{x^2 - 4}}{2}\right)$

25. $\dfrac{y_1 - y_2}{x_2 - x_1} = -\dfrac{y_2 - y_1}{x_2 - x_1} = -m_2;\ \dfrac{x_2 - x_1}{y_3 - y_1} = \dfrac{1}{m_1}$. Hence, $-m_2 = \dfrac{1}{m_1}$, or $m_1 m_2 = -1$.

26. Assume that $P_1P = m_2$ and $PP_2 = m_1$, so that $P_1P_2 = m_1 + m_2$. Note that $\triangle P_1QP$ is similar to $\triangle P_1AP_2$ and therefore

$$\frac{P_1Q}{P_1P} = \frac{P_1A}{P_1P_2}.$$

But $P_1Q = x - x_1$ and $P_1A = x_2 - x_1$. Thus we obtain

$$\frac{x - x_1}{m_2} = \frac{x_2 - x_1}{m_1 + m_2},$$

or

$$x = x_1 + \frac{m_2}{m_1 + m_2}(x_2 - x_1)$$

$$= \left(1 - \frac{m_2}{m_1 + m_2}\right)x_1 + \frac{m_2}{m_1 + m_2}x_2$$

$$= \frac{m_1}{m_1 + m_2}x_1 + \frac{m_2}{m_1 + m_2}x_2.$$

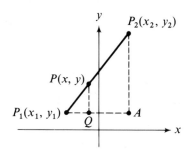

A similar argument gives

$$y = \frac{m_1}{m_1 + m_2}y_1 + \frac{m_2}{m_1 + m_2}y_2.$$

27. $\log_a x = \dfrac{\log_b x}{\log_b a}$

28. From Exercise 27 we have

$$\log_a b = \frac{\log_b b}{\log_b a} = \frac{1}{\log_b a}.$$

Exercise 8.1 (page 246) **1.** $\{(3, 2)\}$ **3.** $\{(2, 1)\}$ **5.** $\{(-5, 4)\}$ **7.** $\left\{\left(0, \dfrac{3}{2}\right)\right\}$ **9.** $\left\{\left(\dfrac{2}{3}, -1\right)\right\}$
11. $\{(1, 2)\}$ **13.** $\left\{\left(-\dfrac{19}{5}, -\dfrac{18}{5}\right)\right\}$ **15.** $\left(x, -\dfrac{3}{2}x + 3\right), x \in R$ **17.** No solution
19. $a = 1, b = -1$ **21.** $a = 0, b = 2$ **23.** $y = -\dfrac{10}{3}x + 2$ **25.** $C = \dfrac{5}{9}(F - 32)$
27. 32 pounds **29.** $14,400 at 9% and $4800 at 8% **31.** 32 and 64 mph **33.** $\{(-1, 1)\}$
35. $\{(2, 1)\}$

37. **a.** From $\dfrac{a_1}{a_2} = \dfrac{b_1}{b_2}$, it follows that $\dfrac{a_1}{b_1} = \dfrac{a_2}{b_2}$. Hence the slopes are equal; the two lines are the same or they are parallel. The equations are either consistent or inconsistent, respectively. Assume that the equations are consistent and the two lines are the same. Then, the y-intercepts $\dfrac{-c_1}{b_1} = \dfrac{-c_2}{b_2}$, from which $\dfrac{b_1}{b_2} = \dfrac{c_1}{c_2}$; but this contradicts the hypothesis $\dfrac{b_1}{b_2} \neq \dfrac{c_1}{c_2}$; hence the equations are inconsistent. **b.** From (a) above, $\dfrac{a_1}{b_1} = \dfrac{a_2}{b_2}$, and it follows that $\dfrac{a_1}{a_2} = \dfrac{b_1}{b_2}$. Now, either $\dfrac{b_1}{b_2} = \dfrac{c_1}{c_2}$ or $\dfrac{b_1}{b_2} \neq \dfrac{c_1}{c_2}$. Assume $\dfrac{b_1}{b_2} = \dfrac{c_1}{c_2}$; then $\dfrac{c_1}{b_1} = \dfrac{c_2}{b_2}$; the graphs would be the same straight line, and the equations would be consistent. This contradicts the hypothesis that states that the equations are inconsistent; hence $\dfrac{b_1}{b_2} \neq \dfrac{c_1}{c_2}$.

Answers to Odd-Numbered Exercises

Exercise 8.2 (page 253) 1. $\{(1, 2, -1)\}$ 3. $\{(2, -2, 0)\}$ 5. $\{(2, 2, 1)\}$ 7. $\{(0, 1, 2)\}$
9. Dependent 11. $\{(4, -2, 2)\}$ 13. 3, 6, 6 15. 60 nickels, 20 dimes, 5 quarters
17. $x = 40$ cm., $y = 60$ cm., $z = 55$ cm. 19. $a = -6, b = -8, c = 0$ 21. $a = 3, b = 1, c = -2$
23. $a = \frac{1}{4}, b = \frac{1}{2}, c = 0$ 25. $\{(1, 1, 1)\}$ 27. $\{(0, 1, 2)\}$
29. $\left(\frac{5}{3} - \frac{1}{3}z, -\frac{5}{3}z + \frac{1}{3}, z\right), z \in R; \left\{\left(\frac{5}{3}, \frac{1}{3}, 0\right), (2, 2, -1)\right\}$

Exercise 8.3 (page 260) 1. $\frac{4}{x} - \frac{4}{x+1}$ 3. $\frac{-1}{x+1} + \frac{2}{x+2}$ 5. $\frac{-4}{x+2} + \frac{5}{x+3}$
7. $\frac{1/2}{x} + \frac{1}{x+1} + \frac{-3/2}{x+2}$ 9. $\frac{1/2}{x} + \frac{-3}{x+1} + \frac{7/2}{x+2}$ 11. $\frac{1/2}{x} + \frac{-1}{x+1} + \frac{1/2}{x+2}$
13. $\frac{-3/4}{x} + \frac{-1/2}{x^2} + \frac{3/4}{x-2}$ 15. $\frac{1/2}{x-1} + \frac{1/4}{(x-1)^2} + \frac{1/2}{x+1} + \frac{-1/4}{(x+1)^2}$ 17. $\frac{-1}{x} + \frac{-1}{x^2} + \frac{1}{x-1}$
19. $x^2 + \frac{2}{x} + \frac{-1}{x+1}$ 21. $x + 2 + \frac{3}{x-1} + \frac{1}{(x-1)^2}$ 23. $a = 1, b = -1, c = 0$
25. $a = -\frac{3}{4}, b = \frac{3}{4}, c = \frac{1}{2}$

Exercise 8.4 (page 268) 1. $\{(-1, -4), (5, 20)\}$ 3. $\{(2, 3), (3, 2)\}$ 5. $\{(4, -3), (-3, 4)\}$
7. $\{(-1, 3), (-1, -3), (1, 3), (1, -3)\}$ 9. $\{(-3, \sqrt{2}), (-3, -\sqrt{2}), (3, \sqrt{2}), (3, -\sqrt{2})\}$
11. $\{(\sqrt{3}, 4), (\sqrt{3}, -4), (-\sqrt{3}, 4), (-\sqrt{3}, -4)\}$ 13. $\{(1, -2), (-1, 2), (2, -1), (-2, 1)\}$
15. $\{(3, 1), (-3, -1), (-2\sqrt{7}, \sqrt{7}), (2\sqrt{7}, -\sqrt{7})\}$
17. $\left\{\left(2\sqrt{\frac{1}{3}}, \sqrt{\frac{1}{3}}\right), \left(-2\sqrt{\frac{1}{3}}, -\sqrt{\frac{1}{3}}\right), (-6i, 4i), (6i, -4i)\right\}$ 19. $\{(i\sqrt{7}, 4), (-i\sqrt{7}, 4)\}$
21. $\{(2i, -2i), (-2i, 2i), (2\sqrt{2}, \sqrt{2}), (-2\sqrt{2}, -\sqrt{2})\}$ 23. $\left\{\left(-1, \frac{1}{10}\right)\right\}$ 25. $\{(1, 0), (10, -1)\}$
27. $\{(0, 2)\}$ 29. $\left\{(1, 0), \left(\frac{3}{2}, \frac{1}{2}\right)\right\}$

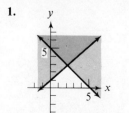

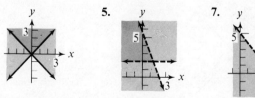

31. $x = 2, y = 3$ 33. $800 at 4\%$ 35. **a.** 1 **b.** 2 **c.** 4 37. $\{(2, 2), (-2, -2)\}$

Exercise 8.5 (page 273)
1. 3. 5. 7.

9.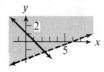
11.
13.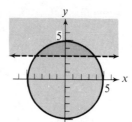
15.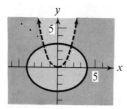
17.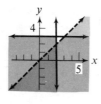
19.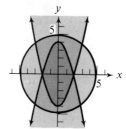
21.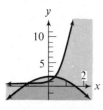
23.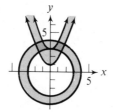

Exercise 8.6 (page 276)

1.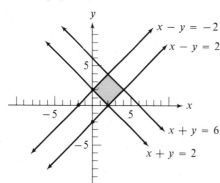

3.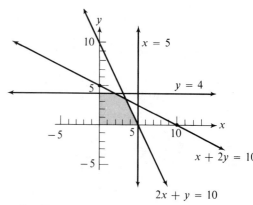

5.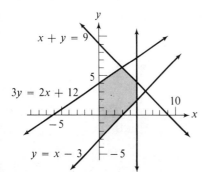

7. No **9.** Yes

Answers to Odd-Numbered Exercises

Exercise 8.7 (page 278) **1.** Maximum: 14; Minimum: 2 **3.** Maximum: 80/3; Minimum: 0 **5.** Maximum: 17; Minimum: -16 **7.** Maximum profit: $1860 **9.** Maximum profit: $790 **11.** 200 standard models and 400 deluxe models

Chapter 8 Review (page 280) **1.** $\left\{\left(\frac{1}{2}, \frac{7}{2}\right)\right\}$ **2.** $\{(1, 2)\}$ **3.** $\left\{\left(\frac{37}{13}, -\frac{10}{13}\right)\right\}$ **4.** $\{(-2, 1)\}$ **5.** $\left(x, \frac{2}{3}x - \frac{1}{3}\right), x \in R$ **6.** No solution **7.** $a = 2, b = 5$ **8.** $a = \frac{2}{3}, b = \frac{5}{3}$ **9.** $\{(2, 0, -1)\}$ **10.** $\{(2, 1, -1)\}$ **11.** $\{(2, -1, 3)\}$ **12.** $a = 2, b = -3, c = 4$ **13.** $\frac{2}{2x-3} - \frac{1}{x+1}$ **14.** $\frac{-1}{x} + \frac{1}{x-1} + \frac{1}{x+1}$ **15.** $\{(1, 2), (4, -13)\}$ **16.** $\{(1, 1), (-1, -1)\}$ **17.** $\{(2, 3), (2, -3), (-2, 3), (-2, -3)\}$ **18.** $\left\{\left(\frac{1}{\sqrt{5}}, -\frac{1}{\sqrt{5}}\right), \left(-\frac{1}{\sqrt{5}}, \frac{1}{\sqrt{5}}\right), \left(\frac{1}{\sqrt{11}}, \frac{-2}{\sqrt{11}}\right), \left(-\frac{1}{\sqrt{11}}, \frac{2}{\sqrt{11}}\right)\right\}$ **19.** $\left\{\left(-\frac{1}{3}, 10^{-2/3}\right)\right\}$ **20.** $\left\{\left(\frac{1}{20}, -1\right)\right\}$ **21.** $\left\{\left(\frac{7}{8}, \frac{17}{8}\right)\right\}$

27. Minimum: 0; Maximum: 8 **28.** Minimum: 0; Maximum: 11 **29.** Maximum profit: $1100 **30.** Maximum profit: $210

Exercise 9.1 (page 287) **1.** 2×2, $\begin{bmatrix} 6 & 2 \\ -1 & 3 \end{bmatrix}$ **3.** 2×3, $\begin{bmatrix} 2 & 1 \\ -7 & 4 \\ 3 & 0 \end{bmatrix}$ **5.** 3×3, $\begin{bmatrix} 2 & 4 & -2 \\ 3 & 0 & 3 \\ -1 & 1 & 1 \end{bmatrix}$

7. 2×4, $\begin{bmatrix} 4 & 2 \\ -3 & 1 \\ -1 & 1 \\ 0 & 6 \end{bmatrix}$ **9.** $\begin{bmatrix} 3 & 1 \\ 3 & 9 \end{bmatrix}$ **11.** $\begin{bmatrix} 9 & -1 & -1 \\ 2 & 3 & 6 \end{bmatrix}$ **13.** $\begin{bmatrix} -2 & -3 \\ 4 & 0 \end{bmatrix}$

15. $\begin{bmatrix} 2 & -9 & -13 \\ 7 & -4 & 1 \\ 9 & 0 & 0 \end{bmatrix}$ **17.** $\begin{bmatrix} 10 \\ 3 \\ -3 \end{bmatrix}$ **19.** $\begin{bmatrix} 2 & 3 & 4 \\ -1 & 6 & 2 \\ 1 & 0 & 3 \end{bmatrix}$

21. $\begin{bmatrix} 2 & -1 \\ 2 & 1 \end{bmatrix}$ **23.** $\begin{bmatrix} 1 & 0 \\ -5 & 3 \end{bmatrix}$

25. Let $A = \begin{bmatrix} a_{11} & a_{12} \\ a_{21} & a_{22} \end{bmatrix}$; then $A^t = \begin{bmatrix} a_{11} & a_{21} \\ a_{12} & a_{22} \end{bmatrix}$ and $[A^t]^t = \begin{bmatrix} a_{11} & a_{12} \\ a_{21} & a_{22} \end{bmatrix} = A$. This result holds for both $n \times n$ and $m \times n$ matrices.

27. The i, j entry of $A + B$ is $a_{ij} + b_{ij}$, which is a real number. Hence $A + B$ is a matrix with real-number entries.

28. The i, j entry of $A + B$ is $a_{ij} + b_{ij}$ and the i, j entry of $B + A$ is $b_{ij} + a_{ij}$. By the commutative law for addition of real numbers $a_{ij} + b_{ij} = b_{ij} + a_{ij}$. Thus $A + B = B + A$.

Exercise 9.2 (page 293) **1.** $\begin{bmatrix} 0 & -5 & 5 \\ -15 & 5 & -10 \end{bmatrix}$ **3.** $[-1]$ **5.** $\begin{bmatrix} 1 & -13 \\ 4 & -7 \end{bmatrix}$ **7.** $\begin{bmatrix} 30 & -39 \\ 29 & 14 \end{bmatrix}$

9. $\begin{bmatrix} -5 & -1 \\ 8 & -1 \end{bmatrix}$ **11.** $\begin{bmatrix} -1 & 0 & -2 \\ 1 & 2 & 8 \\ 0 & 1 & 3 \end{bmatrix}$ **13.** $\begin{bmatrix} 1 & 0 & 0 \\ 0 & 1 & 0 \\ 0 & 0 & 1 \end{bmatrix}$ **15.** $\begin{bmatrix} 1 & 0 \\ -1 & 2 \end{bmatrix}$ **17.** $\begin{bmatrix} 1 & -2 \\ 1 & 2 \end{bmatrix}$

19. $\begin{bmatrix} -2 & 3 \\ 2 & -4 \end{bmatrix}$ **21.** $\begin{bmatrix} -1 & 1 \\ -1 & -1 \end{bmatrix}$ **23.** $\begin{bmatrix} 1 & -1 \\ 1 & 0 \end{bmatrix}$

25. Since $A + B = \begin{bmatrix} 0 & 2 \\ -1 & 3 \end{bmatrix}$, $A - B = \begin{bmatrix} -2 & 2 \\ 1 & -1 \end{bmatrix}$, $A^2 = \begin{bmatrix} 1 & 0 \\ 0 & 1 \end{bmatrix}$, $B^2 = \begin{bmatrix} 1 & 0 \\ -3 & 4 \end{bmatrix}$,

and $AB = \begin{bmatrix} -3 & 4 \\ -1 & 2 \end{bmatrix}$, we have **a.** $(A + B)(A + B) = \begin{bmatrix} 0 & 2 \\ -1 & 3 \end{bmatrix} \cdot \begin{bmatrix} 0 & 2 \\ -1 & 3 \end{bmatrix} = \begin{bmatrix} -2 & 6 \\ -3 & 7 \end{bmatrix}$

and $A^2 + 2AB + B^2 = \begin{bmatrix} 1 & 0 \\ 0 & 1 \end{bmatrix} + 2\begin{bmatrix} -3 & 4 \\ -1 & 2 \end{bmatrix} + \begin{bmatrix} 1 & 0 \\ -3 & 4 \end{bmatrix} = \begin{bmatrix} -4 & 8 \\ -5 & 9 \end{bmatrix}$;

hence $(A + B)(A + B) \neq A^2 + 2AB + B^2$. **b.** $(A + B)(A - B) = \begin{bmatrix} 0 & 2 \\ -1 & 4 \end{bmatrix} \cdot \begin{bmatrix} -2 & 2 \\ 1 & -1 \end{bmatrix} = \begin{bmatrix} 2 & -2 \\ 6 & -6 \end{bmatrix}$

and $A^2 - B^2 = \begin{bmatrix} 1 & 0 \\ 0 & 1 \end{bmatrix} - \begin{bmatrix} 1 & 0 \\ -3 & 4 \end{bmatrix} = \begin{bmatrix} 0 & 0 \\ 3 & -3 \end{bmatrix}$; hence $(A + B)(A - B) \neq A^2 - B^2$.

For Exercises 27 and 29, let $A = \begin{bmatrix} a_{11} & a_{12} \\ a_{21} & a_{22} \end{bmatrix}$, $B = \begin{bmatrix} b_{11} & b_{12} \\ b_{21} & b_{22} \end{bmatrix}$, and $C = \begin{bmatrix} c_{11} & c_{12} \\ c_{21} & c_{22} \end{bmatrix}$.

27. Since $AB = \begin{bmatrix} a_{11}b_{11} + a_{12}b_{21} & a_{11}b_{12} + a_{12}b_{22} \\ a_{21}b_{11} + a_{22}b_{21} & a_{21}b_{12} + a_{22}b_{22} \end{bmatrix}$,

$$(AB)C = \begin{bmatrix} (a_{11}b_{11} + a_{12}b_{21})c_{11} + (a_{11}b_{12} + a_{12}b_{22})c_{21} & (a_{11}b_{11} + a_{12}b_{21})c_{12} + (a_{11}b_{12} + a_{12}b_{22})c_{22} \\ (a_{21}b_{11} + a_{22}b_{21})c_{11} + (a_{21}b_{12} + a_{22}b_{22})c_{21} & (a_{21}b_{11} + a_{22}b_{21})c_{12} + (a_{21}b_{12} + a_{22}b_{22})c_{22} \end{bmatrix}.$$

The element in the first row, first column is given by $a_{11}b_{11}c_{11} + a_{12}b_{21}c_{11} + a_{11}b_{12}c_{21} + a_{12}b_{22}c_{21} = (a_{11}b_{11}c_{11} + a_{11}b_{12}c_{21}) + (a_{12}b_{21}c_{11} + a_{12}b_{22}c_{21}) = a_{11}(b_{11}c_{11} + b_{12}c_{21}) + a_{12}(b_{21}c_{11} + b_{22}c_{21})$.
When the remaining elements are treated in a similar manner, we get

$$(AB)C = \begin{bmatrix} a_{11}(b_{11}c_{11} + b_{12}c_{21}) + a_{12}(b_{21}c_{11} + b_{22}c_{21}) & a_{11}(b_{11}c_{12} + b_{12}c_{22}) + a_{12}(b_{21}c_{12} + b_{22}c_{22}) \\ a_{21}(b_{11}c_{11} + b_{12}c_{21}) + a_{22}(b_{21}c_{11} + b_{22}c_{21}) & a_{21}(b_{11}c_{12} + b_{12}c_{22}) + a_{22}(b_{21}c_{12} + b_{22}c_{22}) \end{bmatrix}$$

$$= \begin{bmatrix} a_{11} & a_{12} \\ a_{21} & a_{22} \end{bmatrix} \cdot \begin{bmatrix} b_{11}c_{11} + b_{12}c_{21} & b_{11}c_{12} + b_{12}c_{22} \\ b_{21}c_{11} + b_{22}c_{21} & b_{21}c_{12} + b_{22}c_{22} \end{bmatrix} = \begin{bmatrix} a_{11} & a_{12} \\ a_{21} & a_{22} \end{bmatrix} \cdot \left(\begin{bmatrix} b_{11} & b_{12} \\ b_{21} & b_{22} \end{bmatrix} \cdot \begin{bmatrix} c_{11} & c_{12} \\ c_{21} & c_{22} \end{bmatrix} \right)$$

$$= A(BC).$$

29. $B + C = \begin{bmatrix} b_{11} + c_{11} & b_{12} + c_{12} \\ b_{21} + c_{21} & b_{22} + c_{22} \end{bmatrix}$; hence $(B + C)A$

$$= \begin{bmatrix} (b_{11} + c_{11})a_{11} + (b_{12} + c_{12})a_{21} & (b_{11} + c_{11})a_{12} + (b_{12} + c_{12})a_{22} \\ (b_{21} + c_{22})a_{11} + (b_{22} + c_{22})a_{21} & (b_{21} + c_{22})a_{12} + (b_{22} + c_{22})a_{22} \end{bmatrix}.$$

The element in the first row, first column is given by

$$b_{11}a_{11} + c_{11}a_{21} + b_{12}a_{21} + c_{12}a_{21} = (b_{11}a_{11} + b_{12}a_{21}) + (c_{11}a_{21} + c_{12}a_{21}).$$

When the remaining elements are treated in a similar manner, we get

$$(B + C)A = \begin{bmatrix} (b_{11}a_{11} + b_{12}a_{21}) + (c_{11}a_{11} + c_{12}a_{21}) & (b_{11}a_{12} + b_{12}a_{22}) + (c_{11}a_{12} + c_{12}a_{22}) \\ (b_{21}a_{11} + b_{22}a_{21}) + (c_{21}a_{11} + c_{22}a_{21}) & (b_{21}a_{12} + b_{22}a_{22}) + (c_{21}a_{12} + c_{22}a_{22}) \end{bmatrix}$$

$$= \begin{bmatrix} b_{11}a_{11} + b_{12}a_{21} & b_{11}a_{12} + b_{12}a_{22} \\ b_{21}a_{11} + b_{22}a_{21} & b_{21}a_{12} + b_{22}a_{22} \end{bmatrix} + \begin{bmatrix} c_{11}a_{11} + c_{12}a_{21} & c_{11}a_{12} + c_{12}a_{22} \\ c_{21}a_{11} + c_{22}a_{21} & c_{21}a_{12} + c_{22}a_{22} \end{bmatrix}$$

$$= \begin{bmatrix} b_{11} & b_{12} \\ b_{21} & b_{22} \end{bmatrix} \cdot \begin{bmatrix} a_{11} & a_{12} \\ a_{21} & a_{22} \end{bmatrix} + \begin{bmatrix} c_{11} & c_{12} \\ c_{21} & c_{22} \end{bmatrix} \cdot \begin{bmatrix} a_{11} & a_{12} \\ a_{21} & a_{22} \end{bmatrix} = BC + CA.$$

31. $(A_{2 \times 2} \cdot B_{2 \times 2}) = \begin{bmatrix} a_{11}b_{11} + a_{12}b_{21} & a_{11}b_{12} + a_{12}b_{22} \\ a_{21}b_{11} + a_{22}b_{21} & a_{21}b_{12} + a_{22}b_{22} \end{bmatrix}$

Hence, $(A_{2 \times 2} \cdot B_{2 \times 2})^t = \begin{bmatrix} a_{11}b_{11} + a_{12}b_{21} & a_{21}b_{11} + a_{22}b_{21} \\ a_{11}b_{12} + a_{12}b_{22} & a_{21}b_{12} + a_{22}b_{22} \end{bmatrix}.$

By the commutative property of multiplication for real numbers,

$$(A_{2 \times 2} \cdot B_{2 \times 2})^t = \begin{bmatrix} b_{11}a_{11} + b_{21}a_{12} & b_{11}a_{21} + b_{21}a_{22} \\ b_{12}a_{11} + b_{22}a_{12} & b_{12}a_{21} + b_{22}a_{22} \end{bmatrix} = \begin{bmatrix} b_{11} & b_{21} \\ b_{12} & b_{22} \end{bmatrix} \cdot \begin{bmatrix} a_{11} & a_{21} \\ a_{12} & a_{22} \end{bmatrix} = B_{2 \times 2}^t \cdot A_{2 \times 2}^t.$$

33. Let $A = \begin{bmatrix} a_{11} & a_{12} \\ a_{21} & a_{22} \end{bmatrix}$ and c be a scalar; then by Definition 9.6 and the closure property of multiplication of real numbers, $c \begin{bmatrix} a_{11} & a_{12} \\ a_{21} & a_{22} \end{bmatrix} = \begin{bmatrix} ca_{11} & ca_{12} \\ ca_{21} & ca_{22} \end{bmatrix}$, which is a 2×2 matrix.

35. By Definition 9.6 and the distributive property for real numbers, and by Definition 9.3,

$$(c + d) \begin{bmatrix} a_{11} & a_{12} \\ a_{21} & a_{22} \end{bmatrix} = \begin{bmatrix} (c + d)a_{11} & (c + d)a_{12} \\ (c + d)a_{21} & (c + d)a_{22} \end{bmatrix} = \begin{bmatrix} ca_{11} + da_{11} & ca_{12} + da_{12} \\ ca_{21} + da_{21} & ca_{22} + da_{22} \end{bmatrix}$$

$$= \begin{bmatrix} ca_{11} & ca_{12} \\ ca_{21} & ca_{22} \end{bmatrix} + \begin{bmatrix} da_{11} & da_{12} \\ da_{21} & da_{22} \end{bmatrix} = c \begin{bmatrix} a_{11} & a_{12} \\ a_{21} & a_{22} \end{bmatrix} + d \begin{bmatrix} a_{11} & a_{12} \\ a_{21} & a_{22} \end{bmatrix} = cA + dA.$$

37. By Definitions 9.6 and 9.5, $(-1)A = -1\begin{bmatrix} a_{11} & a_{12} \\ a_{21} & a_{22} \end{bmatrix} = \begin{bmatrix} -a_{11} & -a_{12} \\ -a_{21} & -a_{22} \end{bmatrix} = -A.$

39. $\mathbf{0}_{2 \times 2} = \begin{bmatrix} 0 & 0 \\ 0 & 0 \end{bmatrix}$; hence, by Definition 9.6, $c \cdot \mathbf{0}_{2 \times 2} = c\begin{bmatrix} 0 & 0 \\ 0 & 0 \end{bmatrix} = \begin{bmatrix} c \cdot 0 & c \cdot 0 \\ c \cdot 0 & c \cdot 0 \end{bmatrix} = \begin{bmatrix} 0 & 0 \\ 0 & 0 \end{bmatrix} = \mathbf{0}_{2 \times 2}.$

Exercise 9.3 (page 302) **1.** $\{(2, -1)\}$ **3.** $\{(5, 1)\}$ **5.** $\{(8, 1)\}$ **7.** $\{(-2, 2, 0)\}$

9. $\left\{\left(\frac{5}{4}, \frac{5}{2}, -\frac{1}{2}\right)\right\}$ **11.** $\left\{\left(-\frac{77}{27}, -\frac{8}{27}, \frac{29}{27}\right)\right\}$

13. $\{(1, 1-z, z)\}$ **15.** $(3, 3, -1)$ **17.** No solution **19.** No solution

21. $\begin{bmatrix} k & 0 \\ 0 & 1 \end{bmatrix}\begin{bmatrix} a & b \\ c & d \end{bmatrix} = \begin{bmatrix} k \cdot a + 0 \cdot c & k \cdot b + 0 \cdot d \\ 0 \cdot a + 1 \cdot c & 0 \cdot b + 1 \cdot d \end{bmatrix} = \begin{bmatrix} ka & kb \\ c & d \end{bmatrix}$

23. $\begin{bmatrix} 0 & 1 \\ 1 & 0 \end{bmatrix}\begin{bmatrix} a & b \\ c & d \end{bmatrix} = \begin{bmatrix} 0 \cdot a + 1 \cdot c & 0 \cdot b + 1 \cdot d \\ 1 \cdot a + 0 \cdot c & 1 \cdot b + 0 \cdot d \end{bmatrix} = \begin{bmatrix} c & d \\ a & b \end{bmatrix}$

25. $\begin{bmatrix} 1 & 0 \\ k & 1 \end{bmatrix}\begin{bmatrix} a & b \\ c & d \end{bmatrix} = \begin{bmatrix} 1 \cdot a + 0 \cdot c & 1 \cdot b + 0 \cdot d \\ k \cdot a + 1 \cdot c & k \cdot b + 1 \cdot d \end{bmatrix} = \begin{bmatrix} a & b \\ ka + c & kb + d \end{bmatrix}$

Exercise 9.4 (page 307) **1.** 3 **3.** 4 **5.** 0 **7.** -11

9. $M_{11} = \begin{vmatrix} 0 & 3 & -1 \\ 1 & 2 & 2 \\ -1 & 3 & 1 \end{vmatrix},\quad A_{11} = \begin{vmatrix} 0 & 3 & -1 \\ 1 & 2 & 2 \\ -1 & 3 & 1 \end{vmatrix}$

11. $M_{23} = \begin{vmatrix} 2 & 1 & 0 \\ -2 & 1 & 2 \\ 1 & -1 & 1 \end{vmatrix},\quad A_{23} = -\begin{vmatrix} 2 & 1 & 0 \\ -2 & 1 & 2 \\ 1 & -1 & 1 \end{vmatrix}$

13. $M_{31} = \begin{vmatrix} 1 & -2 & 0 \\ 0 & 3 & -1 \\ -1 & 3 & 1 \end{vmatrix},\quad A_{31} = \begin{vmatrix} 1 & -2 & 0 \\ 0 & 3 & -1 \\ -1 & 3 & 1 \end{vmatrix}$

15. $M_{44} = \begin{vmatrix} 2 & 1 & -2 \\ 1 & 0 & 3 \\ -2 & 1 & 2 \end{vmatrix},\quad A_{44} = \begin{vmatrix} 2 & 1 & -2 \\ 1 & 0 & 3 \\ -2 & 1 & 2 \end{vmatrix}$

17. 16 **19.** -4 **21.** 0 **23.** -24 **25.** 4

27. Let $A = \begin{vmatrix} x & y & 1 \\ x_1 & y_1 & 1 \\ x_2 & y_2 & 1 \end{vmatrix} = 0.$

Expanding about the first row gives $\delta(A) = x\begin{vmatrix} y_1 & 1 \\ y_2 & 1 \end{vmatrix} - y\begin{vmatrix} x_1 & 1 \\ x_2 & 1 \end{vmatrix} + 1\begin{vmatrix} x_1 & y_1 \\ x_2 & y_2 \end{vmatrix} = 0$; hence $(y_1 - y_2)x + (x_2 - x_1)y + (x_1 y_2 - y_1 x_2) = 0$. Further, y_1, y_2, x_1, x_2 are real numbers; hence there exist real numbers, a, b, c such that $(y_1 - y_2) = a, (x_2 - x_1) = b$, and $(x_1 y_2 - y_1 x_2) = c$ with a and b not both 0 since $(x_1, y_1) \neq (x_2, y_2)$. Substituting yields $ax + by + c = 0$, which is the equation of a straight line. By Theorem 9.9, the line passes through (x_1, y_1) and (x_2, y_2).

Answers to Odd-Numbered Exercises

29. If the ith row of A is identically 0, then expanding about the ith row we have
$$\delta(A) = 0 \cdot A_{i1} + 0 \cdot A_{i2} + \cdots + 0 \cdot A_{in} = 0.$$
Similarly if the ith column is identically 0, then $\delta(A) = 0$.

31. Let $A = \begin{bmatrix} a_{11} & a_{12} \\ a_{21} & a_{22} \end{bmatrix}$; by Definition 9.6 and the fact that $\delta(A) = a_{11}a_{22} - a_{12}a_{21}$, it follows that

$$aA = \begin{bmatrix} aa_{11} & aa_{12} \\ aa_{21} & aa_{22} \end{bmatrix} \text{ and } \delta(aA) = a^2 a_{11}a_{22} - a^2 a_{12}a_{21} = a^2(a_{11}a_{22} - a_{12}a_{21}) = a^2 \delta(A).$$

Exercise 9.5 (page 313) **1.** Theorem 9.7 **3.** Theorem 9.9 **5.** Theorem 9.10 **7.** Theorem 9.10 **9.** Theorem 9.11 **11.** Theorem 9.11

13. $\begin{vmatrix} 1 & 3 \\ 0 & -4 \end{vmatrix}$ **15.** $\begin{vmatrix} 1 & -2 & 1 \\ 0 & 7 & 1 \\ 0 & 2 & 1 \end{vmatrix}$ **17.** $\begin{vmatrix} 0 & 1 & -3 & -2 \\ 0 & 2 & 1 & 2 \\ 1 & 1 & 2 & 3 \\ 0 & 1 & 1 & 1 \end{vmatrix}$ **19.** $-1\begin{vmatrix} 2 & 1 \\ -1 & 2 \end{vmatrix} = -5$

21. $\begin{vmatrix} -1 & -5 \\ 2 & -2 \end{vmatrix} = 12$ **23.** $\begin{vmatrix} 4 & 4 \\ 3 & 7 \end{vmatrix} = 16$ **25.** $\begin{vmatrix} 6 & 1 \\ 0 & 3 \end{vmatrix} = 18$ **27.** $-16\begin{vmatrix} 1 & 2 \\ 2 & 3 \end{vmatrix} = 16$

29. $\begin{vmatrix} 4 & -4 \\ 3 & -9 \end{vmatrix} = -24$

31. Multiply column 1 by $(-a)$ and add result to column 2; also, multiply column 1 by $(-a^2)$ and add result to column 3, to obtain $\begin{vmatrix} 1 & a & a^2 \\ 1 & b & b^2 \\ 1 & c & c^2 \end{vmatrix} = \begin{vmatrix} 1 & 0 & 0 \\ 1 & b-a & b^2-a^2 \\ 1 & c-a & c^2-a^2 \end{vmatrix}$. Expand about the first row to obtain

$$1 \begin{vmatrix} b-a & b^2-a^2 \\ c-a & c^2-a^2 \end{vmatrix} = (b-a)[c^2-a^2] - (c-a)[b^2-a^2]$$
$$= (b-a)[(c-a)(c+a)] - (c-a)[(b-a)(b+a)]$$
$$= -(a-b)[(c-a)(c+a)] + (c-a)[(a-b)(a+b)]$$
$$= (a-b)(c-a)[-(c+a) \cdot (a+b)]$$
$$= (a-b)(c-a)(b-c) = (b-c)(c-a)(a-b).$$

33. If Rows i and j of the matrix A are identical, then interchanging the two rows results in the same matrix. Thus, by Theorem 9.8, $\delta(A) = -\delta(A)$ and hence, $\delta(A) = 0$. A similar argument shows $\delta(A) = 0$ if two columns are identical.

Exercise 9.6 (page 320) **1.** $\begin{bmatrix} 3 & -2 \\ -1 & 1 \end{bmatrix}$ **3.** $\dfrac{1}{5}\begin{bmatrix} 1 & 3 \\ -1 & 2 \end{bmatrix}$ **5.** $|A| = 0$; no inverse

7. $-1\begin{bmatrix} 4 & -7 \\ -3 & 5 \end{bmatrix}$ **9.** $\dfrac{1}{2}\begin{bmatrix} -2 & -4 \\ 4 & 7 \end{bmatrix}$ **11.** $|A| = 0$; no inverse

13. $\dfrac{1}{6}\begin{bmatrix} 2 & 2 & -5 \\ -4 & 2 & 1 \\ 0 & 0 & 3 \end{bmatrix}$ **15.** $\dfrac{1}{3}\begin{bmatrix} -2 & 3 & -1 \\ -1 & 0 & 1 \\ 6 & -6 & 3 \end{bmatrix}$ **17.** $|A| = 0$; no inverse

19. $|A| = 0$; no inverse **21.** $\begin{bmatrix} 2 & -3 & 11 \\ -2 & 4 & -13 \\ 1 & -2 & 7 \end{bmatrix}$ **23.** $-1\begin{bmatrix} 0 & 0 & -1 \\ 0 & -1 & 0 \\ -1 & 0 & 0 \end{bmatrix}$

25. Let $A \cdot B = \begin{bmatrix} 2 & 3 \\ 1 & -1 \end{bmatrix} \cdot \begin{bmatrix} 0 & 1 \\ 3 & 1 \end{bmatrix}$; $A \cdot B = \begin{bmatrix} 9 & 5 \\ -3 & 0 \end{bmatrix}$ and $\delta(AB) = 15$,

so $(A \cdot B)^{-1} = \dfrac{1}{15}\begin{bmatrix} 0 & -5 \\ 3 & 9 \end{bmatrix}$; also, since $\delta(A) = -5$ and $\delta(B) = -3$,

$B^{-1} = -\dfrac{1}{3}\begin{bmatrix} 1 & -1 \\ -3 & 0 \end{bmatrix}$ and $A^{-1} = \dfrac{1}{5}\begin{bmatrix} -1 & -3 \\ -1 & 2 \end{bmatrix}$; $B^{-1} \cdot A^{-1} = \dfrac{1}{15}\begin{bmatrix} 0 & -5 \\ 3 & 9 \end{bmatrix}$; hence $(A \cdot B)^{-1} = B^{-1} \cdot A^{-1}$.

27. Let $A = \begin{bmatrix} 3 & 0 & 1 \\ 2 & 1 & 0 \\ 0 & 1 & 2 \end{bmatrix}$; then $\delta(A) = 8$ and $A^{-1} = \dfrac{1}{8}\begin{bmatrix} 2 & 1 & -1 \\ -4 & 6 & 2 \\ 2 & -3 & 3 \end{bmatrix}$.

Let $B = \begin{bmatrix} 2 & 1 & 0 \\ 1 & 1 & 2 \\ 0 & 1 & 0 \end{bmatrix}$; then $\delta(B) = -\dfrac{1}{4}$ and $B^{-1} = -\dfrac{1}{4}\begin{bmatrix} -2 & 0 & 2 \\ 0 & 0 & -4 \\ 1 & -2 & 1 \end{bmatrix}$;

$A \cdot B = \begin{bmatrix} 6 & 4 & 0 \\ 5 & 3 & 2 \\ 1 & 3 & 2 \end{bmatrix}$, $\delta(A \cdot B) = -32$, and $(A \cdot B)^{-1} = -\dfrac{1}{32}\begin{bmatrix} 0 & -8 & 8 \\ -8 & 12 & -12 \\ 12 & -14 & -2 \end{bmatrix}$;

$B^{-1} \cdot A^{-1} = -\dfrac{1}{4}\begin{bmatrix} -2 & 0 & 2 \\ 0 & 0 & -4 \\ 1 & -2 & 1 \end{bmatrix} \cdot \dfrac{1}{8}\begin{bmatrix} 2 & 1 & -1 \\ -4 & 6 & 2 \\ 2 & -3 & 3 \end{bmatrix} = -\dfrac{1}{32}\begin{bmatrix} 0 & -8 & 8 \\ -8 & 12 & -12 \\ 12 & -14 & -2 \end{bmatrix}$;

hence $(A \cdot B)^{-1} = B^{-1} \cdot A^{-1}$.

29. Let $A = \begin{bmatrix} a_{11} & a_{12} \\ a_{21} & a_{22} \end{bmatrix}$; then $\delta(A) = (a_{11}a_{22} - a_{12}a_{21}) \neq 0$, since A is nonsingular, and hence $\dfrac{1}{\delta(A)}$ is defined and A^{-1} exists.

$$A^{-1} = \dfrac{1}{\delta(A)}\begin{bmatrix} a_{22} & -a_{12} \\ -a_{21} & a_{11} \end{bmatrix} = \begin{bmatrix} \dfrac{a_{22}}{\delta(A)} & \dfrac{-a_{12}}{\delta(A)} \\ \dfrac{-a_{21}}{\delta(A)} & \dfrac{a_{11}}{\delta(A)} \end{bmatrix}$$

and $\delta(A^{-1}) = \dfrac{a_{22}a_{11}}{[\delta(A)]^2} - \dfrac{a_{12}a_{21}}{[\delta(A)]^2} = \dfrac{a_{22}a_{11} - a_{12}a_{21}}{[\delta(A)]^2} = \dfrac{\delta(A)}{[\delta(A)]^2} = \dfrac{1}{\delta(A)}$.

31. From the results of Exercise 9.4–33, and Exercise 29 above,

$$\delta[B^{-1}AB] = \delta[B^{-1}(AB)] = \delta(B^{-1}) \cdot \delta(AB) = \delta(B^{-1}) \cdot \delta(A) \cdot \delta(B) = \dfrac{1}{\delta(B)} \cdot \delta(A) \cdot \delta(B) = \delta(A).$$

Exercise 9.7 (page 325)   **1.** $\{(1, 1)\}$ **3.** No solution   **5.** $\{(6, 4)\}$ **7.** $\left\{\left(-\dfrac{18}{7}, \dfrac{19}{7}\right)\right\}$

9. $\left(x, -\dfrac{1}{3}x + \dfrac{2}{9}\right), x \in R$ **11.** $\{(2, 2)\}$ **13.** $\{(1, 1, 1)\}$ **15.** $\{(1, 1, 0)\}$ **17.** $\{(1, -2, 3)\}$

19. $\{(3, -1, -2)\}$ **21. a.** $\left\{\left(\dfrac{11}{6}, \dfrac{5}{3}, -\dfrac{1}{2}\right)\right\}$ **b.** $\left\{\left(-\dfrac{13}{12}, -\dfrac{1}{6}, -\dfrac{3}{4}\right)\right\}$ **c.** $\left\{\left(\dfrac{5}{6}, -\dfrac{1}{3}, \dfrac{1}{2}\right)\right\}$

Answers to Odd-Numbered Exercises

Exercise 9.8 (page 329) 1. $\{(\frac{13}{5}, \frac{3}{5})\}$ 3. $\{(\frac{22}{7}, \frac{20}{7})\}$ 5. $\{(6, 4)\}$ 7. Inconsistent
9. $\{(4, 1)\}$ 11. $\{(\frac{1}{a+b}, \frac{1}{a+b})\}$ $(a \neq -b)$ 13. $\{(1, 1, 0)\}$ 15. $\{(1, -2, 3)\}$
17. $\{(3, -1, -2)\}$ 19. $\{(-\frac{1}{3}, -\frac{25}{24}, -\frac{5}{8})\}$ 21. $\{(1, -\frac{1}{3}, \frac{1}{2})\}$
23. $\{(w, x, y, z)\} = \{(2, -1, 1, 0)\}$
25. $|A| = \begin{vmatrix} a_1 & b_1 \\ a_2 & b_2 \end{vmatrix}$ and $\delta(A) = a_1 b_2 - a_2 b_1$;

$|A_y| = \begin{vmatrix} a_1 & c_1 \\ a_2 & c_2 \end{vmatrix}$ and $\delta(A_y) = a_1 c_2 - a_2 c_1 = 0$, so $a_1 c_2 = a_2 c_1$;

$|A_x| = \begin{vmatrix} c_1 & b_1 \\ c_2 & b_2 \end{vmatrix}$ and $\delta(A_x) = b_2 c_1 - b_1 c_2 = 0$, so $b_1 c_2 = b_2 c_1$;

hence, $\frac{a_1 c_2}{b_1 c_2} = \frac{a_2 c_1}{b_2 c_1}$; $\frac{a_1}{b_1} = \frac{a_2}{b_2}$ and $a_1 b_2 = a_2 b_1$, so $a_1 b_2 - a_2 b_1 = 0$; therefore $\delta(A) = 0$.

Chapter 9 Review (page 330)

1. $\begin{bmatrix} 1 & -1 \\ 1 & 1 \end{bmatrix}$ 2. $\begin{bmatrix} 2 & 5 & -2 \\ 14 & -1 & 12 \end{bmatrix}$ 3. $\begin{bmatrix} -2 & -3 & 1 \\ 3 & -7 & 6 \\ -5 & 2 & 5 \end{bmatrix}$

4. $\begin{bmatrix} -8 & -3 & -6 \\ 6 & -3 & 0 \\ -9 & 5 & 5 \end{bmatrix}$ 5. $\begin{bmatrix} -21 & 7 \\ -14 & 0 \\ -7 & -7 \end{bmatrix}$ 6. $[13]$ 7. $\begin{bmatrix} -13 & 3 \\ -19 & 27 \end{bmatrix}$ 8. $\begin{bmatrix} 18 & 7 & 25 \\ 8 & -1 & 11 \\ 3 & 0 & 3 \end{bmatrix}$

9. $\{(2, -1)\}$ 10. $\{(2, 1, 1)\}$ 11. -3 12. 7 13. -3 14. 14 15. 0 16. -1
17. 15 18. -1578 19. $\frac{1}{34}\begin{bmatrix} -3 & 2 \\ 11 & 4 \end{bmatrix}$ 20. $\frac{1}{6}\begin{bmatrix} 1 & 3 & -2 \\ -3 & -3 & 6 \\ 1 & -3 & 4 \end{bmatrix}$ 21. $\{(-2, 1)\}$
22. $\{(3, -1, 1)\}$ 23. a. $\{(\frac{1}{2}, -\frac{1}{2})\}$ b. $\{(\frac{1}{4}, -\frac{7}{4})\}$ c. $\{(1, -2)\}$ 24. $\{(-\frac{16}{7}, -\frac{13}{7})\}$
25. $\{(2, -1, 0)\}$

Supplemental Exercises, Chapters 8 and 9 (page 332) 1. $a = -\frac{3}{4}, b = \frac{3}{4}, c = \frac{1}{2}$
2. $a = 4, b = -4, c = 7$ 3. $a = 1, b = -2, c = 2, d = 1$
4. $a = 2, b = 3, c = -12, d = -3, e = 29$ 5. $\{(-\frac{28}{3}, \frac{20}{3}, 2, -\frac{4}{3})\}$ 6. $\{(2, -1, 3, -1)\}$
7. $\{(1, -1, 4), (-1, -1, 4)\}$ 8. $\{(2, \sqrt{3}, \frac{5}{4}), (-2, \sqrt{3}, \frac{5}{4}), (2, -\sqrt{3}, \frac{5}{4}), (-2, -\sqrt{3}, \frac{5}{4})\}$
9. $(-7z - 1, 5z + 1, z), z \in R$ 10. $(2y - 4, y, 2y - 1), y \in R$ 11. $\{(\frac{\sqrt{2}}{2} + \frac{\sqrt{2}}{2}i, -\frac{\sqrt{2}}{2} - \frac{\sqrt{2}}{2}i)\}$

12. Let $A = \begin{bmatrix} a_1 & 0 & 0 & 0 \\ 0 & a_2 & 0 & 0 \\ 0 & 0 & a_3 & 0 \\ 0 & 0 & 0 & a_4 \end{bmatrix}$; then expanding all minors about the first row we have

$$\delta(A) = a_1 \begin{vmatrix} a_2 & 0 & 0 \\ 0 & a_3 & 0 \\ 0 & 0 & a_4 \end{vmatrix} = a_1 a_2 \begin{vmatrix} a_3 & 0 \\ 0 & a_4 \end{vmatrix} = a_1 a_2 a_3 a_4.$$

This result generalizes to $n \times n$ diagonal matrices.

13. This proof is identical to Exercise 12 except we expand all minors about the first column instead of the first row. The result generalizes to $n \times n$ upper-triangular matrices.

14. Let $A = \begin{bmatrix} a_{11} & a_{12} & a_{13} \\ 0 & a_{22} & a_{23} \\ 0 & 0 & a_{33} \end{bmatrix}$, then $A_{12} = A_{13} = A_{23} = 0$ because each of the minors used to compute these cofactors has a zero row or zero column. Now

$$A^{-1} = \frac{1}{\delta(A)} \begin{bmatrix} A_{11} & A_{12} & A_{13} \\ A_{21} & A_{22} & A_{23} \\ A_{31} & A_{32} & A_{33} \end{bmatrix}^t = \frac{1}{\delta(A)} \begin{bmatrix} A_{11} & 0 & 0 \\ A_{21} & A_{22} & 0 \\ A_{31} & A_{32} & A_{33} \end{bmatrix}^t$$

$$= \frac{1}{\delta(A)} \begin{bmatrix} A_{11} & A_{21} & A_{31} \\ 0 & A_{22} & A_{32} \\ 0 & 0 & A_{33} \end{bmatrix},$$

which is upper triangular. The result generalizes to nonsingular, $n \times n$, upper-triangular matrices.

15. $a_1 \neq 0$, $a_2 \neq 0$, $a_3 \neq 0$, $a_4 \neq 0$; $A^{-1} = \begin{bmatrix} \frac{1}{a_1} & 0 & 0 & 0 \\ 0 & \frac{1}{a_2} & 0 & 0 \\ 0 & 0 & \frac{1}{a_3} & 0 \\ 0 & 0 & 0 & \frac{1}{a_4} \end{bmatrix}$;

This result generalizes to $n \times n$ diagonal matrices.

16. No. Let $A = \begin{bmatrix} a \\ b \end{bmatrix}$ and $B = [c \ d]$. If $AB = I_{2 \times 2}$, then $ac = 1$ (and therefore $a \neq 0$, $c \neq 0$) and $bd = 1$ (and therefore $b \neq 0$, $d \neq 0$). But $ad = 0$ and $bc = 0$; thus $a = 0$ or $d = 0$, and $b = 0$ or $c = 0$. This is a contradiction. Thus, AB could not be $I_{2 \times 2}$.

17. Note $\delta(A)\delta(B) = \delta(AB) = \delta(I) = 1$; thus $\delta(A) \neq 0$ and A is nonsingular. Thus, $B = A^{-1}$ and $BA = I$.

18. Note $1 = \delta(I_{4 \times 4}) = \delta(A_{4 \times 4}^{-1} A_{4 \times 4}) = \delta(A_{4 \times 4}^{-1}) \delta(A_{4 \times 4})$. Hence, $\delta(A_{4 \times 4}^{-1}) = \frac{1}{\delta(A_{4 \times 4})}$.

Exercise 10.1 (page 338) 1. $-4, -3, -2, -1$ 3. $-\frac{1}{2}, 1, \frac{7}{2}, 7$ 5. $2, \frac{3}{2}, \frac{4}{3}, \frac{5}{4}$ 7. $0, 1, 3, 6$

Answers to Odd-Numbered Exercises

9. $-1, 1, -1, 1$ **11.** $1, 0, -\frac{1}{3}, \frac{1}{2}$ **13.** $11, 15, 19$ **15.** $2, 5, 8$ **17.** $x + 2, x + 3, x + 4$

19. $2x + 7, 2x + 10, 2x + 13$ **21.** $32, 128, 512, 2048$ **23.** $\frac{8}{3}, \frac{16}{3}, \frac{32}{3}, \frac{64}{3}$ **25.** $\frac{x}{a}, -\frac{x^2}{a^2}, \frac{x^3}{a^3}, -\frac{x^4}{a^4}$

27. $\frac{x^2}{-4}, \frac{x^3}{16}, \frac{x^4}{-64}$ **29.** $4n + 3, 31$ **31.** $-\frac{1}{3}(-3)^{n-1}, -243$ **33.** $(4n+1)x - 4x, 29x$

35. $48(2)^{n-1}, 1536$ **37.** $-\frac{1}{3}(-3)^{n-1}, -243$ **39.** $x + (n-1)(x^3 - x); \ -3x + 4x^3$

41. $x + (n-1)(y - x); \ 5y - 4x$ **43.** $\frac{y^{n-1}}{x^{n-2}}; \ \frac{y^7}{x^6}$ **45.** $2; 3; 41$ **47.** 28th **49.** 3

Exercise 10.2 (page 344) **1.** $1 + 4 + 9 + 16$ **3.** $-\frac{1}{2} + \frac{1}{4} - \frac{1}{8}$ **5.** $-5 - 3 - 1 + 1 + 3 + 5 + 7$

7. $1 + \frac{1}{2} + \frac{1}{4} + \cdots$ **9.** $\sum_{j=1}^{4} x^{2j-1}$ **11.** $\sum_{j=1}^{5} j^2$ **13.** $\sum_{j=1}^{\infty} j(j+1)$ **15.** $\sum_{j=1}^{\infty} \frac{j+1}{j}$ **17.** 63

19. 806 **21.** -6 **23.** 10,000 **25.** 1092 **27.** $\frac{31}{32}$ **29.** $\frac{364}{2187}$ **31.** 2816 **33.** 180

35. 2040 **37.** 196 **39.** \$2333.39 **41.** $\frac{3}{4}, \frac{7}{8}, \frac{15}{16}, \frac{31}{32}; \ 1$ **43.** $p = 4, q = -3$

Exercise 10.3 (page 351) **1.** $\lim_{n \to \infty} s_n = 0$ **3.** $\lim_{n \to \infty} s_n = 1$ **5.** $\lim_{n \to \infty} s_n$ is undefined. **7.** $\lim_{n \to \infty} s_n = 0$

9. Convergent, $\lim_{n \to \infty} \left| 0 - \frac{1}{2^n} \right| = 0$ **11.** Divergent, $\lim_{n \to \infty} n$ is undefined.

13. Convergent, $\lim_{n \to \infty} \left| 0 - (-1)^{n+1} \frac{1}{2^{n-1}} \right| = 0$ **15.** 24 **17.** No sum **19.** 2 **21.** $\frac{8}{9}$

23. $\frac{31}{99}$ **25.** $2\frac{410}{999}$ **27.** $\frac{29}{225}$ **29.** 20 cm.

Exercise 10.4 (page 357) **1.** $8 \cdot 7 \cdot 6 \cdot 5 \cdot 4 \cdot 3 \cdot 2 \cdot 1$ **3.** $6 \cdot 5 \cdot 4 \cdot 3 \cdot 2 \cdot 1$ **5.** $5 \cdot 4 \cdot 3 \cdot 2 \cdot 1 = 120$

7. $\frac{9 \cdot 8 \cdot 7!}{7!} = 72$ **9.** $\frac{5 \cdot 4 \cdot 3 \cdot 2 \cdot 1 \cdot 7!}{8 \cdot 7!} = 15$ **11.** $\frac{8 \cdot 7 \cdot 6!}{2 \cdot 1 \cdot 6!} = 28$ **13.** $3!$ **15.** $\frac{6!}{2!}$ **17.** $\frac{8!}{5!}$

19. $\frac{6!}{5!1!} = 6$ **21.** $\frac{3!}{3!0!} = 1$ **23.** $\frac{7!}{0!7!} = 1$ **25.** $\frac{5!}{2!3!} = 10$

27. $(n)(n-1)(n-2) \cdot \cdots \cdot 3 \cdot 2 \cdot 1$ **29.** $(3n)(3n-1)(3n-2) \cdot \cdots \cdot 3 \cdot 2 \cdot 1$

31. $(n-2)(n-3)(n-4) \cdot \cdots \cdot 3 \cdot 2 \cdot 1$ **33.** $(n+2)(n+1)$ **35.** $\frac{n+1}{n+3}$ **37.** $\frac{2n-1}{2n-2}$

39. $x^5 + 15x^4 + 90x^3 + 270x^2 + 405x + 243$ **41.** $x^4 - 12x^3 + 54x^2 - 108x + 81$

43. $8x^3 - 6x^2y + \frac{3}{2}xy^2 - \frac{1}{8}y^3$ **45.** $\frac{1}{64}x^6 + \frac{3}{8}x^5 + \frac{15}{4}x^4 + 20x^3 + 60x^2 + 96x + 64$

47. $x^{20} + 20x^{19}y + \dfrac{20 \cdot 19}{2!}x^{18}y^2 + \dfrac{20 \cdot 19 \cdot 18}{3!}x^{17}y^3$, or $\binom{20}{0}x^{20} + \binom{20}{1}x^{19}y + \binom{20}{2}x^{18}y^2 + \binom{20}{3}x^{17}y^3$

49. $a^{12} + 12a^{11}(-2b) + \dfrac{12 \cdot 11}{2!}a^{10}(-2b)^2 + \dfrac{12 \cdot 11 \cdot 10}{3!}a^9(-2b)^3$, or

$\binom{12}{0}a^{12} + \binom{12}{1}a^{11}(-2b) + \binom{12}{2}a^{10}(-2b)^2 + \binom{12}{3}a^9(-2b)^3$

51. $x^{10} + 10x^9(-\sqrt{2}) + \dfrac{10 \cdot 9}{2!}x^8(-\sqrt{2})^2 + \dfrac{10 \cdot 9 \cdot 8}{3!}x^7(-\sqrt{2})^3$, or

$\binom{10}{0}x^{10} + \binom{10}{1}x^9(-\sqrt{2}) + \binom{10}{2}x^8(-\sqrt{2})^2 + \binom{10}{3}x^7(-\sqrt{2})^3$ 53. 1.22   55. \$1480

57. $-3003a^{10}b^5$ 59. $3360x^6y^4$ 61. **a.** $1 - x + x^2 - x^3 + \cdots$ **b.** $1 - x + x^2 - x^3 + \cdots$

63. $\binom{k}{r} + \binom{k}{r-1} = \dfrac{k!}{r!(k-r)!} + \dfrac{k!}{(r-1)!(k-r+1)!}$

$= \dfrac{k!(k-r+1) + k!r}{r!(k-r+1)!} = \dfrac{k!(k+1)}{r!(k-r+1)!}$

$= \dfrac{(k+1)!}{r![(k+1)-r]!} = \binom{k+1}{r}$

Exercise 10.5 (page 364)   **1.**   **a.**   For $n = 1$: $\dfrac{n}{2} = \dfrac{1}{2}$; $\dfrac{n(n+1)}{4} = \dfrac{1(1+1)}{4} = \dfrac{1}{2}$.

b.   For $n = k$: $\dfrac{1}{2} + \dfrac{2}{2} + \dfrac{3}{2} + \cdots + \dfrac{k}{2} = \dfrac{k(k+1)}{4}$ and $(k+1)$st term $= \dfrac{k+1}{2}$;

hence $\dfrac{1}{2} + \dfrac{2}{2} + \dfrac{3}{2} + \cdots + \dfrac{k}{2} + \dfrac{k+1}{2} = \dfrac{k(k+1)}{4} + \dfrac{k+1}{2} = \dfrac{k^2 + k + 2k + 2}{4}$

$= \dfrac{k^2 + 3k + 2}{4} = \dfrac{(k+1)(k+2)}{4}$.

3. **a.**   For $n = 1$: $2n = 2(1) = 2$; $n(n+1) = 1(1+1) = 2$.
    b.   For $n = k$: $2 + 4 + 6 + \cdots + 2k = k(k+1)$ and $(k+1)$st term is $2(k+1)$;
hence $2 + 4 + 6 + \cdots + 2k + 2(k+1) = k(k+1) + 2(k+1)$
$= (k+1)(k+2)$.

5. **a.**   For $n = 1$: $6n - 4 = 6(1) - 4 = 2$; $n(3n-1) = 1(3(1)-1) = 2$.
    b.   For $n = k$: $2 + 8 + 14 + \cdots + (6k-4) = k(3k-1)$ and $(k+1)$st term is $6(k+1) - 4$;
hence $2 + 8 + 14 + \cdots + (6k-4) + [6(k+1) - 4] = k(3k-1) + 6k + 2 = 3k^2 + 5k + 2$
$= (k+1)(3k+2) = (k+1)[3(k+1) - 1]$.

7.   **a.**   For $n = 1$: $3n + 4 = 3(1) + 4 = 7$; $\dfrac{n(3n+11)}{2} = \dfrac{1(3(1)+11)}{2} = 7$.

b.   For $n = k$: $7 + 10 + 13 + \cdots + (3k+4) = \dfrac{k(3k+11)}{2}$ and $(k+1)$st term is $3(k+1) + 4$;

hence $7 + 10 + 13 + \cdots + (3k+4) + [3(k+1) + 4] = \dfrac{k(3k+11)}{2} + 3k + 7 = \dfrac{3k^2 + 17k + 14}{2}$

$= \dfrac{(k+1)(3k+14)}{2} = \dfrac{(k+1)[3(k+1) + 11]}{2}$.

9. **a.** For $n = 1$: $n^2 = 1^2 = 1$; $\dfrac{n(n+1)(2n+1)}{6} = \dfrac{1(2)(3)}{6} = 1$.

   b. For $n = k$: $1^2 + 2^2 + 3^2 + \cdots + k^2 = \dfrac{k(k+1)(2k+1)}{6}$ and $(k+1)$st term is $(k+1)^2$;

hence $1^2 + 2^2 + 3^2 + \cdots + k^2 + (k+1)^2$

$= \dfrac{k(k+1)(2k+1)}{6} + (k+1)^2$

$= \dfrac{k(k+1)(2k+1) + 6(k+1)^2}{6} = \dfrac{(k+1)[k(2k+1) + 6(k+1)]}{6} = \dfrac{(k+1)(2k^2 + 7k + 6)}{6}$

$= \dfrac{(k+1)(k+2)(2k+3)}{6} = \dfrac{(k+1)[(k+1)+1][2(k+1)+1]}{6}$.

11. **a.** For $n = 1$: $(2n-1)^3 = (2-1)^3 = 1^3 = 1$; $n^2(2n^2 - 1) = 1(2-1) = 1(1) = 1$.

    b. For $n = k$: $1^3 + 3^3 + 5^3 + \cdots + (2k-1)^3 = k^2(2k^2 - 1)$ and the $(k+1)$st term is $[2(k+1) - 1]^3 = (2k+1)^3$; hence

$1^3 + 3^3 + 5^3 + \cdots + (2k-1)^3 + (2k+1)^3 = k^2(2k^2 - 1) + (2k+1)^3 = 2k^4 + 8k^3 + 11k^2 + 6k + 1$;

by use of the factor theorem and synthetic division,

$$2k^4 + 8k^3 + 11k^2 + 6k + 1 = (k+1) \times (k+1)(2k^2 + 4k + 1);$$

also, $2k^2 + 4k + 1 = 2(k^2 + 2k + 1) - 2 + 1 = 2(k+1)^2 - 1$;
hence, $2k^4 + 8k^3 + 11k^2 + 6k + 1 = (k+1)^2[2(k+1)^2 - 1]$.

13. **a.** For $n = 1$: $n(n+1) = 1(2) = 2$; $\dfrac{n(n+1)(n+2)}{3} = \dfrac{1(2)(3)}{3} = 2$.

    b. For $n = k$: $1 \cdot 2 + 2 \cdot 3 + 3 \cdot 4 + \cdots + k(k+1) = \dfrac{k(k+1)(k+2)}{3}$ and the $(k+1)$st term is

$(k+1)[(k+1)+1] = (k+1)(k+2)$; hence

$1 \cdot 2 + 2 \cdot 3 + 3 \cdot 4 + \cdots + k(k+1) + [(k+1)(k+2)]$

$= \dfrac{k(k+1)(k+2)}{3} + (k+1)(k+2) = \dfrac{[k(k+1)(k+2)] + [3(k+1)(k+2)]}{3}$

$= \dfrac{(k+1)(k+2)(k+3)}{3} = \dfrac{(k+1)[(k+1)+1][(k+1)+2]}{3}$.

15. **a.** For $n = 1$: $1^3 + 2 \cdot 1 = 3$; 3 is divisible by 3.

    b. For $n = k$: assume $k^3 + 2k$ is divisible by 3; for $n = k + 1$:

$$(k+1)^3 + 2(k+1) = k^3 + 3k^2 + 3k + 1 + 2k + 2$$
$$= (k^3 + 2k) + (3k^2 + 3k + 3).$$

Since by hypothesis $k^3 + 2k$ is divisible by 3 and since $3k^2 + 3k + 3$ is divisible by 3, $(k+1)^3 + 2(k+1)$ is divisible by 3.

17. For $n = k$: $2 + 4 + 6 + \cdots + 2k = k(k+1) + 2$ and $(k+1)$st term is $2(k+1)$; hence

$2 + 4 + 6 + \cdots + 2k + 2(k+1) = k(k+1) + 2 + 2(k+1) = (k^2 + 3k + 2) + 2$

$= (k+1)(k+2) + 2 = (k+1)[(k+1)+1] + 2$.

However, for $n = 1$: $2n = 2(1) = 2$; $n(n+1) + 2 = 1(2) + 2 = 4$. Hence not true for every $n \in N$.

19. Since $s_1 = a$ and $s_{n+1} = s_n + d$, it follows that:
   a. For $n = 1$: $s_n = s_1 = a = a + 0 \cdot d = a + (n - 1)d$.
   b. For $n = k$: $s_k = a + (k - 1)d$; now
   $$s_{k+1} = s_k + d = a + (k - 1)d + d = a + kd = a + (k + 1 - 1)d.$$

21. Since $s_n = ar^{n-1}$ and $S_{n+1} = S_n + s_{n+1}$, it follows that:
   a. For $n = 1$: $S_1 = s_1 = a = a\dfrac{1-r}{1-r} = \dfrac{a - ar^1}{1-r}$.
   b. For $n = k$: $S_k = \dfrac{a - ar^k}{1-r}$ and $s_{k+1} = ar^k$; now
   $$S_{k+1} = S_k + s_{k+1} = \dfrac{a - ar^k}{1-r} + ar^k = \dfrac{a - ar^k + ar^k - ar^{k+1}}{1-r} = \dfrac{a - ar^{k+1}}{1-r}.$$

Chapter 10 Review (page 365) 1. 2, 5, 10   2. $\dfrac{1}{2}, \dfrac{1}{3}, \dfrac{1}{4}$ 3. 13, 16, 19   4. $a - 4, a - 6, a - 8$

5. $-18, 54, -162$ 6. $\dfrac{3}{2}, \dfrac{9}{4}, \dfrac{27}{8}$ 7. $s_n = 5n - 8$; $s_7 = 27$ 8. $s_n = (-2)\left(\dfrac{-1}{3}\right)^{n-1}$; $s_5 = \dfrac{-2}{81}$

9. 25   10. 6th term   11. $2 + 6 + 12 + 20$ 12. $\sum\limits_{k=1}^{\infty} x^{k+1}$ 13. 119   14. $\dfrac{121}{243}$ 15. 3

16. $\dfrac{8}{3}$ 17. $\dfrac{1}{2}$ 18. $\dfrac{4}{9}$ 19. $5 \cdot (2 \cdot 1)$ 20. 48   21. 21   22. $\dfrac{1}{n(n+1)!}$

23. $x^{10} - 20x^9 y + 180x^8 y^2 - 960x^7 y^3$ 24. $-15{,}360 x^3 y^7$

25. Formula holds for 1. Assume formula holds for n; test for $n + 1$:
$$3 + 6 + 9 + \cdots + 3n + 3(n + 1) = \dfrac{3n(n+1)}{2} + 3(n + 1).$$

Right-hand member is equivalent to:
$$\dfrac{3n^2 + 3n}{2} + \dfrac{6(n+1)}{2} = \dfrac{3n^2 + 9n + 6}{2} = \dfrac{3(n^2 + 3n + 2)}{2} = \dfrac{3(n+1)(n+2)}{2} = \dfrac{3(n+1)((n+1)+1)}{2}.$$

26. Formula holds for 1. Assume formula holds for n; test for $n + 1$:
$$\dfrac{1}{2} + \dfrac{1}{4} + \dfrac{1}{8} + \cdots + \dfrac{1}{2^n} + \dfrac{1}{2^{n+1}} = 1 - \dfrac{1}{2^n} + \dfrac{1}{2^{n+1}}.$$

Right-hand member is equivalent to: $\dfrac{2^{n+1} - 2 + 1}{2^{n+1}} = \dfrac{2^{n+1} - 1}{2^{n+1}} = 1 - \dfrac{1}{2^{n+1}}.$

Exercise 11.1 (page 371) 1. 1, 5, 8   3. 2, 6, 16   5. 2, 2, 4   7. 4   9. 24   11. 16
13. 64   15. 24   17. 216   19. 375   21. 30   23. 10   25. 48   27. $\dfrac{5!}{2!}$, or 60

29. $\dfrac{8!}{3!}$, or 6720   31. $P_{5,3} = \dfrac{5!}{2!} = \dfrac{5 \cdot 4!}{2!} = 5\left(\dfrac{4!}{2!}\right) = 5(P_{4,2})$

33. $P_{n,3} = \dfrac{n!}{(n-3)!} = \dfrac{n(n-1)!}{(n-3)!} = n\left[\dfrac{(n-1)!}{(n-3)!}\right] = n(P_{n-1,2})$ 35. 9   37. 24   39. 48

Answers to Odd-Numbered Exercises

Exercise 11.2 (page 376) **1.** 7 **3.** 15 **5.** $\binom{52}{5}$ **7.** $\binom{13}{5} \cdot \binom{13}{5} \cdot \binom{13}{3}$ **9.** $4 \cdot \binom{13}{5}$
11. 164 **13.** 10 **15.** 210 **17.** 12

Exercise 11.3 (page 379) **1.** $\{1, 2, 3, 4, 5, 6\}, \{3, 4, 5, 6\}, \dfrac{2}{3}$
3. $\{(H, H), (H, T), (T, H), (T, T)\}, \{(H, H), (T, T)\}, \dfrac{1}{2}$ **5.** $\dfrac{1}{6}$ **7.** $\dfrac{1}{18}$ **9.** $\dfrac{5}{9}$ **11.** $\dfrac{1}{52}$
13. $\dfrac{3}{26}$ **15.** $\dfrac{1}{17}$ **17.** $\dfrac{11}{221}$ **19.** $\dfrac{1}{190}$ **21.** $\dfrac{3}{38}$

Exercise 11.4 (page 382) **1.** $\dfrac{1}{6}$ **3.** $\dfrac{7}{18}$ **5.** $\dfrac{13}{18}$ **7.** $\dfrac{5}{33}$ **9.** $\dfrac{1}{11}$ **11.** $\dfrac{5}{22}$ **13.** $\dfrac{14}{33}$
15. $\dfrac{15}{22}$ **17.** $\dfrac{3}{13}$ **19.** $2.14; less **21.** 5.3 cents **23.** $3.29

Exercise 11.5 (page 386) **1.** $\dfrac{20}{91}$; no **3. a.** $\dfrac{2}{45}$ **b.** $\dfrac{28}{75}$ **c.** $\dfrac{4}{225}$ **d.** $\dfrac{1}{9}$
5. a. $\dfrac{11}{36}$ **b.** $\dfrac{5}{36}$ **c.** $\dfrac{2}{11}$ **d.** No **7. a.** $\dfrac{1}{2}$ **b.** $\dfrac{1}{2}$ **c.** $\dfrac{1}{4}$ **d.** $\dfrac{1}{4}$ **e.** $\dfrac{1}{8}$
9. a. $\dfrac{71}{72}$ **b.** $\dfrac{5}{9}$ **c.** $\dfrac{5}{36}$ **d.** $\dfrac{61}{72}$ **11. a.** $\dfrac{1}{210}$ **b.** $\dfrac{29}{210}$ **c.** $\dfrac{29}{70}$ **d.** $\dfrac{29}{30}$; yes
13. $\dfrac{1}{4}$

Chapter 11 Review (page 388) **1.** 16 **2.** 64 **3.** 128 **4.** 360 **5.** 792
6. 1,033,885,600 **7.** 200 **8.** 84 **9.** $\dfrac{1}{17}$ **10.** $\dfrac{1}{221}$ **11.** $\dfrac{10}{17}$ **12.** $\dfrac{25}{102}$ **13.** $\dfrac{26}{51}$
14. $\dfrac{80}{221}$ **15.** $\dfrac{41}{663}$ **16.** $\dfrac{12}{25}$ **17.** $\dfrac{25}{1326}$ **18.** $\dfrac{20}{221}$ **19.** $\dfrac{25}{221}$ **20.** $\dfrac{95}{663}$

Supplemental Exercises, Chapters 10 and 11 (page 390) **1.** $\dfrac{2}{3}$ **2.** 0 **3.** 2 **4.** 1 **5.** 2
6. $\dfrac{1}{12}$ **7.** $P[\text{exactly } k \text{ successes}] = \binom{n}{k} p^k (1 - p)^{n-k}$ **8.** $P[\text{exactly } k \text{ failures}] = \binom{n}{k} p^{n-k} (1 - p)^k$
9. $P[\text{at least } k \text{ successes}] = \sum_{j=k}^{n} \binom{n}{j} p^j (1 - p)^{n-j}$ **10.** $P[\text{at least } k \text{ failures}] = \sum_{j=k}^{n} \binom{n}{j} p^{n-j} (1 - p)^j$
11. $\sum_{k=0}^{n} P[\text{exactly } k \text{ successes}] = \sum_{k=0}^{n} \binom{n}{k} p^k (1 - p)^{n-k} = [p + (1 - p)]^n = 1$

12. Expectation: $\sum_{k=0}^{n} k\binom{n}{k}p^k(1-p)^{n-k} = \sum_{k=1}^{n} \frac{n!}{(k-1)!(n-k)!} p^k(1-p)^{n-k}$

$= \sum_{j=0}^{n-1} \frac{n!}{j!(n-j-1)!} p^{j+1}(1-p)^{n-j-1} = np \sum_{j=0}^{n-1} \frac{(n-1)!}{j!(n-1-j)!} p^j(1-p)^{(n-1)-j}$

$= np[p + (1-p)]^{n-1} = np$

13. N **14.** $\dfrac{1}{2^{n-1}}$

15.   a. For $n = 1$: $n(n+1)(n+2) = 6$ and 3 is a factor of 6.   **b.** For $n = k$:
3 is a factor of $k(k+1)(k+2)$, i.e., there is $p \in N$ so that $k(k+1)(k+2) = 3p$. For $n = k+1$:
$(k+1)(k+1+1)(k+1+2) = (k+1)(k+2)(k+3) = k(k+1)(k+2) + 3(k+1)(k+2)$
$= 3p + 3(k+1)(k+2) = 3[p + (k+1)(k+2)]$.
Since $p + (k+1)(k+2) \in N$, 3 is a factor of $(k+1)(k+2)(k+3)$.

16.   a. For $n = 2$: $I_{2 \times 2} = \begin{bmatrix} 1 & 0 \\ 0 & 1 \end{bmatrix}$ and $\delta(I_{2 \times 2}) = (1)(1) - (0)(0) = 1$.

b. For $n = k$: $\delta(I_{k \times k}) = 1$. For $n = k+1$, we expand the determinant of $I_{(k+1) \times (k+1)}$ about the first row and note that $\delta(I_{(k+1) \times (k+1)}) = (1)(-1)^{1+1}\delta(I_k) = 1$.

17.   a. For $n = 1$: X_0 satisfies $AX = \lambda X$ by hypothesis.

b. For $n = k$: X_0 satisfies $A^k X = \lambda^k X$, i.e., $A^k X_0 = \lambda^k X_0$. For $n = k+1$,

$A^{k+1} X_0 = A(A^k X_0) = A(\lambda^k X_0) = \lambda^k A X_0 = \lambda^k (\lambda X_0) = \lambda^{k+1} X_0.$

Thus, X_0 satisfies $A^{k+1} X = \lambda^{k+1} X$.

18.   a. For $n = 1$: $(x+y)^1 = \sum_{j=0}^{1} \binom{1}{j} x^j y^{1-j}$.

b. For $n = k$: $(x+y)^k = \sum_{j=0}^{k} \binom{k}{j} x^j y^{k-j}$. For $n = k+1$:

$(x+y)^{k+1} = (x+y)(x+y)^k$

$= (x+y) \sum_{j=0}^{k} \binom{k}{j} x^j y^{k-j}$

$= x \sum_{j=0}^{k} \binom{k}{j} x^j y^{k-j} + y \sum_{j=0}^{k} \binom{k}{j} x^j y^{k-j}$

$= x \left[x^k + \sum_{j=0}^{k-1} \binom{k}{j} x^j y^{k-j} \right] + y \left[y^k + \sum_{j=1}^{k} \binom{k}{j} x^j y^{k-j} \right]$

$= x^{k+1} + x \sum_{j=0}^{k-1} \binom{k}{j} x^j y^{k-j} + y \sum_{j=1}^{k} \binom{k}{j} x^j y^{k-j} + y^{k+1}$

$= x^{k+1} + \sum_{j=0}^{k-1} \binom{k}{j} x^{j+1} y^{k-j} + \sum_{j=1}^{k} \binom{k}{j} x^j y^{k+1-j} + y^{k+1}$

$= x^{k+1} + \sum_{r=1}^{k} \binom{k}{r-1} x^r y^{k+1-r} + \sum_{r=1}^{k} \binom{k}{r} x^r y^{k+1-r} + y^{k+1}$

$= x^{k+1} + \sum_{r=1}^{k} \left[\binom{k}{r-1} + \binom{k}{r} \right] x^r y^{k+1-r} + y^{k+1}$

$= x^{k+1} + \sum_{r=1}^{k} \binom{k+1}{r} x^r y^{k+1-r} + y^{k+1}$ (by Exercise 10.4-63)

$= \sum_{r=0}^{k+1} \binom{k+1}{r} x^r y^{k+1-r}.$

Answers to Odd-Numbered Exercises

Exercise A.1 (page 396)

1.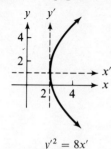
$y'^2 = 8x'$

3.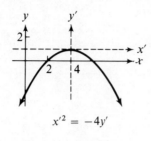
$x'^2 = -4y'$

5.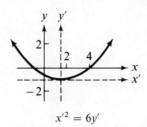
$x'^2 = 6y'$

7. $4x'^2 + 9y'^2 = 36$; ellipse
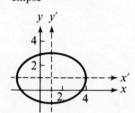

9. $9x'^2 + y'^2 = 9$; ellipse
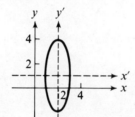

11. $4x'^2 - 9y'^2 = 36$; hyperbola
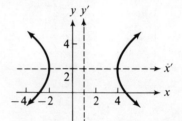

13. $3x'^2 - 2y'^2 = -6$; hyperbola
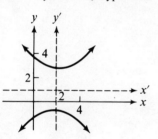

15. $2x'y' - 3 = 0$

17. $x'^2 + 3x'y' = 0$

19. $x'^2 - 3x'y' + 2y'^2 - 3 = 0$

21. $x'^2 + y'^2 - r^2 = 0$

Exercise A.2 (page 399)

1. a. **b.** $x + y - 1 = 0$ **3. a.** **b.** $2x + 3y - 5 = 0$

5. a. **b.** $x = 1 - 2y + y^2$ **7. a.** **b.** $x - y - 1 = 0$

9. $x = \dfrac{4}{3-t}$, $y = tx = \dfrac{4t}{3-t}$ **11.** $x = \dfrac{-1}{2+5t}$, $y = \dfrac{-t}{2+5t}$ **13.** $x = \dfrac{4}{\pm\sqrt{1+t^2}}$, $y = \dfrac{4t}{\pm\sqrt{1+t^2}}$

15. Note that $(y_1 + bt - y_1) = \dfrac{b}{a}(x_1 + at - x_1)$. Thus $x = x_1 + at$ and $y = y_1 + bt$ satisfy $y - y_1 = \dfrac{b}{a}(x - x_1)$.

Index

A

Abscissa, 135
Absolute value:
   definition of, 17
   equations involving, 119
   functions involving, 156
Addition, of matrices, 284
Additive identity law, 8
Additive inverse:
   of a matrix, 285
   of a real number, 8
Algebra, fundamental theorem of, 191
Algebraic expression(s):
   definition of, 24
   rational, 46
   terms of, 25
Antilogarithm, 223
Arithmetic, fundamental theorem of, 193
Arithmetic progression:
   definition of, 336
   nth term of an, 336
   sum of n terms of an, 341
Associative law:
   of addition, 8
   of multiplication, 8
Asymptote:
   as an aid of graphing, 204
   horizontal, 202
   of a hyperbola, 176
   oblique, 205
   vertical, 202
Augmented matrix, 298
Axes, translation of, 394
Axioms, 7, 401
Axis of symmetry, 167

B

Base:
   of a logarithm, 215
   of a power, 24
Binary operation, 7
Binomial expansion:
   coefficients as combinations, 375
   rth term of a, 357
Binomial theorem, 355
Bracket function, 158

C

Cartesian coordinate system:
   in the plane, 135
   in three dimensions, 252
Cartesian product, 134, 368
Characteristic of a logarithm, 221
Circle, equation of a, 174
Circular permutation, 373
Closed convex polygon, 274
Closure:
   for addition, 8
   for multiplication, 8
   for positive real numbers, 16
Coefficient, 25
Coefficient matrix, 298
Cofactor, 305
Column matrix, 282
Column vector, 282
Combination(s), 374
Common difference, 336
Common logarithm, 220
Common ratio, 337
Commutative law:
   of addition, 8
   of multiplication, 8
Complement of a set, 3
Complementary events, 377
Completeness property, 412
Completing the square in a quadratic equation, 97
Complex fractions, 54
Complex number(s), 79
Component of an ordered pair, 134
Composite number, 193
Compound interest, 229
Conditional inequality, solution by sign graph, 112
Conditional probability, 385
Conformable matrices, 291
Conic section(s), 173
Conjugate:
   of a binomial, 76
   of a complex number, 83
   zeros of a polynomial function, 190
Consistent equations, 241
Constant, 3
Constant function, 143
Constant of variation, 181
Convergent sequence, 348
Convex polygon, 274
Convex set, 274
Coordinate(s), 14, 134
Counting properties, 367
Cramer's rule, 327
Critical number, 114
Cube root, 67

D

Decreasing function, 212
Degree, 25
Dependence, linear, 241

Dependent events, 384
Determinant(s):
    cofactor of an element of a, 305
    in Cramer's rule, 327
    definition of, 304
    expansion of a, 304, 306
    function, 304
    of a matrix, 304
    minor of an element of a, 305
    order of a, 304
    properties of, 310
    solution of system of equations by, 326
    value of a, 306
Diagonal matrix, 292
Diagonal of a matrix, 292
Difference:
    of matrices, 284
    of real numbers, 9
Dimension of a matrix, 282
Direct variation, 180
Directed distance, 144
Discriminant, 100
Disjoint sets, 3
Distance:
    directed, 144
    between two points, 145
Distributive law, 8
Division:
    algorithm, 40
    of polynomials, 39
    of real numbers, 9
    synthetic, 42
    by zero, 9, 109
Domain:
    of an exponential function, 212
    of a function, 137
    of a logarithmic function, 214
    of a relation, 135

E

e:
    decimal approximation for, 213
    logarithms to the base, 223
    table for e^x, 413
    table for $\log_e$, 416
Element:
    of a determinant, 304
    of a set, 1
Elementary transformation:
    of an equation, 90
    of an inequality, 108
    of matrices, 295
Ellipse, equation of an, 175

Empty set, 2
Equality:
    addition law for, 11
    of complex numbers, 80
    definition of, 7
    of matrices, 283
    multiplication law for, 11
    properties of, 7
    of sets, 2
Equation(s):
    of a circle, 174
    consistent, 241
    dependent, 241
    of an ellipse, 175
    equivalent, 89
    exponential, 228
    first-degree, 91, 142
    of a hyperbola, 176
    inconsistent, 241
    involving absolute value, 119
    involving radicals, 102
    linear, 91, 142
    matrix form for linear system of, 322
    of a parabola, 166
    parametric, 397
    quadratic, 95
    second-degree, 95
    solution of by substitution, 105, 245, 263
    of a straight line, 142
    systems of linear, in three variables, 249
    systems of linear, in two variables, 241
    systems of nonlinear, 262
    for word problems, 122
Equivalent equations, 89
Equivalent expressions, 24
Equivalent inequalities, 108
Equivalent systems of equations, 242
Event(s), 377
Expansion of a binomial, 357, 375
Expansion of a determinant:
    by cofactors, 306
    meaning of, 306
Exponent(s):
    definition of, 24
    laws of, 30, 38
    as a logarithm, 215
    rational, 66, 211
    real, 212
Exponential equation, 228
Exponential function, 212
Extraction of roots, 96
Extraneous solutions, 102

F

Factor theorem, 188
Factorial notation, 353
Factoring polynomials, 34, 188, 191
Field(s):
    meaning of, 7
    ordered, 16
    real numbers as a, 7
Finite set, 2
First-degree equation(s), 91, *see also* Linear equations
First-degree function, *see* Function(s), linear
First-degree relation, 160
Fraction(s), 46
    partial, 257
Function(s):
    bracket, 158
    constant, 143
    decreasing, 212
    definition of, 137
    domain of a, 137
    exponential, 212
    graph of a, 137, 165
    increasing, 212
    inverse of a, 153
    involving absolute value, 156
    linear, 144
    logarithmic, 215
    notation, 137
    one-to-one, 153
    polynomial, 186
    quadratic, 165
    range of a, 137
    rational, 201
    sequence, 334
    step, 158
    zeros of a, 169
Fundamental principle of fractions, 13, 46, 83
Fundamental theorem:
    of algebra, 191
    of arithmetic, 193

G

Geometric progression:
    definition of, 337
    infinite, 350
    nth term of a, 337
    sum of n terms of a, 342
Graph(s):
    of an absolute-value function, 157
    of a bracket function, 158

Graph(s) (*continued*):
    of conic sections, 173
    of a constant function, 143
    of an exponential function, 212
    of inequalities, 110
    intercepts of a, 168
    of a linear equation in three variables, 252
    of a linear function, 142
    of a linear inequality, 160
    of a logarithmic function, 214
    of an ordered pair, 134
    of a polynomial function, 197
    of a quadratic function, 165
    of a quadratic inequality, 169
    of a rational function, 201
    of a real number, 15
    of a relation, 137
    of a second-degree equation, 165
    of systems of equations, 263
    of systems of inequalities, 274
    turning point of a, 198

H

Horizontal asymptote, 202
Hyperbola:
    asymptotes of a, 176
    equation of a, 176

I

i, 81
Identity element:
    for addition, 8
    for multiplication, 8
Imaginary numbers, 81
Inclination of a line segment, 144
Inconsistent equations, 241
Increasing function, 212
Independence, linear, 241
Independent events, 384
Index:
    of a radical, 72
    of summation, 344
Inequalities:
    equivalent, 108
    involving absolute value, 119
    linear, 107
    quadratic, 169
    solution of, 108
    solution sets of, 108
    systems of, 272
    transformations of, 108

Infinite geometric progression, 350
Infinite sets, 2
Integers, set of, 6
Integral exponents, 60
Intercept:
    form for a linear equation, 150
    of a graph of a function, 142
Interest, 229
Intersection:
    of sets, 2
    of solution sets of linear equations, 242, 249
Interval notation, 18
Inverse:
    additive, for matrices, 285
    additive, for real numbers, 8
    of an exponential function, 214
    of a function, 153
    multiplicative, for a real number, 8
    multiplicative, for a square matrix, 315
Inverse relations, 153
Inverse variation, 180
Irrational numbers, 6

J

Joint variation, 182

L

Laws of signs, 12
Leading coefficient, 26
Leading term, 26
Least common denominator, 50
Least common multiple, 50
Length of a line segment, 144
Less than, 16
Limit of a sequence, 348
Line graph, 110
Linear combination, 243, 264
Linear dependence, 241
Linear equation(s):
    definition of, 91
    intercept form of a, 150
    point-slope form of a, 149
    slope-intercept form of a, 149
    standard form of a, 142
    systems of, 241, 249
    two-point form, 152
Linear function, 144
Linear independence, 241
Linear programming, 277

Linear system(s):
    coefficient matrix of a, 298
    solution of, by determinants, 326
    solution of, by linear combination, 243, 264
    in three variables, 249
    in two variables, 241
Location theorem, 186
Logarithm(s):
    applications of, 228
    to base e, 223
    to base 10, 220
    characteristic of a, 221
    common, 220
    as an exponent, 215
    laws of, 216
    mantissa of a, 221
    natural, 223
    table of $\log_e$, 416
    table of $\log_{10}$, 414
Logarithmic function(s), 215

M

Mantissa of a logarithm, 221
Mathematical induction, 361
Matrices, 282
Matrix:
    augmented, 298
    coefficient, 298
    column, 282
    definition of, 282
    determinant of a, 304
    diagonal, 292
    elementary transformation of, 295
    entries of a, 282
    identity, for addition, 284
    identity, for multiplication, 293
    inverse of a square, 315
    negative of a, 284
    nonsingular, 297, 317
    order of a, 282
    principal diagonal of a, 292
    product with a real number, 288
    row, 282
    row-equivalent, 296
    singular, 297
    square, 291
    transpose of a, 283
    zero, 284
Maximum point:
    local, 198
    of a parabola, 166
Member of a set, 1

Minimum point:
  local, 198
  of a parabola, 166
Monic polynomial, 26
Monomial, 25
Multiplication:
  of complex numbers, 80
  identity element for, in R, 8
  identity matrix for, 293
  of matrices, 288
  of a matrix and a real number, 288
  of polynomials, 31
Multiplicative identity, 8
Multiplicative inverse, 8
Multiplicity of solutions, 96
Mutually exclusive events, 382

N

Natural number, 1, 6
Negative:
  of a matrix, 284
  of a number, 8
  of a real number, 8, 15
Nonsingular matrix, 297
Notation:
  factorial, 353
  interval, 18
  scientific, 62
  set-builder, 4
  summation, 344
nth term:
  of an arithmetic progression, 336
  of a geometric progression, 337
Null set, 2
Number(s):
  complex, *see* Complex number(s)
  composite, 193
  negative of a, 8
  prime, 193
Number line, 15

O

Oblique asymptote, 205
Operations:
  binary, 7
  on sets, 2
Order:
  of a determinant, 304
  of a matrix, 282
  of a radical, 72
  symbols for, 17
Ordered field, 16

Ordered pair(s):
  in Cartesian products, 134
  components of an, 134
  definition of, 134
  in functions, 137
  graph of an, 134
  in relations, 135
Ordinate, 135
Origin, 15
Outcome, 377

P

Parabola, 166
  axis of symmetry, 167
  equation of a, 166
  vertex of, 166
Parameter, 397
Partial fractions, 257
Partial sum, 349
Permutation(s):
  circular, 373
  definition of, 369
  distinguishable, 370
  of n things n at a time, 369
  of n things r at a time, 370
pH, 235
Plane, equation of a, 252
Point-slope form of a linear equation, 149
Polynomial(s):
  completely factored, 34
  definition of a, 25
  degree of a, 25
  factoring, 34, 188, 191
  leading coefficient of a, 26
  leading term of a, 26
  linear combination of, 242
  linearly dependent, 241
  monic, 26
  over R, 26
  prime, 34
  products of, 31
  quotients of, 39
  real, 26
  standard form, 26
  sums of, 26
Polynomial equation:
  complex zeros of a, 190
  rational zeros of a, 193
  real zeros of a, 188
Polynomial function:
  complex zeros of a, 190
  definition of, 197
  graph of a, 197
  rational zeros of a, 193

real zeros of a, 188
turning points of graph of a, 198
Positive real number, 15
Postulates, 7
Power(s):
  definition of, 24
  with integral exponents, 60
  with rational exponents, 211
  with real exponents, 212
Prime number, 193
Prime polynomial, 34
Probability:
  a posteriori, 379
  a priori, 378
  conditional, 385
  function, 377
  meaning of, 378
  of mutually exclusive events, 382
Product(s):
  of complex numbers, 80
  involving fractions, 53
  of matrices, 289
  of a matrix and a real number, 288
  of polynomials, 30
  of rational expressions, 53
Programming, linear, 277
Proof, by mathematical induction, 361
Proportion, 182
Pure imaginary number, 81

Q

Quadratic equation(s), 95
  with complex solutions, 100
  discriminant of a, 100
  graph of a, 165
  solution of, 95
  standard form of a, 95
  systems involving, 262
  in two variables, 173
Quadratic formula, 99
Quadratic function, 165
Quadratic inequalities, 113, 169
Quadratic relation, 173
Quotient(s):
  of complex numbers, 83
  definition of, 9
  involving fractions, 54
  of polynomials, 39
  of rational expressions, 53

R

Radical expression(s), 71
Radicand, 72

Index

Range:
  of an exponential function, 212
  of a function, 137
  of a logarithmic function, 215
  of a relation, 135
  of summation, 344
Rational exponents, 211
Rational expression, 46
Rational function, 201
Rational number(s), 6
Rational zeros, of a polynomial equation, 193
Rationalizing:
  denominators, 73
  numerators, 73
Real number(s):
  completeness of, 412
  equality of, 7
  field properties of, 11
  order in the set of, 15
  product, with a matrix, 288
  set of, 6
Reciprocal, 8
Rectangular coordinate system, *see* Cartesian coordinate system
Recursive definition, 336
Reflexive law of equality, for real numbers, 7
Relation(s):
  definition of, 135
  domain of a, 135
  first-degree, 160
  graph of a, 134
  ordered pairs in a, 134
  quadratic, 165
  range of a, 135
Relatively prime numbers, 194
Remainder theorem, 187
Replacement set, 3
Ring, 412
Rise of a line segment, 145
Root(s), 99, *see also* Solution(s)
  cube, 67
  nth, 67
  square, 67
Row-equivalent matrices, 296
Row matrix, 282
Row vector, 282
Run of a line segment, 145

S

Sample point, 377
Sample space, 377
Second-degree equation, *see* Quadratic equation(s)

Sequence(s):
  alternating, 349
  arithmetic, 336
  convergent, 348
  definition of, 334
  divergent, 349
  finite, 334
  function, 334
  general term of a, 335
  geometric, 337
  infinite geometric, 350
  limit of a, 348
  strictly increasing, 348
Series:
  arithmetic, 341
  definition of, 341
  geometric, 343
  infinite geometric, 350
Set(s):
  complement of a, 3
  of complex numbers, 79
  convex, 274
  designation of a, 1
  disjoint, 3
  element of a, 1, 3
  empty, 2
  equality of, 2
  finite, 2
  of imaginary numbers, 80
  infinite, 2
  of integers, 6
  intersection of, 2
  of irrational numbers, 6
  member of a, 1, 3
  of natural numbers, 1, 6
  null, 2
  operations on, 2
  of rational numbers, 6
  of real numbers, 6
  replacement, 3
  union of, 2
  universal, 3
Set-builder notation, 4
Set function, probability, 377
Sigma notation, 344
Sign graph, 113
Singular matrix, 297
Slope of a line segment, 145
Slope-intercept form of a linear equation, 149
Solution(s), 89
  of equations by substitution, 105, 245, 263
  of exponential equations, 228
  of a linear inequality, 107
  of polynomial equations, 190

  of quadratic inequalities, 112, 169
  of a system by substitution, 105, 245, 263
  of systems in three variables, 249
  of systems in two variables, 241, 322
  of systems of inequalities, 272
  of systems of nonlinear equations, 262
Solution set(s):
  of an inequality, 107
  of a system, 241
Square matrix, 291
Square root:
  of a negative number, 81
  of a positive number, 67
Standard form:
  of a linear equation, 142
  of a quadratic equation, 95
Step function, 158
Subset:
  as a combination, 374
  definition of, 2
  as an event, 377
Substitution:
  axiom, 7
  solution of equations by, 105, 245, 263
Subtraction:
  of matrices, 284
  of real numbers, 9
Sum(s):
  of complex numbers, 80
  of an infinite series, 350
  involving fractions, 50
  of matrices, 283
  of polynomials, 26
  of real numbers, 8
Summation:
  index of, 344
  range of, 344
Summation notation, 344
Symmetric law of equality, for real numbers, 7
Synthetic division:
  definition, 42
  in locating zeros of polynomial equations, 195
System(s):
  equivalent, 242
  of linear equations in three variables, 249
  of linear equations in two variables, 241
  of linear inequalities, 272
  of nonlinear equations, 262

T

Table(s):
  of e^x, 413
  of $\log_e$, 416
  of $\log_{10}$, 414
  of squares, square roots, and prime factors, 417
Term, 25
Theorem, 11
Transformations, elementary:
  of equations, 90
  of inequalities, 108
  of matrices, 295
Transitive law of equality, for real numbers, 7
Transpose of a matrix, 283
Trichotomy law, 16
Turning point of a graph, 198

U

Union of sets, 2
Universal set, 3

V

Value:
  of a determinant, 306
  of a function, 137
Variable, 3
Variation:
  constant of, 181
  direct, 180
  inverse, 180
  joint, 182
Vector(s):
  column, 282
  row, 282
Vertex, 166
Vertical asymptote, 202

W

Word problems, 122

Z

Zero(s):
  complex, of functions, 190
  division by, 9, 109
  of a function, 169
  locating, of a polynomial equation, 180
  matrix, 284
  polynomial, 25
  rational, of a polynomial equation, 193

STUDENT QUESTIONNAIRE

Your chance to rate **College Algebra, Fifth Edition** (*Beckenbach/Drooyan/Wooton*)

In order to keep this text responsive to your needs, it would help us to know what you, the student, thought of this text. We would appreciate it if you would answer the following questions. Then cut out the page, fold, seal, and mail it; no postage is required. Thank you for your help.

Which chapters did you cover?   (circle)   1   2   3   4   5   6   7   8   9   10   11   A   B   All

Does the book have enough worked-out examples?   Yes _____   No _____

enough exercises?   Yes _____   No _____

Which helped most?

Explanations _____   Examples _____   Exercises _____   All three _____   Other _____
                                                                                                (fill in)

Were the answers at the back of the book helpful?   Yes _____   No _____

Did the answers have any typos or misprints? If so, where?

Did you know about the *Student Guide to College Algebra, Fifth Edition*?   Yes _____   No _____
Did you use the guide?   Yes _____   No _____

For you, was the course elective? _____   Required? _____

Do you plan to take more mathematics courses?   Yes _____   No _____

If yes, which ones?

How much algebra did you have before this course?   Terms in high school   (circle)   1   2   3   4

Courses in college   1   2   3

If you had algebra before, how long ago?

Last 2 years _____   3–5 years ago _____   5 years or longer _____

What is your major or your career goal? _____   Your age? _____

What did you like the most about *College Algebra, Fifth Edition*?

Can we quote you?   Yes _____   No _____

What did you like least about the book?

College _____   State _____

-- FOLD HERE --

	First Class PERMIT NO. 34 Belmont Ca.

BUSINESS REPLY MAIL
No postage necessary if mailed in United States

Postage will be paid by
WADSWORTH PUBLISHING COMPANY, INC.
10 Davis Drive
Belmont, California 94002

        ATTN: Rich Jones, Mathematics editor

[9.1] $\begin{bmatrix} a_1 & b_1 & c_1 \\ a_2 & b_2 & c_2 \end{bmatrix}$, etc.   matrix

$A_{m \times n}$ m by n matrix

a_{ij} the element in the ith row and jth column of the matrix A

A^t the transpose of the matrix A

$\mathbf{0}_{m \times n}$ the m by n zero matrix

$-A_{m \times n}$ the negative of $A_{m \times n}$

[9.2] $I_{n \times n}$ the identity matrix for all n by n matrices

[9.3] $A \sim B$ A is row-equivalent to B (for matrices)

[9.4] $\begin{vmatrix} a_1 & b_1 & c_1 \\ a_2 & b_2 & c_2 \\ a_3 & b_3 & c_3 \end{vmatrix}$, etc.   determinant

$\delta(A)$ the determinant of A

M_{ij} the minor of the element a_{ij}

A_{ij} the cofactor of a_{ij}

[9.6] A^{-1} the inverse of A

[10.1] $s(n)$, or s_n the nth term of a sequence

[10.2] S_n the sum of the first n terms in a sequence

$\sum$ the sum

S_∞ the sum of an infinite sequence

[10.3] $\lim_{n \to \infty} s_n$ the limit of a sequence

[10.4] $n!$ n factorial, or factorial n

$0!$ zero factorial

$\binom{n}{r}$ $\dfrac{n!}{r!(n-r)!}$

[11.1] $n(A)$ the number of elements in the set A

$A \times B$ the Cartesian product of A and B

$P_{n,n}$ the number of permutations of n things taken n at a time

$P_{n,r}$ the number of permutations of n things taken r at a time